METHODS IN MOLECULAR BIOLOGY

Series Editor
John M. Walker
School of Life and Medical Sciences
University of Hertfordshire
Hatfield, Hertfordshire, AL10 9AB, UK

For further volumes:
http://www.springer.com/series/7651

Tau Protein

Methods and Protocols

Edited by

Caroline Smet-Nocca

UMR CNRS 8576, Villeneuve d'Ascq, France

Humana Press

Editor
Caroline Smet-Nocca
UMR CNRS 8576
Villeneuve d'Ascq, France

ISSN 1064-3745 ISSN 1940-6029 (electronic)
Methods in Molecular Biology
ISBN 978-1-4939-6596-0 ISBN 978-1-4939-6598-4 (eBook)
DOI 10.1007/978-1-4939-6598-4

Library of Congress Control Number: 2016961536

Cover illustration: Negative stain electron microscopy with a 200,000 × magnification of PHF-like fibrils made of recombinant Tau protein (longest human isoform) induced by heparin.

Printed on acid-free paper

This Humana Press imprint is published by Springer Nature
The registered company is Springer Science+Business Media LLC
The registered company address is: 233 Spring Street, New York, NY 10013, U.S.A.

Preface

Isolated from a heat stable fraction of tubulin-bound proteins in 1975, Tau protein was associated a decade later with the paired helical filaments (PHFs) found in neurofibrillary tangles (NFTs) of Alzheimer's disease (AD) brain. As the main constituent of PHFs, Tau has thus become a pharmacological target of interest. Although phosphorylation is involved in the regulation of Tau microtubule binding activity allowing for neuronal plasticity, Tau in AD brain is found in an abnormally hyperphosphorylated form which combines high level with abnormal phosphorylation (i.e., phosphorylation sites that are not detected in normal adult brains). Tau is mainly found in the axonal compartment of adult mature neurons under physiological conditions, separated from the somatodendritic compartment by a diffusion barrier in the axon initial segment. In contrast, Tau missorting is an early marker of Tau dysfunction in AD together with hyperphosphorylation and aggregation. Furthermore, decreasing Tau levels has a positive effect on AD cellular and in vivo models suggesting that at least part of Tau effect in neurodegeneration comes from a gain of toxic function and/or accumulation. Hence, the current model of Tau pathogenesis in AD proposes that (hyper)phosphorylated Tau detaches from microtubules, accumulates in diverse neuronal compartments, and aggregates into intraneuronal PHFs.

Together Tau protein and β-amyloid (Aβ) peptide are AD molecular biomarkers that both form amyloid assemblies. Aβ peptide as the main component of the extracellular amyloid plaques—which is provided by the enzymatic cleavage of the amyloid precursor protein (APP) by the presenilins—has been in the forefront over the past decades as a promising target for the development of AD therapeutics according to the amyloid cascade hypothesis. The etiological role of APP in familial AD (that represents 1 % of total cases) related to autosomal dominant mutations of either *APP*, *presenilin-1*, or *presenilin-2* genes, together with the extraneuronal localization of amyloid deposits, has first motivated attempts to target Aβ peptide in a therapeutic strategy aiming at reducing amyloid plaques, however with poor outcome up to now. Hence, new strategies directed against Tau protein have emerged. The role of Tau in AD has been highlighted by the finding that mutations of the *MAPT* gene encoding for Tau have been linked to the etiology of other dementia called tauopathies indicating that Tau itself is able to trigger neurodegeneration in the absence of Aβ. Furthermore, the neurofibrillary degeneration is, unlike Aβ deposits, in good correlation with AD progression through the brain and severity of cognitive decline. The number and extent of NFT lesions are used in *postmortem* definite diagnosis of AD. In addition, phospho-epitope staining allows for defining AD stages and progression of neurofibrillary degeneration through the brain, referred to as Braak staging. A spreading of neurofibrillary degeneration in AD brain in a specific temporal and regional pattern is observed, but molecular mechanisms of transmission between neurons and the involved pathogenic specie(s) remain to be defined.

It has been suggested that abnormal phosphorylation could be accompanied by conformational change(s) that would turn the native into an aggregation-prone conformation as first suggested by AD-specific, conformation-dependent antibodies. This is even more intriguing for the disorder structure of Tau protein—classified as an intrinsically disordered protein (IDP)—that seemingly escapes the dogma of structure-function relationships. However, the molecular details still have to be described, a difficult task given the complexity of Tau post-translational modifications. Tau oligomers seem to be the most toxic species and have also been involved in the spreading of Tau pathology. Hence, transformation of Tau protein involves progressive conformational changes from a monomeric form into soluble, prefibrillar oligomers that constitute the nuclei of the fibrillization process and finally into *bona fide* PHFs containing β-structures. Furthermore, truncated forms of Tau, other post-translational modifications such as lysine ubiquitination, acetylation or methylation, or other proteins found in PHF extracted from AD brain account as potential factors involved in the acquisition of the aggregation capacity.

This volume of *Methods in Molecular Biology* dedicated to Tau protein covers basic and advanced methods and protocols from in vitro assays to in vivo models to address the molecular and functional aspects of Tau physiopathology, and deals with many related technical issues. Hereafter are described protocols for the conformational studies of native Tau protein and investigation of its physiological function in microtubule binding and tubulin polymerization; in vitro methods of formation of Tau oligomers and PHF-like fibrils, the study of fibrillization kinetics and a screening assay of Tau aggregation inhibitors; protocols for the characterization and in vitro introduction of post-translational modifications on Tau protein for further functional studies; analytical tools for the detection of Tau proteins, their modifications and cellular interactions, and *MAPT* gene mutations in various biological samples; and cellular and in vivo models for the investigations of Tau physiopathology.

Part I

Although it has been studied for several decades, Tau protein has still not finished revealing its secrets. Several studies have suggested that distinct conformations could drive Tau function and toxicity, but a description of diverse Tau species is still required at the molecular level. The flexibility and structural dynamics of Tau and other IDPs enlarge the concept of "structure-function relationships" requiring the development of new structural methodologies to decipher the fascinating and intriguing world of IDPs. Detection of transient folding and secondary structure elements in the monomeric form of Tau requires a combination of biophysical methodologies including, amongst others, circular dichroism and small-angle X-ray scattering described in Chapter 1 and Raman Spectroscopy in Chapter 2. Detection of subtle conformational changes of Tau protein upon phosphorylation or other post-translational modifications is challenging for most experimental methods. Atomistic molecular dynamics simulations described in Chapter 3, combined with experimental NMR data, provide structural and dynamic information on folding of a short phospho-peptide encompassing the AT8 epitope (pS202/pT205). The study of Tau:tubulin or Tau:microtubule interactions is of major interest in the context of Tau physiopathology when one wants to decipher the role of post-translational modifications or cofactors on microtubule binding and tubulin polymerization, or the intricate molecular mechanism by which Tau promotes tubulin assembly. Microtubules constitute a dynamic protein machinery while Tau protein keeps a high degree of flexibility even at the microtubule surface. Together these behaviors hamper the full structural description of the Tau:tubulin and Tau:microtubule complexes at the molecular level. Furthermore, negatively charged patches at the microtubule surface could promote Tau self-association when studying interactions of Tau with tubulin or microtubules leading to misinterpretation of dissociation constants and kinetic parameters.

Chapter 4 describes tubulin preparation from lamb or pig brains for in vitro polymerization assays. Various experimental conditions are proposed to study Tau:tubulin and Tau:microtubule interactions in vitro and in living cells using turbidimetry, co-sedimentation, and FRET assays. Pitfalls and cautions related to data interpretation are emphasized.

Part II

In vitro aggregation assays both aim to decipher the fibrillization process at the molecular level and to screen small-molecule inhibitors interfering with fibril formation in a perspective of AD cure. Part II focuses on in vitro methods and protocols for the study of Tau fibrillar structure, fibrillization process, and screening of small-molecule inhibitors of Tau aggregation. In addition, protocols using Tau oligomers as seeds in in vitro aggregation assays and in in vivo model of Tau spreading are included. Due to its high solubility, external aggregation inducers are required to stimulate Tau fibrillization in vitro that otherwise would not be of any practical use, i.e., in a few day timescale. Molecules enabling Tau aggregation are polyanions such as heparin, polyglutamate or RNA, anionic micelles and vesicles made up of fatty acids, as well as small-molecule anionic compounds or negatively charged gold nanorods. Inducers stimulate Tau fibrillization into PHF-like fibrils with morphologies resembling fibrils extracted from AD brains, hence modeling an authentic pathological process. A combination of biophysical techniques are used for the specific detection of amyloid-type fibril formation, conformational changes upon aggregation, quantification of fibril amounts and monitoring of fibrillization kinetics, as well as the description of fibril morphology. Chapter 5 describes the co-crystallization procedure of a short Tau peptide encompassing the minimal motif of fibrillization—the PHF6 peptide—with small-molecule compounds, collection of X-ray diffraction data from microcrystals and data processing. This model has provided structures of a pharmacophore for rational design of small-molecule compounds targeting Tau amyloid fibrils. Chapter 6 presents protocols for in vitro fibrillization of full-length Tau protein from the monomeric form using either Geranine G, ODS, or anionic microspheres as external aggregation inducers combined with three complementary methods of fibril detection, transmission electron microscopy (TEM) imaging, filter-trap immunoassay, and Thioflavin dye fluorescence. The advantages and disadvantages of each method are emphasized. Chapter 7 describes three alternative methods for the detection of Tau fibrils formed with heparin and the study of fibrillization kinetics. Vibrational spectroscopy such as Fourier transform infrared (FTIR) and ultraviolet resonance Raman (UVRR) spectroscopy can discriminate between different classes of secondary structure, therefore allowing for the detection of specific β-structure formation upon fibrillization. These techniques are combined with atomic force microscopy (AFM) for fibril imaging. Chapter 8 presents a screening method of Tau aggregation inhibitors (TAI) combining a cell-free Tau-Tau binding immunoassay and a cell-based assay involving a stable cell line expressing an aggregation-prone Tau fragment and inducible for the expression of full-length Tau. This method allows investigation of Tau aggregation inhibition with small molecules together with evaluation of their properties of cell penetration and toxicity. Chapter 9 provides a method for the in vitro preparation of Tau oligomers with Aβ42 or α-synuclein seeds, methods for isolation of Tau oligomers from biological sources, and their detection using Tau oligomer-specific antibodies in ELISA, dot blot, Western blot, and immunohistochemistry assays. This method provides important tools for the early detection of Tau pathogenic species in vivo and the study of Tau pathology propagation.

Part III

The disordered nature of Tau induces a high level and diversity of post-translational modifications. Therefore, the characterization of such a complex array of post-translational modifications is very challenging, and deciphering their role in modulation of Tau functions has been rendered even more complicated. In particular, phosphorylation has focused attention due to its close link with pathogenesis. The number of potential phosphorylatable Ser/Thr residues (80 sites) as well as the number of involved kinases has complicated the description of pathological patterns of phosphorylation and the sequence of events. While phosphorylation has been extensively studied, other post-translational modifications have only been recently described and their characterization is thus less comprehensive. Part III deals with methods for the characterization of Tau post-translational modifications generated either in vivo or in vitro and methods of production of Tau proteins with specific pattern of post-translational modifications as tools for the study of their role in Tau functions. Chapter 10 presents identification and quantification of lysine-directed modifications in Tau protein by liquid-chromatography-tandem mass spectrometry. A method for the characterization of Tau phosphorylation and acetylation patterns by high-resolution NMR spectroscopy is described in Chapter 11. Protocols in this chapter include the recombinant production of the activated ERK2 kinase and a functional fragment of the Creb-binding protein in *E. coli* for the in vitro modification of isotopically labeled Tau protein that fulfills the amount required for NMR analyses. Additionally, a protocol for the preparation of a rat brain extract is provided for the production of phosphorylated Tau samples. As an alternative to the enzymatic strategy that affords heterogeneous patterns of modification, a protocol of Expressed Protein Ligation (EPL) is described in Chapter 12 for the introduction of selective and quantitative phosphorylation or *O*-β-N-acetylglucosaminylation (*O*-GlcNAcylation) in the C-terminal region of Tau protein. This method is complemented by a protocol for traceless purification of the ligation product. As in vitro enzymatic *O*-GlcNAcylation of Tau is challenging, Chapter 13 describes an alternative method for high-level incorporation of *O*-GlcNAc moieties in Tau protein with co-expression of Tau and *O*-GlcNAc transferase (OGT) in *E. coli* and enrichment of the *O*-GlcNAc-modified Tau fraction, as well as selective detection of *O*-GlcNAc-Tau in brain lysates with the S400-*O*-GlcNAc antibody.

Part IV

Part IV describes basic and advanced analytical tools to detect Tau proteins, Tau post-translational modifications, and Tau interacting partners from various biological sources. Additional protocols are presented to identify mutations of the *MAPT* gene encoding for Tau protein. Although mainly found in neurons Tau protein is also detected in the extracellular space. An increase of Tau and phospho-Tau levels is detected in the cerebrospinal fluid (CSF) of AD patients. Synaptic release and postsynaptic uptake of Tau species are responsible, at least in part, for interneuronal spreading of Tau pathology. Hence, release and diffusion of Tau in the interstitial fluid (ISF) account for the extracellular Tau content. In addition, intraneuronal aggregates of hyperphosphorylated, misfolded PHF-Tau become extraneuronal when neurons die (referred to as "ghost" tangles). Detecting intraneuronal and extracellular Tau species is thus a key issue not only in understanding Tau physiology and pathology but also for (early) diagnosis of neurodegenerative diseases. Challenging characterization of the neuropathological lesions such as neurofibrillary tangles required development of methods for the isolation and identification of their proteinaceous constituents. With a high content of various post-translational modifications coupled to the expression of six isoforms, proteomic analyses of Tau proteins and PHF-Tau become very difficult. Chapter 14 describes a two-dimensional electrophoresis-based protocol for the analysis of Tau isovariants extracted from human or mouse brain and cells. Chapter 15 highlights pitfalls and offers tips to circumvent detection of nonspecific signals when analyzing mouse brain proteins by Western blotting with anti-Tau monoclonal antibodies. Chapter 16 deals with quantitative flow cytometry analysis of Tau in synaptosomes—isolated nerve terminals enriched in synaptic proteins obtained during nerve tissue homogenization—as a model system for the study of Tau release and synaptic Tau pathology. This method includes a protocol for the preparation of crude (P2) synaptosomal fraction from cryopreserved *postmortem* tissue of AD brain and P2 immunolabeling. Flow cytometry allows purification of synaptosomes from P2 fraction based on a particle size selection achieving a high degree of purity and single-cell quantitative analysis of (phospho)Tau. Chapter 17 presents a protocol of in vivo microdialysis in freely moving mice to detect extracellular Tau proteins in brain interstitial fluid. Changes of Tau levels in ISF related to drug administration can be monitored using this protocol. While Tau was first identified as a tubulin-binding protein, it has been recently described as a scaffold protein with several protein partners. Chapter 18 describes the proximity-ligation assay (PLA) method to detect Tau interacting partners into neurons. This procedure is illustrated by the characterization of interactions between Tau and the bridging integrator 1 (BIN1). BIN1 is one of the most important genetic risk factors in late-onset AD identified by genome-wide association study while BIN1 protein was shown to modulate neurofibrillary tangle pathology. Finally, Chapter 19 describes a method for PCR amplification, bidirectional Sanger sequencing of the *MAPT* gene from genomic DNA of patients, and further sequence analyses in order to detect single nucleotide *MAPT* mutations combined with exon trapping. As *MAPT* mutations that affect splicing of exon 10 are commonly associated with tauopathies, exon trapping enables to functionally connect DNA variants to dysregulation of exon 10 splicing of *MAPT*.

Part V

Part V concerns various cellular and in vivo models for investigation of Tau physiopathology. Chapter 20 provides a detailed protocol to grow primary neurons, as well as a cell fixation and staining protocol for immunofluorescence microscopy of endogenous Tau. Chapter 21 proposes a nonviral transfection protocol of primary neurons ensuring low levels of Tau comparable to endogenous levels and thus providing a good model to investigate Tau sorting and transport into neurons. Chapters 22 and 23 provide two protocols of cellular assays based on fluorescence biosensors for the detection of Tau aggregation within the cell. Chapter 22 describes the Tau-BiFC cell-based model that uses Bimolecular Fluorescence Complementation as a convenient tool to investigate intracellular tau aggregation. An example of modulation of Tau-Tau interactions in cell by Forskolin, a kinase activator, is given. Chapter 23 describes a rapid and sensitive FRET-based flow cytometry biosensor assay to quantitatively measure Tau seeding activity. Chapter 24 investigates Tau fibrillar aggregates in the mouse retina as a part of the central nervous system accessible for noninvasive in vivo imaging using confocal scanning laser ophthalmoscopy in combination with fluorophore staining of fibrillar Tau in the retinal ganglion cell layer. This method can be exploited to evaluate the efficiency of potential drugs in preclinical studies with a follow-up of mice over several months of treatment. Chapter 25 presents an in vivo hyperthermic stress model to study the effects of increased oxidative stress on Tau functions in mouse brain. Chapter 26 describes the Drosophila model to study the modifiers of Tau toxicity. This method includes a protocol for the expression of Tau in the Drosophila eye which is used as a readout in a genetic screen of Tau toxicity modifiers. Chapter 27 deals with a protocol of in situ hybridization of neurotrophin mRNA with labeled riboprobes in the THY-Tau22 mouse model exhibiting an AD-like Tau pathology. This method is combined with immunohistochemical and immunofluorescence analysis of brain slices allowing for double labeling of neurotrophin mRNA and Tau phospho-epitopes. Chapter 28 describes Pin1 knockout mouse model and its use for the study of Tau pathology. This chapter includes several behavior tests as well as a protocol to harvest mouse brain and prepare paraffin-embedded sections for immunostaining and brain lysate for immunoblotting.

I am indebted to all authors who contributed, sharing their protocols and technical expertise, and who made this book possible. I would also like to thank Professor John Walker, the series editor, for inviting me to edit this volume of the Methods in Molecular Biology series. I am very grateful to him for this opportunity. Finally, I would like to warmly thank my colleagues for their encouragements and meaningful discussions.

Lille, France ***Caroline Smet-Nocca***

Contents

Part V Cellular and In Vivo Models of Tau Physiopathology

Contributors

Onder Albayram • *Division of Translational Therapeutic, Department of Medicine and the Cancer Center, Beth Israel Deaconess Medical Center, Harvard Medical School, Boston, MA, USA*

Diane Allegro • *Aix-Marseille Université, INSERM, CRO2, UMR-S 911, Marseille, France*

Pascale Barbier • *Aix-Marseille Université, INSERM, CRO2, UMR-S 911, Marseille, France*

Tiphany De Bessa • *Aix-Marseille Université, INSERM, CRO2, UMR-S 911, Marseille, France*

Tina Bilousova • *UCLA School of Nursing and Mary S. Easton Center for Alzheimer's Research at UCLA, Los Angeles, CA, USA*

Alexis Bretteville • *Santé publique et épidémiologie moléculaire des maladies liées au vieillissement, INSERM, UMR 1167, Lille, France; Institut Pasteur de Lille, Lille, France; Université de Lille, Lille, France*

Gilles Breuzard • *Aix-Marseille Université, INSERM, CRO2, UMR-S 911, Marseille, France*

Luc Buée • *Inserm UMR-S1172, Jean-Pierre Aubert Research Centre (JPARC), Alzheimer & Tauopathies, Lille, France*

Valérie Buée-Scherrer • *Inserm UMR-S1172, Jean-Pierre Aubert Research Centre (JPARC), Alzheimer & Tauopathies, Lille, France*

François-Xavier Cantrelle • *Université de Lille, Sciences et Technologies, Unité de Glycobiologie Structurale et Fonctionnelle (UMR CNRS 8576), Lille, France; CNRS, UMR 8576, CNRS, UMR 8576*

Béatrice Chambraud • *INSERM, Paris-Saclay University, UMR 1195 Paris, France*

Julien Chapuis • *Santé publique et épidémiologie moléculaire des maladies liées au vieillissement, INSERM, UMR 1167, Lille, France; Institut Pasteur de Lille, Lille, France; Université de Lille, Lille, France*

Alban Chauderlier • *Inserm UMR-S1172, Jean-Pierre Aubert Research Centre (JPARC), Alzheimer & Tauopathies, Lille, France*

Subashchandrabose Chinnathambi • *Neurobiology Group, Division of Biochemical Sciences, National Chemical Laboratory (CSIR-NCL), Pune, India; Academy of Scientific and Innovative Research (AcSIR), New Delhi, India*

Lucie Delattre • *Inserm UMR-S1172, Jean-Pierre Aubert Research Centre (JPARC), Alzheimer & Tauopathies, Lille, France*

Florie Demiautte • *Santé publique et épidémiologie moléculaire des maladies liées au vieillissement, INSERM, UMR 1167, Lille, France; Institut Pasteur de Lille, Lille, France; Université de Lille, Lille, France*

Clément Despres • *Université de Lille, Sciences et Technologies, Unité de Glycobiologie Structurale et Fonctionnelle (UMR CNRS 8576), Lille, France; CNRS, UMR, Lille, France*

François Devred • *Aix-Marseille Université, INSERM, CRO2, UMR-S 911, Marseille, France*

Marc I. Diamond • *University of Texas- Southwestern Medical Center, Dallas, TX, USA; Center for Alzheimer's and Neurodegenerative Diseases, Dallas, TX, USA*

Carol Dobson-Stone • *Neuroscience Research Australia, Sydney, NSW, Australia; School of Medical Sciences, University of New South Wales, Sydney, NSW, Australia*

Pierre Dourlen • *Institut Pasteur de Lille, Lille, France; INSERM UMR1167, Lille, France; Université de Lille, Lille, France*

SABIHA EDDARKAOUI • *Inserm UMR-S1172, Jean-Pierre Aubert Research Centre (JPARC), Alzheimer & Tauopathies, Lille, France*
FRANCISCO-JOSE FERNANDEZ-GOMEZ • *Inserm UMR-S1172, Jean-Pierre Aubert Research Centre (JPARC), Alzheimer & Tauopathies, Lille, France*
JENNIFER L. FURMAN • *University of Texas- Southwestern Medical Center, Dallas, TX, USA; Center for Alzheimer's and Neurodegenerative Diseases, Dallas, TX, USA*
MARIE-CHRISTINE GALAS • *Inserm UMR-S1172, Jean-Pierre Aubert Research Centre (JPARC), Alzheimer & Tauopathies, Lille, France*
NEHA S. GANDHI • *School of Biomedical Sciences, CHIRI Biosciences and Curtin Institute for Computation, Curtin University, Perth, WA, Australia*
JULIA E. GERSON • *George P. and Cynthia Woods Mitchell Center for Neurodegenerative Diseases, University of Texas Medical Branch, Galveston, TX, USA; Departments of Neurology, and Neuroscience and Cell Biology, University of Texas Medical Branch, Galveston, TX, USA*
NALINI VIJAY GORANTLA • *Neurobiology Group, Division of Biochemical Sciences, National Chemical Laboratory (CSIR-NCL), Pune, India; Academy of Scientific and Innovative Research (AcSIR), New Delhi, India*
JEREMY GUNAWARDENA • *Department of Systems Biology, Harvard Medical School, Boston, MA, USA*
KAREN HOPPENS GYLYS • *UCLA School of Nursing and Mary S. Easton Center for Alzheimer's Research at UCLA, Los Angeles, CA, USA*
CHRISTIAN P.R. HACKENBERGER • *Leibniz-Institut für Molekulare Pharmakologie (FMP), Berlin, Germany; Department Chemie, Humboldt Universität zu Berlin, Berlin, Germany*
CHARLES R. HARRINGTON • *TauRx Therapeutics Ltd., Aberdeen, Scotland, UK; School of Medicine, Medical Sciences and Nutrition, University of Aberdeen, Scotland, UK*
SÉBASTIEN S. HÉBERT • *Département de Psychiatrie et Neurosciences, Université Laval, Québec, QC, Canada; Centre de Recherche du CHU de Québec, Neurosciences, Québec, QC, Canada*
JOCHEN HERMS • *Department of Translational Brain Research, German Centre for Neurodegenerative Diseases (DZNE), Munich, Germany*
SAMANTHA H. HINCKLEY • *Ohio State Biochemistry Program, The Ohio State University, Columbus, OH, USA*
DAVID HORSLEY • *School of Medicine, Medical Sciences and Nutrition, University of Aberdeen, Scotland, UK; TauRx Therapeutics Ltd., Aberdeen, Scotland, UK*
RAKEZ KAYED • *George P. and Cynthia Woods Mitchell Center for Neurodegenerative Diseases, University of Texas Medical Branch, Galveston, TX, USA; Departments of Neurology, and Neuroscience and Cell Biology, University of Texas Medical Branch, Galveston, TX, USA; University of Texas Medical Branch, Galveston, TX, USA*
PUNEET KHANDELWAL • *Academy of Scientific and Innovative Research (AcSIR), New Delhi, India; Physical and Materials Chemistry Division, CSIR-National Chemical Laboratory (CSIR-NCL), Pune, Maharashtra, India*
DOHEE KIM • *Convergence Research Center for Diagnosis, Treatment and Care System of Dementia, Korea Institute of Science and Technology (KIST), Brain Science Institute, Seoul, South Korea; Department of Biotechnology, Translational Research Center for Protein Function Control, College of Life Science and Biotechnology, Yonsei University, Seoul, South Korea*
DONG JIN KIM • *Korea Institute of Science and Technology (KIST), Brain Science Institute, Center for Neuro-medicine, Seoul, South Korea*
YUN KYUNG KIM • *Convergence Research Center for Diagnosis, Treatment and Care System of Dementia, Korea Institute of Science and Technology (KIST), Brain Science Institute,*

Seoul, South Korea; Biological Chemistry, University of Science and Technology (UST), Daejeon, South Korea

ASAMI KONDO • *Division of Translational Therapeutic, Department of Medicine and the Cancer Center, Beth Israel Deaconess Medical Center, Harvard Medical School, Boston, MA, USA*

PREDRAG KUKIC • *Department of Chemistry, University of Cambridge, Cambridge, UK*

JEFF KURET • *Ohio State Biochemistry Program, The Ohio State University, Columbus, OH, USA; Department of Biological Chemistry and Pharmacology, Ohio State Biochemistry Program, The Ohio State University, Columbus, OH, USA*

JOHN B.J. KWOK • *Neuroscience Research Australia, Sydney, NSW, Australia; School of Medical Sciences, University of New South Wales, Sydney, NSW, Australia*

MEYTAL LANDAU • *Department of Biology, Technion-Israel Institute of Technology, Haifa, Israel*

ISABELLE LANDRIEU • *UMR8576 CNRS-Université de Lille, Sciences et Technologies, Résonance Magnétique Nucléaire (RMN), IRI Building (CNRS), Campus CNRS Parc de la Haute Borne50, avenue de Halley, Lille, France*

SUNGSU LIM • *Convergence Research Center for Diagnosis, Treatment and Care System of Dementia, Korea Institute of Science and Technology (KIST), Brain Science Institute, Seoul, South Korea*

LAUREN LIN • *Neuroscience Signature Program, The Ohio State University, Columbus, OH, USA*

GUY LIPPENS • *Université de Lille, Sciences et Technologies, Unité de Glycobiologie Structurale et Fonctionnelle (UMR CNRS 8576), Lille, France; CNRS, UMR 8576, Lille, France*

KUN PING LU • *Division of Translational Therapeutic, Department of Medicine and the Cancer Center, Beth Israel Deaconess Medical Center, Harvard Medical School, Boston, MA, USA*

JULIA LÜDTKE • *German Center for Neurodegenerative Diseases (DZNE), Bonn, Germany*

RICARDO L. MANCERA • *School of Biomedical Sciences, CHIRI Biosciences and Curtin Institute for Computation, Curtin University, Perth, WA, Australia*

EVA-MARIA MANDELKOW • *German Center for Neurodegenerative Diseases (DZNE), Bonn, Germany; Max-Planck-Institute for Metabolism Research, Hamburg Outstation, c/o DESY, Hamburg, Germany; Center for Advanced European Studies and Research (Caesar), Bonn, Germany*

NIKI NANAVATY • *Neuroscience Signature Program, The Ohio State University, Columbus, OH, USA*

SAMANTHA B. NICHOLLS • *Department of Neurology, Alzheimer's Disease Research Laboratory, MassGeneral Institute for Neurodegenerative Disease, Massachusetts General Hospital, Harvard Medical School, Charlestown, MA, USA*

HÉLÈNE OBRIOT • *Inserm UMR-S1172, Jean-Pierre Aubert Research Centre (JPARC), Alzheimer & Tauopathies, Lille, France*

FRANCK R. PETRY • *Département de Psychiatrie et Neurosciences, Université Laval, Québec, QC, Canada*

VINCENT PEYROT • *Aix-Marseille Université, INSERM, CRO2, UMR-S 911, Marseille, France*

EMMANUEL PLANEL • *Département de Psychiatrie et Neurosciences, Université Laval, Québec, QC, Canada; Centre de Recherche du CHU de Québec, Neurosciences, Québec, QC, Canada*

PANKAJ PODDAR • *Academy of Scientific and Innovative Research (AcSIR), New Delhi, India; Physical and Materials Chemistry Division, CSIR-National Chemical Laboratory, Pune, India*

Sudhakaran Prabakaran • *Department of Systems Biology, Harvard Medical School, Boston, MA, USA*

Haoling Qi • *Université de Lille, Sciences et Technologies, Unité de Glycobiologie Structurale et Fonctionnelle (UMR CNRS 8576), Lille, France; CNRS, UMR, Lille, France*

Gayathri Ramachandran • *Department of Molecular Biosciences, Northwestern University, Evanston, IL, USA*

Oliver Reimann • *Leibniz-Institut für Molekulare Pharmakologie (FMP), Berlin, Germany; Institut für Chemie und Biochemie, Freie Universität Berlin, Berlin, Germany*

Janet E. Rickard • *School of Medicine, Medical Sciences and Nutrition, University of Aberdeen, Scotland, UK; TauRx Therapeutics Ltd., Aberdeen, Scotland, UK*

Katharina Schindowski-Zimmermann • *Institute of Applied Biotechnology, University of Applied Sciences Biberach, Biberach/Riss, Germany*

Christian Schön • *Center for Neuropathology, Ludwig-Maximilian-University, Munich, Germany; Department of Translational Brain Research, DZNE – German Centre for Neurodegenerative Diseases, Munich, Germany*

Urmi Sengupta • *George P. and Cynthia Woods Mitchell Center for Neurodegenerative Diseases, University of Texas Medical Branch, Galveston, TX, USA; Departments of Neurology, and Neuroscience and Cell Biology, University of Texas Medical Branch, Galveston, TX, USA*

Nicolas Sergeant • *Inserm UMR-S1172, Jean-Pierre Aubert Research Centre (JPARC), Alzheimer & Tauopathies, Lille, France*

Alexander V. Shkumatov • *Structural Biology Research Centre, VIB, Brussel, Belgium*

Caroline Smet-Nocca • *Université de Lille, Sciences et Technologies, Unité de Glycobiologie Structurale et Fonctionnelle (UMR CNRS 8576), Lille, France; CNRS, UMR 8576, Lille, France*

Einav Tayeb-Fligelman • *Department of Biology, Technion-Israel Institute of Technology, Haifa, Israel*

Stefani N. Thomas • *Department of Pathology, Johns Hopkins University School of Medicine, Baltimore, MD, USA*

David J. Vocadlo • *Department of Molecular Biology and Biochemistry, Simon Fraser University, Burnaby, BC, Canada; Department of Chemistry, Simon Fraser University, Burnaby, BC, Canada*

Claude M. Wischik • *TauRx Therapeutics Ltd., Aberdeen, Scotland, UK; School of Medicine, Medical Sciences and Nutrition, University of Aberdeen, Scotland, UK*

Kaoru Yamada • *Department of Neuropathology, Graduate School of Medicine, The University of Tokyo, Tokyo, Japan*

Austin J. Yang • *Department of Anatomy and Neurobiology, University of Maryland, Baltimore, MD, USA; National Institute on Aging, National Institutes of Health, Bethesda, MD, USA*

Scott A. Yuzwa • *Department of Molecular Biology and Biochemistry, Simon Fraser University, Burnaby, BC, Canada; Neurosciences & Mental Health Program, The Hospital for Sick Children, Toronto, ON, Canada*

Hans Zempel • *German Center for Neurodegenerative Diseases (DZNE), Bonn, Germany; Max-Planck-Institute for Metabolism Research, Hamburg Outstation, c/o DESY, Hamburg, Germany*

Xiao Zhen Zhou • *Division of Translational Therapeutic, Department of Medicine and the Cancer Center, Beth Israel Deaconess Medical Center, Harvard Medical School, Boston, MA, USA*

Part I

Native Tau Protein: Conformational Studies and Tubulin Interactions

Chapter 1

Conformational Dynamics of Intracellular Tau Protein Revealed by CD and SAXS

Nalini Vijay Gorantla, Alexander V. Shkumatov, and Subashchandrabose Chinnathambi

Abstract

A native conformation of a protein is essential for its biological role. In certain conditions, some proteins show non-native conformations, leading to aggregation, which in turn may produce severe pathologies. Such physiological conditions are classified as protein misfolding diseases. Alzheimer's disease (AD) is the most common form of dementia. Extracellular senile plaques formed by Amyloid β and intracellular aggregates formed by microtubule-associated protein Tau (MAPT) are the hallmarks of AD. Physiological role of MAPT is to maintain the integrity and stability of microtubules, however it tends to self-aggregate forming intracellular paired helical filaments (PHFs) during AD. MAPT is also subjected to various post-translational modifications such as phosphorylation, glycosylation, truncation, and acetylation. Being natively unfolded, MAPT is prone to full characterization at atomic level. Small-angle X-ray scattering (SAXS) is often applied in combination with other biophysical methods, like nuclear magnetic resonance (NMR), circular dichroism (CD), fluorescence spectroscopy, analytical ultracentrifugation (AUC), or dynamic light scattering (DLS) to characterize natively unfolded systems. Here we describe the practical aspects of MAPT characterization by SAXS and CD in detail as well as outline the inferred structural and functional implications.

Key words Tau protein, Protein conformation, Tau aggregation, Alzheimer disease, Circular dichroism, Small angle X-ray scattering

1 Introduction

Microtubules are the structural components within cells. Many cellular processes such as mitosis and vesicular transport require the involvement of microtubules. Microtubules serve as the tracks for motor proteins, which enable the intracellular transport of vesicles, organelles, and protein complexes [1–3]. During mitosis, microtubules act as spindle fibers attached to opposite ends of the cell and pull apart the aligned chromosomes [4]. During these processes, the dynamics of the microtubules (assembly and

Nalini Vijay Gorantla and Alexander V. Shkumatov contributed equally with all other contributors

Caroline Smet-Nocca (ed.), *Tau Protein: Methods and Protocols*, Methods in Molecular Biology, vol. 1523,
DOI 10.1007/978-1-4939-6598-4_1, © Springer Science+Business Media New York 2017

disassembly of tubulin) is modulated by microtubule-associated proteins (MAPs) that bind to the lattice surface of microtubules.

Tau protein is present both in the central and peripheral nervous system [5, 6]. Tau is one of the major MAPs in neurons and is predominantly present in axons [7, 8]. A single pre-mRNA of the Tau gene is alternatively spliced into six isoforms in the central nervous system. In fetal brain, the smallest isoform (hTau23-lacking exons 2, 3, and 10) is the predominant one, whereas in adult brain, all isoforms can be found in roughly equal amounts [6]. The full-length isoform is hTau40, whereas the other isoforms differ by the presence or absence of insert domains (I1 and I2) in N-terminal part and the repeat domain 2 (R2) in the C-terminal half of Tau protein (*see* Fig. 1). The alternative splicing of Tau is indeed regulated according to the developmental stages [9]. The repeat domains (R1–R4) together with the proline-rich flanking regions carry out the physiological role of Tau, i.e., the microtubule binding and assembly (*see* Fig. 1). On the other hand, the repeat domains are also involved in Tau-induced pathology as they form the core of the PHFs [10, 11]. The hexapeptide motifs in the repeats R2 and R3 can undergo a conformational change from a typical random coil to a β-sheet conformation and thus act as seeds for Tau aggregation [10, 12–15]. The N-terminal domain of Tau (~200 residues) projects away from the microtubule surface [16] and may serve as an anchor for other cellular components such as kinases, membranes, or motor components [17]. As Tau mainly

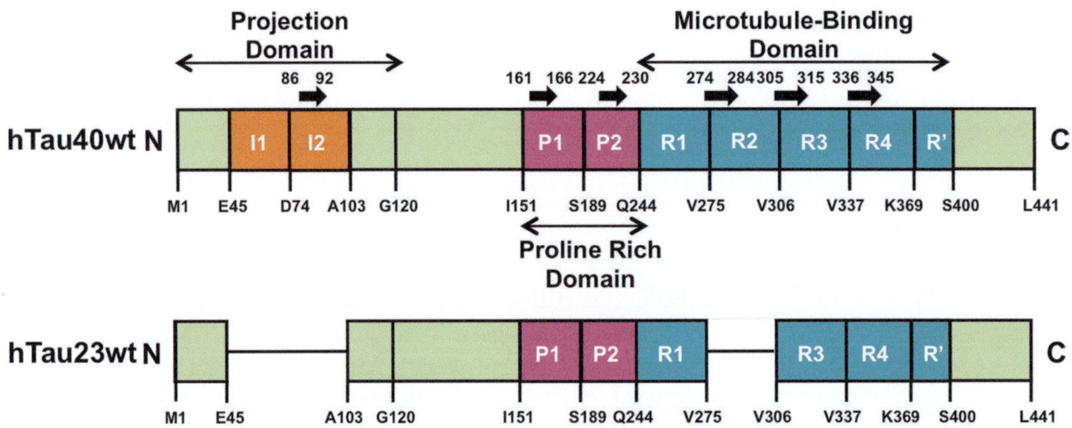

Fig. 1 Schematic diagram of human Tau isoforms. The bar diagram shows the longest and the shortest isoforms of Tau (hTau40 and hTau23, respectively). hTau40 has 441 amino acids and comprises two inserts, I1 and I2, encoded by exon 2 and 3 towards the acidic N-terminal end. This is followed by proline rich region, P1 and P2. Towards the C-terminus, Tau has four imperfect repeats: R1, R2, R3, and R4, each consisting of approximately 30 amino acids. Repeat R2 is encoded by exon 10. These repeats are responsible for microtubule binding and also aggregates formation (PHFs-paired helical filament) in Alzheimer's disease condition. The black arrows indicate the region on Tau, which can form a transient β-sheet. The alternative splicing of exon 2, exon 3, and exon 10 leads to six different isoforms of Tau

stabilizes microtubules, it is expected that the binding to microtubules could induce the structuration of Tau. Indeed, some reports suggest that Tau compaction occurs upon binding to microtubules [18], whereas other studies point out that Tau retains much of its disordered state even when interacting with microtubules [19, 20].

Structural investigations, in particular at atomic level, of Tau are hindered by its highly flexible nature. Attempts to analyze Tau by high-resolution methods, e.g., by NMR, have so far confirmed its highly dynamic properties and importance of transient long-range contacts [21]. Additional structural information on Tau can be obtained by small-angle X-ray scattering (SAXS) [22, 23]. In contrast to other methods, SAXS is well suited to study intrinsically disordered proteins (IDPs), allowing investigations under various conditions (pH, temperature, ionic strength, etc.). Moreover the information obtained by SAXS can be readily combined with NMR chemical shifts (CSs), residual dipolar couplings (RDCs), and paramagnetic relaxation enhancements (PREs) as recently demonstrated by Schwalbe et al. [24], where a combination of the state of the art ensemble calculation procedures and experimental restraints obtained by NMR and SAXS provided an accurate quantitative ensemble description of Tau at atomic level.

Buffer-subtracted SAXS curves not only provide a rapid assessment of overall parameters, such as molecular mass (MM), radius of gyration (R_g), and the overall shape of a macromolecule, but also allow to qualitatively estimate compactness and provide a quantitative description of the conformational ensembles for flexible macromolecules, sampling astronomical number of conformations in solution [25]. Traditionally, Kratky plots ($I(s)*s^2$ as a function of s) have been used to identify disordered states and distinguish them from globular ones [26]. A more accurate variation, the so-called dimensionless Kratky plot ($s^2R_g{}^2I(s)/I_0$ as a function of sR_g), allows a comparative estimate of the degree of disorder of different macromolecules independent of their masses and sizes. A dumbbell-shaped curve is expected for a compact, ordered globule and a plateau for a fully disordered protein [27] (*see* Fig. 2). A typical procedure to quantitatively describe flexibility based on SAXS data consists of two steps: (1) computational sampling of possible conformations of a macromolecule by modeling missing parts or exploring the conformational space using molecular dynamics-like approaches in case full atomic structure is available and (2) selection of a model subset that would satisfy (fit) the experimental SAXS data [28–32]. One of the most commonly used approaches is ensemble optimization method (EOM) [33]. If available, EOM treats partial high-resolution structures as rigid bodies, whereas missing parts are modeled as Cα traces with dihedral angles selected randomly from a quasi-Ramachandran plot [33]. A theoretical scattering curve is then calculated for each model in the random pool (10,000 models or more) representing the conformational space of a macromolecule. In the second step, a genetic algorithm

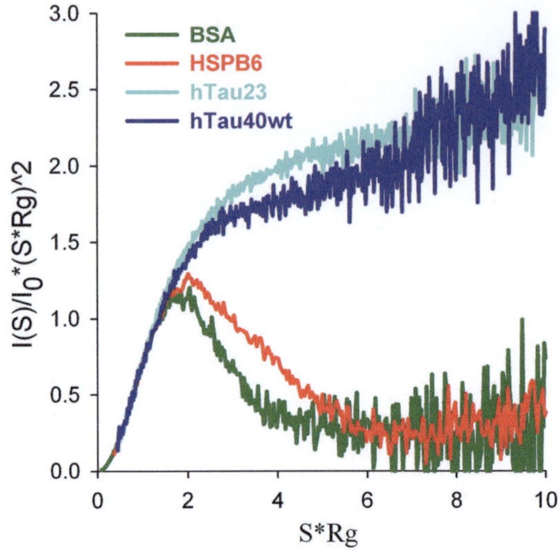

Fig. 2 hTau40 is highly flexible as revealed by small-angle X-ray scattering. Dimensionless Kratky plot showing globular bovine serum albumin (BSA) drawn as a *green line*, natively unfolded hTau40 protein [22] drawn as a *blue line* and a partially disordered (about 45 % of the residues are presumably flexible) human small heat shock protein, HSPB6 [59] drawn as a *red line*. Highly flexible hTau40 shows no peak as opposed to folded BSA, showing almost ideal bell-shaped peak. For partially flexible proteins, the peak becomes asymmetric and its position is shifted more to the right. Scattering data for hTau40 (PED7AAD) and HSPB6 (PED7AAA) are taken from pE-DB database [60]

repeatedly selects the subsets of theoretical curves, which if averaged, fit the scattering data. After certain number of steps, the corresponding ensemble of models, having the lowest discrepancy to the experimental data, is reported. Such strategy not only provides a set of models but also reflects the distribution of R_g and D_{max}, which in turn may suggest a preferential conformational state of the macromolecule: opened, closed or highly dynamic (sampling both closed and opened conformations) (*see* Fig. 3).

In the following sections, we describe the practical aspects of the application of solution scattering for analysis of natively disordered Tau protein in combination with CD measurements. For more general aspects of SAXS-based description of IDPs, we refer the reader to a recent review by Kihney and Svergun [34].

1.1 Residual Secondary Structure of Tau Mapped by CD Spectroscopy

Various biophysical techniques including FRET [35–39], NMR [21], electron microscopy [15], X-ray diffraction [40], FTIR [39, 41], fluorescence spectroscopy [38, 42], and attenuated total reflection-Infrared spectroscopy (ATR-IR) [38, 42, 43] have confirmed that Tau is an IDP and that the protein does not contain a significant amount of secondary structure in solution. In its native form Tau has random coil structure (*see* Fig. 4), whereas upon aggregation it attains a cross β structure. Even the aggregated form

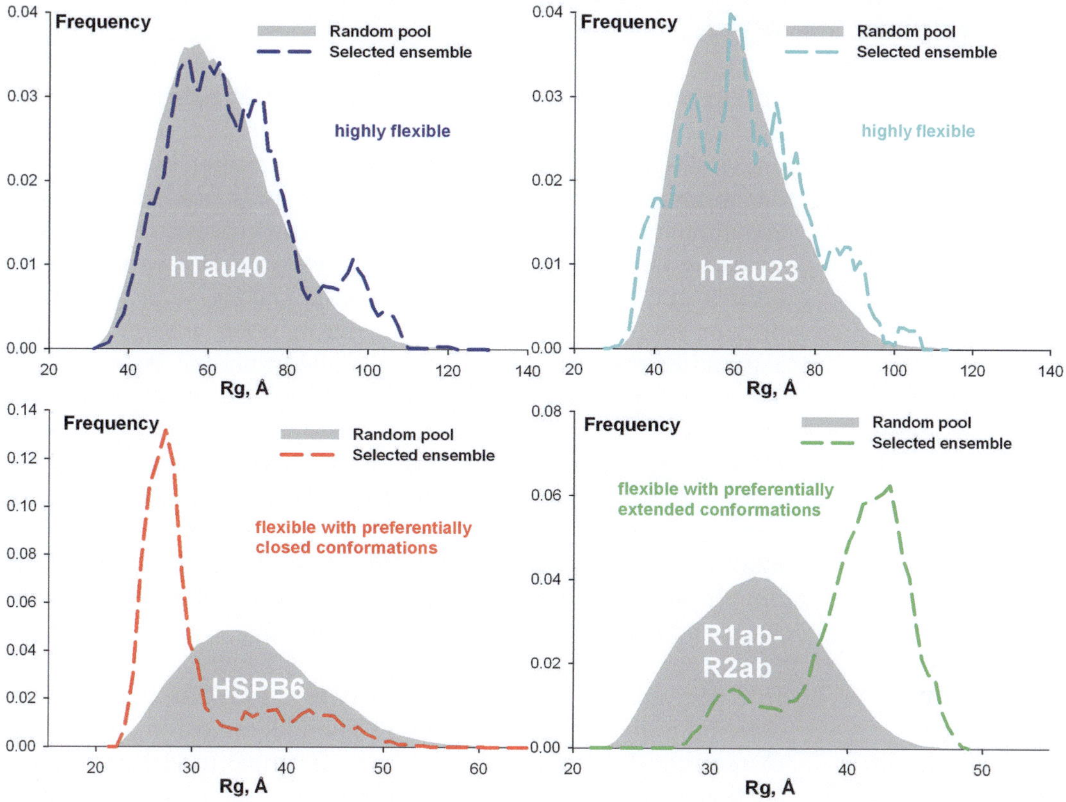

Fig. 3 Radii of gyration (R_g) distribution for flexible systems derived from ensemble modeling with EOM. The area filled with *gray* shows the distribution for the random pool of models; *dashed lines* of different colors correspond to ensembles selected by EOM. Highly flexible constructs, like hTau40, would typically sample a broad range of conformations [22] and the ensemble R_g distribution would be as broad as the random pool (*blue dashed line* vs. *gray filled area*). Less flexible macromolecules show a narrower ensemble R_g distribution compared to the random pool. If there is a certain typically function-related preference towards more closed or opened conformations, the ensemble R_g distribution peak would be either shifted to the left (*red line*; HPSB6 [59]) or to the right (*green line*; R1ab-R2ab [61]), respectively

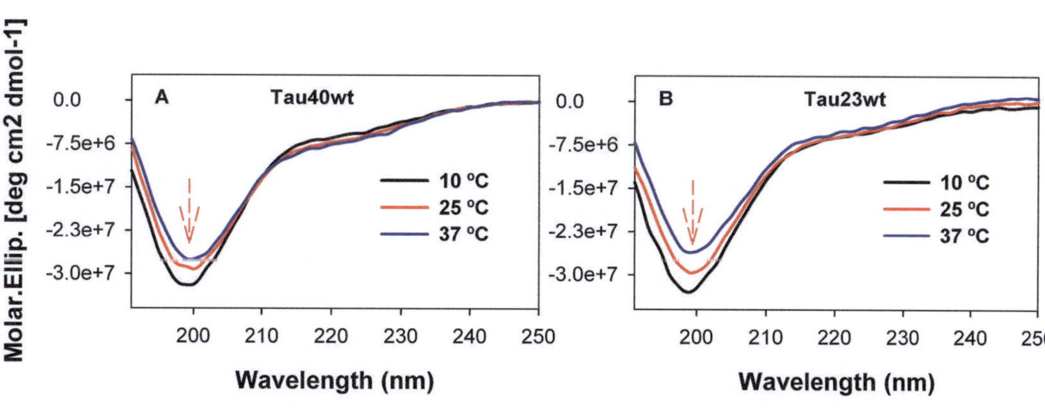

Fig. 4 Conformation of Tau monitored by circular dichroism spectroscopy. CD spectroscopy determines the secondary structural composition with signatures reflecting the molar ellipticities of α-helical, β-pleated sheets, disordered and extended helical conformations. (**a**) and (**b**) CD Spectra of hTau40wt and hTau23wt are performed at three different temperatures; 10, 25 and 37 °C. Here it is noticed that as the temperature increases the absorbance of the circularly polarized light decreases. But there is no change in the conformation of the protein. Note that the dip at 200 nm corresponds to random coil and this dip fills in as the temperature rises, which is consistent throughout the varied temperatures

of Tau cannot be crystallized, leaving molecular details at atomic level unresolved. Nevertheless, the changes in Tau's conformation under different conditions like pH, temperature, ionic strength, denaturation conditions can be studied by CD [44–46]. Our CD experiments (*see* Fig. 4) did not show any changes in secondary structure but at the same time clearly suggested that the negative peak at 200 nm became less pronounced and the value at 217 nm became more negative with an isosbestic point around 210 nm, suggesting that the strengthening of hydrophobic interaction of Tau in solution [23, 44]. Similarly CD revealed the change in the structure of Tau repeat region upon binding to anionic lipids [47]. During aggregation Tau undergoes various structural transitions; from misfolded to transiently folded intermediate, to either oligomers or protofilaments. The transition from oligomers to fibril formation is proven by CD spectroscopy in combination with immunostaining using conformation-dependent antibodies like MC1 and Alz-50 [48].

1.2 Tau Conformation Revealed by Small-Angle X-Ray Scattering (SAXS)

In a first comprehensive SAXS study of various Tau isoforms and deletion mutants, Mylonas et al. [22] highlighted a high potential of recently developed (at that time) ensemble modeling towards analysis of flexible proteins. In particular, data suggested that all constructs are unfolded as judged by high values of radii of gyration and Kratky plots [22]. The EOM results showed that Tau protein is quite loose and highly flexible with the repeat domain, which is considered the core constituent of the PHFs, being more extended than the terminal parts. Interestingly, aggregation-promoting mutations have little effect on the overall shape of the protein in solution, indicating that any changes in aggregation behavior or in the display of abnormal Alzheimer phospho-epitopes appear to be based on local rather than global structural changes [22]. Taken together, these results favor the "paper-clip" model of Tau, proposed previously based on FRET data, which also assumes some possible compaction of Tau [37].

In the following studies, reversible compaction of hTau40 and a mutant that mimics a phosphorylated state is observed upon rapid temperature increase or decrease using SAXS and DLS [23]. Interestingly, the compaction is not observed in the constructs missing one of the repeat domains, R2 [23]. Similar thermal collapse is observed using SAXS although upon gradual temperature increase of hTau40 sample [49]. However, it remained unclear whether the observed compaction is stable and reversible and why the results obtained by Ciasca et al. [49] contradicted the observations made by Shkumatov et al. [23]. In another study using hTau40, anionic DMPG lipid monolayers and X-ray/neutron scattering techniques it is shown that strong interaction of Tau with anionic lipids induces its structural compaction, similar to that of a folded protein, and a membrane disruption [50]. This membrane-induced structural compaction may render the otherwise soluble and stable Tau aggregation-prone and cause seed fibrillation under certain conditions [50]. Taken together, all these results suggest, on one hand, high plasticity of Tau protein and, on the other hand, responsiveness towards changes in external conditions (e.g., temperature or interactions with lipids). Interestingly, the observed compaction agrees well with the reactivity of certain antibodies (Alz-50, MC1, Tau-66, and SMI34) that detect abnormal Tau in early stages of Alzheimer's disease [45, 51, 52], suggesting that a stabilization of the folded state could have pathological consequences.

Recent work by Schwalbe et al. [24] used a hybrid approach, combining various information derived from NMR experiments (CSs, RDCs, PREs) and SAXS, to present amino acid-specific ensemble descriptions of hTau40. Used restraints could describe both long- and short-range conformational sampling of hTau40 and confirm previous findings, suggesting an extended nature of the repeat domain [22, 37]. Moreover, the key finding of the study by Schwalbe et al. [24] is that hTau40 exhibits an enhanced propensity to sample the polyproline II (βP) region of Ramachandran space compared with the statistical coil.

2 Materials

1. Phosphate buffered saline (PBS): 137 mM sodium chloride (NaCl), 3 mM potassium chloride (KCl), 10 mM disodium hydrogen phosphate (Na_2HPO_4), 2 mM potassium dihydrogen phosphate (KH_2PO_4). To make 10× PBS, weigh 80 g NaCl, 2 g KCl, 14.4 g Na_2HPO_4, and 2.4 g KH_2PO_4, dissolve and adjust to 1000 mL with water, filter and store at room temperature.
2. *E. coli* BL21(DE3) competent cells.
3. LB medium containing 100 μg/mL ampicillin.
4. Isopropyl β-D-1-thiogalactopyranoside (IPTG) stock solution: 1 M IPTG in deionized water.
5. Cell lysis buffer: 20 mM sodium 2-(*N*-morpholino)ethanesulfonate (Na-MES) pH 6.8, 1 mM ethylene glycol tetraacetic acid (EGTA), 2 mM $MgCl_2$, 5 mM dithiothreitol (DTT), 1 mM phenylmethylsulfonyl fluoride (PMSF), and protease inhibitor cocktail.
6. 5 M NaCl solution.
7. 1 M DTT solution.
8. Loading buffer: 20 mM Na-MES pH 6.8, 50 mM NaCl, 1 mM EGTA, 1 mM $MgCl_2$, 2 mM DTT, 0.1 mM PMSF, and protease inhibitor cocktail.
9. SP-Sepharose 16/10 column (GE Healthcare).

10. SP-Sepharose buffer A: 20 mM Na-MES pH 6.8, 50 mM NaCl, 1 mM EGTA, 1 mM $MgCl_2$, 2 mM DTT, 0.1 mM PMSF.
11. SP-Sepharose buffer B: 20 mM Na-MES pH 6.8, 1 M NaCl, 1 mM EGTA, 1 mM $MgCl_2$, 2 mM DTT, 0.1 mM PMSF, and protease inhibitor cocktail.
12. Amicon ultracentrifugal filter, molecular weight cutoff 5,000 Da (Millipore).
13. HiLoad Superdex G75, prep grade gel filtration column (GE Healthcare).
14. 2 mM DTT in PBS buffer.
15. Protein sample buffer (CD spectroscopy): 50 mM phosphate buffer, pH 6.8.
16. Bicinchoninic acid (BCA) Protein Assay.
17. 5 μM bovine serum albumin (BSA) in 50 mM phosphate buffer, pH 6.8.
18. 6 M guanidinium hydrochloride.
19. JLA 8.1000 rotor, Avanti J-26 XP centrifuge (Beckmann Coulter).
20. Ti45 rotor, Optima XPN 100 Ultracentrifuge (Beckman Coulter).
21. Cell homogenizer (TS-Series, Constant system).
22. 10–150 mL super loop (GE Healthcare).
23. Jasco 815 CD spectrometer (Jasco Corporation).
24. Nitrogen gas bottle Impurities (in ppm): O_2, 2.0; Moisture, 2.0; Total Hydrocarbon, 0.5 (Vadilal Chemicals Limited).
25. 1 mm path length cuvette Type J/21, Quartz material (Jasco Corporation).
26. SigmaPlot software (Systat Software).

3 Methods

3.1 Expression and Purification of Recombinant Tau Protein Isoforms (See Fig. 1) [13]

1. The glycerol stock of *E. coli* BL21 (DE3) competent cells [53] is inoculated into 100 mL LB medium containing ampicillin (100 μg/mL) and the culture is grown overnight at 37 °C. This primary culture is then used for inoculating a fresh 4 L of LB medium supplemented with ampicillin and grown at 37 °C with shaking until the optical density at 600 nm (OD_{600}) reached 0.6. The same culture is induced with 0.5 mM IPTG (from IPTG stock solution) and further incubated at 37 °C for 4–6 h. The grown cells are harvested by high-speed centrifugation at 6238 × *g* for 20 min and resuspended in cell lysis buffer.

2. The resuspended cells are subjected to mechanical cell disruption by high shear force (15,000–20,000 psi) using a homogenizer. NaCl and DTT are added to disrupted cells to final concentrations of 0.5 M and 5 mM, respectively, followed by boiling at 95 °C for 20 min. The total cell lysate is centrifuged at 164,700 × *g* for 45 min at 4 °C, and the supernatant is collected. The collected supernatant is dialyzed against loading buffer required for further purification.
3. Tau is a positively charged protein at around pH 7.0, thus cationic exchangers are used. The protein is eluted with a gradient of NaCl (varying the ionic strength) from a SP-Sepharose 16/10 column. After the equilibration with eight column volumes of the SP-Sepharose buffer A, the dialyzed protein sample obtained by heating is loaded on the column using a 10–150 mL super loop and the column is washed with five to ten column volumes of SP-Sepharose buffer A to remove unbound proteins. The elution is performed with a linear gradient of SP-Sepharose buffer B in two steps: first from 0 to 60% in six to ten column volumes and then to 100% in two to four column volumes. The eluted fractions are checked on SDS-PAGE and fractions containing the protein of interest are pooled together and concentrated using an ultracentrifugal filter devices.
4. The concentrated protein solution from SP Sepharose column is injected into pre-equilibrated gel filtration column using a 2 mL loop with an injection needle. Isocratic elution is performed with PBS buffer containing 2 mM DTT at a flow rate of 0.5 mL/min. The eluted fractions are analyzed on SDS-PAGE and pure protein fractions are used for further experiments (CD and light scattering).

3.2 Circular Dichroism Spectroscopy

CD spectroscopy is a popular technique for the determination of the secondary structure content and folding properties of peptides and proteins. Moreover, CD can be employed to determine protein stability and detect conformational changes. CD works only for chiral molecules where the difference in the absorption (ΔA) of the left (A_L) and right (A_R) components of polarized light is measured as a function of wavelength.

$$\Delta A = A_L - A_R$$

1. CD is operated in the UV region (350–180 nm). In far UV region, both peptide bond and disulfide generate signals and in near UV region these are due to aromatic amino acids such as tyrosine, tryptophan, and phenylalanine. Spectra vary from protein to protein based on the content of peptide bond, disulfide bond, and aromatic amino acids.
2. The spectrum is generated by plotting the ellipticity (θ) (deg. $cm^2/dmol^{-1}$) on the y-axis against the wavelength (nm) on x-axis.

3. A spectrum in the positive scale indicates absorption of right circularly polarized light and on the negative scale indicates absorption of left circularly polarized light.
4. The β-sheet of a protein has spectra on negative scale between 210 and 220 nm and on positive scale between 195 and 200 nm.
5. The α-helix of the protein has spectra in the negative scale with two crusts of equal magnitude at 222 and 208 nm and on positive scale at 190 nm, whereas the random coil has spectra of large magnitude on the negative scale at 200 nm.
6. Tau protein concentration is estimated by using BCA Protein Assay.
7. The concentration of hTau40wt and hTau23wt is estimated as 26.27 and 18.5 mg/mL, respectively. They are aliquoted and stored in −80 °C (*see* **Note 1**).
8. The soluble Tau protein is centrifuged at 169,000 × *g* for 30 min to remove aggregates (*see* **Note 2**).
9. Spectra for 5 μM BSA are collected at different temperatures (10, 25 and 37 °C) as a standard.
10. The working concentration of hTau40wt and hTau23wt is 5 μM, which is prepared and measured in 50 mM phosphate buffer pH 6.8 (*see* **Note 3**).
11. 1 mm path length cuvette is used. Prior to experiment cuvette is thoroughly rinsed using 6 M guanidinium hydrochloride followed by water and phosphate buffer (*see* **Note 4**).
12. Nitrogen purging is done in order to remove the oxygen present. If present, oxygen will absorb the far-UV and produce ozone, which will cause harm to the optics of the instrument (*see* **Note 5**).
13. Parameters are adjusted as following: photometric mode (CH1), CD; photometric mode (CH2), HT (high tension); sensitivity, standard; DIT (data integration time), 1 s; band width, 1.0 nm; measure range, 250–190 nm; data pitch, 1.0 nm; scanning speed, 100 nm/min; base line correction, none; shutter control, auto; CD detector, PMT (photomultiplier tube); accumulations, ten times.
14. A Peltier element is attached to control the given temperature during the measurement. After adding the sample in the cuvette it is incubated for 5 min at particular temperature to make sure that the sample reached the temperature and then the spectrum reading is performed.
15. The samples are placed into the cell at 10 °C, incubated for 5 min and measured. The same procedure is followed for 25 and 37 °C.

16. The measurement of protein sample buffer is performed at 10, 25, and 37 °C. These values are subtracted from the spectra of hTau40wt and hTau23wt protein to obtain base line corrected spectra (*see* **Note 6**).
17. The spectra raw data is obtained in ellipticity (in milli degrees). This is converted to mean residual ellipticity (MRE), measured in terms of degrees squared centimeter per decimole ($deg{\cdot}cm^2{\cdot}dmol^{-1}$) (*see* **Note 7**).

$$[\theta] = \frac{100(\text{Signal in millideg})}{Cnl}$$

18. The obtained values are plotted using SigmaPlot software.

3.3 Small Angle X-Ray Scattering

Below we provide a concise description of the major steps used for SAXS data collection and analysis of hTau40 and its deletion mutants [23]. This strategy can be used for any other protein construct, in particular if it is partially or fully unfolded. For a more comprehensive description on how to avoid pitfalls during SAXS data collection and analysis, we refer the reader to more general reviews [54, 55].

3.3.1 Data Collection and Processing (See Note 8)

1. Measure standard (BSA or any other standard used at the beamline). Using forward scattering of the standard, molecular mass (MM) of the measured protein can be calculated, based on the relative intensity of the forward scattering, $I(0)$. For example:

$$MM_{\text{sample}} = \frac{I(0)_{\text{sample}}}{I(0)_{\text{BSA}}} \times MM_{\text{BSA}}.$$

2. SAXS relies on a higher contrast of a sample compared to its buffer (background). The first step in SAXS data processing is buffer subtraction, and hence, it is recommended to use buffer from the last step of sample preparation (e.g., flow-through from concentrator) to avoid mismatch between sample and buffer curves at high scattering angles (*see* **Note 8**).
3. Make test measurements to see if the sample is prone to radiation-induced aggregation. In case it is, shorten exposure time or use radioprotectant (glycerol, DTT, Tris (2-carboxymethyl) phosphine (TCEP), etc.).
4. Typically, a dedicated SAXS beamline is equipped with a liquid handling robot. The user can adjust the temperature of the sample storage chamber and measurement cell (*see* **Note 9**).
5. For each sample it is recommended to collect data at three to four different concentration points to evaluate the effect of the concentration (e.g., use 1, 3, and 9 mg/mL aliquots). Values

in concentration series depend on the MM of the sample—higher MM samples give higher signal-to-noise ratio (SNR) and hence lower concentrations can be used.

6. Each sample measurement should be intermittent with the corresponding buffer.
7. Keep an eye on the sample loading to avoid bubbles in the measurement cell. This is especially important if the temperature in the measurement cell is more than 40 °C or if the sample is viscous.
8. Typically, there is an automatic data processing pipeline at dedicated SAXS beamlines. Keep track of any warning (loading problems, bubbles, radiation damage, etc.) and adjust your collection strategy accordingly.
9. Automatic pipelines usually perform initial data processing: radial averaging of individual data frames, their comparison, buffer subtraction, estimation of invariants, etc. Bear in mind that these are preliminary results to mainly assess the sample quality (aggregated or not). It allows one to perform initial quality check and discard all bad curves, typically resulting from radiation-induced aggregation or air bubbles.

3.3.2 Data Evaluation

1. If subtracted curves show negative signal or too high intensity at high angle part of the scattering curve, check for possible sample/buffer mismatch by plotting unsubtracted buffer and sample curves (*see* **Note 10**).
2. Estimate invariants from final buffer-subtracted curve. Radius of gyration, (R_g), maximum dimensions (D_{max}) and forward scattering (I_0) can be evaluated using the Guinier approximation [56], assuming that at very small angles ($s < 1.3/R_g$) the intensity is represented as $I(s) = I(0)\exp(-(sR_g)2/3)$. However, for flexible systems a more reliable estimate of invariants can be obtained from an indirect Fourier transformation of experimental data [57], as it takes the whole scattering curve into account rather than a short region at small angles (*cfr.* Guinier region).
3. For each sample, plot R_g vs. concentration to determine how the data should be processed further. In case of concentration dependency in the collected data, an extrapolated zero-concentration curve (*see* **Note 11**) should be derived and used for further analysis. If there is no concentration effect, the single concentration data set with the highest SNR should be employed for further analysis.
4. Plot the final scattering curve as a Kratky plot ($I(s)*s^2$ as a function of s) to identify the state of the measured construct (folded/fully disordered). The dimensionless Kratky plot ($s^2R_g^2I(s)/I_0$ as a function of sR_g), allows a comparative estimate of the degree of flexibility regardless of a construct size.

Use the dimensionless Kratky plot for the measured construct along with well characterized folded (BSA or lysozyme) and fully unfolded constructs (hTau40 or any other chemically denatured proteins) to assess the degree of flexibility (*see* Fig. 2).

3.3.3 Ensemble Modeling

1. Given that flexible proteins sample a large number of conformations in solution, the best description of SAXS data can be achieved using an ensemble approach. Algorithms, searching for the optimal dummy atom/rigid body arrangement to fit the scattering data best, will not typically produce a converging solution, resulting in different shapes after independent reconstructions. In fact single-model methods should be used for rather rigid constructs, as judged by Kratky plot; else the resulting model (which not necessary will fit the data well) will represent an average over a number of possible conformations in solution and would not fully reflect the dynamic property of the system.
2. The first step in ensemble modeling is to generate a random pool. Commonly used approach—EOM consists of two programs: RANCH and GAJOE. The former can generate *N* models (where *N* is typically 10,000 or more), given a file with a primary sequence of a construct in FASTA format. If a partial structure is available, it can be used as a rigid body with a possibility of applying symmetry operation should the oligomer interface be known. At this stage, it is necessary to specify the Cα-angle distribution used for modeling the missing regions (e.g., "Random-coil" uses a Cα-angle distribution consistent with chemically denatured proteins, while "native-like" uses a Cα-angles distribution consistent with disordered proteins). During this step, a theoretical scattering curve is calculated for each generated model (*see* **Note 12**).
3. In the second step, a genetic algorithm (GAJOE) and subtracted scattering data are used to select subsets of theoretical curves, which if averaged, produce the fit with the lowest discrepancy to the subtracted experimental scattering data. At this stage, an ensemble of models and R_g distribution file are reported.
4. Typically, the genetic algorithm should be run five to ten times. The discrepancy and selected models may somewhat vary from run to run, but the distribution of R_g for selected ensemble vs. random pool should be consistent. In other words, the algorithm should pick the models with similar descriptors (R_g and D_{max}).
5. Inspect the R_g distribution file. Broad R_g distribution of selected ensemble similar to distribution of a random pool indicates high flexibility of a construct, whereas R_g distribution shifted to the left or right from the random pool indicates preference towards more compact or extended conformations, respectively (*see* Fig. 3).

4 Notes

1. Tau is a soluble protein, which does not form any precipitates, but it aggregates if not handled properly.
2. Prior to the end of pelleting only supernatant is collected without touching the bottom of the tube, in order to not disturb the pellet, if any formed, otherwise the particle settled.
3. 5 μM hTau40wt (0.23 mg/mL) and hTau23wt (0.184 mg/mL) protein is prepared in 50 mM phosphate buffer pH 6.8. Initially 20 μM stock is prepared, from which 5 μM working solution is made.
4. Cuvettes of path length ranging from 1 mm to 1 cm are used based on the concentration of the sample. Concentrations ranging from 0.05 to 0.2 mg/mL for 1 mm path length cuvettes, 0.2–1 mg/mL proteins in 0.01–0.02 cm path length cuvettes and 0.005–0.01 mg/mL proteins in 1 cm path length cuvettes are recommended [58].
5. Oxygen absorbs the light in the far-UV region; this prevents the sample from absorbing the light and thus, interferes with the signal.
6. Base line correction is performed manually by subtracting the spectra of phosphate buffer at different temperatures with hTau40wt and hTau23wt protein spectrum.
7. Mean Residual Ellipticity (MRE) is calculated by using the above mentioned formula. The raw output value in mill degrees is multiplied by 100 and this whole value is divided by product of C (concentration of protein in mM; 0.005), n (no. of amino acid residues; 441), and l (path length; 1 mm).
8. Prior to SAXS experiment check for sample homogeneity. A single symmetric peak during size exclusion chromatography and single band on SDS-PAGE are good indicators of samples homogeneity. Additionally, one can detect polydispersity or aggregation using dynamic light scattering (DLS). A polydispersity value of 20% or less signifies a very-well-behaved sample.
9. Check for sample/buffer mismatch, which is manifested by over- or under-subtraction, when buffer scatters more or less than the sample, respectively, at high angle part of the curve. It is a good practice to use buffer from the last sample preparation step (note that even residual glycerol can result in buffer mismatch problems as it scatters very well). Mismatch can also stem from air bubbles present in the sample or a dirty measuring cell window. In the latter case, contact beamline personnel to clean or change the measuring cell window.

10. Heating of sample storage or measurement chamber is much faster compared to cooling.
11. In case there is a concentration effect observed as judged by differences in low angle part of SAXS curves from a concentration series, avoid using very high (>10 mg/mL) or very low concentrations during extrapolation due to possible strong interparticle interactions affecting the curve beyond Guinier region or very low SNR, respectively. To avoid concentration effect, one can consider dialyzing sample in a different buffer.
12. One random pool can be generated using RANCH program from EOM suite and used multiple times for the sample construct recorded at different conditions (temperature, salinity, pH, etc.). It is also possible to generate several random pools and use them together in the genetic algorithm.

Acknowledgments

The authors thank Dr. Hirekodathakallu V. Thulasiram (CSC0130) for his excellent Molecular Biology Laboratory (MBL) facility at the CSIR-NCL, Pune. Tau constructs were kindly gifted by Prof. Roland Brandt from University of Osnabruck, Germany and Prof. Jeff Kuret from Ohio State University College of Medicine, USA. N.V.G. acknowledges research fellowship from University of Grant Commission (UGC) by government of India. This project is supported in part by grants from the DST-SERB SB/YS/LS-355/2013 and 12th 5-year plan CSIR-Network Project BSC0115. The authors acknowledge Dr. Ranjan Singh, Shweta Sonawane, Dr. Yann Stercks, and Dr. Ritika Sethi for proofreading the manuscript and for useful comments. The authors have declared no conflict of interest.

References

1. Hirokawa N (1988) Tau proteins: the molecular structure and mode of binding on microtubules. J Cell Biol 107(4):1449–1459
2. Hirokawa N, Takemura R (2005) Molecular motors and mechanisms of directional transport in neurons. Nat Rev Neurosci 6(3):201–214
3. Mandelkow E, von Bergen M, Biernat J et al (2007) Structural principles of tau and the paired helical filaments of Alzheimer's disease. Brain Pathol 17(1):83–90
4. Illenberger S, Zheng-Fischhofer Q, Preuss U et al (1998) The endogenous and cell cycle-dependent phosphorylation of tau protein in living cells: implications for Alzheimer's disease. Mol Biol Cell 9(6):1495–1512
5. Couchie D, Mavilia C, Georgieff IS et al (1992) Primary structure of high molecular weight tau present in the peripheral nervous system. Proc Natl Acad Sci 89(10):4378–4381
6. Goedert M, Wischik CM, Crowther RA et al (1988) Cloning and sequencing of the cDNA encoding a core protein of the paired helical filament of Alzheimer disease: identification as the microtubule-associated protein tau. Proc Natl Acad Sci 85(11):4051–4055
7. Binder LI, Guillozet-Bongaarts AL, Garcia-Sierra F et al (2005) Tau, tangles, and Alzheimer's disease. Biochim Biophys Acta 1739(2–3):216–223

8. Garcia ML, Cleveland DW (2001) Going new places using an old MAP: tau, microtubules and human neurodegenerative disease. Curr Opin Cell Biol 13(1):41–48
9. Glatz DC, Rujescu D, Tang Y et al (2006) The alternative splicing of tau exon 10 and its regulatory proteins CLK2 and TRA2-BETA1 changes in sporadic Alzheimer's disease. J Neurochem 96(3):635–644
10. von Bergen M, Barghorn S, Li L et al (2001) Mutations of tau protein in frontotemporal dementia promote aggregation of paired helical filaments by enhancing local beta-structure. J Biol Chem 276(51):48165–48174
11. von Bergen M, Friedhoff P, Biernat J et al (2000) Assembly of tau protein into Alzheimer paired helical filaments depends on a local sequence motif (306VQIVYK311) forming beta structure. Proc Natl Acad Sci 97(10):5129–5134
12. Mukrasch MD (2005) Sites of tau important for aggregation populate-structure and bind to microtubules and polyanions. J Biol Chem 280(26):24978–24986
13. von Bergen M, Barghorn S, Biernat J et al (2005) Tau aggregation is driven by a transition from random coil to beta sheet structure. Biochim Biophys Acta (BBA) 1739(2–3):158–166
14. von Bergen M, Barghorn S, Jeganathan S et al (2006) Spectroscopic approaches to the conformation of tau protein in solution and in paired helical filaments. Neurodegener Dis 3(4–5):197–206
15. von Bergen M, Barghorn S, Müller SA et al (2006) The core of tau-paired helical filaments studied by scanning transmission electron microscopy and limited proteolysis. Biochemistry 45(20):6446–6457
16. Chen J, Kanai Y, Cowan NJ et al (1992) Projection domains of MAP2 and tau determine spacings between microtubules in dendrites and axons. Nature 360(6405):674–677
17. Magnani E, Fan J, Gasparini L et al (2007) Interaction of tau protein with the dynactin complex. EMBO J 26(21):4546–4554
18. Butner KA (1991) Tau protein binds to microtubules through a flexible array of distributed weak sites. J Cell Biol 115(3):717–730
19. Al-Bassam J, Ozer RS, Safer D et al (2002) MAP2 and tau bind longitudinally along the outer ridges of microtubule protofilaments. J Cell Biol 157(7):1187–1196
20. Santarella RA, Skiniotis G, Goldie KN et al (2004) Surface-decoration of microtubules by human tau. J Mol Biol 339(3):539–553
21. Mukrasch MD, Bibow S, Korukottu J et al (2009) Structural polymorphism of 441-residue tau at single residue resolution. PLoS Biol 7(2):e34
22. Mylonas E, Hascher A, Bernadó P et al (2008) Domain conformation of tau protein studied by solution small-angle x-ray scattering. Biochemistry 47(39):10345–10353
23. Shkumatov AV, Chinnathambi S, Mandelkow E et al (2011) Structural memory of natively unfolded tau protein detected by small-angle x-ray scattering. Proteins 79(7):2122–2131
24. Schwalbe M, Ozenne V, Bibow S et al (2014) Predictive atomic resolution descriptions of intrinsically disordered hTau40 and α-synuclein in solution from NMR and small angle scattering. Structure 22(2):238–249
25. Bernadó P, Svergun DI (2012) Analysis of intrinsically disordered proteins by small-angle x-ray scattering. In: Intrinsically disordered protein analysis. Springer Science + Business Media, Berlin. doi:10.1007/978-1-4614-3704-8_7
26. Doniach S (2001) Changes in biomolecular conformation seen by small angle x-ray scattering. Chem Rev 101(6):1763–1778
27. Durand D, Vivès C, Cannella D et al (2010) NADPH oxidase activator p67phox behaves in solution as a multidomain protein with semi-flexible linkers. J Struct Biol 169(1): 45–53
28. Ozenne V, Bauer F, Salmon L et al (2012) Flexible-meccano: a tool for the generation of explicit ensemble descriptions of intrinsically disordered proteins and their associated experimental observables. Bioinformatics 28(11): 1463–1470
29. Pelikan M, Hura G, Hammel M (2009) Structure and flexibility within proteins as identified through small angle x-ray scattering. Gen Physiol Biophys 28(2):174–189
30. Różycki B, Kim YC, Hummer G (2011) SAXS ensemble refinement of ESCRT-III CHMP3 conformational transitions. Structure 19(1): 109–116
31. Yang S, Blachowicz L, Makowski L et al (2010) Multidomain assembled states of Hck tyrosine kinase in solution. Proc Natl Acad Sci 107(36):15757–15762
32. Zheng W, Tekpinar M (2011) Accurate flexible fitting of high-resolution protein structures to small-angle x-ray scattering data using a coarse-grained model with implicit hydration shell. Biophys J 101(12):2981–2991
33. Bernadó P, Mylonas E, Petoukhov MV et al (2007) Structural characterization of flexible proteins using small-angle x-ray scattering. J Am Chem Soc 129(17):5656–5664
34. Kikhney AG, Svergun DI (2015) A practical guide to small angle x-ray scattering (SAXS) of flexible and intrinsically disordered proteins. FEBS Lett 589(19 Pt A):2570–2577

35. Jeganathan S, Chinnathambi S, Mandelkow E-M et al (2012) Conformations of microtubule-associated protein tau mapped by fluorescence resonance energy transfer. In: Amyloid proteins, Methods in molecular biology. Springer, Berlin. doi:10.1007/978-1-61779-551-0_7
36. Jeganathan S, Hascher A, Chinnathambi S et al (2008) Proline-directed pseudo-phosphorylation at AT8 and PHF1 epitopes induces a compaction of the paperclip folding of tau and generates a pathological (MC-1) conformation. J Biol Chem 283(46):32066–32076
37. Jeganathan S, von Bergen M, Brutlach H et al (2006) Global hairpin folding of tau in solution. Biochemistry 45(7):2283–2293
38. Li L, von Bergen M, Mandelkow EM et al (2002) Structure, stability, and aggregation of paired helical filaments from tau protein and FTDP-17 mutants probed by tryptophan scanning mutagenesis. J Biol Chem 277(44): 41390–41400
39. von Bergen M, Li L, Mandelkow E (2004) Intrinsic fluorescent detection of tau conformation and aggregation. In: Amyloid proteins. Springer Science + Business Media, Berlin. doi:10.1385/1-59259-874-9:175
40. Schweers O, Schonbrunn-Hanebeck E, Marx A et al (1994) Structural studies of tau protein and Alzheimer paired helical filaments show no evidence for beta-structure. J Biol Chem 269(39):24290–24297
41. Frost B, Ollesch J, Wille H et al (2008) Conformational diversity of wild-type tau fibrils specified by templated conformation change. J Biol Chem 284(6):3546–3551
42. Antzutkin ON, Iuga D, Filippov AV et al (2012) Hydrogen bonding in Alzheimer's amyloid-beta fibrils probed by 15N{17O} REAPDOR solid-state NMR spectroscopy. Angew Chem Int Ed Engl 51(41):10289–10292
43. López Deber MP, Hickman DT, Nand D et al (2014) Engineering amyloid-like assemblies from unstructured peptides via site-specific lipid conjugation. PLoS One 9(9):e105641
44. Jeganathan S, von Bergen M, Mandelkow E-M et al (2008) The natively unfolded character of tau and its aggregation to Alzheimer-like paired helical filaments. Biochemistry 47(40):10526–10539
45. Lichtenberg-Kraag B, Mandelkow EM, Biernat J et al (1992) Phosphorylation-dependent epitopes of neurofilament antibodies on tau protein and relationship with Alzheimer tau. Proc Natl Acad Sci 89(12): 5384–5388
46. Skrabana R, Kontsek P, Mederlyova A et al (2004) Folding of Alzheimer's core PHF subunit revealed by monoclonal antibody 423. FEBS Lett 568(1–3):178–182
47. Künze G, Barré P, Scheidt HA et al (2012) Binding of the three-repeat domain of tau to phospholipid membranes induces an aggregated-like state of the protein. Biochim Biophys Acta 1818(9):2302–2313
48. Maeda S, Sahara N, Saito Y et al (2007) Granular tau oligomers as intermediates of tau filaments. Biochemistry 46(12):3856–3861
49. Ciasca G, Campi G, Battisti A et al (2012) Continuous thermal collapse of the intrinsically disordered protein tau is driven by its entropic flexible domain. Langmuir 28(37): 13405–13410
50. Jones EM, Dubey M, Camp PJ et al (2012) Interaction of tau protein with model lipid membranes induces tau structural compaction and membrane disruption. Biochemistry 51(12):2539–2550
51. Ghoshal N, Garcia-Sierra F, Fu Y et al (2001) Tau-66: evidence for a novel tau conformation in Alzheimer's disease. J Neurochem 77(5): 1372–1385
52. Jicha GA, Bowser R, Kazam IG et al (1997) Alz-50 and MC-1, a new monoclonal antibody raised to paired helical filaments, recognize conformational epitopes on recombinant tau. J Neurosci Res 48(2):128–132
53. Hanahan D, Jessee J, Bloom FR (1991) Plasmid transformation of Escherichia coli and other bacteria. Methods Enzymol 204: 63–113
54. Jacques DA, Trewhella J (2010) Small-angle scattering for structural biology–expanding the frontier while avoiding the pitfalls. Protein Sci 19(4):642–657
55. Skou S, Gillilan RE, Ando N (2014) Synchrotron-based small-angle x-ray scattering of proteins in solution. Nat Protoc 9(7): 1727–1739
56. Guinier A (1939) La diffraction des rayons X aux très petits angles; application à l'étude de phénomènes ultramicroscopiques. Ann Phys (Paris) 1939(12):161–237
57. Perez J, Vachette P, Russo D et al (2001) Heat-induced unfolding of neocarzinostatin, a small all-beta protein investigated by small-angle x-ray scattering. J Mol Biol 308(4): 721–743
58. Greenfield NJ (2007) Using circular dichroism spectra to estimate protein secondary structure. Nat Protoc 1(6):2876–2890
59. Weeks SD, Baranova EV, Heirbaut M et al (2014) Molecular structure and dynamics of the dimeric human small heat shock protein HSPB6. J Struct Biol 185(3):342–354

60. Varadi M, Kosol S, Lebrun P et al (2014) pE-DB: a database of structural ensembles of intrinsically disordered and of unfolded proteins. Nucleic Acids Res 42(Database issue):D326–D335

61. Zoll S, Schlag M, Shkumatov AV et al (2012) Ligand-binding properties and conformational dynamics of autolysin repeat domains in staphylococcal cell wall recognition. J Bacteriol 194(15):3789–3802

Chapter 2

Global Conformation of Tau Protein Mapped by Raman Spectroscopy

Nalini Vijay Gorantla, Puneet Khandelwal, Pankaj Poddar, and Subashchandrabose Chinnathambi

Abstract

Alzheimer's disease (AD) is one of the neurodegenerative disease characterized by progressive neuronal loss in the brain. Its two major hallmarks are extracellular senile plaques and intracellular neurofibrillary tangles (NFTs), formed by aggregation of amyloid β-42 (Aβ-42) and Tau protein respectively. Aβ-42 is a transmembrane protein, which is produced after the sequential action of β- and γ-secretases, thus obtained peptide is released extracellularly and gets deposited on the neuron forming senile plaques. NFTs are composed of microtubule-associated protein-Tau (MAPT). Tau protein's major function is to stabilize the microtubule that provides a track on which the cargo proteins are shuttled and the stabilized microtubule also maintains shape and integrity of the neuronal cell. Tau protein is subjected to various modifications such as phosphorylation, ubiquitination, glycation, acetylation, truncation, glycosylation, deamination, and oxidation; these modifications ultimately lead to its aggregation. Phosphorylation is the major modification and is extensively studied with respect to Tau protein. Tau protein, however, undergoes certain level of phosphorylation and dephosphorylation, which regulates its affinity for microtubule and ultimately leading to microtubule assembly and disassembly. Our main aim was to study the native state of longest isoform of Tau (hTau40WT-4R2N) and its shortest isoform, (hTau23WT-3R0N), at various temperatures such as 10, 25, and 37 °C. Raman spectroscopic results suggested that the proportion of random coils or unordered structure depends on the temperature of the protein environment. Upon increase in the temperature from 10 to 37 °C, the proportion of random coils or unordered structures increased in the case of hTau40WT. However, we did not find a significant effect of temperature on the structure of hTau23WT. This current approach enables one to analyze the global conformation of soluble Tau in solution.

Key words Tau protein, Tau conformation, Tau aggregation, Alzheimer disease, Sedimentation assay, Raman spectroscopy

1 Introduction

Tau belongs to a class of proteins called microtubule-associated proteins (MAPs) whose major function is to stabilize microtubules (MT) that serve as cytoskeleton. Microtubules are important for cell viability, cell polarity, and particularly for development of neurons. The main biological function of Tau is to stimulate

Caroline Smet-Nocca (ed.), *Tau Protein: Methods and Protocols*, Methods in Molecular Biology, vol. 1523,
DOI 10.1007/978-1-4939-6598-4_2, © Springer Science+Business Media New York 2017

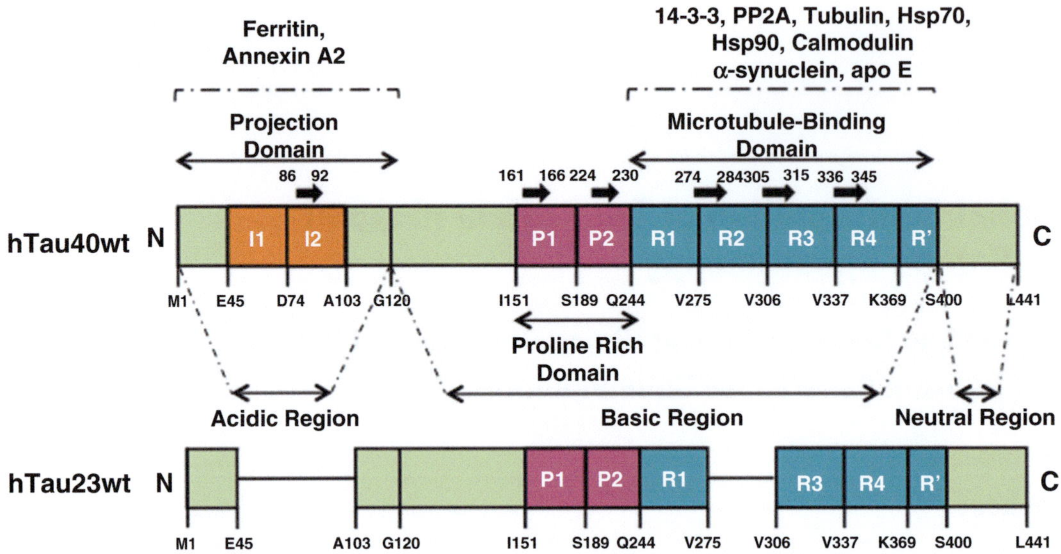

Fig. 1 Schematic presentation of Human Tau protein. Tau protein has an overall basic charge with N-terminal acidic region, basic repeat region and neutral C-terminal. Bar diagrams of hTau40WT and hTau23WT, the longest and the shortest isoform respectively are depicted. hTau40WT has 441 amino acids that include two inserts, I1 and I2, encoded by exon 2 and 3, located at the N-terminal end. This is followed by P1 and P2, the proline-rich regions. R1, R2, R3, and R4 are the four imperfect repeats present towards the C-terminal end. These repeats are binds to tubulin and also responsible for its assembly into microtubule. Many proteins interact with Tau in its N-terminal as well as in the repeat region. Tau protein is highly soluble with no secondary structures, but however it forms transient β-sheet, indicated by black arrows. hTau23WT is the shortest isoform that lacks I1, I2, and R2, which are encoded by exon 2, 3, and 10

microtubule assembly and to stabilize the structure of microtubules (Fig. 1). Tau is a phosphoprotein and its phosphorylation state is developmentally regulated [1–7]. Tau protein is one of the most soluble, natively unfolded, which does not adapt any secondary structure (Fig. 1). However, it forms transient α-helix or β-sheet structures on binding to its interacting partners. Tau protein in Alzheimer's disease (AD) condition forms toxic β-rich aggregates, leading to neurodegeneration. Tau protein has been studied using different spectroscopic techniques in order to understand its structural conformations. The analysis of sedimentation assay and circular dichroism (CD) showed Tau to be a highly asymmetric molecule with very little secondary structure in solution [8–10]. The extensive investigation of Tau structure in solution using X-ray scattering and CD revealed that Tau behaves as a random Gaussian coil with persistence length of ~2 nm [11–15]. Intrinsic fluorescence analysis of tryptophan mutants of Tau confirmed that residues along the polypeptide chain were indeed completely solvent exposed, supporting the lack of structure [16]. All these observations clearly demonstrate that Tau in solution does not contain any secondary structure and can be regarded as 'natively unfolded protein' [11, 17–19]. In addition, nuclear magnetic resonance (NMR) was performed for Tau protein in its longest as well as shortest isoform; Mandelkow [16, 20–25] and Lippens group in 2006 [26, 27] elucidated the use of NMR to study the residual structures not only in the soluble form but also in the aggregates of Tau protein. This revealed the highly dynamic property of Tau and also showed that transient long-range interactions are necessary for aggregate formation [16, 20–25]. Further efforts to visualize Tau in electron microscope suffered from its low contrast due to its low-z number as well as hydrophilic nature (carbon coated TEM grids being hydrophobic-resulting into poor coating) [28], but the glycerol-spray technique clearly revealed the structure of Tau to be an elongated and flexible rod, about 35 nm in length [29]. The global hairpin conformation of Tau protein was explained using fluorescence resonance energy transfer (FRET) technique and electron paramagnetic resonance (EPR) [16, 30–32]. EPR showed the importance of second and third repeat in the formation of protease resistant core region of aggregates [32]. Tau adopts a "paperclip" conformation, whereby the N- and C-terminal domains approach each other and also the repeat domain [30, 32]. Tau protein is subjected to various insults before aggregating and one such is hyperphosphorylation by many cellular kinases. There are many distinct sites of phosphorylation; these were answered using matrix-assisted laser desorption/ionization-Fourier transform ion cyclotron resonance-mass spectroscopy (MALDI-FTICR-MS) [33]. In addition, small-angle X-ray scattering (SAXS) and FRET were used to obtain insight into the structure of the Tau protein [16, 18, 30, 31, 34–37]. These methods revealed that 441-residue Tau is highly dynamic in solution with a distinct domain character and an intricate network of transient long-range contacts important for pathogenic aggregation [28, 37]. Several attempts were made to understand the structure of Tau in soluble as well as aggregated form, but still it remains partially unresolved. The plausible reason is that being natively unfolded, Tau protein may not have static structures (α-helix or β-sheet), but possibly it may adapt an ensemble of dynamically interchanging secondary structure conformations, which limits the application of standard structural biology tools for structural determination [38]. Raman spectroscopy has the potential to characterize the change in the conformational states induced by environmental changes.

Raman spectroscopy has been widely used to noninvasively investigate the changes in secondary structure at all stages of protein aggregation and amyloid fibril formation. Raman spectroscopy is an inelastic light scattering technique that scatters the monochromatic light on interacting with the sample (Fig. 2). The wavelength of the light emitted from the sample will be higher (Stokes scattering) or lower (anti-Stokes scattering) depending on the loss or gain of the energy by the light photon. Nevertheless, in Raman spectroscopy, Stokes scattering is considered in the spectrum due to its higher intensity. But only 1 in 10^7 photons exhibits Stokes scattering due to its feeble scattering cross-section, anti-stoke

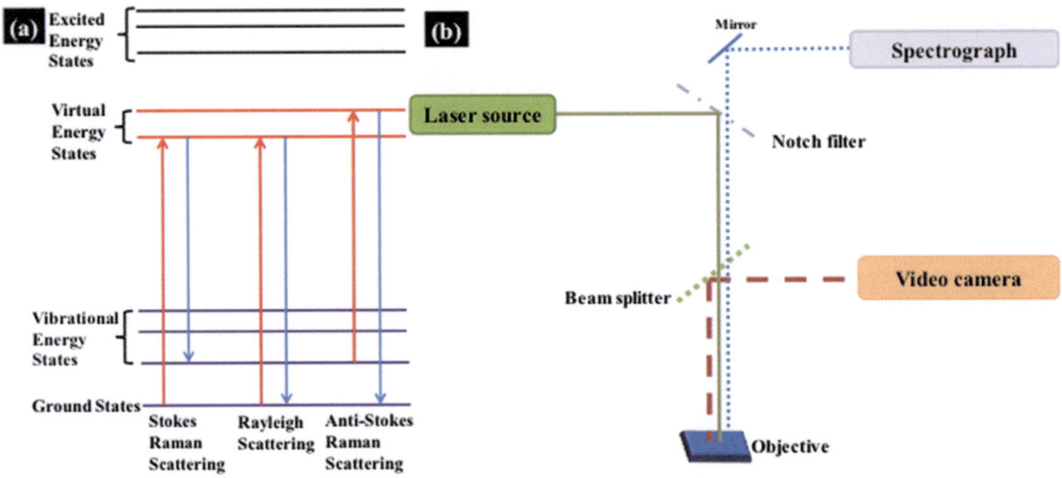

Fig. 2 (**a**) Principle of Raman effect and (**b**) instrumentation. When the sample illuminated with monochromatic laser beam (*green color*), the exchange of quantum vibrational energy occurs, which results into the change in the energy of scattered light. However, most of the scattered light has the same energy (elastic scattering) as the incident light and referred as Rayleigh scattering. Only the very small amount (1 in 10^7 photons) of scattered light (elastic scattering) can be of higher or lower vibrational energies depending on the vibrational state of the molecule and can be referred as Stokes and anti-Stokes Raman scattering. The Raman shifted light (*blue color*) will filter from notch filter and detected by CCD detector

being still lesser and Rayleigh's scattering being more prominent. In order to increase the scattering cross-section, there are several enhancements in Raman spectroscopy, such as surface-enhanced Raman spectroscopy (SERS), tip-enhanced Raman spectroscopy (TERS), Raman optical activity (ROA), resonance Raman, coherent anti-Stokes Raman, and deep UV resonance Raman spectroscopy (DUVRR), etc. The β-sheet conformation of poly-(L-lysine) was studied using vibrational ROA and the structural transitions were also monitored against the function of time [39]. The Tau protein before forming stable aggregates goes through several intermediates structures, such as oligomers, protofilaments, and fibrils. Raman spectroscopy can be used to study these transitions in structure and trace their conformational path [40]. Raman is also coupled to techniques such as atomic force microscopy (AFM), scanning electron microscopy (SEM), and vibrational circular dichroism (VCD) to study the morphology of the amyloid fibrils [41–43]. Elucidation of the structural details of these aggregated proteins is challenging. The morphology of aggregates are different based on different interacting partners, such as metal ions [42] and lipids [44]. Similarly, the effect of temperature, pH, ionic strength, etc. on the protein can be studied using Raman spectroscopy [43]. Recently, Ramachandran et al., employed the UV resonance Raman spectroscopy to look at the signatures of changes in secondary structure and side-chain packing during fibril formation by the four repeat functional domain of Tau in the presence of the inducer heparin [45]. However, there is a lack of Raman study on the natively unfolded Tau protein and its temperature dependence. In this chapter, hTau40WT and hTau23WT are studied in their native condition using a combination of sedimentation analysis and Raman spectroscopy as a function of temperature such as 10, 25, and 37 °C.

The Raman spectroscopic technique is established as a very useful probe to study the structure of natively unfolded proteins (one of the example is Tau protein) [46, 47]. In view of this fact, the Raman spectroscopic measurements were performed on hTau40WT and hTau23WT at three different temperatures such as 10, 25, and 37 °C (Fig. 3a) to analyze the effect of temperature on the secondary structural content of these proteins. Figure 3a shows the Raman spectra of hTau40WT and hTau23WT proteins and the Fig. 3b shows the enlarged area from 1600 to 1700 cm^{-1}, which compares the amide I band for all the cases. Amide I band is very important in assigning the peptide backbone conformation in protein structure [45]. Generally, the presence of an amide I band centered between 1650 and 1658 cm^{-1} shows the presence of a high α-helical content, the Raman band in the range 1660–1665 cm^{-1} shows high proportions of random coil, and the Raman band in the range 1665–1680 cm^{-1} indicated unordered structure, and β-sheet structures. It can be easily noticed from the Fig. 3b that the intensity of amide I band was more in the case of hTau40WT in comparison to hTau23WT. This observation supports the fact that hTau23WT is a shorter construct of hTau40WT. Raman spectra for hTau40WT at different temperatures show that the structure of the protein is dynamic in nature and can be changed upon change in the environment temperature. The amide I band of the hTau40WT was broad, which suggest the presence of various

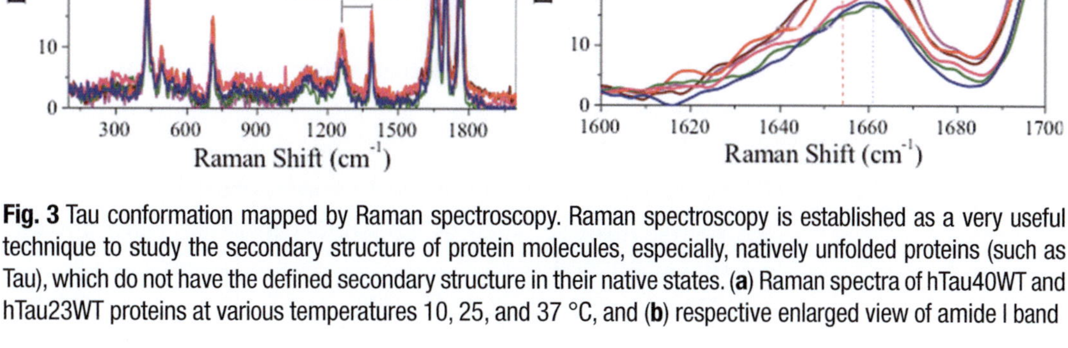

Fig. 3 Tau conformation mapped by Raman spectroscopy. Raman spectroscopy is established as a very useful technique to study the secondary structure of protein molecules, especially, natively unfolded proteins (such as Tau), which do not have the defined secondary structure in their native states. (**a**) Raman spectra of hTau40WT and hTau23WT proteins at various temperatures 10, 25, and 37 °C, and (**b**) respective enlarged view of amide I band

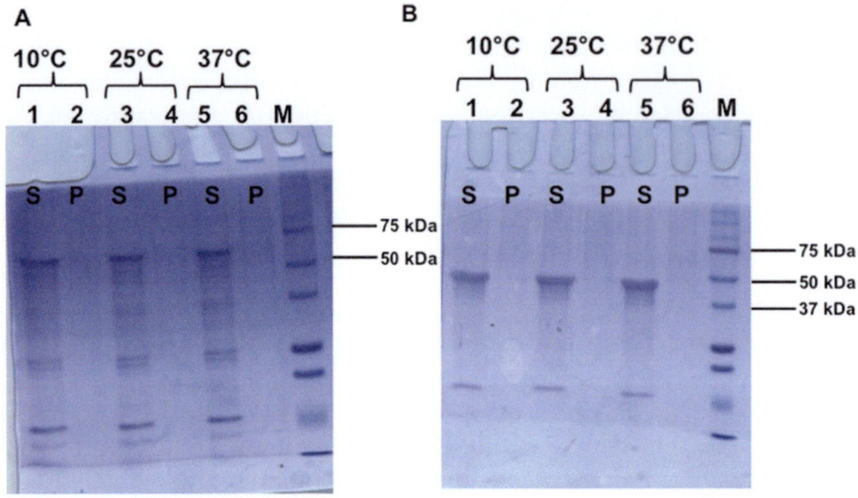

Fig. 4 Sedimentation assay of soluble hTau40WT (**a**) and hTau23WT (**b**) at different temperatures (10, 25, and 37 °C). Supernatant and pellet fractions obtained after sedimentation assay at different temperatures were run on 10 % SDS-polyacrylamide gel. Notation 1 and 2; 3 and 4; 5 and 6 denotes supernatant and pellet at 10, 25, and 37 °C, respectively on the SDS-PAGE. Soluble protein was found in supernatant fraction. These results clearly suggest that the structural changes solely come from temperature

secondary structures in the protein molecule, and can be considered as a doublet. The spectra recorded at 10 °C for hTau40WT shows the most prominent peak ~1654 cm^{-1}, and the less intense peak ~1664 cm^{-1}, related to a high proportion of α-helix and a small proportion of random coils as well as β-sheets. However, upon further increase in the temperature to 25 and 37 °C, leads to the clear shift of the most prominent peak to ~1661 cm^{-1} which indicates the presence of a high proportion of random coils or unordered structures, apparently, with a small proportion of α-helix and β-sheets. However, the significant changes in the structure of hTau23WT, as a function of environmental temperature was not observed. The presence of amide I band ~1661 cm^{-1} shows the presence of a high proportion of random coils or unordered structure at all the three temperatures (10, 25, and 37 °C).

2 Materials

1. Protein sample buffer (Raman spectroscopy): 50 mM phosphate buffer, pH 6.8.
2. BCA Protein Assay.
3. 5 μM BSA in 50 mM phosphate buffer, pH 6.8.
4. Microcentrifuge, such as Eppendorf 5418 R; rotor FA-45-18-11 (Eppendorf).
5. SDS Laemmli sample buffer 5X.

6. Coomassie Brilliant Blue R-250 staining solution: 45 % methanol, 10 % glacial acetic acid, 45 % water, 2.5 g/L Coomassie Brilliant Blue R-250; dissolve 2.5 g Coomassie Brilliant Blue R-250 in 450 mL methanol, mix 100 mL acetic acid to 450 mL water and add to the Coomassie dye solution, and filter.
7. HR-800 Raman spectrophotometer (Jobin Yvon-Horiba, France).

3 Methods

3.1 Protein Preparation

Expression of Tau isoforms (*see* Fig. **1**) in *E. coli* and purification is described in Chapter 1 and in [48] (*see* **Notes 1** and **2**).

3.2 Tau Conformation Monitored by Sedimentation Assay

1. The protein concentrations are measured by using bicinchoninic acid (BCA) assay and concentrations of hTau40WT and hTau23WT are estimated as 26.3 and 18.5 mg/mL, respectively. Bovine serum albumin (BSA) is used as standard. The proteins are aliquoted and stored them in −80 °C (*see* **Note 1**).
2. The protein aliquot at −80 °C is allowed to thaw on ice and is centrifuged at 4 °C and 20,800 × *g* for 1 h (*see* **Notes 2** and **3**).
3. 50 μL of 1 mg/mL protein dilutions of soluble Tau (hTau40WT and hTau23WT) are prepared and incubated at 10, 25, and 37 °C for 30 min. After incubation samples are centrifuged at 20,800 × *g* for 30 min at 10, 25, and 37 °C respectively (*see* **Notes 4** and **5**).
4. 50 μL supernatant is carefully transferred into other tube after centrifugation. The pellet is resuspended in 50 μL of Protein sample buffer (*see* **Note 6**).
5. SDS sample buffer is added to supernatant and pellet fraction and these are resolved by 10 % SDS-polyacrylamide gel electrophoresis (Fig. 4a, b).
6. The percentage of Tau isoforms (hTau40WT and hTau23WT) in the supernatants and pellets are quantified by densitometry of Coomassie Brilliant Blue R-250 stained gels (Fig. 4a, b). These results clearly suggest that the structural changes solely come from temperature.

3.3 Tau Conformation Monitored by Raman Spectroscopy

1. Raman spectra are recorded on a HR-800 Raman spectrophotometer equipped with achromatic Czerny-Turner type monochromator with silver treated mirrors and 800 mm focal length. The monochromatic radiation emitted by a He–Ne laser (633 nm), operating at 20 mW is used as a source. The instrument has Raman-shift detection accuracy of ±1 cm^{-1} between 100 and 2000 cm^{-1}. The instrument is equipped with thermoelectrically cooled (with Peltier junctions), multi-channel, spectroscopic-grade charge coupled device (CCD) detec-

tor (1024 × 256 pixels of 26 μm) with dark current lower than 0.002 electrons pixel^{-1} s^{-1} using a 50× objective.

2. In order to record the temperature dependent Raman spectra using cooling-heating stage, the liquid protein samples are mounted on a LINKAM THMS 600 heating/freezing stage to which a temperature programmer TP 94 is connected with a temperature controller. Liquid nitrogen is used to maintain the stage at low temperature (*see* **Note 7**).
3. Temperature dependent Raman spectra of hTau40WT and hTau23WT protein samples are recorded at 10, 25, and 37 °C. The Raman spectra are recorded only after incubating the sample at specific temperature for at least 5 min (*see* **Note 8**). All the parameters such as RTD exposure time = 15 s, exposure time = 15 s, accumulation number = 2, filter = D2, hole size = 400 μm, slit = 100 μm, grating = 1800 g/mm, are kept fixed while recording the Raman spectra for all the measurements.
4. The background correction and smoothening are performed for all the recorded spectra using instrumental software.
5. All the experiments are performed for at least two different occasions to check the reproducibility.

4 Notes

1. Tau is a soluble protein, which does not form any precipitates, but it aggregates if not handled properly.
2. Prior to the end of pelleting only supernatant is collected without touching the bottom of the tube, in order to not disturb the pellet, if any formed, otherwise the particles settled.
3. After purification and estimating the concentration, proper labeling is done and the protein samples are aliquoted and stored in −80 °C as repeated freeze–thaw may lead to protein aggregation.
4. Centrifugation is done to remove the aggregated protein to ensure that only soluble protein is retained for sedimentation assay.
5. Incubation and pelleting are done at identical temperatures, i.e., if the protein is incubated at 10 °C, it is pelleted at 10 °C.
6. 1 mg/mL of hTau40WT and hTau23WT would correspond to 21.81 and 27.25 μM, respectively.
7. The protein samples should be transferred carefully from the icebox to the Raman temperature stage so that the temperature of protein environment should not increase more than 10 °C.

8. Raman measurements should start from 10 °C which would be beneficial for precise liquid state measurements and secondary structure prediction.

Acknowledgements

The authors thank Dr. Hirekodathakallu V. Thulasiram for his excellent Molecular Biology Laboratory (MBL) facility at the CSIR-NCL, Pune. Tau constructs were kindly gifted by Prof. Roland Brandt from University of Osnabruck, Germany and Prof. Jeff Kuret from Ohio State University College of Medicine, USA. Nalini Vijay Gorantla and Puneet Khandelwal acknowledge the fellowship from University Grant Commission and the Department of Biotechnology, India, respectively. This project is supported in part by grants from the DST-SERB (Young Investigator Grant): SB/YS/LS-355/2013, In-house NCL grant MLP029526 and CSIR-Network Project CSC0406. PP acknowledges support from CSIR-network project NANO-SHE. The authors have declared no conflict of interest.

References

1. Chen J, Kanai Y, Cowan NJ, Hirokawa N (1992) Projection domains of MAP2 and tau determine spacings between microtubules in dendrites and axons. Nature 360:674–677. doi:10.1038/360674a0
2. Butner KA, Kirschner MW (1991) Tau protein binds to microtubules through a flexible array of distributed weak sites. J Cell Biol 115:717–730
3. Kopke E, Tung YC, Shaikh S, Alonso AC, Iqbal K, Grundke-Iqbal I (1993) Microtubule-associated protein tau. Abnormal phosphorylation of a non-paired helical filament pool in Alzheimer disease. J Biol Chem 268: 24374–24384
4. Hirokawa N (1993) Axonal transport and the cytoskeleton. Curr Opin Neurobiol 3:724–731
5. Hirokawa N (1994) Microtubule organization and dynamics dependent on microtubule-associated proteins. Curr Opin Cell Biol 6:74–81
6. Hirokawa N, Shiomura Y, Okabe S (1988) Tau proteins: the molecular structure and mode of binding on microtubules. J Cell Biol 107:1449–1459
7. Hirokawa N, Takemura R (2005) Molecular motors and mechanisms of directional transport in neurons. Nat Rev Neurosci 6:201–214. doi:10.1038/nrn1624
8. Cleveland DW, Hwo SY, Kirschner MW (1977) Physical and chemical properties of purified tau factor and the role of tau in microtubule assembly. J Mol Biol 116:227–247
9. Garcia ML, Cleveland DW (2001) Going new places using an old MAP: tau, microtubules and human neurodegenerative disease. Curr Opin Cell Biol 13:41–48
10. Hoffman PN, Cleveland DW (1988) Neurofilament and tubulin expression recapitulates the developmental program during axonal regeneration: induction of a specific beta-tubulin isotype. Proc Natl Acad Sci U S A 85:4530–4533
11. Schweers O, Schönbrunn-Hanebeck E, Marx A, Mandelkow E (1994) Structural studies of tau protein and Alzheimer paired helical filaments show no evidence for beta-structure. J Biol Chem 269:24290–24297
12. Mandelkow E, Song YH, Schweers O, Marx A, Mandelkow EM (1995) On the structure of microtubules, tau, and paired helical filaments. Neurobiol Aging 16:347–354
13. Mandelkow EM, Schweers O, Drewes G, Biernat J, Gustke N, Trinczek B, Mandelkow E (1996) Structure, microtubule interactions, and phosphorylation of tau protein. Ann N Y Acad Sci 777:96–106
14. Schweers O, Mandelkow EM, Biernat J, Mandelkow E (1995) Oxidation of cysteine-322 in the repeat domain of microtubule-associated protein tau controls the in vitro

assembly of paired helical filaments. Proc Natl Acad Sci U S A 92:8463–8467

15. von Bergen M, Barghorn S, Li L, Marx A, Biernat J, Mandelkow EM, Mandelkow E (2001) Mutations of tau protein in frontotemporal dementia promote aggregation of paired helical filaments by enhancing local beta-structure. J Biol Chem 276:48165–48174. doi:10.1074/jbc.M105196200
16. Li L, von Bergen M, Mandelkow E-M, Mandelkow E (2002) Structure, stability, and aggregation of paired helical filaments from tau protein and FTDP-17 mutants probed by tryptophan scanning mutagenesis. J Biol Chem 277:41390–41400. doi:10.1074/jbc.M206334200
17. Mylonas E, Hascher A, Bernadó P, Blackledge M, Mandelkow E, Svergun DI (2008) Domain conformation of tau protein studied by solution small-angle X-ray scattering. Biochemistry 47:10345–10353. doi:10.1021/bi800900d
18. Shkumatov AV, Chinnathambi S, Mandelkow E, Svergun DI (2011) Structural memory of natively unfolded tau protein detected by small-angle X-ray scattering. Proteins 79:2122–2131. doi:10.1002/prot.23033
19. Mandelkow E, von Bergen M, Biernat J, Mandelkow EM (2007) Structural principles of tau and the paired helical filaments of Alzheimer's disease. Brain Pathol 17:83–90. doi:10.1111/j.1750-3639.2007.00053.x
20. von Bergen M, Friedhoff P, Biernat J, Heberle J, Mandelkow EM, Mandelkow E (2000) Assembly of tau protein into Alzheimer paired helical filaments depends on a local sequence motif ((306)VQIVYK(311)) forming beta structure. Proc Natl Acad Sci U S A 97:5129–5134
21. Mukrasch MD, Biernat J, von Bergen M, Griesinger C, Mandelkow E, Zweckstetter M (2005) Sites of tau important for aggregation populate {beta}-structure and bind to microtubules and polyanions. J Biol Chem 280:24978–24986. doi:10.1074/jbc.M501565200
22. von Bergen M, Barghorn S, Biernat J, Mandelkow EM, Mandelkow E (2005) Tau aggregation is driven by a transition from random coil to beta sheet structure. Biochim Biophys Acta 1739:158–166. doi:10.1016/j.bbadis.2004.09.010
23. von Bergen M, Barghorn S, Jeganathan S, Mandelkow EM, Mandelkow E (2006) Spectroscopic approaches to the conformation of tau protein in solution and in paired helical filaments. Neurodegener Dis 3:197–206. doi:10.1159/000095257
24. von Bergen M, Barghorn S, Muller SA, Pickhardt M, Biernat J, Mandelkow EM, Davies P, Aebi U, Mandelkow E (2006) The core of tau-paired helical filaments studied by scanning transmission electron microscopy and limited proteolysis. Biochemistry 45:6446–6457. doi:10.1021/bi052530j
25. Mukrasch MD, Bibow S, Korukottu J, Jeganathan S, Biernat J, Griesinger C, Mandelkow E, Zweckstetter M (2009) Structural polymorphism of 441-residue tau at single residue resolution. PLoS Biol 7:e34. doi:10.1371/journal.pbio.1000034
26. Lippens G, Sillen A, Smet C, Wieruszeski JM, Leroy A, Buee L, Landrieu I (2006) Studying the natively unfolded neuronal Tau protein by solution NMR spectroscopy. Protein Pept Lett 13:235–246. doi:10.2174/092986606775338461
27. Sibille N, Sillen A, Leroy A, Wieruszeski JM, Mulloy B, Landrieu I, Lippens G (2006) Structural impact of heparin binding to full-length Tau as studied by NMR spectroscopy. Biochemistry 45:12560–12572. doi:10.1021/bi060964o
28. Zingsheim HP, Herzog W, Weber K (1979) Differences in surface morphology of microtubules reconstituted from pure brain tubulin using two different microtubule-associated proteins: the high molecular weight MAP 2 proteins and tau proteins. Eur J Cell Biol 19:175–183
29. Wille H, Drewes G, Biernat J, Mandelkow EM, Mandelkow E (1992) Alzheimer-like paired helical filaments and antiparallel dimers formed from microtubule-associated protein tau in vitro. J Cell Biol 118:573–584
30. Jeganathan S, Chinnathambi S, Mandelkow E-M, Mandelkow E (2012) Conformations of microtubule-associated protein Tau mapped by fluorescence resonance energy transfer. Methods Mol Biol 849:85–99. doi:10.1007/978-1-61779-551-0_7
31. Jeganathan S, von Bergen M, Brutlach H, Steinhoff HJ, Mandelkow E (2006) Global hairpin folding of tau in solution. Biochemistry 45:2283–2293. doi:10.1021/bi0521543
32. Siddiqua A, Margittai M (2010) Three- and four-repeat Tau coassemble into heterogeneous filaments: an implication for Alzheimer disease. J Biol Chem 285:37920–37926. doi:10.1074/jbc.M110.185728
33. Becker JS, Przybylski M (2007) Studies of structure and phosphorylation of tau protein using high resolution mass spectrometry. J Anal At Spectrom 22:761. doi:10.1039/b701440f

34. Schwalbe M, Ozenne V, Bibow S, Jaremko M, Jaremko L, Gajda M, Jensen MR, Biernat J, Becker S, Mandelkow E, Zweckstetter M, Blackledge M (2014) Predictive atomic resolution descriptions of intrinsically disordered hTau40 and α-synuclein in solution from NMR and small angle scattering. Structure 22:238–249. doi:10.1016/j.str.2013.10.020
35. Bernadó P, Svergun DI (2012) Analysis of intrinsically disordered proteins by small-angle X-ray scattering. Methods Mol Biol 896:107–122. doi:10.1007/978-1-4614-3704-8_7
36. Jeganathan S, Hascher A, Chinnathambi S, Biernat J, Mandelkow EM, Mandelkow E (2008) Proline-directed pseudo-phosphorylation at AT8 and PHF1 epitopes induces a compaction of the paperclip folding of Tau and generates a pathological (MC-1) conformation. J Biol Chem 283:32066–32076. doi:10.1074/jbc.M805300200
37. von Bergen M, Li L, Mandelkow E (2005) Intrinsic fluorescent detection of tau conformation and aggregation. Methods Mol Biol 299:175–184
38. Herrero AM, Cambero MI, Ordóñez JA, de la Hoz L, Carmona P (2008) Raman spectroscopy study of the structural effect of microbial transglutaminase on meat systems and its relationship with textural characteristics. Food Chem 109:25–32. doi:10.1016/j.foodchem.2007.12.003
39. McColl IH, Blanch EW, Gill AC, Rhie AGO, Ritchie MA, Hecht L, Nielsen K, Barron LD (2003) A new perspective on beta-sheet structures using vibrational Raman optical activity: from poly(L-lysine) to the prion protein. J Am Chem Soc 125:10019–10026. doi:10.1021/ja021464v
40. Arya S, Kumari A, Dalal V, Bhattacharya M, Mukhopadhyay S (2015) Appearance of annular ring-like intermediates during amyloid fibril formation from human serum albumin. Phys Chem Chem Phys 17:22862–22871. doi:10.1039/C5CP03782D
41. Kurouski D, Lu X, Popova L, Wan W, Shanmugasundaram M, Stubbs G, Dukor RK, Lednev IK, Nafie LA (2014) Is supramolecular filament chirality the underlying cause of major morphology differences in amyloid fibrils? J Am Chem Soc 136:2302–2312. doi:10.1021/ja407583r
42. Zhu F, Davies P, Thompsett AR, Kelly SM, Tranter GE, Hecht L, Isaacs NW, Brown DR, Barron LD (2008) Raman optical activity and circular dichroism reveal dramatic differences in the influence of divalent copper and manganese ions on prion protein folding. Biochemistry 47:2510–2517. doi:10.1021/bi7022893
43. Silva JG, Arruda LM, Pinheiro GS, Lima CL, Melo FEA, Ayala AP, Filho JM, Freire PTC (2015) The temperature-dependent single-crystal Raman spectroscopy of a model dipeptide: L-Alanyl-L-alanine. Spectrochim Acta A Mol Biomol Spectrosc 148:244–249. doi:10.1016/j.saa.2015.04.010
44. Accardo A, Shalabaeva V, Cotte M, Burghammer M, Krahne R, Riekel C, Dante S (2014) Amyloid β peptide conformational changes in the presence of a lipid membrane system. Langmuir 30:3191–3198. doi:10.1021/la500145r
45. Ramachandran G, Milán-Garcés EA, Udgaonkar JB, Puranik M (2014) Resonance Raman spectroscopic measurements delineate the structural changes that occur during tau fibril formation. Biochemistry 53:6550–6565. doi:10.1021/bi500528x
46. Maiti NC, Apetri MM, Zagorski MG, Carey PR, Anderson VE (2004) Raman spectroscopic characterization of secondary structure in natively unfolded proteins: alpha-synuclein. J Am Chem Soc 126:2399–2408. doi:10.1021/ja0356176
47. Ishizaki H, Balaram P, Nagaraj R, Venkatachalapathi YV, Tu AT (1981) Determination of beta-turn conformation by laser Raman spectroscopy. Biophys J 36:509–517. doi:10.1016/S0006-3495(81)84749-2
48. Barghorn S, Biernat J, Mandelkow E (2005) Purification of recombinant tau protein and preparation of Alzheimer-paired helical filaments in vitro. Methods Mol Biol 299:35–51

Chapter 3

Molecular Dynamics Simulation of Tau Peptides for the Investigation of Conformational Changes Induced by Specific Phosphorylation Patterns

Neha S. Gandhi, Predrag Kukic, Guy Lippens, and Ricardo L. Mancera

Abstract

The Tau protein plays an important role due to its biomolecular interactions in neurodegenerative diseases. The lack of stable structure and various posttranslational modifications such as phosphorylation at various sites in the Tau protein pose a challenge for many experimental methods that are traditionally used to study protein folding and aggregation. Atomistic molecular dynamics (MD) simulations can help around deciphering relationship between phosphorylation and various intermediate and stable conformations of the Tau protein which occur on longer timescales. This chapter outlines protocols for the preparation, execution, and analysis of all-atom MD simulations of a 21-amino acid-long phosphorylated Tau peptide with the aim of generating biologically relevant structural and dynamic information. The simulations are done in explicit solvent and starting from nearly extended configurations of the peptide. The scaled MD method implemented in AMBER14 was chosen to achieve enhanced conformational sampling in addition to a conventional MD approach, thereby allowing the characterization of folding for such an intrinsically disordered peptide at 293 K. Emphasis is placed on the analysis of the simulation trajectories to establish correlations with NMR data (i.e., chemical shifts and NOEs). Finally, in-depth discussions are provided for commonly encountered problems.

Key words Scaled molecular dynamics simulation, Tau protein, Antibody, Phosphorylation, AMBER, Protein folding, Free energy surface

1 Introduction

Recent advances in the research of "tauopathies," a group of central nervous system disorders such as Alzheimer's disease (AD), frontotemporal lobar degeneration, Pick's disease, and progressive supranuclear palsy, have highlighted the importance of microtubule associated protein Tau in their pathogenesis and its importance as a target for the development of disease-modifying therapeutics [1]. The Tau protein polymerizes tubulin into microtubules (MT) and stabilizes the MT network. In the case of neurodegenerative tauopathies, the MT-binding domain of the Tau

Caroline Smet-Nocca (ed.), *Tau Protein: Methods and Protocols*, Methods in Molecular Biology, vol. 1523,
DOI 10.1007/978-1-4939-6598-4_3, © Springer Science+Business Media New York 2017

protein oligomerizes to yield fibrillar polymers (aggregates) while destabilizing the MT network. In tauopathies, this abnormal disengagement of the Tau protein from the MTs is linked to the imbalance in the regulation of Tau protein kinases and phosphatases depending on its phosphorylation state. Hyperphosphorylation can lead to the formation of intraneuronal aggregates [2].

The Tau protein represents an ideal substrate for many kinases with 85 phosphorylatable residues (45 serines, 35 threonines and 5 tyrosines) within the full-length Tau 441 isoform [3, 4]. Hyperphosphorylated Tau protein (often referred to as PHF-tau) is the major constituent of paired helical filaments (PHFs) and neurofibrillary tangles (NFTs) found in AD brains, as evidenced by studies based on the creation and characterization of monoclonal antibodies raised against PHFs which recognize phosphorylation-dependent epitopes on Tau [5, 6]. There are many Ser/Thr residues in the N-terminus of the Tau protein positioned next to a Pro residue, which might provide a further phosphorylation-induced conformational switching mechanism for the backbone through cis/trans isomerization of the peptidyl-prolyl (ω) bond [7]. Phosphorylation of Ser/Thr residues in Pro-rich sequences often is not an individual event, and patterns of phosphorylation events separated by 3–4 residues might induce a conformational change [8]. These patterns are commonly associated by their recognizing monoclonal antibody (mAb), such as AT8 (Ser202/Thr205) [9–11], AT100 (Thr212/Ser214) [12], AT180 (Thr231/Ser235) [13, 14], and PHF-1 (Ser396/Ser404) [15], among others.

Among these monoclonal antibodies, AT8 is the most widely applied PHF-tau specific mAb to stain NFTs of the Tau protein in research and in clinical settings, and was shown to recognize PHF-Tau around positions 199–205. The epitope of AT8 was first mapped exclusively to Ser202 and later to doubly phosphorylated sites pSer202 and pThr205 using purified recombinant Tau protein or mutated Tau or synthetic peptides phosphorylated in vitro by different kinases [9, 11, 16, 17]. Pseudophosphorylation of the AT8 epitope alone or in combination with various other epitopes is shown to either open or tighten the conformation of the paperclip model of the Tau protein, which has consequences for the polymerization of MTs and for the aggregation behavior of the Tau protein [18]. Studies using in vitro phosphorylation of AT8 and AT180 by the cyclin-dependent kinase 2 (CDK2)/cyclin A3 (CycA3) kinase complex and NMR spectroscopy have indicated that sole phosphorylation of the AT8/AT180 epitopes only affects the capacity to assemble tubulin into MT and is not sufficient for dissociation of the Tau protein from the MT surface, narrowing the role of phosphorylation in the Pro-rich region of the Tau protein [19]. The AT8 epitope on the Tau F4 fragment (residues 208–324) consists of the repetitive appearance of Pro residues with a characteristic PxxPxxP motif, which may well constitute a recognition pattern for the many aromatic residues exposed at the surface of both α- and β-tubulin [20] but that also might be involved in the aggregation of Tau into tangles. Therefore it is necessary to characterize the conformational changes induced by phosphorylation of the AT8 epitope (^{192}GDRSGYSSPG*p*SPG*p* TPGSRSRT^{212}; henceforth referred to as the AT8 peptide) to gain further understanding of the function of this Pro-rich region of the Tau protein. NMR and molecular modeling approaches are best suited to address this.

The conformational energy landscape of a globular, folded protein usually contains a well-defined energy minimum corresponding to its stable, folded 3D structure. Intrinsically disordered proteins (IDPs), on the other hand, exhibit a number of local minima and hence lack a uniquely defined structure [21, 22]. The soluble phosphorylated Tau protein is an IDP, adopting numerous different conformations during its biological lifetime (it switches conformation within nanoseconds) [23], making it difficult to identify the structure actually associated during the detachment of the Tau protein from the MT, as identified by various antibodies during early and development stages of AD, as well as aggregated forms which are the hallmark of tauopathies [2]. The scaled (SMD) [24] and accelerated MD (aMD) [25, 26] sampling methods developed by McCammon's group can overcome high energy barriers in the protein folding energy landscape with very little prior knowledge of the underlying conformational space and transition states (*see* **Note 1**). These methods have been applied successfully to study the *cis/trans* proline isomerization (which typically occurs in the millisecond time scale), the effects of phosphorylation of Ser residues and to allow the Ser/Thr prolyl bond to interconvert easily between the *trans* and *cis* isomers in phosphorylated peptides [27, 28]. Therefore, we have used SMD as a method of choice in this paper [29]. We provide here a detailed account of the SMD simulation protocol (*see* **Note 2**) that we used to study the structure of the AT8 peptide with two phosphorylated sites, i.e., pSer202 and pThr205.

Phosphorylation of Ser or Thr residues can lead to a significant downfield shift of the corresponding amide proton chemical shift in their NMR spectrum, which can be used to map in detail the in vitro phosphorylation of the Tau protein by different kinases [30]. Chemical shifts and NOEs are available for the AT8 peptide and most previous structural interpretations of chemical shifts have focused on the sensitivity of shifts to backbone Psi and Phi angles and the nature of the hydrogen bonds in the vicinity of the phosphorylation sites [29]. These parameters can be used to validate the dynamically averaged values of backbone chemical shifts obtained from the SMD simulation of this fragment. Phosphorylation appears to induce a specific structure, reducing the IDP character of the AT8 peptide.

2 Materials

2.1 Hardware

Both the AMBER (Assisted Model Building with Energy Refinement) and GROMACS (GROningen MAchine for Chemical Simulations) MD simulation packages are designed to work on Linux systems with a GPU/CPU implementation (any NVIDIA GPU supporting hardware version 2.0, 3.0, 3.5, 5.0 or later) on serial or multiple nodes (parallel). However, there are some considerations that need to be borne in mind for maximizing performance, both in serial and parallel runs. Further information on hardware configurations is available on the AMBER website (http://ambermd.org/gpus/recommended_hardware.htm).

2.2 Software

1. AMBER [31] is a molecular dynamics simulation package which is currently distributed in two parts: AmberTools15 and AMBER14. AmberTools15 consists of several independently developed packages, and includes force fields for macromolecules, organic molecules, lipids and carbohydrates (including most versions of AMBER force fields, the GLYCAM and CHARMM force fields). AmberTools15 is used to prepare input files for a simulation, post-process trajectories and to perform MD simulations and energy minimization (Sander code). The Amber14 package consists of the pmemd program, which resembles the Sander (molecular dynamics) code in AmberTools, but provides better performance on multiple CPUs, and dramatic speed improvements on GPUs. AMBER14 includes support for both CPU and GPU versions and support for additional interaction points, multidimensional replica exchange, a Monte Carlo barostat, scaled MD, accelerated MD, Jarzynski sampling, implicit and explicit solvent constant pH, GBSA, and hydrogen mass repartitioning. AMBER14 is available for academic/nonprofit/government organizations for a modest fee and is distributed electronically. Please refer to the user manual of AMBER14 for installation instructions on how to configure AMBER in serial and parallel in both CPUs and GPUs using either Intel or gnu compilers. Upon installation, the AMBERHOME environment variable should be set:

   ```
   export AMBERHOME=/installation_directory/
   amber14 # (for bash, zsh, ksh, etc.)
   setenv AMBERHOME /installation_directory/
   amber14 # (for csh, tcsh)
   ```

2. GROMACS [32] is a versatile package used to perform molecular dynamics simulations of biomolecules like proteins, lipids and nucleic acids. GROMACS comes with a large selection of flexible tools with a simple interface with command line options for trajectory analyses. For illustration in this chapter, we use GROMACS version 4.6.7.

3. Molecular viewers like VMD [33] (http://www.ks.uiuc.edu/Research/vmd/) and UCSF Chimera [34] (https://www.cgl.ucsf.edu/chimera/) are frequently used for the visualization of simulation trajectories. For example, a movie of the folding of a protein or peptide observed during a simulation can be generated for presentation. We recommend downloading and installing 64-bit latest Linux executable binary files of these programs. This will enable handling visualization and analyses of large simulation trajectories, which require a large amount of memory.
4. A Perl script written by Thomas Cheatham III is available from within the AMBER mailing list which allows calculation of the number of ions (NaCl, KCl, etc.) needed beyond those for used for charge neutralization based on the simulation box size and the required ionic strength.
5. A perl script is available for automatic processing of output log files from AMBER simulations. The script process_mdout.pl script can be downloaded from the AMBER tutorial website (http://ambermd.org/tutorials/basic/tutorial3/files/process_mdout.perl). This script will provide summary and average values of temperature, pressure, density, total energy, kinetic energy and potential energy from the output file. Examining these properties allows to check for convergence of various properties during an MD simulation.
6. Several Python scripts are available for analyses of free energy surfaces (1D and 2D histograms) of trajectories obtained from scaled MD and conventional MD simulations. These scripts can be downloaded from http://scaledmd.ucsd.edu/download/reweighting_plotting.tar.gz and http://mccammon.ucsd.edu/computing/amdReweighting/.

   These scripts require installation of dependent libraries, such as numpy, scripy, Argparse and matlabplotlib.
7. Dictionary of Secondary Structure of Proteins (DSSP) [35] executable files and scripts for secondary structure analyses can be downloaded for various platforms from ftp://ftp.cmbi.ru.nl/pub/software/dssp/. An environment variable DSSP should be set pointing to the location of the dssp executable if using GROMACS:

   ```
   setenv DSSP /opt/dssp/bin/dssp
   ```

   Secondary structure analyses can be performed using the do_dssp script in GROMACS or AMBER; however, we have a customized Shell and Perl script to include the identification of polyproline-II (PP-II) secondary structure when performing DSSP analyses (*see* **Note 3**).

8. ShiftX2 [36] is a chemical shift predictor that calculates chemical shifts of all backbone and side chain atoms for a provided protein structure (in PDB format). SHIFTX2 is written in C, Java and Python and exists as a standalone program, as an online web server and as a VMWare version. All of these versions are available at http://www.shiftx2.ca. Currently, the program cannot predict chemical shifts for proteins/peptides with phosphorylated amino acids.

3 Methods

The AT8 peptide described here is meant to be simulated under conditions close to experiment, i.e., at a temperature of 293 K, an ionic strength of 25 mM and with phosphorylated amino acid residues in di-anionic form (at pH 6.8). The peptide is acetylated at the N-terminus and amidated at the C-terminus to avoid artefactual effects from the charged termini and to better mimic the environment of the fragment within the full-length Tau protein. Peptide bonds to Pro can populate both the *cis* and *trans* isomers but we have considered only the *trans* isomer in this case.

3.1 System Preparation

1. The amino acid sequence of the AT8 peptide is Ac-GDRSGYS SPG*p*SPG*p*TPGSRSRT-NH_2. The tleap module in AMBER is used to build a linear chain structure from which to start the folding simulation. The *ff99SBildn* force field for proteins and phosaa10 parameters for phosphorylated residues is used (*see* **Note 4** regarding choice of force field). Setting the default PBRadii to the "mbondi3" set should be done as part of specifying this force field. AMBER renames the residues pS and pT to SEP and TPO, respectively. AMBER numbers the residues starting with one. $AMBERHOME/exe/tleap -s -f $AMBERHOME/dat/leap/cmd/oldff/leaprc.ff99SBildn > source leaprc. phosaa10 > set default PBRadii mbondi3 > m1 = sequence {ACE GLY ASP ARG SER GLY TYR SER SER PRO GLY SEP PRO GLY TPO PRO GLY SER ARG SER ARG THR NHE} > saveamberparm m1 minvac.prmtop minvac.inpcrd > savepdb m1 minvac.pdb
2. Unrestrained energy minimization is then carried out using an implicit Generalized Born (GB) solvent model. AMBER supports a number of different GB models (http://ambermd.org/doc12/Amber14.pdf), and here igb=8 is chosen. Refer to **Note 5** for generating different starting conformations of the peptide. minvac.in (energy minimization input script) &cntrl imin=1, maxcyc=1000, (invoke energy minimisation) ncyc=500, (switch from steepest descents (SD) to conjugate gradients (CG) after 500 cycles.) cut=10, (non-bonded cutoff, in Angstroms) ntb=0, (no periodicity is applied and PME

is off) igb=8, ntpr=5, (implicit solvent model and print frequency) / $AMBERHOME/bin/sander -O -i minvac.in -o minvac.out -r minvac.rst -p minvac.prmtop -c minvac.inpcrd -ref minvac.inpcrd

3. The resulting PDB file from the above minimized disordered conformation is extracted before solvating the system.

```
$AMBERHOME/bin/ambpdb -p minvac.prmtop <min-
vac.rst> input.pdb -aatm -no ter
```

4. The system is solvated to a margin of at least 12.0 Å from any atom in the peptide. It is important to consider the shape of the peptide and the possible conformational changes that it may experience. If the system is not grossly spherical, such as with a disordered peptide, over the course of the simulation periodic artifacts may occur as an asymmetric molecule rotates within the periodic simulation box. To avoid through-box contacts and through-box disruptions of the solvation shells, a cubic solvation box should be created with dimensions determined from the longest dimension of a periodic box defined by a 12.0 Å solvent margin. The dimensions of the new simulation box obtained after solvation should be noted for the subsequent concentration (molarity) calculations.

```
$AMBERHOME/exe/tleap -s -f $AMBERHOME/dat/
leap/cmd/oldff/leaprc.ff99SBildn
> source leaprc. phosaa10
> loadamberparams frcmod. ionsjc_tip3p
> m1=loadpdb input.pdb
> solvatebox m1 TIP3PBOX 12 iso
```

5. Before quiting the tleap module, the total charge of the system (−2) should be neutralized. This is done by adding counterions as well as additional ions to take into account a salt concentration of 25 mM. In a new terminal window, the molarity.pl script can be executed to calculate the number of ions required to take into account the desired ionic strength.

```
> perl molarity.pl ionic strength in molar
x box dimensions
(e.g., perl molarity.pl 0.025 74.81 74.81
74.81 // this should give approximate 6 mol-
ecules)
```

6. Ions can be added by returning to the tleap terminal window.

```
> addions m1 Na+ 0 // add ions to neutral-
ise charge
```

```
> addions m1 Na+ 6 Cl- 6 // add ionic
strength i.e. 6 molecules
```

7. The solvated system and its respective AMBER topology files are then created for the subsequent execution of the MD simulation.

```
> saveamberparm m1 solvated.prmtop solvated.
inpcrd
> savepdb m1 solvated.pdb
> quit
```

8. To maximize peptide–water interactions during the initial stages of equilibration, the atomic coordinates of the peptide should be fixed followed by (a) energy minimization of water molecules and ions (500 steps of SD followed by 1000 steps of CG) with harmonic positional restraints with a force constant of 10 kcal/mol/Å^2 applied to the peptide backbone atoms, (b) MD simulation in the NPT ensemble at 1 atm for 50 ps, retaining the harmonic positional restraints on the heavy atoms of the peptide, and (c) energy minimization of the entire system (500 steps of SD and 5000 steps of CG) without positional restraints on the peptide.
9. Heat the system from 150 to 293 K (or other chosen temperature) in the isothermal–isovolumetric (NVT) ensemble at 1 atm for 500 ps.

```
heat.in //equilibration in NVT ensemble
&cntrl irest=1, ntx=5, (restart MD)
nstlim=250000, dt=0.002, (run for 500 ps)
ntt=1, tautp=1.0, (temperature control)
ntb=1, ntc=2, ntf=2, (SHAKE, periodic bound-
ary condition)
cut=10.0, (cutoff for non-bonded interac-
tions)
temp0=150, tempi=293,
ntpr=5000,ntwx=5000,ntwv=1000, (output fre-
quency for log and trajectory file)
ig=-1, iwrap=1, (random number seed, wrap
the coordinates in trajectory file)
ipol=0, igb=0, (GBSA model)
ntr=0, nmropt=1, (positional restraints)
/
&wt type='TEMP0', istep1=0, istep2=250000,
value1=150.,
value2=293.0
/
&wt type='END'
/
> mpirun $AMBERHOME/bin/pmemd.cuda_SPFP.
MPI -O -i heat.in -o heat.out -r heat.rst -p
solvated.prmtop -c minimisation_step8c.rst
-x heat.mdcrd
```

10. The system is then equilibrated in full by conducting an isothermal-isobarometric (NPT) simulation for at least 5 ns to

ensure that the system reaches its appropriate average density at the desired temperature.

```
npt.in (equilibration in NPT to stabilize
density)
Title
&cntrl
irest=1, ntx=5, (restart MD)
nstlim=2500000, dt=0.002, (run for 5 ns)
temp0=293.0, ntt=1, tautp=1.0, (temperature
control)
ntc=2, ntf=2, (SHAKE, periodic boundary con-
dition)
cut=10.0, (cutoff for non-bonded interac-
tions)
ntpr=5000, ntwx=5000, ntwv=0, ntwe=0, (output
frequency for log and trajectory file)
ntb=2, ntp=1, taup=1.0, (pressure control)
ipol=0, igb=0,
ig=-1, iwrap=1, (random number seed, wrap the
coordinates in trajectory file)
ntr=0 (positional restraints)
&end
> mpirun $AMBERHOME/bin/pmemd.cuda_SPFP.MPI
-O -i npt.in -o npt.out -r npt.rst -p solvat-
ed.prmtop -c heat.rst -x heat.mdcrd
```

11. The .out file from the previous NPT run can be processed to monitor the average density and volume per frame. These properties can be used in any future NVT simulations.

```
> perl process_mdout.pl npt.out
```

12. For simulations in the NVT and NPT ensembles, bonds containing hydrogen atoms are restrained with the SHAKE algorithm. Weak coupling (Berendsen) to external temperature and pressure baths is used to control both temperature and pressure [37]. The electrostatic interactions are calculated using the particle mesh Ewald summation (PME) method using a fourth order spline for interpolation, with a cutoff of 10.0 Å for long-range interactions.

13. The primary benefit of using the NVT ensemble for production runs is the computational throughput, particularly when compared to the Berendsen barostat as implemented in AMBER pmemd for GPUs. However, when changing ensemble from NPT to NVT, one might see a drift in the average pressure if the restarting frame has a volume that is at either edge of the histogram of the volume calculated from the previous NPT run (**step 9**). To avoid this drift in pressure, a frame should be selected with the volume closest to the average obtained during the NPT simulation as starting point for the

production simulations in the NVT ensemble. The average box edge length can be obtained by taking the cube-root of the average VOLUME from the NPT simulation.

```
> grep VOL npt.out | head -n -2 | awk '{sum+=$9}
END{printf %10.7f\n",(sum/NR)^(1/3)}'
```

Finally, the latest npt restart file (npt.rst) should be modified such that the box lengths are set to the equilibrated average values obtained by running the above command. Using the latest modified restart file, an additional equilibration in the NVT ensemble is run with the chosen volume for 20 ns.

```
nvt.in
&cntrl
irest=1, ntx=5, (restart MD)
nstlim=10000000, dt=0.002, (run for 10 ns)
temp0=293.0, ntt=3, gamma_ln=2.0, (temperature
control)
ntb=1, ntc=2, ntf=2, (SHAKE, periodic boundary
condition)
cut=10.0, (cutoff for non-bonded interactions)
ntpr=2500, ntwx=2500, ntwv=0, ntwe=0, (output
frequency for log and trajectory file)
ipol=0, igb=0,
ntr=0, iwrap=1, ig=-1, (random number seed, wrap
the coordinates in trajectory file)
/
> mpirun $AMBERHOME/bin/pmemd.cuda_SPFP.MPI -O
-i nvt.in -o equilibration_nvt.out -r npt.rst
-p solvated.prmtop -c equilibration_nvt.rst -x
equilibration_nvt.mdcrd
```

14. Finally, 1 μs long production simulations for both conventional MD and SMD are performed in the NVT ensemble at 293 K using Langevin dynamics [38] with a collision frequency $\gamma = 2.0$ ps^{-1}. The coordinates were saved every 5 ps, giving a total of 2,000,000 frames for each 1 μs simulation. In addition to the run control parameters for conventional MD, SMD can be turned on by adding two variables to the &cntrl list (see below). In SMD simulations, the forces on each atom are calculated and then scaled by $\lambda = 0.7$ at each time step to accelerate sampling of the potential energy landscape. When running production simulations in AMBER, especially when using Anderson or Langevin thermostats, it is essential to change the random number generator seed from the default value for every restart. This can be done by setting ig = −1 in the &cntrl list. If the trajectory from multiple runs needs to be reproduced then the setting ig = −1 must be turned off. Depending on the availability of supercomputing power, storage space, memory and processing wall time, a 1 μs simulation

may be split into multiple trajectories of 10 ns each. This should then be run 100 times to obtain 1 μs of simulation time. Periodic boundary PME simulation can be used with the CUDA version of PMEMD for better performance and parallel scaling. Below is an example script used on an NVIDIA Tesla C2075 GPU-based system to run this simulation using two GPUs on a single node. It takes approximately 8.5 h to compute 10 ns of SMD simulation for this peptide with this sytem.

```
smd.in // scaled MD input file
&cntrl
irest=1, ntx=5, (restart MD)
nstlim=5000000, dt=0.002, (run for 10 ns)
temp0=293.0, ntt=3, gamma_ln=2.0, (temperature
control)
ntb=1, ntc=2, ntf=2, (SHAKE, periodic boundary
condition)
cut=10.0, (cutoff for non-bonded interactions)
ntpr=2500, ntwx=2500, ntwv=0, ntwe=0, (output
frequency for log and trajectory file)
ipol=0, igb=0,
ntr=0, iwrap=1, ig=-1, (random number seed, wrap
the coordinates in trajectory file)
scaledMD=1, (turn on scaled MD)
scaledMD_lambda=0.70, (scaling factor)
/
Job1.pbs // input script to submit jobs to GPU
cluster
#PBS -l walltime=12:00:00
#PBS -l select=1:ncpus=12:mpiprocs=2:ngpus=2:me
m=64GB
#PBS -W group_list=XXX
#PBS -q develq
module load openmpi
module load cuda
module load fftw/3.3.3
module load amber/14
cd $PBS_O_WORKDIR
mpirun $AMBERHOME/bin/pmemd.cuda_SPFP.MPI -O -i
smd.in -o smd1.out2 -r smd1.rst -p solvated.
prmtop -c equilibration_nvt.rst -x smd1.mdcrd -
scaledMD smd1.log
gzip -9 smd1.mdcrd
```

3.2 Analyses of Trajectory

1. Basic and common types of analysis can be performed using CPPTRAJ, a successor to PTRAJ in AmberTools. All trajectory files need to be first concatenated and water molecules and ions removed to produce a single trajectory file containing the peptide fragment. This processing will generate a trajectory

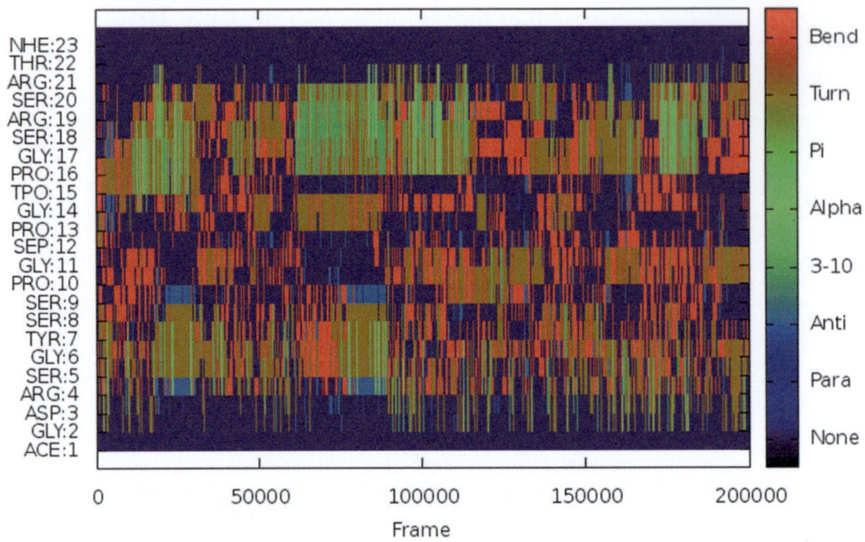

Fig. 1 Secondary structure propensity analysis of the SMD simulation trajectory of the AT8 peptide. Secondary structure content was assessed with the DSSP algorithm in AMBER

file of smaller size but with all snapshots enables quick execution of other commands for analyses. An input file containing a list of all trajectories for analysis is then prepared.

```
ptraj.in // input file for combining trajecto-
ries
trajin smd1.mdcrd.gz
trajin smd2.mdcrd.gz
..
trajin smd100.mdcrd.gz
strip :WAT, Cl-, Na+
trajout trajout.netcdf netcdf nobox
> $AMBERHOME/bin/cpptraj -p solvated.prmtop -i
ptraj.in
```

2. Secondary structure propensities for all residues of the peptide can be computed using the DSSP method of Kabsch and Sander implemented in AMBER, which assigns secondary structure types for residues based on backbone amide (N–H) and carbonyl (C=O) atom positions. By default cpptraj assumes that these atoms are named “N,” “H,” “C,” and “O,” respectively. Cpptraj generates a 2D map of secondary structure versus time, with different colors representing different secondary structure types for each residue, where numbering from 0 to 6 is indicative of random coil, parallel β-strand, antiparallel β-strand, 3_{10}-helix, α-helix, π-helix, and turns, respectively. This 2D map can be visualized in a graphing utility such as gnuplot and saved as an image file. Cpptraj can also output average structural pro-

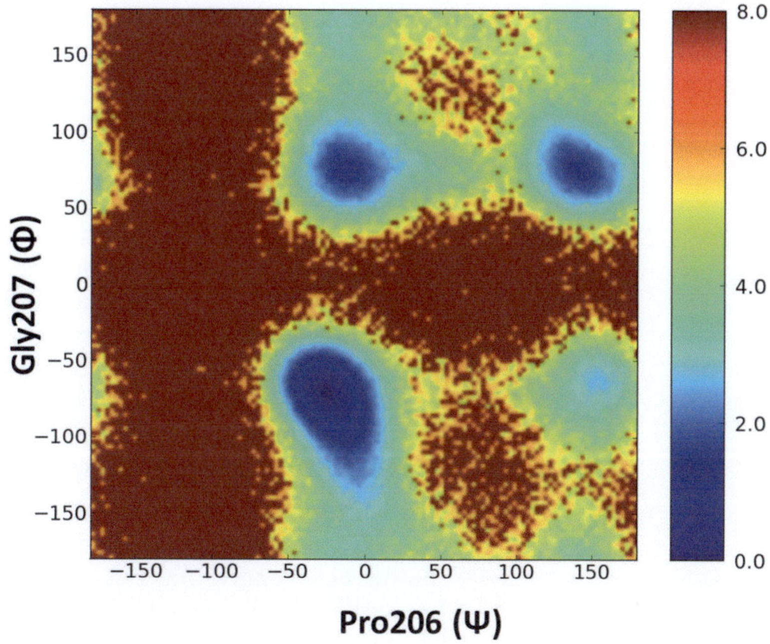

Fig. 2 Free energy plots of the Psi angle of Pro206 and Phi angle of Gly207 from the SMD simulation at $\lambda = 0.7$

pensities over all residues for each frame along with a 2D map. Of particular further interest is the calculation of phi and psi angles for Pro206 (Pro16 in AMBER numbering) and Gly207 (Gly17 in AMBER numbering) in the same cpptraj input script (choice of variables clarified in **Note 6**).

```
ptraj1.in // input file for simpler analysis
such as secondary structure and torsions tra-
jin trajout.netcdf (trajectory with only peptide
coordinates) secstruct :1-23 out dssp.gnu sumout
dssp.dat dihedral phi16 :15@C :16@N :16@CA :16@C
out phi16 type phi dihedral psi16 :16@N :16@CA
:16@C :17@N out psi16 type psi stat psi16 out
stat_proline_psi.newlist dihedral phi17 :16@C
:17@N :17@CA :17@C out phi17 type phi dihe-
dral psi17 :17@N :17@CA :17@C :18@N out psi17
type psi stat phi17 out stat_glycine_phi.newl-
ist > $AMBERHOME/bin/cpptraj -p input.prmtop -i
ptraj1.in (input.prmtop is used, which does not
contain ions or water)
```

The resulting file can be visualized with gnuplot and saved as an image file (e.g., in gif format) (Fig. 1). By default, AMBER uses residue numbering starting from one. Similarly, the sumout file can be easily visualized using Microsoft Excel. Residue numbering can be changed by editing dssp.gnu file with a text editor.

```
> gnuplot gnuplot> set terminal gif size 800,600
enhanced gnuplot> set oiiutput 'dssp.gif' gnu-
plot> load 'dssp.gnu' gnuplot> quit
```

3. SMD alters the potential energy surface to achieve rapid sampling and, therefore, it is necessary to perform reweighting of conformational populations obtained from scaled MD in order to plot the free energy surface. The first step here involves combining the Psi angle distribution of Pro16 and the Phi angle distribution of Gly17 (*see* **Note 6**) into one single file. > paste psi16 phi17 | awk '{print $2,$4}' > Psi_Phi The header line from the file Psi_Phi needs to be removed using a text editor before saving the file again.

   The reweighting scripts from the SMD website (*see* software section) should be downloaded and untarred. File permissions should be changed for binstring_multiproc.py and 2DFree EnergyPlot.py. A weight for each frame in the trajectory should be assigned and outputted to weights.csv

```
> python binstring_multiproc.py -input Psi_Phi
-weight weights.csv -lambda 0.7 -disc 3 -nproc
4 (disc=discretization creates uniform bins, in
this case at 3° spacing, nproc is the number of
processors) > python 2DFreeEnergyPlot.py -in-
put Psi_Phi -job weighthist -Xdim -180 180 -Ydim
-180 180 -disc 3 -weight weights.csv
```

   Two image files should be obtained: one with the free energy (Fig. 2) and another one with the weights.

4. Based on the free energy surface, the trajectory can be recast into four conformational ensembles classified as lower_left, lower_right, upper_right, and upper_left clusters. Trjconv (refer to **Note 7**) can be used to extract coordinates from the trajectory file based on the values of the *X*- and *Y*-axes, as shown in Fig. 2.

```
e.g., trjconv -f input_trajectory -s input.
pdb -conect -dropunder 1 -drop test.xvg -o
lower_left.pdb (test.xvg consists of the file
with the time and binary value if the frame
needs to be dropped at that time)
```

   The respective conformations from these four clusters can be visualized using molecular visualization programs such as VMD or UCSF Chimera (explained in **Note 8**).

5. The time evolution of the Phi angle of Gly207 and the Psi angle of Pro206 can be monitored during the SMD simulation of the AT8 peptide by plotting their values (corresponding to

phi17 and psi16) in Microsoft Excel. Alternatively, the convergence of the sampling can be assessed by monitoring the weighted free energies (in kcal/mol) versus the Phi angle of Gly207 at 100 ns intervals during the simulations. For this purpose PyReweighting-1D.py script obtained from the aMD reweighting tutorial (*see* Subheading 2.2) can be used to calculate the free energies.

```
> python PyReweighting-1D.py -input phi17 -job
noweight -Xdim -180 180 -Ydim -180 180 -disc
3 -T 293 (this outputs the pmf-phi17.xvg file)
PyReweighting-1D.py does not calculate SMD
weights and hence the free energies (column 2 in
pmf-phi17.xvg file) need to be scaled by 1/λ=1.43
to recover the original PES. > awk '{print $1,
$2*1.43}' pmf-phi17_100ns.xvg > phi_FES.xvg
```

6. The presence of salt bridges within the peptide structure in each simulation can be established using the Salt Bridges Plugin of the VMD molecular visualization software, which identifies acidic and basic residues within a distance of 0.6 nm of each other, which is the distance criterion set to define a salt bridge during the simulation. In this case TPO, SEP, ASP and GLU are defined as anionic residues whereas ARG and LYS are the basic residues for these analyses. The following commands should be used in the vmd tk console window:

```
> atomselect macro acidic "resname TPO SEP ASP
GLU" > atomselect macro basic "resname ARG LYS"
> saltbr -sel [atomselect
top all] -condist 6 -log saltbr.log -outdir path
to output_
directory
```

3.3 Validation of the Ensemble Obtained

In the case of the AT8 peptide, its predicted conformational preferences can be verified using the outlying NMR chemical shift of the Gly207 amide proton and the strong NOE peaks observed in the NOESY spectra (*see* **Note 9**). Because the outlying chemical shift of the Gly207 amide proton is expected to depend primarily on the Psi angle of the preceding residue and on its own Phi angle, the low-energy conformations were initially clustered into four clusters depending on these two torsional angles (see above). Chemical shifts of each conformation were initially determined using ShiftX2, and then ensemble averaged within each cluster.

ShiftX2 combines structure-based chemical shift predictions, provided by the module ShiftX+, with sequence based predictions that are obtained from databases of experimental chemical shifts (*see* **Note 10**). However, the software does not support the phosphorylated amino acids. Therefore, the structures were first stripped of the existing phosphate groups and passed on to ShiftX2.

The electrostatic component of the Gly207 amide proton chemical shift arising from the presence of the phosphate group of pThr205 is then predicted using Buckingham's equation [39]:

$$\delta_{\mathrm{EF}} = A_{\mathrm{II}} E \cos\theta$$

where A_{II} is the nuclear shielding polarizability tensor parallel to the polarized H–N bond, E is the electric field at the position of the amide proton and θ is the angle between the H–N bond vector and the vector connecting the phosphorus atom and the amide proton. The value of $A_{\mathrm{II}} = 188$ pmau (1 ppmau $= 1.9447 \times 10^{-18}$ m/V) is used in this work [40, 41].

The electric field at the position of Gly207 amide proton can be calculated using either Coulomb's law or Poisson Boltzmann equation. In this work, we used Coulomb's law:

$$E = \frac{1}{4\pi\varepsilon_0\varepsilon_{\mathrm{r}}}\frac{q}{r^2}$$

where q is the charge of the dianionic pThr205, $\varepsilon_{\mathrm{r}} = 4$ is the relative dielectric constant, ε_0 is the vacuum permittivity, and r is the distance between the phosphorus atom of pThr205 and the amide proton of Glu207. δ_{EF} values of each conformation were calculated first and then ensemble averaged within each cluster.

In order to validate the measured NOE peaks, the distances within individual conformations were computed considering all hydrogen atom pairs. The distance between two hydrogen atoms i and j was calculated using the following equation:

$$d_{\mathrm{NOE}} = \left(\sum r_{ij}^{-6}\right)^{-1/6}$$

where the sum is taken over all equivalent atoms. A NOE peak was assigned as present in the conformation if the distance between the atom pair was less than 5.0 Å. This distance is usually used as a threshold value for the detection of cross peaks in NOESY spectra. As an alternative criterion, the NOE peak was assigned as present in the cluster if the distance between both protons involved was inferior to 5 Å in more than 50% of the conformations. Even though we used such a relaxed criterion, agreement between both was excellent, and almost all conformations within the same cluster predicted either presence or absence of a certain NOE peak.

4 Notes

1. Choice of advanced sampling methods: Temperature replica exchange MD (T-REMD), accelerated MD (aMD) and scaled MD (SMD).

Atomistic molecular dynamics simulations and advanced sampling techniques has helped to characterize the states and transitions of IDPs on longer timescales [42]. In the case of the Tau protein, T-REMD [43] has been used to study folding and aggregation of its PHF6 and PHF6* hexapeptide motifs in the presence of explicit solvent [44–46]. Higher temperature trajectories are necessary in T-REMD in order to enhance sampling with neighboring replicas and, therefore, the number of replicas required increases with the square root of the number of particles in the system (to cover a desired temperature range). This limits the applicability of the method to smaller systems (e.g., systems greater than 16 residues in explicit solvent) because of the computational load. aMD can provide enhanced sampling compared to conventional MD in a number of biological systems, as has been the case with the repeat domains of the Tau protein [47], with hundreds-of-nanosecond aMD simulations being able to capture millisecond-timescale events [48]. Unlike metadynamics, no prior information on reaction coordinates is required in aMD. aMD explores phase space by applying a non-negative additional potential (boost potential; $\Delta V(\vec{r})$) when the system potential $V(\vec{r})$ is lower than a reference energy. aMD trajectories need to be reweighted to recover the canonical ensemble and the original free energy profile of functional biomolecules. Reweighting of aMD trajectories involves evaluating the exponential average of potential changes, resulting in an inherently noisy potential energy surface (PES) for systems larger than small polypeptides. Many studies have focused on improving the accuracy of reweighting for aMD simulations but these methods are system specific and must be used with care [49, 50]. To harness the power of GPUs and the scalability of AMBER14 [51], we used the SMD [24] method, which—has a reweighting protocol to recover the canonical ensemble that is not biased by the fluctuations of energy but instead relies solely on the populations of conformations to reweight. In SMD, the biomolecular PES is modified by scaling $V(\vec{r})$ by a factor of λ that ranges from 0 to 1. A value of λ close to 0 may explore very high energy states. It is thus recommended to use values of $\lambda > 0.5$ for typical biomolecular simulations to retain the balance between enhanced sampling and a physically relevant conformational ensemble.

2. Ensemble methods.

IDPs can explore multiple conformations on fast timescales on a rugged potential energy landscape. Therefore, they can be represented as an ensemble of structures that collectively span their conformational space -. Two software packages, ASTEROIDS [52] that uses the iterative Flexible-Meccano refinement and TRaDES [53, 54], in conjunction with

ENSEMBLE [55], have been developed for generating IDP structures using NMR constraints such as chemical shifts, residual dipolar couplings and data obtained from paramagnetic relaxation enhancement experiments. A discussion of the algorithms used in these programs is beyond the scope of this chapter. However, these packages do not have provisions for generating ensembles from sequences consisting of phosphorylated residues. When sufficient computing power is available, one attractive alternative for generating directly ensembles of the AT8 peptide is through the use of extensive MD simulations.

3. Polyproline (PP-II) content.

   PP-II is a left-handed helical conformation that usually lacks conventional stabilizing internal hydrogen bonds, has average dihedral angle values of Phi = −75° and Psi = +145° and a residue length of approximately 3 [56, 57]. This secondary structure is found in polymers of Pro in aqueous solution, although most PP-II conformations might contain non-Pro residues and some even contain no Pro residues at all. PP-II content is found in particular constructs from the Pro-rich regions of the Tau protein. The extended nature of PP-II may lead to potential interactions of the Tau protein with other molecular partners. The AT8 and AT180 epitopes of the Tau protein have significant numbers of Pro residues and, therefore, it is important to identify this secondary structure type in the analyses of MD trajectories. PP-II assignment is not done with standard packages for secondary structure assignment, such as DSSP and STRIDE and, therefore, newly solved protein structures are not assigned with PP-II in the PDB. Currently programs such as XTLSSTR [58], PROSS [59] and SEGNO [60] assign PP-II secondary structures solely on the basis of the values of the backbone angles, mainly involving the Phi and Psi dihedral angles. For the analyses of PP-II structure in Tau peptides, we recommend using $PPII^{DSSP}$ [61], based on the most popular secondary assignment tool DSSP. DSSP assigns eight types of secondary structures: α-helix (H), extended β-strand in parallel and or anti-parallel β-sheet conformation (E), isolated β-strand (B), 3_{10} helix (G), π-helix (I), bend (S), and coil (−), based on the identification of precise hydrogen bonding patterns corresponding to regular secondary structures, in addition to dihedral angles. In addition to these eight structures, $PPII^{DSSP}$ assigns PP-II structure based solely on the coil region for at least two consecutive amino acids in coil conformation with Phi = $-75° \pm \varepsilon$ and Psi = $+145° \pm \varepsilon$, with $\varepsilon = 29°$. This program can be obtained from the authors.

4. Choice of force fields for IDPs phosphorylated and proline residues, and water potential.

A variety of force fields have been developed and validated for simulations of biomolecules which reliably reproduced several different NMR observables, including—chemical shifts, *J*-couplings and order parameters [62–64]. Several force fields and water potentials have been reparameterized for the study of disordered peptides that had been previously characterized by NMR spectroscopy [65]. Firstly, revised Phi backbone dihedral potentials in the AMBER *ff99SB* force field in conjunction with the TIP4P-Ew water potential have been shown to yield improved conformational ensembles for a variety of small disordered peptides [66]. This force field and water potential combination has also been used to successfully reproduce NMR observables of Aβ peptides [67]. Since then, significant corrections have been made to the AMBER protein force fields, which often lead to structures that are too stable to properly model the conformational behavior of IDPs. The *ff99IDPs* [68, 69] AMBER force field has been developed with the addition of grid-based energy correction maps (CMAPs) on the standard *ff99SBildn* force field to reproduce the main-chain torsional distributions of eight disorder-promoting residues (A, G, P, R, Q, S, E, and K). Most of these force fields have not been tested to reproduce NMR observables for full length IDPs but rather smaller peptides. Another AMBER force field, ff03w [70], together with TIP4P/2005 gives a good description of protein folding thermodynamics and a more accurate description of α or β-proteins like chignolin, WW Pin1 domain, Trp cage and GB1 hairpin, but there is still insufficient evidence for its use in the successful folding of PP-II and other turn like structures which might be dominant in IDP-ordered protein interactions. The water potentials TIP4P-Ew [71] and TIP4P/2005 [72] capture reasonably well the temperature-dependence of the solubility of hydrophobic solutes; however, the combination of these water potentials with AMBER force fields for IDPs has resulted in overly compact disordered/unfolded conformations, with radii of gyration that are in strong disagreement with experimental observations. This disagreement is a consequence of water models that severely underestimate solvent-solvent and solvent-solute dispersion interactions. Narenberg et al. [73], Best et al. [74] and Piana and coworkers [75] have worked on similar aspects of optimization of the solute–solvent dispersion by either scaling the Lennard-Jones interaction between the water oxygen and all protein atoms (*ff99SBws* force field), or by increasing the water dispersion coefficient (TIP4P-D water potential).

Several force fields parameters are available for simulating systems with phosphorylated residues [76–78] starting with an ordered conformation. To our knowledge, no benchmark calculations of IDP force fields have been carried out for IDPs with phosphorylated residues. The *ff99SB* force field and its variants can be used for simulations since efforts had been made to correct glycine sampling to reproduce β-turn structures, which had been especially problematic with most previous AMBER force field variants. e.g. *ff14SB* has revised side chain and backbone dihedral parameters, and particularly designed to improve the agreement between experiments and simulations in explicit water in particular for folding of villin headpiece [79]. Most variants of the AMBER force fields have modified sets of parameters for the ω torsion for Pro-containing peptide bonds, giving equilibrium structures (*cis/trans*) that are consistent with experimental results [80]. The presence of consecutive Pro, Gly and phosphorylated residues in AT8 gives an arena open for the scientific community to seek to improve the performance and accuracy of the specific force fields for IDPs with prominent post-translational modifications.

5. Building a starting structure from its sequence.

   There are a variety of programs that facilitate defining the initial coordinates of a structure from a given sequence. Here an initial structure was generated using the sequence command in the tleap module in AMBER. It is recommended to perform an energy minimization to remove steric clashes after generating starting coordinates in AMBER. The structure generated by tleap is linear in nature and hence may require a much bigger simulation box. The outcomes of a simulation may be affected by the initial conformation of the peptide. It is thus common practice to execute multiple simulations of the same peptide with different starting conformations or using a random seed to ensure consistency of the results, or to look at convergence of NMR observables from an MD simulation. Alternatively, the CONCOORD algorithm [81] can be applied to the structure obtained from tleap to generate an ensemble of new conformations of the peptide. CONCOORD (http://www3.mpibpc.mpg.de/groups/de_groot/concoord/) will generate several hundred structures. Conformations that are similar to the structure generated by AMBER (based on the root mean square difference reported by CONCOORD) can be safely discarded. One of the remaining structures can be solvated to perform MD simulations. In this chapter, we restrict our analyses to a single simulation run.

6. Selecting variables for free energy plots.

   It is common to employ principal component analysis (PCA) to reveal the most important motions in proteins. This method is commonly available in most popular MD simulation packages to analyze protein trajectories. PCA analysis can help to filter global, collective (often slow) motions from local, fast motions [82]. This analysis is usually carried out on the backbone atoms of the ensemble of conformations. The resulting principal components (PCs) are sorted according to their contribution to the total fluctuation along the ensemble of conformations. Usually only a small subset of these PCs (first six or ten) describes the great majority of the total atomic displacement. In this case, the DSSP plot indicated the presence of interesting conformations around pThr205-Arg211. Therefore, the focus was on analyzing the conformations around pThr205 by focussing on the Phi and Psi of neighboring residues.

7. Trajectory conversion from AMBER to GROMACS format.

   The first necessary step is the installation of VMD version ≥1.8.6 on a system that supports dynamic loading. The next step is to install GROMACS version ≥4.6. CMake in newer versions of GROMACS can detect VMD and automatically set the VMDDIR and VMD_PLUGIN_PATH environment variables to point to VMD installations and/or their molfile folders. The VMD plugin allows the trjconv tool in GROMACS to read various trajectory file formats.
   e.g., trjconv –f amber_trajectory_file –o gromacs_trajectory_format –s input.pdb –n index.ndx –timestep 5 –t0 0 // timestep is the frequency in picoseconds

8. Visualization of secondary structure using VMD and Chimera.

   The secondary structure definitions for the molecules in VMD do not change during an animation of the trajectory even if a protein changes its secondary structure. For this purpose a tcl script 'sscache' (http://www.ks.uiuc.edu/Research/vmd/script_library/) is available which will calculate and store the secondary structure assignment for each frame. Similarly, while playing a trajectory using an MD movie in Chimera, the secondary structure assignments are not recomputed along the trajectory. This is relevant when visualizing large conformational changes. To recompute secondary structure at each frame, the command ksdssp in Chimera should be used in the MD movie.

9. Limitation of using NMR observables in restrained MD simulations of IDPs.

   Calculation of protein structure and dynamics based on NMR constraints (NOEs, paramagnetic relaxation enhancements (PREs), chemical shifts, coupling constants, RDCs, etc.) has been used successfully since its inception. These NMR observables are used both as restraints in MD simulations and for the validation of ensembles of conformers aimed at representing the population of unfolded proteins in solution [83]. Solution NMR spectroscopy has remained a powerful tool for studying the dynamic structural ensembles of IDPs compared to other biophysical characterization methods [83–85], such as small angle X-ray scattering (SAXS) and single molecule FRET because of the wealth of local structural information encoding in the NMR observables. However, studies of IDPs by NMR has been hampered by the extremely poor dispersion of chemical shifts along all nuclei dimensions [83]. Moreover, an IDP peptide such as AT8, which is enriched in Pro composition compared to cooperatively folded proteins, restricts detection of its amide proton. This limitation can be overcome by using a ^{13}C-CON spectrum, which contains a resonance for each peptide bond involving a Pro ^{15}N (easily identified by the downfield shift in the ^{15}N dimension), further enhancing the practical resolution of the experiment [86]. Approaches that additionally utilize the advantages of ^{13}C-direct detection have allowed the study of phosphorylation patterns in the AT8 peptide.

Secondary chemical shifts, which represent deviations of the experimentally determined chemical shifts from random coil values, are influenced by the local environment of a nucleus and therefore depend explicitly on the presence of secondary structure (e.g., α-helix, PPII-helix, and β-strand) [87–90]. Programs are now available to determine secondary structure populations in IDPs if chemical shift measurements of backbone atoms are available [91, 92], but for amide proton or nitrogen, the predicted ^{1}H,^{15}N HSQC spectrum by most programs does not resemble experimental data for IDPs. So using these programs to predict secondary structure content of IDPs remains far off. As for phosphorylated residues, we do not have a measure of the true random coil shift yet, especially as the intramolecular hydrogen bond (phosphate moiety—amide proton) seems to dominate the conformational sampling. In solution, IDPs exist as interchanging conformers and the observed chemical shifts are consequently a population-weighted average of interconverting conformers over a time-scale of up to microseconds. A notably smaller population of one of the secondary structures in the ensemble could introduce systematic errors in secondary chemical shifts calculated from a single nucleus. We were able to detect very few nonlocal NOEs with which the structure of the AT8 epitope can be defined. The factors discussed above greatly hampered the prediction of the structure of the AT8 peptide using restrained MD simulations based on NMR observables.

10. Accuracy of NMR chemical shift prediction.

A substantial number of structure-based predictors of NMR chemical shifts have emerged recently. These include SHIFTS [93], CheShift [94], CamShift [95], and 4DSPOT [96, 97]. These programs are able to predict protein chemical shifts with reasonably high accuracy using only protein coordinates as input. Here SHIFTX2 was used. SHIFTX2 is a hybrid method that combines both sequence- and structure-based parameters. SHIFTX2 is fast, accurate, and suitable for calculations of chemical shifts from MD trajectories. Its reported correlation coefficient and RMS error for amide proton shifts are 0.9714 and 0.1711 ppm, respectively [36].

SHIFTX2, similar to other chemical shift predictors is only able to predict protein chemical shifts for naturally occurring amino acids. Therefore, prior to chemical shift calculations, the phosphate groups in the AT8 peptide were substituted with hydroxyl groups. The missing electrostatic component of the Gly207 chemical shift was then calculated using Buckingham's equation [39] and added to the value predicted by SHIFTX2 [98]. The uncertainty of the value of the electrostatic component of the amide proton chemical shift is estimated to be 0.2 ppm.

Acknowledgment

N.S.G. acknowledges the award of a Curtin Early Career Research Fellowship. This work was also supported by resources provided by the Pawsey Supercomputing Centre.

References

1. Iqbal K, del C. Alonso A, Chen S et al (2005) Tau pathology in Alzheimer disease and other tauopathies. Biochim Biophys Acta 1739(2–3): 198–210
2. Noble W, Hanger DP, Miller CCJ et al (2013) The importance of tau phosphorylation for neurodegenerative diseases. Front Neurol 4:83
3. Buee L, Bussiere T, Buee Scherrer V et al (2000) Tau protein isoforms, phosphorylation and role in neurodegenerative disorders. Brain Res Brain Res Rev 33(1):95–130
4. Hanger DP, Anderton BH, Noble W (2009) Tau phosphorylation: the therapeutic challenge for neurodegenerative disease. Trends Mol Med 15(3):112–119
5. Hasegawa M, Morishima-Kawashima M, Takio K et al (1992) Protein sequence and mass spectrometric analyses of tau in the Alzheimer's disease brain. J Biol Chem 267(24):17047–17054
6. Hoffmann R, Lee VM, Leight S et al (1997) Unique Alzheimer's disease paired helical filament specific epitopes involve double phosphorylation at specific sites. Biochemistry 36(26):8114–8124
7. Lim J, Ping Lu K (2005) Pinning down phosphorylated tau and tauopathies. Biochim Biophys Acta 1739(2–3):311–322
8. Morishima-Kawashima M, Hasegawa M, Takio K et al (1995) Proline-directed and non-proline-directed phosphorylation of PHF-tau. J Biol Chem 270(2):823–829
9. Biernat J, Mandelkow EM, Schroter C et al (1992) The switch of tau protein to an Alzheimer-like state includes the phosphorylation of two serine-proline motifs upstream of

the microtubule binding region. EMBO J 11(4):1593–1597
10. Goedert M, Jakes R, Vanmechelen E (1995) Monoclonal antibody AT8 recognises tau protein phosphorylated at both serine 202 and threonine 205. Neurosci Lett 189(3): 167–169
11. Mercken M, Vandermeeren M, Lubke U et al (1992) Monoclonal antibodies with selective specificity for Alzheimer tau are directed against phosphatase-sensitive epitopes. Acta Neuropathol 84(3):265–272
12. Zheng-Fischhofer Q, Biernat J, Mandelkow EM et al (1998) Sequential phosphorylation of TAU by glycogen synthase kinase-3beta and protein kinase A at Thr212 and Ser214 generates the Alzheimer-specific epitope of antibody AT100 and requires a paired-helical-filament-like conformation. Eur J Biochem 252(3): 542–552
13. Goedert M, Jakes R, Crowther RA et al (1994) Epitope mapping of monoclonal antibodies to the paired helical filaments of Alzheimer's disease: identification of phosphorylation sites in tau protein. Biochem J 301(Pt 3):871–877
14. Jicha GA, Lane E, Vincent I et al (1997) A conformation- and phosphorylation-dependent antibody recognizing the paired helical filaments of Alzheimer's disease. J Neurochem 69(5):2087–2095
15. Otvos L Jr, Feiner L, Lang E et al (1994) Monoclonal antibody PHF-1 recognizes tau protein phosphorylated at serine residues 396 and 404. J Neurosci Res 39(6):669–673
16. Goedert M, Cohen ES, Jakes R et al (1992) p42 MAP kinase phosphorylation sites in microtubule-associated protein tau are dephosphorylated by protein phosphatase 2A1. Implications for Alzheimer's disease [corrected]. FEBS Lett 312(1):95–99
17. Porzig R, Singer D, Hoffmann R (2007) Epitope mapping of mAbs AT8 and Tau5 directed against hyperphosphorylated regions of the human tau protein. Biochem Biophys Res Commun 358(2):644–649
18. Jeganathan S, Hascher A, Chinnathambi S et al (2008) Proline-directed pseudo-phosphorylation at AT8 and PHF1 epitopes induces a compaction of the paperclip folding of Tau and generates a pathological (MC-1) conformation. J Biol Chem 283(46): 32066–32076
19. Amniai L, Barbier P, Sillen A et al (2009) Alzheimer disease specific phosphoepitopes of Tau interfere with assembly of tubulin but not binding to microtubules. FASEB J 23(4): 1146–1152
20. Ball LJ, Kuhne R, Schneider-Mergener J et al (2005) Recognition of proline-rich motifs by protein-protein-interaction domains. Angew Chem Int Ed Engl 44(19):2852–2869
21. Uversky VN (2013) Unusual biophysics of intrinsically disordered proteins. Biochim Biophys Acta 1834(5):932–951
22. Papoian GA (2008) Proteins with weakly funneled energy landscapes challenge the classical structure–function paradigm. Proc Natl Acad Sci U S A 105(38):14237–14238
23. Structure of IDPs (2009) In: Structure and function of intrinsically disordered proteins. Chapman and Hall/CRC, pp 121–142
24. Sinko W, Miao Y, de Oliveira CAF et al (2013) Population based reweighting of scaled molecular dynamics. J Phys Chem B 117(42): 12759–12768
25. Hamelberg D, de Oliveira CA, McCammon JA (2007) Sampling of slow diffusive conformational transitions with accelerated molecular dynamics. J Chem Phys 127(15):155102
26. Hamelberg D, Mongan J, McCammon JA (2004) Accelerated molecular dynamics: a promising and efficient simulation method for biomolecules. J Chem Phys 120(24): 11919–11929
27. Hamelberg D, Shen T, McCammon JA (2005) Phosphorylation effects on cis/trans isomerization and the backbone conformation of serine-proline motifs: accelerated molecular dynamics analysis. J Am Chem Soc 127(6): 1969–1974
28. Velazquez HA, Hamelberg D (2015) Dynamical role of phosphorylation on serine/threonine-proline Pin1 substrates from constant force molecular dynamics simulations. J Chem Phys 142(7):075102
29. Gandhi NS, Landrieu I, Byrne C et al (2015) A phosphorylation-induced turn defines the Alzheimer's disease AT8 antibody epitope on the tau protein. Angew Chem Int Ed Engl 54(23):6819–6823
30. Lippens G, Amniai L, Wieruszeski JM et al (2012) Towards understanding the phosphorylation code of tau. Biochem Soc Trans 40(4):698–703
31. Case DA, Babin V, Berryman JT et al (2014) AMBER14. University of California, San Francisco, CA
32. Abraham M, Apol E, Apostolov R et al (2014) GROMACS user manual version 4.6.7.
33. Humphrey W, Dalke A, Schulten K (1996) VMD: visual molecular dynamics. J Mol Graph 14(1):33–38, 27-38
34. Pettersen EF, Goddard TD, Huang CC et al (2004) UCSF chimera – a visualization system

for exploratory research and analysis. J Comput Chem 25(13):1605–1612
35. Kabsch W, Sander C (1983) Dictionary of protein secondary structure: pattern recognition of hydrogen-bonded and geometrical features. Biopolymers 22(12):2577–2637
36. Han B, Liu Y, Ginzinger SW et al (2011) SHIFTX2: significantly improved protein chemical shift prediction. J Biomol NMR 50(1):43–57
37. Berendsen HJ, Postma JPM, van Gunsteren WF et al (1984) Molecular dynamics with coupling to an external bath. J Chem Phys 81(8):3684–3690
38. Loncharich RJ, Brooks BR, Pastor RW (1992) Langevin dynamics of peptides: the frictional dependence of isomerization rates of N-acetylalanyl-N′-methylamide. Biopolymers 32(5): 523–535
39. Buckingham A (1960) Chemical shifts in the nuclear magnetic resonance spectra of molecules containing polar groups. Can J Chem 38(2):300–307
40. Kukic P, Farrell D, Søndergaard C, Bjarnadottir U, Bradley J, Pollastri G, Nielsen JE (2010) Improving the analysis of NMR spectra tracking pH-induced conformational changes: removing artefacts of the electric field on the NMR chemical shift. Proteins 78:971–984
41. Hass MAS, Ringkjøbing Jensen M, Led JJ (2008) Probing electric fields in proteins in solution by NMR spectroscopy. Proteins 72:333–343
42. Stanley N, Esteban-Martín S, De Fabritiis G (2015) Progress in studying intrinsically disordered proteins with atomistic simulations. Prog Biophys Mol Biol 119(1):47–52
43. Sugita Y, Okamoto Y (1999) Replica-exchange molecular dynamics method for protein folding. Chem Phys Lett 314(1–2):141–151
44. Larini L, Gessel MM, LaPointe NE et al (2013) Initiation of assembly of tau (273–284) and its ΔK280 mutant: an experimental and computational study. Phys Chem Chem Phys 15(23): 8916–8928
45. Levine ZA, Larini L, Shea J-E (2014) Tau (273–284): a molecular dynamics study of intrinsically disordered protein conformations in the presence of osmolytes. Biophys J 106(2): 483
46. Ganguly P, Do TD, Larini L et al (2015) Tau assembly: the dominant role of PHF6 (VQIVYK) in microtubule binding region repeat R3. J Phys Chem B 119(13):4582–4593
47. Mukrasch MD, Markwick P, Biernat J et al (2007) Highly populated turn conformations in natively unfolded tau protein identified from residual dipolar couplings and molecular simulation. J Am Chem Soc 129(16):5235–5243
48. Pierce LCT, Salomon-Ferrer R, de Augusto F. Oliveira C et al (2012) Routine access to millisecond time scale events with accelerated molecular dynamics. J Chem Theory Comput 8(9):2997–3002
49. Miao Y, Sinko W, Pierce L et al (2014) Improved reweighting of accelerated molecular dynamics simulations for free energy calculation. J Chem Theory Comput 10(7): 2677–2689
50. Jing Z, Sun H (2015) A comment on the reweighting method for accelerated molecular dynamics simulations. J Chem Theory Comput 11(6):2395–2397
51. Götz AW, Williamson MJ, Xu D et al (2012) Routine microsecond molecular dynamics simulations with AMBER on GPUs. 1. Generalized born. J Chem Theory Comput 8(5): 1542–1555
52. Markwick PR, Bouvignies G, Salmon L et al (2009) Toward a unified representation of protein structural dynamics in solution. J Am Chem Soc 131(46):16968–16975
53. Feldman HJ, Hogue CW (2000) A fast method to sample real protein conformational space. Proteins 39(2):112–131
54. Feldman HJ, Hogue CW (2002) Probabilistic sampling of protein conformations: new hope for brute force? Proteins 46(1):8–23
55. Marsh JA, Forman-Kay JD (2009) Structure and disorder in an unfolded state under nondenaturing conditions from ensemble models consistent with a large number of experimental restraints. J Mol Biol 391(2):359–374
56. Adzhubei AA, Sternberg MJ (1993) Left-handed polyproline II helices commonly occur in globular proteins. J Mol Biol 229(2): 472–493
57. Berisio R, Loguercio S, De Simone A et al (2006) Polyproline helices in protein structures: a statistical survey. Protein Pept Lett 13(8):847–854
58. King SM, Johnson WC (1999) Assigning secondary structure from protein coordinate data. Proteins 35(3):313–320
59. Srinivasan R, Rose GD (1999) A physical basis for protein secondary structure. Proc Natl Acad Sci U S A 96(25):14258–14263
60. Cubellis MV, Cailliez F, Lovell SC (2005) Secondary structure assignment that accurately reflects physical and evolutionary characteristics. BMC Bioinformatics 6(Suppl 4):S8
61. Mansiaux Y, Joseph AP, Gelly J-C et al (2011) Assignment of polyproline II conformation

and analysis of sequence – structure relationship. PLoS One 6(3), e18401

62. Cino EA, Choy W-Y, Karttunen M (2012) Comparison of secondary structure formation using 10 different force fields in microsecond molecular dynamics simulations. J Chem Theory Comput 8(8):2725–2740
63. Lindorff-Larsen K, Maragakis P, Piana S et al (2012) Systematic validation of protein force fields against experimental data. PLoS One 7(2), e32131
64. Beauchamp KA, Lin Y-S, Das R et al (2012) Are protein force fields getting better? A systematic benchmark on 524 diverse NMR measurements. J Chem Theory Comput 8(4): 1409–1414
65. Palazzesi F, Prakash MK, Bonomi M et al (2015) Accuracy of current all-atom force-fields in modeling protein disordered states. J Chem Theory Comput 11(1):2–7
66. Nerenberg PS, Head-Gordon T (2011) Optimizing protein–solvent force fields to reproduce intrinsic conformational preferences of model peptides. J Chem Theory Comput 7(4):1220–1230
67. Ball KA, Phillips Aaron H, Wemmer David E et al (2013) Differences in β-strand populations of monomeric Aβ40 and Aβ42. Biophys J 104(12):2714–2724
68. Wang W, Ye W, Jiang C et al (2014) New force field on modeling intrinsically disordered proteins. Chem Biol Drug Des 84(3):253–269
69. Ye W, Ji D, Wang W et al (2015) Test and evaluation of ff99IDPs force field for intrinsically disordered proteins. J Chem Inf Model 55(5): 1021–1029
70. Best RB, Mittal J (2010) Protein simulations with an optimized water model: cooperative helix formation and temperature-induced unfolded state collapse. J Phys Chem B 114(46):14916–14923
71. Horn HW, Swope WC, Pitera JW et al (2004) Development of an improved four-site water model for biomolecular simulations: TIP4P-Ew. J Chem Phys 120(20):9665–9678
72. Abascal JL, Vega C (2005) A general purpose model for the condensed phases of water: TIP4P/2005. J Chem Phys 123(23):234505
73. Nerenberg PS, Jo B, So C et al (2012) Optimizing solute-water van der Waals interactions to reproduce solvation free energies. J Phys Chem B 116(15):4524–4534
74. Best RB, Zheng W, Mittal J (2014) Balanced protein–water interactions improve properties of disordered proteins and non-specific protein association. J Chem Theory Comput 10(11):5113–5124
75. Piana S, Donchev AG, Robustelli P et al (2015) Water dispersion interactions strongly influence simulated structural properties of disordered protein states. J Phys Chem B 119(16): 5113–5123
76. Homeyer N, Horn AH, Lanig H et al (2006) AMBER force-field parameters for phosphorylated amino acids in different protonation states: phosphoserine, phosphothreonine, phosphotyrosine, and phosphohistidine. J Mol Model 12(3):281–289
77. Steinbrecher T, Latzer J, Case D (2012) Revised AMBER parameters for bioorganic phosphates. J Chem Theory Comput 8(11): 4405–4412
78. Khoury GA, Thompson JP, Smadbeck J et al (2013) Forcefield_PTM: ab initio charge and AMBER forcefield parameters for frequently occurring post-translational modifications. J Chem Theory Comput 9(12):5653–5674
79. Miao Y, Feixas F, Eun C et al (2015) Accelerated molecular dynamics simulations of protein folding. J Comput Chem 36(20): 1536–1549
80. Doshi U, Hamelberg D (2009) Reoptimization of the AMBER force field parameters for peptide bond (Omega) torsions using accelerated molecular dynamics. J Phys Chem B 113(52): 16590–16595
81. De Groot B, Van Aalten D, Scheek R et al (1997) Prediction of protein conformational freedom from distance constraints. Proteins 29(2):240–251
82. David C, Jacobs D (2014) Principal component analysis: a method for determining the essential dynamics of proteins. In: Livesay DR (ed) Protein dynamics, vol 1084, methods in molecular biology. Humana, Louisville, KY, pp 193–226
83. Showalter SA (2007) Intrinsically disordered proteins: methods for structure and dynamics studies. eMagRes 3(2). John Wiley & Sons, Ltd
84. Tompa P (2011) Unstructural biology coming of age. Curr Opin Struct Biol 21(3): 419–425
85. Jensen MR, Zweckstetter M, J-r H et al (2014) Exploring free-energy landscapes of intrinsically disordered proteins at atomic resolution using NMR spectroscopy. Chem Rev 114(13): 6632–6660
86. Bastidas M, Gibbs EB, Sahu D et al (2015) A primer for carbon-detected NMR applications to intrinsically disordered proteins in solution. Concept Magnet Reson A 44(1):54–66
87. Bienkiewicz E, Lumb K (1999) Random-coil chemical shifts of phosphorylated amino acids. J Biomol NMR 15(3):203–206

88. Wishart DS, Sykes BD (1994) Chemical shifts as a tool for structure determination. Methods Enzymol 239:363–392
89. Schwarzinger S, Kroon GJ, Foss TR et al (2001) Sequence-dependent correction of random coil NMR chemical shifts. J Am Chem Soc 123(13):2970–2978
90. Tamiola K, Acar B, Mulder FA (2010) Sequence-specific random coil chemical shifts of intrinsically disordered proteins. J Am Chem Soc 132(51):18000–18003
91. Shen Y, Lange O, Delaglio F et al (2008) Consistent blind protein structure generation from NMR chemical shift data. Proc Natl Acad Sci U S A 105(12):4685–4690
92. Camilloni C, De Simone A, Vranken WF et al (2012) Determination of secondary structure populations in disordered states of proteins using nuclear magnetic resonance chemical shifts. Biochemistry 51(11):2224–2231
93. Moon S, Case DA (2007) A new model for chemical shifts of amide hydrogens in proteins. J Biomol NMR 38(2):139–150
94. Vila JA, Arnautova YA, Martin OA et al (2009) Quantum-mechanics-derived 13Cα chemical shift server (CheShift) for protein structure validation. Proc Natl Acad Sci U S A 106(40): 16972–16977
95. Kohlhoff KJ, Robustelli P, Cavalli A et al (2009) Fast and accurate predictions of protein NMR chemical shifts from interatomic distances. J Am Chem Soc 131(39): 13894–13895
96. Lehtivarjo J, Hassinen T, Korhonen S-P et al (2009) 4D Prediction of protein 1H chemical shifts. J Biomol NMR 45(4):413–426
97. Lehtivarjo J, Tuppurainen K, Hassinen T et al (2012) Combining NMR ensembles and molecular dynamics simulations provides more realistic models of protein structures in solution and leads to better chemical shift prediction. J Biomol NMR 52(3):257–267
98. Kukic P, Farrell D, McIntosh LP et al (2013) Protein dielectric constants determined from NMR chemical shift perturbations. J Am Chem Soc 135(45):16968–16976

Chapter 4

Tau Interaction with Tubulin and Microtubules: From Purified Proteins to Cells

Tiphany De Bessa, Gilles Breuzard, Diane Allegro, François Devred, Vincent Peyrot, and Pascale Barbier

Abstract

Microtubules (MTs) play an important role in many cellular processes and are dynamic structures regulated by an important network of microtubules-associated proteins, MAPs, such as Tau. Tau has been discovered as an essential factor for MTs formation in vitro, and its region implicated in binding to MTs has been identified. By contrast, the affinity, the stoichiometry, and the topology of Tau–MTs interaction remain controversial. Indeed, depending on the experiment conditions a wide range of values have been obtained. In this chapter, we focus on three biophysical methods, turbidimetry, cosedimentation assay, and Förster Resonance Energy Transfer to study Tau–tubulin interaction both in vitro and in cell. We highlight precautions that must be taken in order to avoid pitfalls and we detail the nature of the conclusions that can be drawn from these methods about Tau–tubulin interaction.

Key words Tau protein, Microtubules, Tubulin, Turbidimetry, FRET, Cosedimentation essay, Affinity constant

1 Introduction

Microtubules cytoskeleton plays an important role in many cellular processes such as mitosis, cellular integrity, cellular shaping, intracellular trafficking, and cell migration. Microtubules (MTs) are constantly alternating between polymerization and depolymerization phases. This process, called dynamic instability of MT [1], is highly regulated by an important network of microtubules-associated proteins, MAPs, [2] such as Tau. Since Weingarten's discovery of Tau as an essential factor for microtubule formation in vitro [3], the interaction of Tau with tubulin has been intensively studied. However, differential experimental conditions (different buffer, Tau and tubulin peptides, Tau constructs) lead to surprising range of Tau–MT affinity values. Studying this interaction is difficult for at least two reasons: the intrinsically disordered structure of Tau and the ability of αβ-tubulin dimers to self-assemble

Caroline Smet-Nocca (ed.), *Tau Protein: Methods and Protocols*, Methods in Molecular Biology, vol. 1523,
DOI 10.1007/978-1-4939-6598-4_4, © Springer Science+Business Media New York 2017

into different oligomers depending on the temperature. MTs are hollow cylinders formed at 37 °C by a combination of linear protofilaments, which themselves are formed by αβ-tubulin dimers associated heads to tails, forming a pseudohelical lattice [4–6]. Tubulin can also self-assemble into other oligomers than MT. For example, in solution at 20 °C, addition of magnesium to tubulin dimers induces formation of single and doubles rings [7–9]. These rings were also described as depolymerization products resulting of protofilaments peeling at MTs end [10–12].

Human brain expresses six isoforms of Tau protein produced by alternative splicing (exons 2–3 and 10) of a single mRNA [13]. The interaction domain of Tau with microtubules, called MT binding domain (MTBR), is well documented: it is constituted of three or four (depending on the isoform) partially repeated sequences and a part of N-terminal proline-rich region flanking the MTBR [14–17]. By contrast, despite the great number of methods used to characterize this interaction, its affinity, the stoichiometry, and the topology of binding site on tubulin/MT are still under debate.

In this chapter, we focus on three biophysical methods namely turbidimetry, cosedimentation assay, and Förster Resonance Energy Transfer (FRET) in living cells used to study Tau–tubulin interaction; the use of other useful methods such as analytical ultracentrifugation or microcalorimetry has been recently described [18–20]. We highlight precautions that must be taken in order to avoid pitfalls and we detail the nature of the conclusions that can be drawn from these methods about Tau–tubulin interaction.

2 Materials

Prepare all solutions using ultrapure water and analytical grade reagents.

2.1 Tubulin Purification

2.1.1 Reagents and Solutions

1. 1 M magnesium chloride ($MgCl_2$) commercial aqueous solution.
2. 1 M ethylene glycol-bis(2-aminoethyl-ether)-N,N,NP,NP-tetraacetic acid (EGTA): weigh 7.6 g EGTA in a 20 mL glass beaker and add ultrapure water to have just less than 20 mL. Transfer in a 20 mL graduated flask and adjust volume to 20 mL. Put in a plastic tube or glass bottle and store at room temperature.
3. 0.1 M guanosine triphosphate sodium salt hydrate (GTP): weigh 0.0567 g GTP in a microtube and add 970 μL of ultrapure water. Check the concentration by measuring the absorbance at 256 nm with a molecular extinction coefficient $\varepsilon 256$ nm = 13,700 M^{-1} cm^{-1} and store at −20 °C (*see* **Note 1**).

4. 400 mM phosphate buffer stock (1 L): mix 27.6 g of NaH_2PO_4 H_2O with 28.39 g anhydrous Na_2HPO_4 and add ultrapure water to 1 L in graduated flask.
5. 20 mM phosphate buffer (1 L): put 50 mL of 400 mM phosphate buffer in graduated flask and adjust volume to 1 L. pH must be at 6.95 at 25 °C.
6. PM buffer: 20 mM phosphate buffer, 0.5 mM $MgCl_2$. In a 500 mL glass beaker, add 25 mL 400 mM phosphate buffer and 250 μL of 1 M $MgCl_2$ to 450 mL of ultrapure water. Adjust pH to 7.0 with 10 M NaOH. Transfer the buffer in a 500 mL graduated flask and complete volume to 500 mL. Then, transfer to a 500 mL cylinder and store at 4 °C.
7. PMS buffer: 20 mM phosphate buffer, 0.5 mM $MgCl_2$, 0.24 M sucrose. In a 4 L glass beaker, add 200 mL 400 mM phosphate buffer, 2 mL 1 M $MgCl_2$ and 328.6 g sucrose to 3.5 L of ultrapure water. The pH must be at 6.90 without adjustment. Transfer in 4 L graduated flask and complete volume to 4 L. Put in a glass beaker to store at 4 °C.
8. PMG buffer: 20 mM phosphate buffer, 0.5 mM $MgCl_2$, 0.1 mM GTP. In a 2 L glass beaker, add 100 mL 400 mM phosphate buffer, 1 mL of 1 M $MgCl_2$ to 1850 mL of ultrapure water. Adjust pH to 7 with 10 M NaOH. Transfer and fill it up to 2 L in a graduated flask and then transfer 250 mL in a 500 mL glass beaker to be used to prepare PMG 0.4 M KCl and keep the 1750 mL left in the 2 L glass beaker and store at 4 °C. Weigh 0.1056 g of GTP in a microtube and store it at -20 °C. GTP will be added to the 1.750 L of PMG Buffer on purification day.
9. PMG 0.8 M KCl: 20 mM phosphate buffer, 0.5 mM $MgCl_2$, 0.8 M KCl, 0.1 mM GTP. In a 500 mL glass beaker, add 25 mL of 400 mM phosphate buffer, 250 μL of 1 M $MgCl_2$, and 29.84 g KCl to 450 mL of ultrapure water. Adjust pH to 7.0 with 10 M NaOH and transfer in a 500 mL graduated flask to fill it up to 500 mL and transfer 250 mL in another beaker to prepare PMG 0.4 M KCl. Keep the 250 mL left in a glass beaker to store at 4 °C. Weigh 0.01507 g of GTP in a microtube and store it at −20 °C. GTP will be added in 250 mL PMG 0.8 M KCl on purification day.
10. PMG 0.4 M KCl: 20 mM phosphate buffer, 0.5 mM $MgCl_2$, 0.4 M KCl, 0.1 mM GTP. In a 500 mL glass beaker, mix 250 mL PMG buffer and 250 mL of PMG 0.8 M KCl. Verify pH and adjust it to 7 with 10 M NaOH if it is necessary and store at 4 °C. Weigh 0.03016 g of GTP in a microtube and store at −20 °C. GTP will be added in the 250 mL of PMG 0.4 M KCl on purification day.

11. PMG 1 M sucrose: 20 mM phosphate buffer, 0.5 mM $MgCl_2$, 1 M sucrose, 0.1 mM GTP. In a 500 mL glass beaker, add 171.2 g sucrose, 25 mL of 400 mM phosphate buffer, 250 μL of 1 M $MgCl_2$ and ultrapure water to a volume of 450 mL. Adjust pH to 7.0 with 10 M NaOH and transfer in a graduated flask to fill it up to 500 mL. Store at 4 °C. Prepare 500 μL of 1 M GTP as described in **step 3**. It will be added in this buffer on the purification day.
12. Microtubule formation buffer: 20 mM phosphate buffer, 10 mM $MgCl_2$, 3.4 M glycerol, 1 mM EGTA, pH 6.5, 0.1 mM GTP. In a 500 mL glass beaker, add 156.4 g glycerol to 300 mL ultrapure water and mix few minutes to dilute glycerol (*see* **Note 2**). Then add 25 mL 400 mM phosphate buffer, 5 mL of 1 M $MgCl_2$, 500 μL of 1 M EGTA and ultrapure water to 450 mL. Adjust pH to 6.5 and complete volume to 500 mL in a graduated flask. Store at 4 °C. 0.1 mM GTP will be added freshly.
13. Sephadex G25 ultra fine: 50 g of Sephadex G25 beads are preincubated overnight in 1 L water ultrapure before use (*see* **Note 3**).
14. DEAE A50: weigh 10 g of DEAE A50 in 1 L of 20 mM phosphate buffer, pH7.
15. Guanidine-HCl 6 M: weigh 11.46 g of guanidine-HCl and add 20 mL of water (*see* **Note 4**).

2.1.2 Special Equipment

1. For slaughterhouse: ice box, plastic bag to put brains, gloves, lab coat, and disposable paper.
2. For purification: big container to boil water, cold room, blender, centrifuge and ultracentrifuge and matching tubes (18 × 250 mL and 8 × 50 mL), sintered-glass filter and vacuum flask, column 3 cm × 50 cm and collector, dialysis tubing : cellulose molecular weight cutoff 3500 Spectra/Por® molecular porous n°3 18 mm × 50 ft, nitrogen tank.

2.2 Tau Purification

2.2.1 Reagents and Buffers

1. Transformed bacteria *Escherichia coli* BL21DE3 with pET containing Tau DNA.
2. Medium Luria Broth Miller (200 mL) sterilized. Weigh 5 g medium in a 200 mL glass beaker and add ultrapure water to 200 mL. Adjust volume in a graduated flask and transfer in a bottle to sterilize and store at 4 °C.
3. Medium Luria Broth Buffered (1 L) sterilized and supplemented with glucose and ampicillin. Weigh 54.8 g medium in a 1 L glass beaker and add ultrapure water. Adjust volume in a graduated flask and transfer in two bottles to sterilize it. Store it at 4 °C.

4. Ampicillin 100 mg/mL. Weigh 1 g ampicillin in 15 mL tube and add 10 mL ultrapure water. Make 1 mL aliquots and store them at −20 °C.
5. 1 M Isopropyl-beta-D-1-thiogalactopyranoside (IPTG): weigh 2.4 g in a 15 mL tube and add 10 mL ultrapure water. Make 1 mL aliquots and store them at −20 °C.
6. 1 M DL-Dithiothreitol (DTT): weigh 1.54 g in a 15 mL tube and add 10 mL ultrapure water. Make 1 mL aliquots and store them at −20 °C.
7. Lysis Buffer (100 mL): 8 mM Triton X100, 45 mM 2-(N-morpholino) ethanesulfonic acid (MES), 1 mM DTT. Weigh 0.5 g Triton and 0.9 g MES. Mix in a 100 mL glass beaker with 95 mL of ultrapure water and add 100 μL DTT 1 M. pH should be around 3.9. Adjust volume with a graduated flask. Make 10 mL aliquots and store them at −20 °C.
8. Activation Buffer (500 mL): 45 mM MES, 1 M NaCl. Dissolve 4.4 g MES and 29 g NaCl in 450 mL ultrapure water in a 500 mL glass beaker. Increase pH to 6.5 with NaOH and complete volume to 500 mL in a graduated flask (*see* **Note 5**).
9. Equilibration Buffer (500 mL): 45 mM MES. Weigh 4.4 g MES and dissolve in 450 mL ultrapure water in a 500 mL glass beaker. Adjust pH to 6.5 with KOH and complete volume to 500 mL in a graduated flask (*see* **Note 5**).
10. Elution Buffer (500 mL): 45 mM MES, 0.5 M NaCl. Weigh 4.4 g MES and 14.5 g NaCl and dissolve them in 450 mL ultrapure water in a 500 mL glass beaker. Adjust pH to 6.5 with NaOH and complete volume to 500 mL in a graduated flask (*see* **Note 5**).

2.2.2 Special Equipment

Incubator, French press, centrifuge, FPLC apparatus, dialysis tubing: cellulose molecular weight cutoff 3500 Spectra/Por® molecular porous n°3 18 mm × 50 ft, freeze-dryer.

2.3 Tau–Tubulin Interaction Experiments in Solution

2.3.1 Microtubule Formation in Solution

1. 400 mM phosphate buffer, 1 M EGTA, 1 M GTP is prepared as described in Subheading 1.
2. 1 M $MgCl_2$ commercial aqueous solution.
3. 1 M Tris(2-carboxyethyl)phosphine (TCEP): weigh 1.4 g in a 5 mL plastic tube, add 4.5 mL ultrapure water. Use a Vortex to strongly mix the solution for few minutes (*see* **Note 6**). Make 500 μL aliquots and store them at −20 °C.
4. Assembly Condition Buffer : 3.4 M glycerol, 20 mM phosphate buffer, 1 mM EGTA, 10 mM $MgCl_2$, pH 6.5, 0.1 mM GTP, 1 mM TCEP. Microtubule formation buffer is prepared as described in Subheading 2.1.1 and 0.1 mM GTP and 1 mM TCEP are added freshly before used.

5. Non Assembly Condition Buffer: 20 mM phosphate buffer pH 6.5, 0.1 mM GTP, 1 mM TCEP. In a 500 mL glass beaker, add 25 mL of 400 mM phosphate buffer to 450 mL of ultrapure water. Adjust pH to 6.5 with 10 M NaOH. Transfer the buffer in a 500 mL graduated flask and complete volume to 500 mL. Add 0.1 mM GTP and 1 mM TCEP freshly just before used.
6. 10^{-3} M Paclitaxel: weigh 0.8 mg of paclitaxel in 1 mL in dimethyl sulfoxide (DMSO). Its concentration is measured spectrometrically with $\varepsilon_{273\text{nm}} = 1{,}700\ \text{M}^{-1}\ \text{cm}^{-1}$ in ethanol.
7. Gravity column (1 cm × 25 cm).
8. Centrifuge to run at 1,200 × *g* at 4 °C.
9. Spectrophotometer and quartz cuvettes.

2.3.2 Cosedimentation Assay

1. Glycerol purity superior to 99.5 %.
2. All buffers and reagents necessary to make SDS-PAGE (polyacrylamide gel electrophoresis in denaturing conditions) with 12 % acrylamide.
3. SDS PAGE materials.
4. Ultracentrifuge to run at 88,000 × *g* at 37 °C.

2.3.3 Electronic Microscopy

1. Uranyl acetate 2 %: weigh 0.2 g uranyl acetate and add 10 mL ultrapure water. Mix with vortex and check carefully for the dissolution of crystals (*see* **Note 7**).
2. 200 mesh carbon-coated formvar films on copper grids.
3. Thermostated room at 37 °C.
4. Transmission electron microscope.

2.4 Tau–Tubulin Interactions by FRET in Living Cells

2.4.1 Cell Culture, Plasmids, and Transfection Reagent

1. 1.5×10^5 cells (purchased from ATCC, USA).
2. Growth media: media supplemented with 10 % heat-inactivated fetal bovine serum (FBS) (*see* **Note 8**).
3. 8-well Lab-Tek chamber cover glass (Nunc).
4. Media without FBS (*see* **Note 8**).
5. 10× phosphate buffer saline stock solution.
6. 10 mM HEPES buffer.
7. Two plasmid DNA species (1 mg/mL): pEGFP-Tau DNA coding for hTau40 protein, the longest human isoform of Tau [21], coupled to the fluorescent protein EGFP as an energy donor; and pmCherry-α-tubulin DNA coding for α-tubulin protein (wild-type α1B isotype) coupled to the fluorescent mCherry protein [22] as the energy acceptor.
8. Commercial transfection reagent such as LipofectAmine 2000 (Life Technologies, Invitrogen).

2.4.2 Confocal Laser Scanning Microscopy

1. An inverted microscope equipped with a Plan-Apochromat 63× oil immersion objective (Numerical Aperture = 1.4).
2. A Confocal Laser Scanning Microscope (CLSM) equipped with Argon (488 nm) and Helium–Neon (543 nm) laser lines, specific dichroic filter and a spectral mode to select specific domains of the emission spectrum (*see* **Note 9**).
3. A photobleaching module coupled to CLSM including an observation beam, a photo-destruction beam and a time-resolved recording system.
4. The public-domain ImageJ software [23] and the plug-in PixFRET [24] for image analysis.

3 Methods

3.1 Tubulin Purification

Tubulin purification necessitates three consecutive days, one for materials and buffer preparations and two others for the purification.

3.1.1 Tubulin Purification (Day 1)

1. Equilibrate the ionic exchange resin in 20 mM phosphate buffer. From stock solution 400 mM phosphate buffer, prepare 1 L of phosphate buffer 20 mM by adding 50 mL of 400 mM phosphate buffer in graduated flask and adjust volume to 1 L with ultrapure water. Then, weigh 10 g of DEAE-Sephadex A50 in a 2 L flask and add 1 L 20 mM phosphate buffer. Put this flask in boiling water and let DEAE-Sephadex A50 swell for 5 h (*see* **Note 10**).
2. G25 Preparation: pack the Sephadex G25 beads (preincubated in ultrapure water overnight) in a column of 50 cm high and 3 cm diameter (3 cm × 50 cm). Leave 2 cm at the top of the column to add the sample. Equilibrate it with 400 mL PM buffer at a 1 mL/min flow rate (*see* **Note 11**).
3. Chemical preparation: crush ammonium sulfate crystals in powder to dissolve it easier.
4. Weigh GTP samples as described in Subheading 2.1.1 and store it at −20 °C.
5. Material and room preparation: clean and organize the cold storage room. Dispose of materials.

3.1.2 Tubulin Purification (Day 2)

The brains (fifteen lamb or pig brains) should be collected just after the animal death in the slaughterhouse and put them in plastic bag and rapidly in ice box for the transport. From there, conduct all the experiments in a cold room to prevent tubulin denaturation. Nitrile gloves are used for all purification steps and workers must wear mask during brain homogenization.

1. Brain homogenization: remove superficial blood vessels and meninges from brains with dissection pliers. Cut and rinse them with an equal volume of PMS (do not exceed 2.5 L) that will be eliminated through gauze. Repeat this step three times. Add PMS (total volume must not exceed 3 L to limit centrifuge run numbers) to brains and homogenize them in a blender for 30 s.
2. To eliminate membranes and cellular fragments, homogenate is centrifuged in 12 tubes (250 mL) at 13,000 × *g* for 30 min at 4 °C (*see* **Note 12**). During this centrifugation, add GTP to PMG buffer, PMG 0.4 M KCl and PMG 0.8 M KCl. Equilibrate G25 with 400 mL PMG at a flow rate of 1 mL/min. Begin DEAE-Sephadex A50 equilibration: get rid of the buffer by filtration with sintered-glass filter and vacuum flask. Add 250 mL PMG buffer to Sephadex beads and increase pH to 6.7 with few drops of 10 M NaOH. Get rid of the buffer as described above and add 250 mL PMG buffer.
3. Ammonium sulfate fractionation 32 %: discard pellets (*see* **Note 13**) and measure supernatant volume with cylinder. Put it in a 2 L glass beaker and add slowly 177 g of ammonium sulfate per liter of supernatant. When ammonium sulfate is completely dissolved, wait for 10 min.
4. Centrifuge the precipitate at 13,000 × *g* for 30 min at 4 °C in six tubes of 250 mL (*see* **Note 12**). During this second centrifugation, continue to equilibrate DEAE-Sephadex A50: increase pH to 6.8 with 10 M NaOH, throw out buffer and add 250 mL PMG buffer. Increase pH to 6.9 and throw out buffer. Add 250 mL PMG buffer. DEAE-Sephadex A50 is ready to use.
5. Ammonium sulfate fractionation 43 %: discard pellets (*see* **Note 13**) and measure the supernatant volume with cylinder. Put it in a 2 L glass beaker and add slowly 71 g of ammonium sulfate per liter of supernatant. When ammonium sulfate is completely dissolved, wait for 10 min.
6. Centrifuge the precipitate at 13,000 × *g* for 30 min at 4 °C (*see* **Note 12**). Discard the supernatant. Collect pellets with spatula and suspend them by gentle mixing in a 100 mL glass beaker. Wash thoroughly centrifugation tubes with PMG buffer. Do it three times but do not exceed a final volume of 40 mL PMG buffer (*see* **Note 14**).
7. DEAE A50: the protein solution is then mixed with 250 mL of the DEAE-Sephadex A50 which has been previously equilibrated in PMG buffer. Incubate for 10 min with gentle agitation. Get rid of buffer by filtration on sintered-glass filter and vacuum flask. Eliminate filtrate and add 200 mL of PMG 0.4 M KCl on DEAE-Sephadex A50 containing sample.

Incubate with gentle agitation for 5 min. Get rid of buffer by filtration as described above twice. Change vacuum flask to keep filtrate. Add 125 mL PMG 0.8 M KCl, wait 5 min and eluate tubulin by filtration. This step has to be repeated twice.

8. Add slowly 62 g ammonium sulfate to the 250 mL of filtrate in a 500 mL glass beaker and wait for 5 min with gentle agitation after total ammonium sulfate dissolution.
9. Centrifuge at 66,000 × *g* for 25 min at 4 °C (*see* **Note 12**). Discard the supernatant and collect pellets with spatula and suspend them by gentle mixing in a 100 mL glass beaker. Wash thoroughly centrifugation tubes with PMG buffer. Do it three times but do not exceed a final volume of 20 mL PMG buffer (*see* **Note 14**). The totally dissolved pellets will be passed through Sephadex G25 chromatography.
10. Sephadex G25 chromatography: when the PMG buffer completely entered in the resin, load carefully the protein solution. When half of the mix is in the column add buffer on the top without mixing solutions (*see* **Note 15**). Eluate with a 1 mL/min flow rate and collect fractions of 3–3.5 mL. In tubes containing protein where a meniscus is observed (use plastic tubes), add 1 M $MgCl_2$ to have a final concentration of 30 mM (110 μL for 3.5 mL). Mix by inverting and collect all samples in which white precipitate is instantaneously observed. Gather them in centrifugation tubes.
11. Centrifuge at 9,000 × *g* for 10 min at 4 °C (*see* **Note 12**). During this centrifugation, add 500 μL 0.1 M GTP to the 500 mL of PMG sucrose buffer.
12. Collect pellets and dissolve them in a minimum volume of PMG sucrose buffer (do not exceed 5 mL) with gentle agitation (*see* **Note 14**).
13. The solution is placed in a dialysis tubing (cellulose molecular weight cutoff 3500, Spectra/Por® molecular porous n°3 18 mm × 50 ft) and dialyzed for 16 h against PMG sucrose buffer supplemented with 0.1 mM GTP.

3.1.3 Tubulin Purification and Storage (Day 3)

1. Collect tubulin from the dialysis tubing and centrifuge it at 12,000 × *g* for 30 min at 4 °C (*see* **Note 12**). Discard pellets and keep the supernatant. Make 300 μL aliquot in cryotubes and stock them in liquid nitrogen. Store 2 mL of dialysis buffer.
2. Tubulin concentration is determined by spectrophotometry. Put 5 μL of dialysis buffer in 500 μL of 6 M Guanidine-HCl and make the blank. Put in triplicate 5 μL of tubulin sample in 500 μL and measure the absorbance at 275 nm. Determine the concentration using the Lambert–Beer law with a molecular extinction coefficient of $\varepsilon_{275\ \mathrm{nm}} = 1.09\ \mathrm{mL\ mg^{-1}\ cm^{-1}}$ taking into account the dilution factor (here 101).

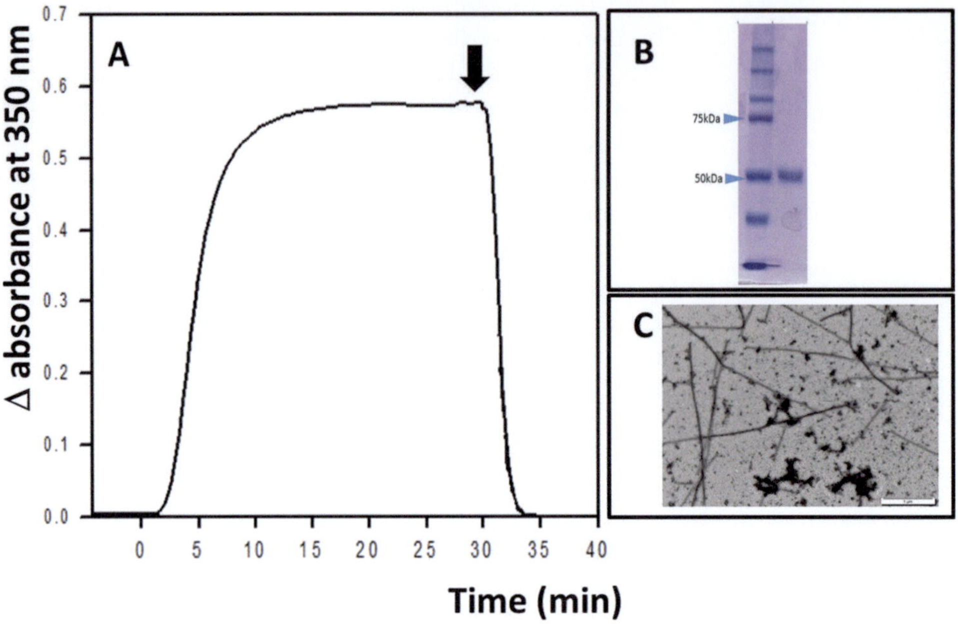

Fig. 1 Tubulin purification: (**A**) MT formation time course at 350 nm of 30 μM of purified tubulin in assembly conditions at 37 °C. At time indicated by *arrow*, the temperature is decreased to 10 °C. (**B**) 12 % SDS-PAGE of Tubulin that migrates in one band at 50 kDa. (**C**) Electronic microscopy of the MTs sample. The *white scale bar* represents 1 μm

3. Test the tubulin capacity to form microtubules using a turbidimetry assay (*see* Subheading 3.3.2).

At the end of the purification, you should have 200–300 mg of pure tubulin at 40–80 mg/mL and an absorbance variation of about 0.5 for 30 μM (3.0 mg/mL) of tubulin corresponding to microtubule formation (Fig. 1).

3.2 Tau Purification

3.2.1 Bacterial Pregrowth

1. Switch on the "Hoffman heater" 30 min before the beginning of the bacterial growth (*see* **Note 16**).
2. Near the "Hoffman heater" in a 500 mL Erlenmeyer flask, mix 100 mL sterilized medium Luria Broth Miller with 100 μL Ampicillin and 100 μL stock bacteria. Place it in an incubator at 37 °C overnight with agitation.

3.2.2 Bacterial Culture

1. Switch on the "Hoffman heater" 30 min before the beginning of bacterial growth (*see* **Note 16**).
2. Near the "Hoffman heater" first, remove 5 mL medium without bacteria to make blank to read absorbance and second, in each two 2 L Erlenmeyer flask, 500 mL sterilized medium Luria Broth Buffered are mixed with 1.85 g glucose, 100 μL ampicillin, and 10 mL pregrowth bacteria. Place them in an incubator at 37 °C with agitation for few hours. Read absorbance at

600 nm every hour until it reaches a value between 0.6 and 1. At this moment add 750 μL of 1 M IPTG to induce protein expression and incubate for 2 h 30 min.

3. Centrifuge at 5,000 × *g* for 10 min at 4 °C and remove supernatant. Pellets are recovered and dissolved in lysis buffer (*see* **Note 17**).

3.2.3 Proteins Extraction

1. Apply twice on the dissolved pellet a 4 t pressure in French press.
2. Boil the bacterial lysate at 90 °C for 10–15 min, to select thermal-resistant proteins like Tau.
3. Centrifuge for 30 min at 30,000 × *g* at 4 °C (*see* **Note 12**). Keep the supernatant (*see* **Note 17**).

3.2.4 Tau Purification

1. Chromatography preparation (*see* **Note 18**): A SP Sephadex® High Performance 5 mL column is activated by passing 5 mL (one column volume) of activation buffer (45 mM MES, 1 M NaCl) with a NaCl gradient of 0–100% (mix with equilibration buffer constituted of 45 mM MES) at a flow rate of 1 mL/min. Beads are rinsed with 3 column volumes (15 mL) of activation buffer. Reverse the NaCl gradient from 100 to 0% in 5 mL. Before sample injection, column is washed with 15 mL of equilibration buffer.
2. Start to collect 5 mL fractions and inject sample with equilibration buffer at a flow rate of 0.5 mL/min. Absorbance increases during sample deposit (resin–protein interaction). Rinse column with equilibration buffer as far as absorbance comes back to baseline level and then begin elution.
3. Apply a 30% NaCl gradient: increase the salt concentration to 30% in two column volumes (10 mL) with a flow rate of 1 mL/min. Apply it as long as absorbance is constant.
4. Increase the salt gradient to 70% in one column volume (5 mL) with a flow rate of 1 mL/min. Tau is released at this step. When Tau protein has been released from the column, a 100% gradient is applied to wash column for few minutes at a flow rate of 1 mL/min.
5. To verify that Tau is present in fractions or to know if there is no Tau in flow through, you may take 20 μL of each fraction, add 5 μL of loading buffer 5× and depose them on SDS PAGE 12% acrylamide (*see* **Note 17**).
6. Fractions containing Tau are pooled in a dialysis tubing (cellulose molecular weight cutoff 3500, Spectra/Por® molecular porous n°3 18 mm × 50 ft) and dialyzed for 2 h, three times against ultrapure water. Then, Tau is collected and frozen at −80 °C to be lyophilized. Once lyophilized, Tau can be stored at −20 °C for several months.

7. Before use, Tau is dissolved in the desired buffer, centrifuged at $3{,}000 \times g$ for 10 min and its concentration is determined by spectrophotometry using the extinction coefficient $\varepsilon_{280nm} = 7{,}700\ M^{-1}\ cm^{-1}$ after light scattering correction (Fig. 2).

3.3 Tau–Tubulin/Microtubule Interaction in Solution

Microtubule formation is a cooperative system that needs in vitro a tubulin minimum concentration called "tubulin critical concentration" (Cr). Cr is defined as the inverse of the apparent MT-elongation constant K_{app} [25]. To determine Cr, we measured microtubule amount by turbidimetry or sedimentation assay (tubulin concentration in the pellet) as a function of total tubulin concentration and fit the results to a linear regression. The Cr corresponds in this case to the value of x (tubulin concentration) found for $y = 0$ (no microtubule formation). More accurately, measuring the concentration found in the supernatant of sedimentation assay of microtubule solution leads directly the Cr value. This value is dependent of the experimental conditions. It is well known that magnesium ions, glycerol, dimethyl sulfoxide, sucrose are positive factors that reduce the Cr. For example, Cr of tubulin is 7 μM

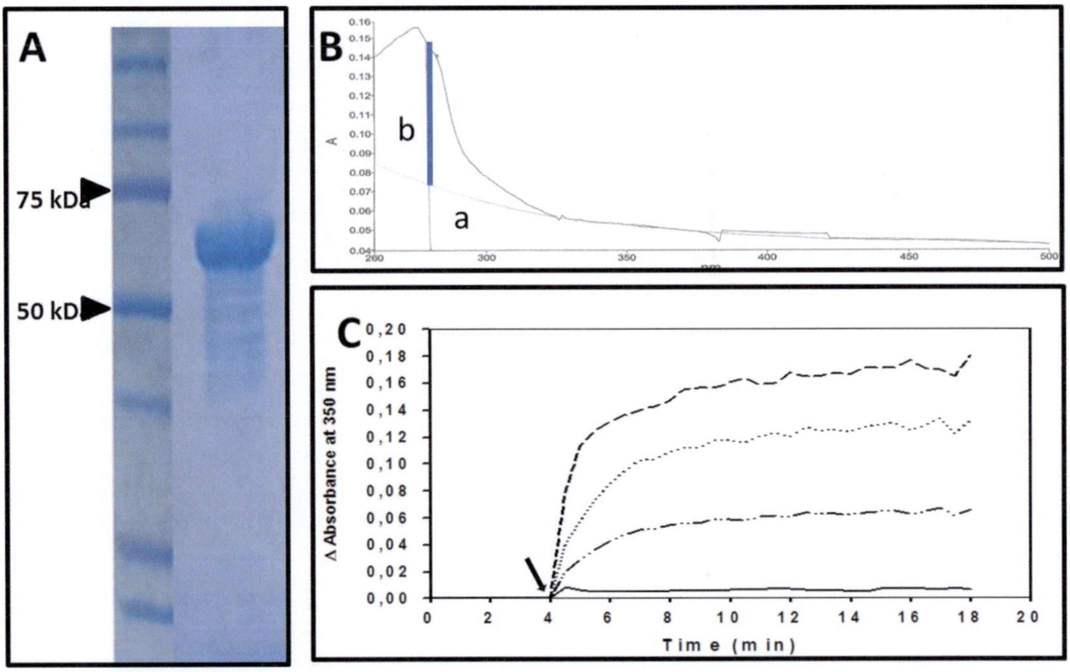

Fig. 2 Tau purification: (**A**) 12 % SDS-page of purified Tau protein. As largely described in literature, the largest 441 amino acid long isoform migrate at an apparent molecular mass of approximately 65 kDa. (**B**) Concentration determination by UV–visible wavelength scan of Tau protein. The *line A* corresponds to the light diffusion that should be subtracted to the scan to not sur-estimate the concentration. The corrected absorbance at 280 nm (*line B*) is then used to determine the Tau concentration. (**C**) Typical time course of MTs formation at 37 °C induced by Tau. Tubulin concentration is 5 μM is mixed with no (*dark line*) or 3.5 (*dash-dot line*), 4.2 (*dot line*) and 5 μM of Tau (*dash line*). The *arrow* indicates the Tau addition

in 20 mM phosphate buffer, 3.4 M glycerol, 1 mM EGTA, 10 mM $MgCl_2$, 0.1 mM GTP, pH 6.5 [26] and 20 μM in 80 mM K-Pipes, 1 mM $MgCl_2$, 1 mM EGTA, GTP 1 mM pH 6.8 [27]. In these conditions, called in this chapter as "assembly conditions buffer," presence of Tau leads to *a microtubule formation facilitated by ligand binding*. In 2004, we found experimental conditions in which the presence of Tau induces the microtubule formation in the absence of any other known polymerization factors (20 mM phosphate buffer pH 6.5, 0.1 mM GTP, 1 mM TCEP) (*see* **Note 6**). In this condition, called in this chapter "non-assembly condition," presence of Tau leads to *a microtubule formation induced by ligand binding* [28].

3.3.1 Tubulin Preparation

Before using tubulin it is necessary to eliminate sucrose. Tubulin is equilibrated (*see* **Note 19**) in the appropriate buffer (assembly or non-assembly condition) on gel filtration Sephadex G25 resin on a gravity column (1 cm*25 cm).

1. Equilibrate the column with cold 40 mL of appropriate buffer at a flow rate of 1 mL/min and defreeze rapidly tubulin sample.
2. When the buffer is removed (without completely drying the resin), put tubulin solution (*see* **Note 20**) on the top. When all the protein entered in the resin, put the buffer and start elution at a flow rate of 1 mL/min. Collect 500 μL fractions and put them on ice. The tubulin concentrations into the tubes with meniscus (*see* **step 10** in Subheading 3.1.2) are determined by spectrophotometry as described in Subheading 3.1.3 with elution as control instead of dialysis buffer.

3.3.2 Turbidimetry Assay

Tubulin prepared as described above is diluted to 7.5 μM in the appropriate buffer to be at a tubulin final concentration of 5 μM (this concentration is under the 7 μM Cr for the assembly condition). Sample volume and number of samples depend on the spectrophotometer (for us, it is 300 μL and six samples).

1. Before starting the time-driven microtubule formation, make blank on the spectrophotometer with buffer alone.
2. Put 200 μL of tubulin at 7.5 μM in the six different cuvettes and put them on ice.
3. Put rapidly the cuvettes in the spectrophotometer thermostated at 37 °C and measure the absorbance time course at 350 nm for 5 min. You may have to wipe condensation off.
4. Add rapidly in each cuvettes 100 μL of buffer for control or different Tau concentrations (from 0.5 to 5 μM). Do not use Tau concentration above the tubulin concentration because it induces microtubule bundling [28]. The signal should increase rapidly indicating microtubule formation. The turbidity is

observed for 30 min (Fig. 2). The samples then can be used for cosedimentation assay and/or electron microscopy.

5. Mix 15 μM of tubulin prepared as described in Subheading 3.3.1 in 20 mM phosphate buffer, 8 mM $MgCl_2$, pH 6.5, 1 mM TCEP with 15 μM of paclitaxel, a microtubule inducer and stabilizing agent.
6. Incubate for 25 min at 37 °C. At this stage the solution is opaque due to the microtubule formation.
7. Dilute the paclitaxel-stabilized microtubules (200 μL) to 7.5 μM.
8. Mix 200 μL of paclitaxel-stabilized MT with 100 μL of buffer or different concentrations of Tau from 0.5 to 20 μM. After 5 min, take 10 μL for transmission electron microscopy and 200 μL for cosedimentation assays.

3.3.3 Cosedimentation Assays: Microtubules Bound Tau, Free Tau, and Free Tubulin Separation

Microtubule and Tau bound to microtubule are separated from free Tau and tubulin by centrifugation. To prevent nonspecific association, centrifugation is conducted on 60% glycerol cushion.

1. Prepare the glycerol cushion by adding 60 mL of glycerol to 40 mL of appropriate buffer and form microtubules as described in the Subheading 3.3.2.
2. Place 200 μL samples on 200 μL glycerol 60% in ultracentrifuge tubes pre-warmed at 37 °C and centrifuge at 88,000 × *g* for 20 min at 37 °C in a pre-warmed rotor.
3. Collect supernatant (200 μL) containing free Tau and tubulin and mix it with 40 μL of 5× SDS-PAGE loading buffer. Turn upside down the tubes containing the pellets to remove the rest of supernatant.
4. Wash the pellets containing MT and bound Tau tactfully with 200 μL of warm (37 °C) buffer to eliminate free Tau and free tubulin. Repeat this step three times.
5. Dissolve the pellet in 200 μL cold buffer to depolymerize microtubules. A manual action may be necessary to suspend correctly pellets. Add 40 μL of 5× SDS-PAGE loading buffer to the 200 μL dissolved pellets.
6. Heat samples at 95 °C before loading it in a 12% SDS-PAGE.
7. 10 μL of pellets and supernatants are loaded on a same 12% SDS-PAGE to eliminate error due to coloration variation. Protein content is revealed by Coomassie Brilliant Blue staining. To quantify free and bound tubulin and free and bound Tau, loading of each sample must be readjusted to obtain comparable and non-saturated spots. Protein quantification is determined with band intensity quantification by ImageJ.

The experiments must be done three times to have a representative average. Because Tau concentration in the pellet represents MT-bound Tau concentration and that in the supernatant the free one, Scatchard plot can be used to determine affinity constant and the stoichiometry of the reaction. For analysis used the following equation:

$$\frac{B}{[Tau]_{\text{free}}} = nK_{\text{a}} - BK_{\text{a}}$$

where B is the ratio of the concentration of bound ligand (here Tau in the pellet) to total available binding sites (total tubulin concentration), $[\text{Tau}]_{\text{free}}$ is the concentration of free Tau, n is the number of binding sites, and K_{a} the affinity constant of Tau for the microtubule. Plotting this data, $B/[\text{Tau}]_{\text{free}}$ versus B, yields the Scatchard plot with a slope $-K_{\text{a}}$ and an x-intercept leads to n. This approach can be used only in the case of studying Tau interaction with paclitaxel-stabilized MT. For the two other conditions, Scatchard plots are curved downward reflecting the fact that ligand-induced tubulin assembly may proceed by a ligand facilitated elongation (in assembly conditions) in which the elongation precedes the binding of Tau (Eq. 1) or a ligand-mediated elongation pathway (in non-assembly conditions), in which the binding of Tau, to either tubulin or microtubule, precedes the elongation pathways (Eq. 2) [28].

$$MT_n + \text{tubu} \overset{K2}{\leftrightarrow} MT_{n+1} + \text{Tau} \overset{K1}{\leftrightarrow} MT_{n+1} - \text{Tau}. \tag{1}$$

$$\text{Tau} + \text{tubu} \overset{K1}{\leftrightarrow} \text{Tubu} - \text{Tau} + MT_n \overset{K2}{\leftrightarrow} MT_{n+1} - \text{Tau}. \tag{2}$$

Equation 1 is characterized by an apparent elongation constant $K_{\text{app}} = K_2(1 + K_1[\text{Tau}]_{\text{free}})$ (for more details, *see* the appendix of ref. [29]) and Eq. 2 is characterized by an apparent elongation constant $K_{\text{app}} = K_1K_2[\text{Tau}]_{\text{free}}/(1 + K_1[\text{Tau}]_{\text{free}})$. As already mentioned in Subheading 3.3, the K_{app} is related to 1/Cr and the Cr corresponds to the free tubulin concentration in the supernatant of the cosedimentation assay. So plot K_{app} in function of free Tau concentration and fit the data with the equation of K_{app} described above. Use Eq. 1 in assembly condition and Eq. 2 in non-assembly condition.

3.3.4 Transmission Electronic Microscopy

1. Removed 10 μL at the end of microtubule formation reaction (*see* **Note 21**).
2. Put down on a Parafilm sheet a 200 mesh carbon-coated formvar films on copper grids
3. Put a little drop of samples (3 μL) on grids and incubate for 15 s.

4. Eliminate the excess of sample by taking the grid with tweezer and absorb sample with filter paper (apply the grid perpendicularly).
5. Add 3 μL uranyl acetate on the grid for 30 s for staining.
6. Eliminate uranyl acetate exceed as describe in **step 4**.
7. Dry grids overnight at 37 °C and observe it using a transmission electron microscope.

3.4 Tau–Tubulin Interactions by FRET in Living Cells

3.4.1 Transfection Protocol with Plasmids DNA

The transfection protocol is performed as recommended by the supplier (LipofectAmine 2000, Life technology, U.S.). For cell transfection, 0.4 μg of DNA (0.4 μg of pEGFP-Tau DNA in well 1, 0.4 μg of pmCherry-α-tubulin DNA in well 2, 0.2 μg of pEGFP-Tau DNA and 0.2 μg of pmCherry-α-tubulin DNA in well 3) are applied on cells. After transfection, the well 1 contains cells expressing EGFP-tau proteins (corresponding to the condition "donor alone"), the well 2 contains cells expressing mCherry-α-tubulin proteins (corresponding to the condition "acceptor alone"), and the well 3 contains cells expressing both EGFP-tau and mCherry-α-tubulin proteins (corresponding to the condition "uFRET"). Cells are observed after 24–48 h post-transfection.

3.4.2 Confocal Imaging: FRET Measurement by Sensitized Emission

1. Place cells in medium supplemented with 1 % FBS and 10 mM HEPES to reduce pH variations (*see* **Note 22**), and maintain cells at 37 °C.
2. Configure the CLSM's settings such as imaging conditions are optimized before data collection (*see* **Note 23**). The donor EGFP-tau is excited sequentially at 488 and 543 nm and the fluorescence emission is collected using the CLSM's spectral mode simultaneously between 496 and 535 nm (channel 1, corresponding to the donor settings: excitation of the donor, detection of the donor) and between 580 nm and 650 nm (channel 2 corresponding to the uncorrected FRET (uFRET) settings: excitation of the donor, detection of the acceptor) (*see* **Note 23**). The acceptor mCherry-α-tubulin is then excited at 543 nm and the fluorescence emissions are collected between 580 and 650 nm (channel 3 corresponding to the acceptor setting: excitation of the acceptor, detection of the acceptor).
3. Adjust gain and offset of the donor and uFRET channels with cells expressing EGFP-tau alone. The pixel intensities must be linear with the total dynamic scale of fluorescence intensities (for example 256-grey scale for an 8-bit image depth) (*see* **Notes 24** and **25**). Collect three images of cells on the field.
4. Adjust gain and offset of the acceptor channel exclusively with cells expressing mCherry-α-tubulin alone (*see* **Note 26**). Collect three images of cells on the field.

5. Image cells expressing the donor and acceptor pairs through the donor, the uFRET and the acceptor settings previously adjusted in **steps 3** and **4**.

3.4.3 Confocal Imaging: FRET Measurement by the Acceptor Photobleaching

An alternative method to quantify the absolute efficiency of FRET between the donor EGFP-tau and the acceptor mCherry-α-tubulin is to measure the EGFP emission before and after selective photobleaching of the mCherry [30].

1. Place cells expressing both the donor and the acceptor in medium supplemented with 1% FBS and 10 mM HEPES to reduce pH variations (*see* **Note 22**), and maintain cells at 37 °C.
2. Configure the CLSM settings such as: EGFP-tau is excited at 488 nm and the fluorescence emission is collected using the CLSM's spectral mode between 496 and 535 nm; and mCherry-α-tubulin is excited at 543 nm and the fluorescence emission is collected between 580 and 650 nm. The donor and acceptor fluorophores are sequentially excited.
3. Adjust gain and offset of the donor channel with cells expressing EGFP-tau alone so that as the fluorescence intensity of EGFP-tau does not saturated PMT (*see* **Note 27**).
4. Adjust gain and offset of mCherry-α-tubulin channel with cells expressing mCherry-α-tubulin alone. The pixel intensities must be linear with the total dynamic scale of fluorescence intensities (for example 256-grey scale for an 8-bit image depth).
5. Collect two images of cells with the 488 and 543 nm excitation lines with donor and acceptor settings as previously described in **steps 3** and **4** (*see* **Note 28**).
6. Irradiate the acceptor on a region of interest (ROI) in the cell using 100% laser power with a 543 nm exciting beam. As reported in ref. [30], mCherry-α-tubulin was photobleached for 31.56 s (25 frames) on a $6 \times 6\ \mu m^2$ ROI (*see* **Note 29**).
7. Recollect two images of cells with the 488 and 543 nm excitation lines after the mCherry-α-tubulin photobleaching in the ROI.

3.4.4 FRET Calculation for Sensitized-Emission FRET

The ImageJ plug-in PixFRET (Center for Integrative Genomics, Faculty of biology and medicine, University of Lausanne, Switzerland) [24] can generate images of sensitized-emission FRET, and hence show where FRET occurs within a cell or a group of cells, by computing pixel by pixel the images of a sample acquired in a three channel setting.

1. For donor Spectral Bleed-Through (SBT) determination, create a stack of donor and uFRET images of cells expressing EGFP-tau alone.

2. Select a ROI manually by outlining cells on the donor image and determine the mean fluorescence intensity of the ROI in the Analyze menu of ImageJ software; do the same measure of the mean fluorescence intensity on the uFRET image (*see* **Note 30**).
3. Calculate the intensity ratio of the uFRET on donor images; the mean value of all intensity ratios corresponds to the donor SBT parameter A (*see* **Note 31**).
4. Repeat **steps 1–3** to calculate the acceptor SBT parameter *B* with images of cells expressing mCherry-α-tubulin alone.
5. Before starting the PixFRET plug-in, create a stack of three images of the cell or the group of cells expressing both the donor EGFP-tau and the acceptor mCherry-α-tubulin: the first one in the uFRET setting, the second one in the donor setting and the third one in the acceptor setting.
6. Launch PixFRET from the plug-ins menu in ImageJ.
7. Enter the value for the donor and acceptor SBTs in the Donor and Acceptor Models bookmarks, respectively (*see* **Note 32**), and click on the FRET bookmark to select a ROI on the stack in order to determine the background outside cells; press Get button.
8. In the parameter box below, select "no Gaussian Blur" and a value of 1.0 for threshold correction factor (*see* **Note 33**), as well as the type of normalization of the FRET values among 4 possibilities: by dividing by the donor or the acceptor intensity [31], by the product of donor and acceptor intensities [32] or by the square root of the product of donor and acceptor intensities [33]; users can also generate an image displaying the FRET efficiencies in the sample. For example, the methodology reported in ref. [30] was to normalize corrected FRET (Fc) to the donor intensity.
9. Click on "Compute FRET" button to generate the normalized FRET image, in addition to the Fc image calculated from the formula described by ref. [34]:

$$\mathrm{Fc} = I_{\mathrm{FRET}} - A \times I_{\mathrm{donor}} - B \times I_{\mathrm{acceptor}}$$

   Where I_{FRET}, I_{donor}, and I_{acceptor} are pixel intensities (subtracted from their background intensity) in cells under the distinct uFRET, donor, and acceptor settings, respectively. The parameters *A* and *B* were calculated in **steps 3** and **4**.
10. To reduce background noise, prefer to use a nonlinear filtering of the computed FRET image with for example a 1-pixel-range median filter than a Gaussian filter [35, 36].

11. In the Image menu of ImageJ, select an adequate LookUp Table (LUT) and adjust brightness and contrast to take over all the dynamic scale of fluorescence intensities (*see* **Note 34**).
12. Save the computed FRET images as a TIFF file.

3.4.5 FRET Calculation for FRET by the Acceptor Photobleaching

1. With ImageJ software [23], create a stack of four images of cells expressing EGFP-tau and mCherry-α-tubulin before and after the acceptor photobleaching.
2. On the post-bleaching image of mCherry-α-tubulin, select manually a ROI overlaying the bleached zone.
3. Place on the pre-bleaching image of the donor with the horizontal scrollbar in the stack and determine the mean fluorescence intensity of the ROI using the Measure function of the Analyze menu in ImageJ.
4. Do the same measure of the fluorescence intensity on the post-bleaching image of the donor.
5. Calculate the FRET efficiency (%E) from fluorescence recovery after the photobleaching kinetic of the donor emission in the presence of the acceptor in ROIs [37]:
6. Return to **step 2** and select manually a ROI in another zone of the cell; determine the mean fluorescence intensity of the ROI on pre- and post-bleaching images of the donor and calculate %E which must be significantly lower than for the bleached zone (*see* **Note 35**). For example, we obtained %E of 8.7 ± 0.2 for EGFP-tau in the mCherry-α-tubulin bleached zone vs. %E of 1.1 ± 0.1 in a control zone (unpublished data) (*see* **Note 36**).

$$\%E = 1 - \left(I_{\text{donor}}\right)_{\text{pre-bleaching}} / \left(I_{\text{donor}}\right)_{\text{post-bleaching}}$$

where $(I_{\text{donor}})_{\text{pre-bleaching}}$ and $(I_{\text{donor}})_{\text{post-bleaching}}$ are the donor emission before and after selective photobleaching of the acceptor, respectively.

7. We recommend completing data with figure showing location of the fluorescence variation of EGFP-tau and mCherry-α-tubulin before and after the acceptor photobleaching, as the example reported in Fig. 3.

4 Notes

1. GTP is dissolved in water. Make Aliquots of 100 μL and store them at −20 °C. GTP should be added freshly to buffer to avoid hydrolysis of GTP into GDP which is inactive for MTs formation.
2. Glycerol is viscous. Precautions should be taking to be sure that it was accurately mixed with buffer or water.

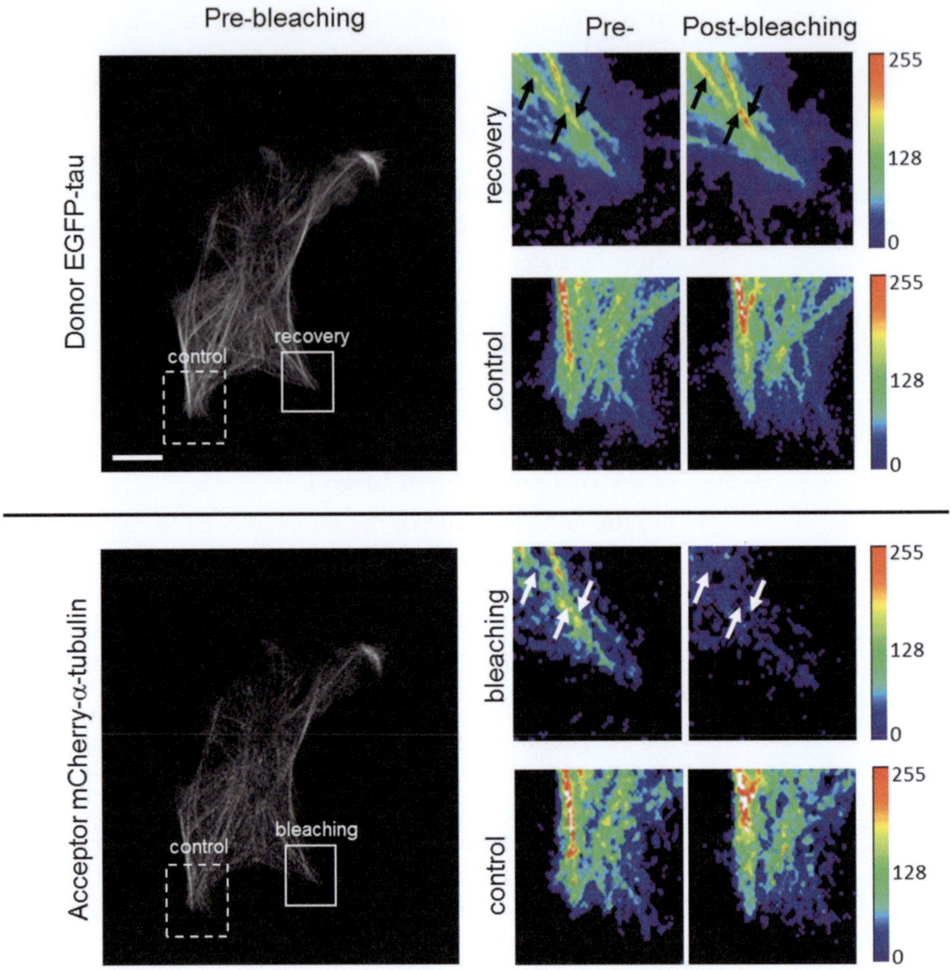

Fig. 3 Energy transfer between EGFP-tau and mCherry-a-tubulin revealed by the acceptor photobleaching in living cells. The quenched and unquenched EGFP-tau emission is compared after specific photobleaching of the mCherry-α-tubulin fluorophore (respectively, *upper* and *lower panels* in false colors under Pre- and Post-bleaching titles). Prior to the acceptor photobleaching, microtubules and uniformly labeled with EGFP-tau; after photobleaching, the fluorescence emission of EGFP-tau increased in spots along microtubules (*black arrows* in *upper panel*) confirming FRET in the initial state; no significant fluorescence intensity fluctuation is observed inside the control region. Scale bar: 10 mm; side bars: color pixel scoring for fluorescence intensities from 0 (*dark purple*) to 255 (*red*); EGFP-tau and mCherry-α-tubulin are excited at 488 and 543 nm respectively, using the multi-tracking mode of CLSM system

3. Add a pinch of azide to prevent bacteria or fungus growth. After use, unpacked resin, wash ten times with ultrapure water. It can be used ten times.
4. In the case of guanidine-HCl, the volume increases when it is dissolved and the reaction is endothermic. Do not put a volume of water larger than the half of the total volume.

5. All FPLC buffers can be stored at room temperature for few months.
6. TCEP powder is very difficult to dissolve at concentration of 1 M. TCEP is preferred to DTT to reduce disulfide bonds because it is more stable. It is used to maintain Tau monomeric and have no effect on MTs formation.
7. This solution must be filtered just before use to remove crystals or possible pollutants which perturb coloration. To use uranyl acetate use gloves.
8. Add 2 mM L-glutamine and nonessential amino acids if absent in media both to stimulate growth and to prolong the viability of cells in culture.
9. We use a Leica DMI6000CS inverted microscope coupled to a CLSM Leica TCS SP5, with the TD488/543/633 dichroic filter equipping the system.
10. During DEAE swelling, you must check there is always enough water and that it is boiling.
11. The resin must not be dried. Keep an eye on column when already 500 mL went through.
12. Think to thermostat centrifuge and rotor and tubes at the good temperature before run.
13. Collect all wastes in a special bag to eliminate them in rules respect.
14. Wash all tubes meticulously with spatula and collect all volumes, as little as they are, to optimize the output. Start using spatula to crash pellet and then use a magnetic stirring bar which has almost the same length as glass beaker diameter and finish by gently by gently pipetting in and out the sample with a 1 mL pipetman without making bubbles. It is very important to suspend every tiny particle.
15. When you add buffer on the top of mixing solution a biphasic layer formed. It is the sign that tubulin is here and concentrated.
16. Do not wear gloves when you work near a heat source but handle carefully bacteria samples and wash hands after experiments.
17. Pellets can be kept in lysis buffer at −20 °C for few weeks or experiment can continue. If sample is frozen, defreeze and homogenize it before use.
18. Make sure that the instrument is ready to start (flow rate, good pressure, no leakages, no blocks).
19. Tubulin is sensitive to temperature. So all steps to prepare it must be made at 4 °C in a cold room or ice box with cold buffer.

20. If you put 150 μL of tubulin solution at 670 μM, that corresponds to 10 mg of protein and you recuperate 6–7 mg into four tubes at an average of approximately 35 μM corresponding to the 2/3 of the amount of protein put initially onto the column.
21. To maintain MT, grids preparation is made in a thermostated room at 37 °C at room temperature, because rings of tubulin might form.
22. Do not add 10 mM Hepes to the medium if the CLSM is equipped with a heat- and CO_2-regulated chamber.
23. Use samples expressing the donor or the acceptor fluorophores alone to maximize uFRET signals, to minimize SBT, to avoid widespread saturation, to avoid unintended photobleaching, and to minimize background noise [34].
24. Avoid images showing pixels with saturated intensity indicating an excessive excitation laser power.
25. The same PMT settings in donor and acceptor channels and low laser power are recommended, in order to generate a robust uFRET signal at donor excitation in the acceptor channel, without at the same time saturating the donor emission signal in the donor channel.
26. Do not modify gain and offset of the uFRET channel, otherwise to calculate wrong donor and acceptor SBT (parameters *A* and *B* in the FRET calculation in the Subheading 3.4.4).
27. We recommend adjusting gain and offset of the donor channel in order to spread fluorescence intensities on two-thirds of the dynamic intensity scale.
28. Users can focus on two cells in the same image, a first cell on which users irradiate a ROI to photobleach the acceptor and another cell as control.
29. On our CLSM with a 63×1.4 NA objective, these conditions destroy mCherry with very little or no direct bleaching of the EGFP donor. The optimal exposure conditions that bleach acceptor maximally with minimal direct effect on donor should be determined for each microscope system and filter set, using for example droplets of donor and acceptor under oil, cells transfected with donor and acceptor fusion proteins alone, or cells co-transfected with non-interacting donor and acceptor fusion proteins.
30. Keep the same selected ROI in the donor image while switching on the uFRET image.
31. The main causes of bleed-through are the emission of the donor in the acceptor channel and the excitation of the acceptor when exciting the donor. To calculate donor (and acceptor) SBTs, average fluorescence intensities in cells expressing only the donor (or the acceptor) should be quantified, both in the uFRET and in donor (or acceptor) channels. The SBT ratios

are the ratios between the fluorescence intensities in each channel after background subtraction, such as:

- For the parameter A: $A = (I_{FRET}/I_{Donor})_{Exc\ 488nm}$

  When the donor alone is expressed and excited at 488 nm;

- For the parameter B: $B = (I_{FRET})_{Exc\ 488nm}/(I_{Acceptor})_{Exc\ 543nm}$

  When the acceptor alone is expressed and excited at both 488 and 543 nm.

32. Keep the value 0 for FRET and Donor/Acceptor boxes in the Background sections; press Accept button; in the Model Donor/Acceptor boxes, select "no Gaussian Blur" to compute FRET images from native uFRET, donor and acceptor images; press on Constant button just below and fill the box with the calculated parameters A (in the Donor Model bookmark) and B (in the Acceptor Model bookmark); press Accept button.
33. The Fc and normalized FRET images are calculated only if pixel values in each image are above a given threshold, set by default to average background values; the threshold correction factor is a multiplication factor applied to background values in order to modify this threshold; the pixels below the threshold are displayed in blue (or black in the new user's LUT) in the computed Fc and normalized FRET images.
34. The computed Fc and normalized FRET images are encoded on 32 bits but visualized on 256 color levels; the minimum value is displayed in black and the maximum value in white. Hence, an artificially amplified FRET signal may be observed in a cell where no FRET occurs (negative control). Users should therefore always:
    - Check the amplitude of the FRET signal by performing a profile plot through the cell of interest and compare this amplitude to that of a negative control (such as with cells expressing the donor–acceptor couple free EGFP and mCherry-α-tubulin for example) and to that a positive control (such as with cells expressing the donor–acceptor couple EGFP-β-tubulin and mCherry-α-tubulin).
    - Stack the image of interest with an image of a negative and a positive control for FRET in order to visualize both images with the same LUT scale.
35. This procedure uses each cell or sub-region as its own internal standard; in addition, users can calculate %E in a neighboring cell that should be similar to the one of the internal standard.
36. The significance of %E is discussed in ref. [38].

References

1. Mitchison T, Kirschner M (1984) Dynamic instability of microtubule growth. Nature 312: 237–242
2. Bhat KM, Setaluri V (2007) Microtubule-associated proteins as targets in cancer chemotherapy. Clin Cancer Res 13:2849–2854
3. Weingarten MD, Lockwood AH, Hwo SY et al (1975) A protein factor essential for microtubule assembly. Proc Natl Acad Sci U S A 72: 1858–1862
4. Tilney LG, Bryan J, Bush DJ et al (1973) Microtubules: evidence for 13 protofilaments. J Cell Biol 59:267–275
5. Chrétien D, Wade RH (1991) New data on the microtubule surface lattice. Biol Cell 71: 161–174
6. Wade RH (2007) Microtubules: an overview. Methods Mol Med 137:1–16
7. Weisenberg RC, Timasheff SN (1970) Aggregation of microtubule subunit protein. Effects of divalent cations, colchicine and vinblastine. Biochemistry 9:4110–4116
8. Frigon RP, Timasheff SN (1975a) Magnesium-induced self-association of calf brain tubulin. I. Stoichiometry. Biochemistry 14:4559–4566
9. Frigon RP, Timasheff SN (1975b) Magnesium-induced self-association of calf brain tubulin. II. Thermodynamics. Biochemistry 14: 4567–4573
10. Mandelkow EM, Mandelkow E, Milligan RA (1991) Microtubule dynamics and microtubule caps: a time-resolved cryo-electron microscopy study. J Cell Biol 114:977–991
11. Müller-Reichert T, Chrétien D, Severin F, Hyman AA (1998) Structural changes at microtubule ends accompanying GTP hydrolysis: information from a slowly hydrolyzable analogue of GTP, guanylyl (alpha, beta)methylenediphosphonate. Proc Natl Acad Sci U S A 95:3661–3666
12. Brouhard GJ, Rice LM (2014) The contribution of αβ-tubulin curvature to microtubule dynamics. J Cell Biol 207:323–334
13. Himmler A, Drechsel D, Kirschner MW, Martin DW Jr (1989) Tau consists of a set of proteins with repeated C-terminal microtubule-binding domains and variable N-terminal domains. Mol Cell Biol 9:1381–1388
14. Butner KA, Kirschner MW (1991) Tau protein binds to microtubules through a flexible array of disturbed weak sites. J Cell Biol 115: 717–730
15. Goode BL, Feinstein SC (1994) Identification of a novel microtubule binding and assembly domain in the developmentally regulated inter-repeat region of tau. J Cell Biol 124: 769–782
16. Gustke N, Trinczek B, Biernat J et al (1994) Domains of tau protein and interactions with microtubules. Biochemistry 33:9511–9522
17. Sillen A, Barbier P, Landrieu I et al (2007) NMR investigation of the interaction between the neuronal protein tau and the microtubules. Biochemistry 46:3055–3064
18. Devred F, Barbier P, Lafitte D et al (2010) Microtubule and MAPs: thermodynamics of complex formation by AUC, ITC, fluorescence, and NMR. Methods Cell Biol 95: 449–480
19. Tsvetkov PO, Makarov AA, Malesinski S et al (2012) New insights into tau-microtubules interaction revealed by isothermal titration calorimetry. Biochimie 94:916–919
20. Tsvetkov PO, Barbier P, Breuzard G et al (2013) Microtubule-associated proteins and tubulin interaction by isothermal titration calorimetry. Methods Cell Biol 115:283–302
21. Goedert M, Wischik CM, Crowther RA et al (1988) Cloning and sequencing of the cDNA encoding a core protein of the paired helical filament of Alzheimer disease: identification as the microtubule-associated protein tau. Proc Natl Acad Sci U S A 85:4051–4055
22. Dompierre JP, Godin JD, Charrin BC et al (2007) Histone deacetylase 6 inhibition compensates for the transport deficit in Huntington's disease by increasing tubulin acetylation. J Neurosci 27:3571–3583
23. Rasband W (1997–2007) ImageJ. US National Institutes of Health, Bethesda, MA, http://rsb.info.nih.gov/ij/plugins.html
24. Feige JN, Sage D, Wahli W et al (2005) PixFRET, an ImageJ plug-in for FRET calculation that can accommodate variations in spectral bleed-throughs. Microsc Res Tech 68: 51–58
25. Oosawa F, Asakura S (1975) Thermodynamics of the polymerization of protein. Academic, London
26. Devred F, Douillard S, Briand C, Peyrot V (2002) First tau repeat domain to growing and taxol-stabilized microtubules, and serine 262 residue phosphorylation. FEBS Lett 523: 247–251
27. Weis F, Moullintraffort L, Heichette C et al (2010) The 90-kDa heat shock protein Hsp90 protects tubulin against thermal denaturation. J Biol Chem 285:9525–9534

28. Devred F, Barbier P, Douillard S et al (2004) Tau induces ring and microtubule formation from alphabeta-tubulin dimers under nonassembly conditions. Biochemistry 43:10520–10531
29. Buey RM, Diaz JF, Andreu JM et al (2004) Interaction of epothilone analogues with the paclitaxel binding site: relationship between binding affinity, microtubule stabilization, and cytotoxicity. Chem Biol 11:225–236
30. Breuzard G, Hubert P, Nouar R et al (2013) Molecular mechanisms of Tau binding to microtubules and its role in microtubule dynamics in live cells. J Cell Sci 126:2810–2819
31. Wouters FS, Verveer PJ, Bastiaens PI (2001) Imaging biochemistry inside cells. Trends Cell Biol 11:203–211
32. Gordon GW, Berry G, Liang XH et al (1998) Quantitative fluorescence resonance energy transfer measurements using fluorescence microscopy. Biophys J 74:2702–2713
33. Xia Z, Liu Y (2001) Reliable and global measurement of fluorescence resonance energy transfer using fluorescence microscopes. Biophys J 81:2395–2402
34. Youvan DC, Silva CM, Bilina E et al (1997) Calibration of fluorescence resonance energy transfer in microscopy using genetically engineered GFP derivatives on nickel chelating beads. Biotechnology 3:1–18
35. Gonzales RC, Woods RE (2007) Digital image processing, 3rd edn. Pearson Education, Upper Saddle River, NJ
36. Pratt WK (2006) Digital image processing, 4th edn. Wiley-Interscience, New York, NY
37. Siegel RM, Chan FK, Zacharias DA et al (2000) Measurement of molecular interactions in living cells by fluorescence resonance energy transfer between variants of the green fluorescent protein. Sci STKE 38:1
38. Nouar R, Devred F, Breuzard G, Peyrot V (2013) FRET and FRAP imaging: approaches to characterise tau and stathmin interactions with microtubules in cells. Biol Cell 105: 149–161

Part II

In Vitro Self-Association of Tau Protein: Oligomer and Fibril Formation

Chapter 5

X-Ray Structural Study of Amyloid-Like Fibrils of Tau Peptides Bound to Small-Molecule Ligands

Einav Tayeb-Fligelman and Meytal Landau

Abstract

Atomic structures of Tau involved in Alzheimer's disease complexed with small molecule binders are the first step to define the Tau pharmacophore, leading the way to a structure-based design of improved diagnostics and therapeutics. Yet the partially disordered and polymorphic nature of Tau hinders structural analyses. Fortunately, short segments from amyloid proteins, which exhibit similar biophysical properties to the full-length proteins, also form fibrils and oligomers, and their atomic structures can be determined using X-ray microcrystallography. Such structures were successfully used to design amyloid inhibitors. This chapter describes experimental procedures used to determine crystal structures of Tau peptide segments in complex with small-molecule binders.

Key words Tau, Alzheimer's disease, Small molecules, Pharmacophore, Microcrystals, Amyloid-like peptides, Microcrystallography

1 Introduction

Amyloid fibrils and oligomers of Tau have been identified in association with Alzheimer's disease, but the development of their diagnostics and therapeutics [1–12] has proceeded in a near vacuum of information about the structure of Tau for many years. While amyloid fibrils might represent one of the most ordered states of proteins in the living cells, they are unamenable to direct structural characterization using X-ray methodologies owing to their partially unstructured and polymorphic nature. Over the last decade, Eisenberg and coworkers demonstrated that this limitation could be circumvented by examining the structured spine of the amyloid, formed by 6–8 residues, whose peptide structure can be determined. These amyloid-like peptides only crystallize as microcrystals, of about 1 μm in width. Therefore, specialized methodology is needed to elucidate the spine structures of the fibrils they contribute forming. Using this approach, the common β-spine architecture of fibril-forming proteins associated with human

Caroline Smet-Nocca (ed.), *Tau Protein: Methods and Protocols*, Methods in Molecular Biology, vol. 1523,
DOI 10.1007/978-1-4939-6598-4_5,

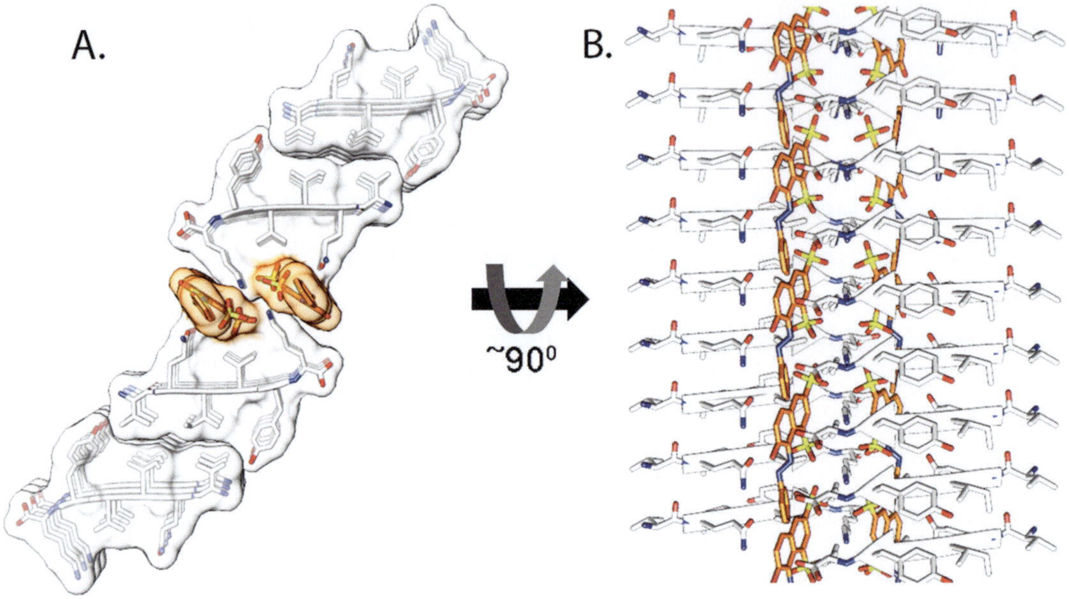

Fig. 1 Structure of the VQIVYK segment from Tau in complexed with orange-G [27]. The VQIVYK segments pack in parallel, in-register β-sheets (*cartoon arrows*) that form steric zippers (two zippers are shown in panel (**a**)). VQIVYK and orange-G are shown as sticks with non-carbon atoms colored by atom type; the carbons of VQIVYK are colored *white*, while that of orange-G are colored *orange*. Two orange-G molecules mediate contacts between two pairs of steric zippers; that is, orange-G is located between the protofilaments composing the fibril. In panel (**a**), the view looks down the fibril axis. Partially transparent surfaces of the molecules show the tight packing of orange-G inside the fibril. In panel (**b**), the view is perpendicular to the fibril axis; nine layers of the fibril are depicted. Only the two sheets that are in contact with orange-G are shown for clarity

aggregation diseases was illuminated [13–24]. These structures were used to design *efficient* peptide and small-molecule inhibitors to the Alzheimer's-disease associated Amyloid-β and Tau proteins [25–27].

The VQIVYK segment of Tau was suggested as the minimal interaction motif for fibril formation [28]. It is possible that other Tau segments can also serve as the amyloid spine of the protein. The crystal structure of VQIVYK was determined in two crystal forms; both show the common steric-zipper motif of amyloid fibril-like structures, consisting of a pair of tightly mated β-sheets [16, 19]. The atomic structure of the VQIVYK spine segment was used to design an all-D-amino-acid inhibitor for the Tau protein fibril formation [25]. Other structures, of the VQIVYK spine segment bound to small molecules, shade light on the Tau pharmacophore [27]. Namely, these structures, of VQIVYK bound to the dye orange-G (Fig. 1), the natural compound curcumin [29], and the Alzheimer's diagnostic compound DDNP [30], revealed the molecular framework of small-molecule binding to the Tau spine segment, within cylindrical cavities running along the β-spines of the fibrils (Fig. 1). A similar structure of the KLVFFA spine segment from amyloid-β in complex with orange-G was used to design small-molecules inhibitors that showed significant reduction of amyloid-β toxicity in cell-based assays [26]. The common feature of the amyloid/small-molecule complexes structures is that the small molecules bind to the fibrils in a similar orientation, along the β-sheets, with their long axes parallel to the fibril axis. This orientation was previously proposed for the binding of thioflavin T to bovine insulin and bovine β-lactoglobulin amyloid fibrils using polarized laser confocal microscopy [31]. A similar mode of binding was seen in co-crystals of oligomer-like β-2-microglubulin with thioflavin T, showing that thioflavin T is bound between β-sheets, orthogonal to the β-strands [32]. The orientation of Congo-red was also suggested to be parallel to the amyloid long axis based on electron diffraction, linear dichroism [33], and an NMR-based model of Congo-red bound to the fungal prion domain HET-s(218–289) [34]. The crystals structures of the complexes also showed that the small molecules bind different Tau peptide polymorphs, revealing that a cocktail of compounds may be required for future amyloid diagnostics and therapies. Such structures are a major milestone in the structure-based design of improved diagnostics and therapeutics targeting Tau.

2 Materials

Prepare all solutions using double deionized water and analytical grade reagents. Perform all experiments at room temperature (unless indicated otherwise). Diligently follow all waste disposal regulations when disposing waste materials. All solutions are made using standard lab protocols. Do not add sodium azide to reagents due to its seeding properties. Note that amyloid segments may pose dangers to health and safety precautions should be taken (*see* **Note 1**).

2.1 Tau Amyloid-Like Peptides Preparation

Peptides are costume synthesized with over 98 % purity. For crystallization experiments leave the termini uncapped. Peptides should be stored in lyophilized form at −20 °C. Preferably, equilibrate peptides to room temperature in a desiccator prior to opening and weighing. *See* **Note 1** for safety measures. For crystallization assays a peptide concentration of 1–40 mg/ml is suitable (*see* **Note 2**). Tau spine peptide VQIVYK is dissolved in water. *See* **Note 2** for dissolving other peptide segments.

2.2 Small-Molecule Preparation

1. Small molecules that were shown to affect the aggregation of Tau (for example *see* refs. 1–12) can be tested for co-crystallization with Tau spine peptides. The small molecules are first tested for their solubility. For water-insoluble small molecules, dimethyl sulfoxide (DMSO) is used (*see* **Note 3**).
2. 0.1 μm Ultrafree-MC centrifugal filter device (AMICON, Bedford, MA, USA).

2.3 Crystallization Experiments

1. Mosquito nanodispenser robot (TTP LabTech).
2. P-97 Flaming/Brown Micropipette Puller (Sutter Instrument) or the PC-10 Puller (NARISHIGE).
3. 5 μl calibrated borosilicate glass micropipettes (VWR).
4. Capillary cutting stone (Hampton Research).
5. 12.8 mm long brass pins (Hampton Research).
6. Adjustable, magnetic, crystal mount (Hampton Research).
7. Standard goniometer key (Hampton Research).
8. Dual-Thickness MicroLoops™ LD, SPINE/18 mm length (e.g., Mitegen LLC), and cap bases suited for the relevant beam-line with suited vials (Hampton Research).
9. Discovery V20 microscope (Zeiss).

3 Methods

3.1 Predicting Amyloid-Like Spine Segments

Fibril-formation propensities of segments of Tau and other amyloid proteins are predicted using computational methods, for example ZipperDB [35], Tango [36], Waltz [37], and Zyggregator [38]. Segments forming amyloid fibrils typically consist of 4–7 residues. Methods for the evaluation of peptide fibril-formation are described elsewhere [39].

3.2 Amyloid-Like Peptide and Small-Molecule Preparation for Co-crystallization Experiments

1. Dissolve *water-soluble peptides* in water and *water-insoluble peptides* in a proper solvent (*see* **Note 2**) to prepare a stock solution of 10–100 mM.
2. Dissolve the small molecules in the proper solvent (*see* **Note 3**) to prepare a stock solution of 10–100 mM.
3. Standard concentrations and molar ratio between peptide and small-molecule used for co-crystallization experiments is between 1:1 and 10:1 mM of peptide and small molecule, respectively. It is recommended to use a small-molecule concentration of up to 1 mM (*see* **Note 4**). Dilute and mix the peptide and small molecules to their final concentration. The final solvent should take into account the solubility of both the peptide and small molecule (*see* **Note 5**).
4. As an optional step, filter the solution through a 0.1 μm Ultrafree-MC centrifugal filter device.

3.3 Crystallization Screens

1. Perform all crystallizations using the hanging drop/vapor diffusion method at room temperature and store the plates at 18–20 °C.
2. Set 96-well plates with crystallization screens (e.g., using the Mosquito nanodispenser robot) (*see* **Note 6**). Drops are typically a mixture of 100 nl peptide solution and 100 nl reservoir

solution in three different ratios: 1:1, 2:1, and 1:2. Other drop volumes can be used.

3. Examine the potential formation and appearance of needle-shaped crystals. In case the examined small molecule is colored, formation of colored crystals (for example *see* Fig. 2a) might indicate the presence of a complex, or of crystals containing simply the small molecule. If the color is washed off during crystal manipulation and mounting (described in the next section) it might indicate that the small molecule is not present inside the crystal.

3.4 Crystal Mounting (See Note 7)

There are two options for mounting crystals for X-ray data collection experiments depending on peptide length:

3.4.1 Mounting Crystals of Peptides Shorter Than Eight Residues on Glass Capillaries (Fig. 2) (See Note 8)

1. Using a micropipette puller instrument (such as the P-97 Flaming/Brown Micropipette Puller or the PC-10 Puller) whet glass capillaries (5 μl borosilicate glass calibrated micropipettes). Depending on the specification of the micropipette puller instrument, adjust the heat and pull velocity to obtain sharp needles with a tip width of about 10–20 μm.
2. Shorten the sharpened capillaries (from the blunt end) using a capillary cutting stone to about 1.3 cm.
3. Super-glue the sharpened capillaries into 12.8 mm long brass pins. Leave about 5–6 mm of the sharp end of the capillary sticking out of the brass pin (Fig. 2).

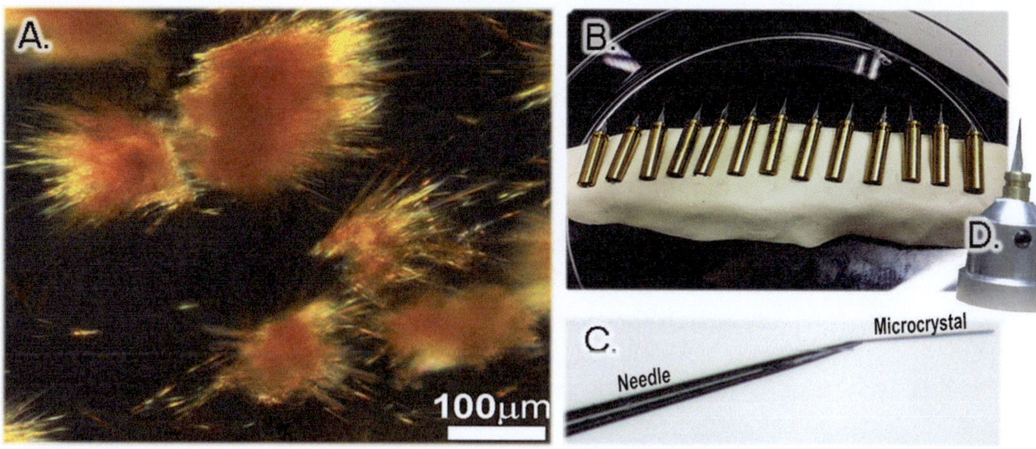

Fig. 2 Mounting microcrystals of Amyloid-like peptides. (**a**) A Crystallization drops is occupied by numerous *orange*-colored microcrystals of VQIVIK in complex with orange-G; a *scale bar* is indicated. (**b**) The sharpened needle capillaries glued to the brass pins (described in Subheading 3.4) are situated on clay stored in a petri dish. The needles are used to mount a single microcrystal. (**c**) A close-up on a microcrystal situated on the tip of the needle capillary. (**d**) The brass pin holding the microcrystal is inserted into the magnetic crystal mount designed to position the brass pin onto a goniometer head for X-ray data collection

4. Mount your needle crystals (*see* **Note 9**) on the glass capillary using a powerful Stereomicroscope (for example Discovery V20 microscope).
5. At the synchrotron, mount the brass pins into adjustable, magnetic, crystal mount (Fig. 2d). The Adjustable Crystal Mount is a magnetic piece designed to position the brass pin with the glass capillary onto a goniometer head for X-ray data collection. You will need a standard goniometer key to adjust the screws (*see* **Note 10**).

3.4.2 Flash Freezing Crystals of Peptides That Are Eight Residues or Longer (Potential Fibrils or Oligomers)

1. Mount the crystals on very small cryogenic loops depending on the crystal size (*see* **Note 11**). You might need to use cryoprotectants.
2. Flash freeze and store the crystals in liquid nitrogen.

3.5 X-Ray Diffraction Data Collection for Microcrystals (See Note 12)

1. For the microcrystals, X-ray diffraction data is usually collected in 5° wedges using a 5 μm beam size. Crystals diffracting to high resolution (above 1.5 Å) should also be collected in 3° wedges.
2. The crystals are cryo-cooled (100 K) for data collection. Transmission intensity and exposure time should be varied according to the quality of the crystal and of the X-ray diffraction (lower intensity should be used for crystals with a fast decay time for instance).

3.6 X-Ray Diffraction Data Processing

Processing the X-ray diffraction of amyloid peptides is often challenging due to the small unit cell and a low number of observed reflections [16]. Previous experience with Tau peptides in complex with small molecules showed that the electron density attributed to the small molecule was undifferentiated (to different extents), which hindered the determination of the structures in atomic detail [27]. It is recommended to validate the presence of the small molecule in the crystal using additional methods (*see* **Note 13**).

1. Most amyloid peptide data indexation, integration and scaling were performed using XDS/XSCALE [40] and DENZO [41].
2. Search for molecular replacement solutions, for example using PHASER [42], with the peptide as search models consisted of geometrically idealized β-strands, or previously determined polymorphs of the same segment.
3. Perform crystallographic refinements with REFMAC5 [43], PHENIX [44], or BUSTER [45]. Model building is performed with COOT [46].
4. After assignment of the peptide segment into the electron density 2Fo-Fc map, the difference Fo-Fc map should show positive density in case the small molecule is indeed forming an

ordered complex with the amyloid peptide (use mesh of +3σ). This density should be located in a void large enough to accommodate the small molecule. Model the small molecule into the density (*see* **Note 14**) and evaluate the results (*see* **Note 15**).

4 Notes

1. Safety measures when handling amyloid-like peptides:

   When handling the peptides in the lyophilized form, use personal protective equipment including a respirator protection mask. Before weighting and dissolving the peptides centrifuge the vial to minimize powder leftovers on the cap and sides of the vial. Open the vials only inside the balance. Carry out all peptide weighing in an analytical balance and keep the doors closed when possible. Add the dissolving solution quickly on the side of the vial (not directly into the powder) and close the lid. Gently vortex the vial.

2. Depending on the solubility of the peptides, final concentration and dissolving buffer vary. Dissolve *water soluble peptides* in water to a final concentration of 1–40 mg/ml (start with 10 mg/ml) [14, 16, 17, 19, 23, 24, 27, 47]. Dissolve *water insoluble peptides* in low or high pH buffers [17], DMSO [23, 27], ethanol [27], 15 % acetonitrile in 15 mM Bis-Tris [22] or 10–20 mM lithium hydroxide.

   The final concentration of the peptide and dissolving conditions used for crystallization assays can be adjusted according to the results obtained. For example: in case of a very soluble peptide, a higher concentration might be needed. The appearance of many small and/or fibrillary-looking crystals might indicate that a lower concentration is needed for obtaining larger nucleation sites. Using different solvents might yield different crystal forms with different X-ray diffraction and packing properties [19].

3. The solubility of the small molecule is often a concern, as many of the compounds are water insoluble at the concentrations needed for co-crystallization (~1 mM) [26, 27]. In some cases the solubility of the small molecule is indicated by the vendor or in general databases, but there is still a need to test the solubility in the final solvent that contains the peptide.

   For preparing small-molecule stock and final solutions, the first step is to test the small molecules for their solubility in water. Water insoluble small molecules can be dissolved in the minimal required percentage of DMSO [23, 27]. The solubility in either water or DMSO should be then evaluated at the final desired concentration and peptide mixture. It is noteworthy that some crystals of peptides with small molecules did form from a solution in which the small molecule was only partially dissolved.

4. Theoretically, the chosen molar ratio between the peptide and small molecule should favor an access of the small molecule, in order to increase probably of binding. However, it is advised not to exceed a concentration of 1 mM of the small molecules, as it increases the chances of formation of crystals that contain only the small molecule, without the presence of the peptide. This limits the range of the molar ratio used for co-crystallization depending on the concentration needed for the crystallization of the peptide.
5. For the dilution and mixing of the peptide and small molecule to their final solution, use the minimal percentage needed of DMSO or other organic solvent since these are highly volatile and can make it difficult setting the crystallization assays.
6. Common crystallization screens used for amyloid-like peptides include: Index, Crystal Screen, PEG/ Ion, and SaltRX by Hampton Research, Wizard by Emerald Biosystems, as well as ComPAS, JCSG+, and PACT by Qiagen.
7. Crystal mounting is the retrieval and transfer of a single crystal from its growth solution into a suitable mounting tool that can be used for X-ray diffraction data collection.
8. Most protein crystals require flash freezing in liquid nitrogen and cryoprotection. Yet, due to the low solvent content of the amyloid peptide microcrystals, they can be stored at room temperature. In addition, due to the especially small size of the peptide microcrystals, they are mounted on glass capillaries, and not on the commonly used cryogenic loop. This enables better centering of the crystal on the goniometer for X-ray diffraction data collection [14, 16, 18, 19, 22, 23, 27].
9. Mounting the micro-crystals is not trivial and requires some expertise. Make sure to mount only one crystal per capillary. It is preferred to have the needle crystal aligned with the capillary, with a protuberant edge of the crystal (Fig. 2c). Crystal mounted with a large certain angle relative to the capillary will be more difficult to center during X-ray data collection.
10. Mounting the brass pins (with glued needle capillaries) into the adjustable crystal mount (Fig. 2d) will results in an overall length of 18 mm standard at nearly all synchrotron beamlines in the world.
11. In crystal mounting, it is generally recommended to use a loop that is slightly smaller than the crystal (for example 20 μm loop for a 25 μm crystal; Robert Thorne (Mitegen LLC), personal communication). For most amyloid peptide microcrystals,

Dual-Thickness MicroLoops™ LD, SPINE/18 mm length (e.g., Mitegen LLC) are suited. The pins are super-glued into cap bases suited for the relevant beam-line with suited vials.

12. Due to the especially small size of the amyloid-like peptide microcrystals, a Microfocus Beamline (5 μm beam size) is required. For example, NE-CAT 24-ID-E of the Advanced Photon Source (APS), Argonne National Laboratory, USA; ID13, ID23-2 or MASSIF-3 at the European Synchrotron Radiation Facility (ESRF), Grenoble, France; I24 at the Diamond light source, Oxfordshire, United Kingdom; P11 and EMBL P14 at PETRA III, Hamburg, Germany; or X06SA at the Swiss Light Source (SLS), Villigen, Switzerland.

13. To assess whether the small molecule is indeed forming a complex with the peptide in the crystal, few approaches can be taken. A first indication is that the crystals show a distinctive color attributed to the small molecule, whereas the control crystals (grown under identical condition without the small molecule) are colorless. Upon solving the structure, the control crystals also lack the additional positive density attributed to the small molecule (*see* Subheading 3.6, **step 4**). Crystals of complexes grown under alternative, nonoverlapping, crystallization conditions that still show a similar positive electron density, further supports its attribution to the small molecule. Strong evidence for the presence of the small molecules in the crystals may also be obtained by mass spectrometric analyses of the crystals [27]. This method can reveal the molar abundance of the small molecule within the crystal as well. In previously determined structures [27], it was conjectured that the lack of an anchoring site for the hydrophobic, uncharged small molecules to specific residues in the fibril, leads to undifferentiated electron density. Furthermore, the nature of the binding site (a narrow tube running along the β-sheets) (Fig. 1) implies that the apolar small molecules are free to drift along the fibril axis.

14. Modeling the small molecule into the structure is enabled using COOT [46]. If available, use an experimentally determined structure of the small molecule (for example via the Cambridge Structural Database). Otherwise, generate three-dimensional (3-D) structures of the small molecule, using for instance Corina (Molecular Networks; http://www.molecular-networks.com/online_demos/corina_demo) and Chemical Identifier Resolver (http://cactus.nci.nih.gov/translate/). Additional 3-D conformations can be generated using OpenEye Omega [48].

    For structures of amyloid peptides in complex with small molecules, the latter constitutes a significant part of the asymmetric unit of the crystal [16]. Therefore, it is anticipated that computational docking [49, 50] could be of help for the correct assignment of the small molecule atoms.

15. Refine and evaluate the generated docked structures based on their free-R value [51]. In case the presence of the small molecule in the crystal was confirmed, yet the refinement with the small molecule did not improve the free-R value, it might be concluded that the X-ray diffraction does not allow the determination of the position of the small molecule in atomic detail [27].

Acknowledgments

ML thanks the U.S.-Israel Binational Science Foundation (BSF), Alon Fellowship from the Israeli Council for Higher Education, David and Inez Mayers Career Advancement Chair in Life Sciences, the J. and A. Tau Biological Research Fund, the I-CORE Program of the Planning and Budgeting Committee and The Israel Science Foundation, Center of Excellence in Integrated Structural Cell Biology; Grant No 1775/12, and the Support for training and career development of researchers (Marie Curie) CIG, Seventh framework program, Single Beneficiary.

References

1. Anand K, Sabbagh M (2015) Early investigational drugs targeting tau protein for the treatment of Alzheimer's disease. Expert Opin Investig Drugs 24(10):1355–1360
2. Brunden KR, Ballatore C, Crowe A et al (2010) Tau-directed drug discovery for Alzheimer's disease and related tauopathies: a focus on tau assembly inhibitors. Exp Neurol 223:304–310
3. Bulic B, Pickhardt M, Khlistunova I et al (2007) Rhodanine-based tau aggregation inhibitors in cell models of tauopathy. Angew Chem Int Ed Engl 119:9375–9379
4. Bulic B, Pickhardt M, Mandelkow E-M et al (2010) Tau protein and tau aggregation inhibitors. Neuropharmacology 59:276–289
5. Bulic B, Pickhardt M, Schmidt B et al (2009) Development of tau aggregation inhibitors for Alzheimer's disease. Angew Chem Int Ed Engl 48:1740–1752
6. Crowe A, Ballatore C, Hyde E et al (2007) High throughput screening for small molecule inhibitors of heparin-induced tau fibril formation. Biochem Biophys Res Commun 358:1–6
7. Fuse S, Matsumura K, Fujita Y et al (2014) Development of dual targeting inhibitors against aggregations of amyloid-β and tau protein. Eur J Med Chem 85:228–234
8. Larbig G, Pickhardt M, Lloyd DG, Schmidt B, Mandelkow E (2007) Screening for inhibitors of tau protein aggregation into Alzheimer paired helical filaments: a ligand based approach results in successful scaffold hopping. Curr Alzheimer Res 4(3):315–323
9. Harrington CR, Storey JMD, Clunas S et al (2015) Cellular models of aggregation-dependent template-directed proteolysis to characterize tau aggregation inhibitors for treatment of Alzheimer disease. J Biol Chem 290:10862–10875
10. Karakani AM, Riazi G, Mahmood GS et al (2015) Inhibitory effect of corcin on aggregation of 1N/4R human tau protein in vitro. Iran J Basic Med Sci 18:485–492
11. Paranjape SR, Riley AP, Somoza AD et al (2015) Azaphilones inhibit tau aggregation and dissolve tau aggregates in vitro. ACS Chem Neurosci 6:751–760
12. Wobst HJ, Sharma A, Diamond MI et al (2015) The green tea polyphenol (−)-epigallocatechin gallate prevents the aggregation of tau protein into toxic oligomers at substoichiometric ratios. FEBS Lett 589:77–83

13. Balbirnie M, Grothe R, Eisenberg DS (2001) An amyloid-forming peptide from the yeast prion Sup35 reveals a dehydrated beta-sheet structure for amyloid. Proc Natl Acad Sci U S A 98:2375–2380
14. Nelson R, Sawaya MR, Balbirnie M et al (2005) Structure of the cross-beta spine of amyloid-like fibrils. Nature 435:773–778
15. Ivanova MI, Thompson MJ, Eisenberg D (2006) A systematic screen of beta(2)-microglobulin and insulin for amyloid-like segments. Proc Natl Acad Sci U S A 103:4079–4082
16. Sawaya MR, Sambashivan S, Nelson R et al (2007) Atomic structures of amyloid cross-beta spines reveal varied steric zippers. Nature 447:453–457
17. Wiltzius JJ, Sievers SA, Sawaya MR et al (2008) Atomic structure of the cross-beta spine of islet amyloid polypeptide (amylin). Protein Sci 17:1467–1474
18. Ivanova MI, Sievers SA, Sawaya MR et al (2009) Molecular basis for insulin fibril assembly. Proc Natl Acad Sci U S A 106: 18990–18995
19. Wiltzius JJ, Landau M, Nelson R et al (2009) Molecular mechanisms for protein-encoded inheritance. Nat Struct Mol Biol 16:973–978
20. Wiltzius JJ, Sievers SA, Sawaya MR et al (2009) Atomic structures of IAPP (amylin) fusions suggest a mechanism for fibrillation and the role of insulin in the process. Protein Sci 18:1521–1530
21. Laganowsky A, Benesch JL, Landau M et al (2010) Crystal structures of truncated alphaA and alphaB crystallins reveal structural mechanisms of polydispersity important for eye lens function. Protein Sci 19:1031–1043
22. Apostol MI, Wiltzius JJ, Sawaya MR et al (2011) Atomic structures suggest determinants of transmission barriers in mammalian prion disease. Biochemistry 50:2456–2463
23. Colletier JP, Laganowsky A, Landau M et al (2011) Molecular basis for amyloid-beta polymorphism. Proc Natl Acad Sci U S A 108: 16938–16943
24. Liu C, Zhao M, Jiang L et al (2012) Out-of-register beta-sheets suggest a pathway to toxic amyloid aggregates. Proc Natl Acad Sci U S A 109:20913–20918
25. Sievers SA, Karanicolas J, Chang HW et al (2011) Structure-based design of non-natural amino-acid inhibitors of amyloid fibril formation. Nature 475:96–100
26. Jiang L, Liu C, Leibly D et al (2013) Structure-based discovery of fiber-binding compounds that reduce the cytotoxicity of amyloid beta. Elife 2:e00857
27. Landau M, Sawaya MR, Faull KF et al (2011) Towards a pharmacophore for amyloid. PLoS Biol 9:e1001080
28. von Bergen M, Friedhoff P, Biernat J et al (2000) Assembly of tau protein into Alzheimer paired helical filaments depends on a local sequence motif ((306)VQIVYK(311)) forming beta structure. Proc Natl Acad Sci U S A 97:5129–5134
29. Yang F, Lim GP, Begum AN et al (2005) Curcumin inhibits formation of amyloid beta oligomers and fibrils, binds plaques, and reduces amyloid in vivo. J Biol Chem 280: 5892–5901
30. Jacobson A, Petric A, Hogenkamp D et al (1996) 1,1-Dicyano-2-[6-(dimethylamino) naphthalen-2-yl]propene (DDNP): a solvent polarity and viscosity sensitive fluorophore for fluorescence microscopy. J Am Chem Soc 118:5572–5579
31. Krebs MR, Bromley EH, Donald AM (2005) The binding of thioflavin-T to amyloid fibrils: localisation and implications. J Struct Biol 149:30–37
32. Wolfe LS, Calabrese MF, Nath A et al (2010) Protein-induced photophysical changes to the amyloid indicator dye thioflavin T. Proc Natl Acad Sci U S A 107: 16863–16868
33. Childers WS, Mehta AK, Lu K et al (2009) Templating molecular arrays in Amyloid's cross-{beta} grooves. J Am Chem Soc 131: 10165–10172
34. Schutz AK, Soragni A, Hornemann S et al (2011) The amyloid-Congo red interface at atomic resolution. Angew Chem Int Ed Engl 50:5956–5960
35. Goldschmidt L, Teng PK, Riek R et al (2010) Identifying the amylome, proteins capable of forming amyloid-like fibrils. Proc Natl Acad Sci U S A 107:3487–3492
36. Fernandez-Escamilla AM, Rousseau F, Schymkowitz J et al (2004) Prediction of sequence-dependent and mutational effects on the aggregation of peptides and proteins. Nat Biotechnol 22:1302–1306
37. Maurer-Stroh S, Debulpaep M, Kuemmerer N et al (2010) Exploring the sequence determinants of amyloid structure using position-specific scoring matrices. Nat Methods 7:237–242

38. Tartaglia GG, Pawar AP, Campioni S et al (2008) Prediction of aggregation-prone regions in structured proteins. J Mol Biol 380:425–436
39. Moshe A, Landau M, Eisenberg D (2016) Preparation of crystalline samples of amyloid fibrils and oligomers. In: Eliezer D (ed) Protein amyloid aggregation, vol 1345, Methods and protocols, methods in molecular biology. Springer, New York, pp 201–210. doi: 10.1007/978-1-4939-2978-8_13
40. Kabsch W (1993) Automatic processing of rotation diffraction data from crystals of initially unknown symmetry and cell constants. J Appl Crystallogr 26:795–800
41. Otwinowski Z, Minor W (1997) Processing of X-ray diffraction data collected in oscillation mode. In: Carter CW Jr, Sweet RM (eds) Macromolecular crystallography, part A, vol 276. Academic, New York
42. Read RJ (2001) Pushing the boundaries of molecular replacement with maximum likelihood. Acta Crystallogr D Biol Crystallogr 57:1373–1382
43. Murshudov GN, Vagin AA, Dodson EJ (1997) Refinement of macromolecular structures by the maximum-likelihood method. Acta Crystallogr D Biol Crystallogr 53:240–255
44. Adams PD, Afonine PV, Bunkoczi G et al (2010) PHENIX: a comprehensive Python-based system for macromolecular structure solution. Acta Crystallogr D Biol Crystallogr 66:213–221
45. Bricogne G, Blanc E, Brandl M et al (2009) BUSTER, Version 2.8.0. Global Phasing Ltd., Cambridge, UK
46. Emsley P, Cowtan K (2004) Coot: model-building tools for molecular graphics. Acta Crystallogr D Biol Crystallogr 60:2126–2132
47. Ivanova MI, Sievers SA, Guenther EL et al (2014) Aggregation-triggering segments of SOD1 fibril formation support a common pathway for familial and sporadic ALS. Proc Natl Acad Sci U S A 111:197–201
48. Bostrom J, Greenwood JR, Gottfries J (2003) Assessing the performance of OMEGA with respect to retrieving bioactive conformations. J Mol Graph Model 21:449–462
49. Davis IW, Baker D (2009) RosettaLigand docking with full ligand and receptor flexibility. J Mol Biol 385:381–392
50. Meiler J, Baker D (2006) ROSETTALIGAND: protein-small molecule docking with full side-chain flexibility. Proteins 65:538–548
51. Brunger AT (1992) Free R value: a novel statistical quantity for assessing the accuracy of crystal structures. Nature 355:472–475

Chapter 6

Detection and Quantification Methods for Fibrillar Products of In Vitro Tau Aggregation Assays

Niki Nanavaty, Lauren Lin, Samantha H. Hinckley, and Jeff Kuret

Abstract

Alzheimer's disease is characterized in part by the intracellular misfolding and aggregation of tau protein. The aggregates, which range in size from small oligomers to large filaments, are markers for disease diagnosis and staging, potential vectors for disease propagation, and candidate sources of neurotoxicity. Here we present protocols for synthesizing large tau aggregates characterized by filamentous morphology and cross-β-sheet structure from monomeric full-length tau precursors in vitro. We also describe their detection and quantification through thioflavin dye binding, filter trap, and transmission electron microscopy methods. These methods cover applications requiring high-throughput capability as well as those requiring high-resolution analysis of aggregation mechanism.

Key words Alzheimer's disease, Tau, Aggregation, Electron microscopy, Filter trap assay, Thioflavin dye saturation assay

1 Introduction

Human brain tau is a monomeric protein that normally functions as a binding partner for the microtubule cytoskeleton (reviewed in ref. [1]). It is expressed as a mixture of six isoforms that arise from alternative splicing of exons 2, 3, and 10 from *MAPT* gene transcripts (reviewed in ref. [2]). Exons 2 and 3 encode 29-residue acidic inserts in the N-terminal projection domain of tau, whereas exon 10 encodes a 31-residue microtubule binding repeat in the C-terminal domain (Fig. 1). As a result, full-length human tau isoforms range in size from 352 to 441 amino acids (prior to N-terminal processing; [3]) and contain either 3 or 4 microtubule binding repeats (Fig. 1). Each tau isoform is enriched in polar, glycine, and proline residues, and contains below-average contents of hydrophobic residues, relative to typical globular proteins [4]. Consistent with this composition, tau isoforms are mostly disordered in aqueous solution, and maintain high water solubility even at boiling temperatures [5] or acidic pH [6]. Disordered structure also

Caroline Smet-Nocca (ed.), *Tau Protein: Methods and Protocols*, Methods in Molecular Biology, vol. 1523,
DOI 10.1007/978-1-4939-6598-4_6, © Springer Science+Business Media New York 2017

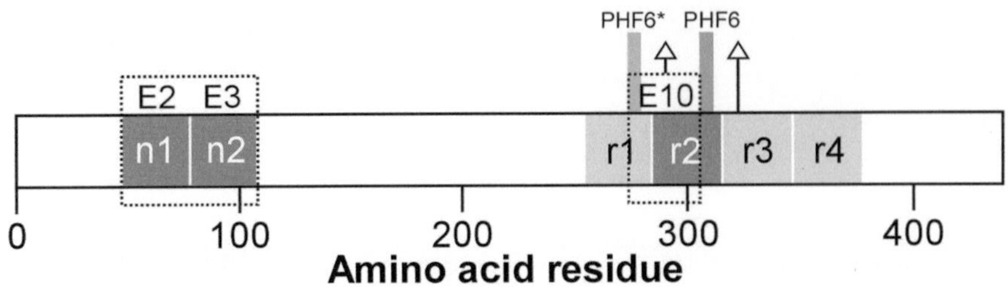

Fig. 1 Tau protein primary structure. Human brain tau is composed of six isoforms derived from alternatively splicing of exons E2, E3, and E10 (*dotted boxes*). Each isoform consists of an N-terminal projection domain containing up to two alternative segments (n1 and n2), and a C-terminal microtubule binding domain containing three or four imperfect repeats (r1–r4; mapped on the basis of sequence homology as in [38]). They also contain one or two cysteine residues (*triangles*) that are prone to oxidation and adduct formation. Tau aggregation leading to filament formation is mediated by hexapeptide motifs (PHF6 and PHF6*) located in the microtubule binding domain. Sequence numbering corresponds to the human 2N4R isoform (National Center for Biotechnology Information accession number NP_005901)

exposes nucleophilic cysteine residues to the surrounding solvent (one in 3R forms, two in 4R forms; Fig. 1) where they can spontaneously engage in disulfide bond formation or react covalently with available electrophiles. Other solvent-exposed hydrophobic sequences, located in the microtubule binding region, self-associate to form the core regions of filamentous aggregates in disease [7].

The goal of in vitro modeling is to convert normally soluble and stable tau monomers into disease-related aggregates over experimentally tractable time periods. Because 4R isoforms contain two cysteine residues capable of forming intramolecular disulfide bonds, it also is desirable to maintain cysteine residues in their reduced form when working with these isoforms for the purposes of investigating aggregation mechanism [8, 9] or minimizing interaction with electrophilic small molecules [10]. Finally, neuronal tau concentrations have been estimated as being low micromolar [11], and so in addition to pH and ionic strength, physiologically relevant aggregation conditions should emulate this concentration.

To meet these constraints, in vitro aggregation reactions involving full-length tau isoforms are frequently initiated by the addition of an exogenous aggregation inducer (reviewed in ref. [12]). Typical inducers are polyanions at neutral pH, and may function by interacting with cationic tau monomers to stabilize aggregation competent conformations [13]. Protocols for three anionic inducers are described below, each with its own advantages and disadvantages depending on purposes of experimentation. Octadecyl sulfate (ODS) is an anionic surfactant that acts in micellar form to rapidly induce aggregation at low micromolar concentrations under reducing conditions [14]. Because this inducer has a low extinction coefficient, it is well suited for experiments involving optical detection methods. Anionic microspheres retain the aggregation promoting activity of ODS, but because of their size, facilitate visualization of nucleation efficiency by transmission electron microscopy (TEM) [15]. Their availability in various charge polarities and densities supports experimentation aiming to investigate the role of charge interactions in promoting aggregation propensity. The third inducer, Geranine G, drives aggregation of four repeat tau isoforms at submicromolar concentrations [16], allowing fine dissection of aggregation mechanism at physiologically relevant tau concentrations. However, because of its high extinction coefficient, it is not well suited for experimentation involving optical methods.

Tau aggregates formed in the presence of inducers can be detected and quantified by multiple approaches. Here, protocols for three complementary assays will be presented: Thioflavin dye fluorescence, filter trap immunoassay, and TEM imaging. Thioflavin dye based assays leverage the strong fluorescence and Stokes shift undergone by the extrinsic fluorescent probes Thioflavin S (ThS) or Thioflavin T (ThT) in the presence of cross-β-sheet structure [17–19]. As a result, it is possible to rapidly and inexpensively detect fibrillary tau aggregate products in the presence of monomeric reactants [20, 21]. This assay is ideal for high throughput applications, such as screens for aggregation inhibitors or dye displacement agents [22–24]. Filter trap assays leverage semipermeable filters to separate large aggregate products from monomeric reactants [25]. Tau aggregates trapped on the filter can be detected with anti-tau antibodies and quantified by highly sensitive fluorescent or chemiluminescent detection assays [26]. This method offers very high sensitivity and also compatibility with optically active additives. TEM, in contrast, is a low-throughput method that offers unparalleled resolution of filament morphology and length distribution without needing reporter molecules [27, 28]. Additional methods for analyzing protein aggregates also can be consulted [29].

2 Materials

2.1 Tau Filament Production

All buffers and reagents are made with ultrapure water (18.2 MΩ-cm at 25 °C) and filtered prior to use (pore size ≤0.22 μm). Use of siliconized pipette tips and microfuge tubes minimizes losses of tau protein on plastic surfaces, especially when studied at submicromolar concentrations.

1. Recombinant tau proteins: these can be prepared by liquid chromatography after *E. coli* expression as described elsewhere [30, 31] (*see* **Note 1**).
2. 5× Aggregation buffer: 50 mM HEPES, pH 7.4, 500 mM NaCl, 5 mM DTT (*see* **Note 2**).

3. 2 mM Geranine G (Alfa Chemistry); prepared in water. Filter stock solutions through 0.22 μm syringe filter and store at −20 °C (*see* **Note 3**).
4. 0.5 mg/ml anionic microsphere (e.g., Product series PC02N, Bangs Laboratories Inc.), suspended in water (*see* **Note 4**).
5. 2 mM Octadecyl sulfate (ODS; ThermoFisher Scientific); prepared in 50% (v/v) isopropanol (*see* **Note 5**).

2.2 Tau Filament Detection

2.2.1 Thioflavin-Dye Binding Assay

1. Thioflavin dye stock solutions: Dissolve ThT (Sigma Aldrich) in water at 100 μM or ThS (Sigma Aldrich) in water at 0.06 mg/ml. Filter stock solutions through 0.22 μm syringe filter and store at −20 °C (*see* **Note 6**).
2. 96-well untreated polypropylene plates for compound dilutions (e.g., ThermoFisher Scientific).
3. 96-well black untreated polystyrene plates for fluorescence detection (e.g., Costar®, Corning Inc. Life Sciences).
4. Adhesive plate seals (e.g., AlumaSeal).
5. Microplate shaker (e.g., Jitterbug microplate shaker).
6. Fluorimeter. Any standard instrument is adequate, but plate reader format is most useful for high-throughput applications.

2.2.2 Filter Trap Assay

1. Microfiltration apparatus (e.g., Bio-Dot).
2. Polyvinylidene fluoride (PVDF) or nitrocellulose membranes, 0.2–0.45 μm pore size (*see* **Note 7**).
3. TBS: 150 mM NaCl in 100 mM Tris–HCl, pH 7.4.
4. Blocking buffer: 3% nonfat dry milk in TBS.
5. Primary anti-tau antibody (*see* **Note 8**).
6. Secondary HRP-conjugated antibody (e.g., HRP-linked goat anti-mouse IgG).
7. HRP detection system (e.g., Enhanced Chemiluminescence).

2.2.3 EM Assay

1. Transmission electron microscope. We use a Tecnai G2 Spirit BioTWIN transmission electron microscope (FEI Company) operated at 80 kV acceleration voltage and equipped with digital image capture.
2. Uranyl acetate (UA) (Electron Microscopy Sciences). Prepare 2% (w/v) solution in water (*see* **Note 9**).
3. Copper grids 300-mesh formvar/carbon-coated (Electron Microscopy Sciences). These commercial grids are supplied with film laid on the shiny side. They can be used directly without glow discharging.

4. Hydrophobic laboratory film (e.g., Parafilm). Cut into 4 × 4 in. squares for easiest handling.
5. Glutaraldehyde (25% w/v in water; Electron Microscopy Sciences).
6. Fine-tipped forceps for handling grids (110 mm, Structure Probe Inc.).
7. Cellulose filter paper (e.g., Whatman Qualitative No. 2 filter paper) for blotting off excess liquids from grids. Cut into small squares for easiest manipulation.
8. Grid box (Electron Microscopy Sciences) for storage and transport of grids (*see* **Note 10**).

3 Methods

3.1 Tau Aggregation

All aggregation experiments are conducted in 1× aggregation buffer at 37 °C for up to 24 h with or without agitation and can be scaled up in volume to meet assay needs. However, they differ in the choice of inducer and concentration of tau protein (*see* **Note 11**).

1. For Geranine G-induced aggregation, incubate recombinant tau protein (0.4–10 μM) in the presence of aggregation buffer containing 100 μM Geranine G (final concentration).
2. For microsphere-induced aggregation, incubate recombinant tau protein (2–4 μM) in the presence of assembly buffer containing 0.05 mg/ml anionic microspheres final concentration.
3. For ODS-induced aggregation, incubate recombinant tau protein (2–4 μM) in the presence of aggregation buffer containing 50 μM ODS.

3.2 Detection Assays

3.2.1 Thioflavin Dye Assay

This assay is best used with aggregates prepared with ODS or microsphere inducers, as Geranine G interferes with Thioflavin dye fluorescence.

1. Remove 90 μl aliquots of aggregation reaction product to a 96-well plate (*see* **Note 12**).
2. Add 10 μl of 100 μM ThT or 0.06 mg/ml ThS to each well, and seal plate with adhesive foil (*see* **Note 13**).
3. Secure plates in micro-plate shaker and shake for 1 min.
4. Incubate plates at 37 °C for 1 h, then shake on Jitterbug for 1 min.
5. Place plates in fluorimeter, and read at monochrometer wavelengths $\lambda_{ex} = 440$ nm and $\lambda_{em} = 490$ nm. For filter-based fluorimeters, read at wavelengths $\lambda_{ex} = 440/30$ nm and $\lambda_{em} = 485/20$ nm.

3.2.2 Filter Trap Assay

This assay is best used with aggregates prepared with ODS or Geranine G inducers, as microspheres confound results by accumulating on the membranes. All steps can be carried out at room temperature.

1. To a 50 μl aliquot of aggregation reaction product, add 150 μl of 2 % SDS and vortex.
2. Pre-equilibrate a PVDF or nitrocellulose membrane in 2 % SDS and mount in a vacuum microfiltration apparatus (*see* **Note 14**).
3. Apply diluted aggregation reaction products to each well of the microfiltration apparatus, then apply vacuum to draw them through the membrane.
4. Wash (with vacuum) each well of the microfiltration apparatus twice with 200 μl 2 % SDS.
5. Remove membrane from filtration apparatus, and incubate in blocking buffer for 2 h.
6. Incubate membranes with primary anti-tau antibody for 1.5 h.
7. Wash membranes (3 × 10 min) in blocking buffer.
8. Incubate membranes with the secondary HRP-conjugated antibody for 1.5 h.
9. Wash membranes (3 × 10 min) in TBS, then develop with high sensitivity detection method, such as the ECL western blotting analysis system.
10. Capture chemiluminescence intensity on film or an imaging system and quantify by densitometry (*see* **Note 15**).

3.2.3 TEM Assay

This assay is compatible with aggregates produced by all inducers. All steps can be carried out at room temperature.

1. To a 50 μl aliquot of aggregation reaction product, add glutaraldehyde to 1 % (w/v) final concentration (*see* **Note 16**).
2. Place 50 μL each of glutaraldehyde-treated aggregation reaction product, two drops of water, and two drops of UA onto a sheet of Parafilm.
3. Carefully pick the grid up by the rim using tweezers, and rest the grid shiny-side down onto the sample drop for 1 min (*see* **Note 17**). The grid will be supported by surface tension while fibrils diffuse and adsorb to the surface of the grid, so do not entirely immerse.
4. Lift the grid from the sample droplet using tweezers and remove extra sample by blotting the grid edge with filter paper.
5. Clean the grid by quickly dipping the grid reflective-side down onto the first water drop, and cautiously blot off extra liquid with filter paper as before.

6. Repeat wash in first UA drop, blot, and set grid down for 1 min in second drop of UA (*see* **Note 18**). Ensure reflective side is still facing down.
7. Take the grid out of the UA, and blot extra UA away with filter paper. Wash once again by placing the grid onto the second drop of water, and blot the excess. Leave the grid face up on top of filter paper until completely dry; stained grids may be stored at room temperature for weeks.
8. Obtain images using the transmission electron microscope, and save images for analysis. Record each image's magnification scale for future reference (*see* **Note 19**).

4 Notes

1. This purification technique has been used on all six tau isoforms [9], missense mutants [32], and posttranslational modifications [33]. Additional purification steps, such as boiling, can be included without altering aggregation performance [14]. Purified tau is aliquoted and stored at −80 °C in the presence of reducing agent (5 mM dithiothreitol or 5 mM Tris(2-carboxyethyl)phosphine hydrochloride).
2. DTT must be added freshly just before initiating aggregation.
3. Geranine G is a member of the Thiazine red family of dyes. Historically, Geranine G was vended as a mixture of isoforms, but in recent years it is available primarily as a pure substance identical to CAS registry number 764584-03-4.
4. Microsphere diameters ranging from 40 to 90 nm with anionic surface charge densities of 12–62 Å^2/equivalent have been tested successfully [15]. In contrast, neutral or positively charged beads are not active as aggregation inducers. Microspheres typically are sold as 100 mg/ml suspensions.
5. ODS is an ionic surfactant that micellizes efficiently in the presence of counterions such as Na^+ or tau proteins at neutral pH [14].
6. ThT is a defined substance and available in a highly purified form, but ThS is a complex mixture of components [19]. Therefore, ThS concentrations are usually reported in units of mass concentration.
7. Although standard 0.45 μm pore membranes used for routine immunoblotting applications perform well in this assay, use of finer pore sizes will increase aggregate trapping and detection sensitivity [26]. Regardless of pore size, high-protein binding capacity is essential, and so cellulose acetate cannot substitute for nitrocellulose or PVDF [26].

8. Many anti-tau antibodies are commercially available. This protocol has been used successfully [26] with anti-tau IgG antibodies Tau 1 [34] and Tau 5 [35], both of which recognize continuous epitopes [30], and IgM antibody Alz50 [36], which is selective for filamentous tau [30].
9. UA is photosensitive and toxic. The solid material should be stored under the hood in a moisture-free dark metal container. Care should be taken to avoid contact with and exposure to the material and subsequent solutions containing UA. Used UA and lab disposables that come in contact with UA are disposed of per institutional guidelines and appropriate Materials Safety Data Sheet.
10. Secure the lid of the grid box with a small piece of labeling tape while transporting grids to prevent the inadvertent opening of the box and losing and/or mixing of the grids.
11. The efficiency of tau aggregation varies with tau/inducer ratio for both ODS and microsphere inducers but not for Geranine G [14, 15]. For this reason, working outside the specified tau concentrations requires adjustment and optimization of tau/inducer ratio for ODS and microspheres.
12. Wells set up without tau protein provide negative controls for nonspecific thioflavin dye fluorescence.
13. Seal plate with adhesive foil seal. It is important to seal each well to prevent spillage from one well to another. Using wax paper from the adhesive foil backing allows for even pressure to be applied.
14. PVDF membranes must be pre-wet in neat methanol before transfer into SDS solution.
15. Filter trap assays developed with chemiluminescence can be nonlinear with respect to aggregate concentration. This can be detected and calibrated by use of a standard curve as described previously [26].
16. Glutaraldehyde is toxic and care should be taken not to breathe the fumes or allow contact with skin. Under the hood with gloves, add a drop of glutaraldehyde to the inside wall of the tube just above the sample. Carefully allow the drop to mix with the sample by gentle flicking of the tube. Aggressive mixing or vortexing can cause clumping of tau filaments.
17. It is necessary to maintain constant adsorption time for all samples due to the time dependent nature of filament adsorption [37].
18. UA most frequently generates a positive staining effect when interacting with tau filaments, where filaments appear dark against a light background. Positive staining is adequate for quantification of filament length. However, UA adopts negative staining in certain areas of the grid. Negative stain is particularly useful for determining filament morphology.

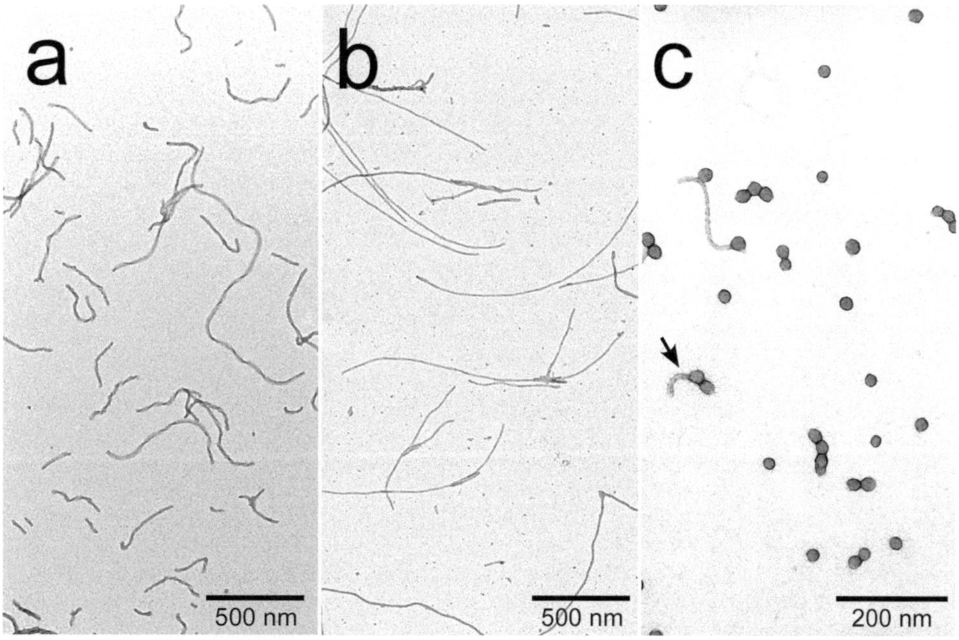

Fig. 2 Electron microcsopy detection of tau aggregates generated by different inducers. Tau filaments induced by Geranine G (**a**), ODS (**b**), and 90 nm anionic microspheres 12 Å²/equivalent (**c**) were prepared and imaged according to the TEM assay protocol (Subheading 3.2.3). The *arrow* identifies a filament nucleated on a microsphere surface

19. Three or more random images from each grid are captured typically at 8000- to 100,000-fold magnification. High magnification better captures morphological features, while lower magnification is necessary for quantitative assessments of filament length. A conventional magnification for quantification of tau filament lengths in the presence of Geranine G ranges 18,000- to 35,000-fold (Fig. 2a).

Acknowledgment

This work was supported by NIH grant NS077441. Electron micrographs were generated using instruments at the Campus Microscopy and Imaging Facility, The Ohio State University, Columbus, OH.

References

1. Morris M, Maeda S, Vossel K, Mucke L (2011) The many faces of tau. Neuron 70:410–426
2. Andreadis A (2012) Tau splicing and the intricacies of dementia. J Cell Physiol 227:1220–1225
3. Hanger DP, Byers HL, Wray S, Leung KY, Saxton MJ, Seereeram A, Reynolds CH, Ward MA, Anderton BH (2007) Novel phosphorylation sites in tau from Alzheimer brain support a

role for casein kinase 1 in disease pathogenesis. J Biol Chem 282:23645–23654

4. Dafforn TR, Smith CJ (2004) Natively unfolded domains in endocytosis: hooks, lines and linkers. EMBO Rep 5:1046–1052
5. Cleveland DW, Hwo SY, Kirschner MW (1977) Purification of tau, a microtubule-associated protein that induces assembly of microtubules from purified tubulin. J Mol Biol 116:207–225
6. Lindwall G, Cole RD (1984) The purification of tau protein and the occurrence of two phosphorylation states of tau in brain. J Biol Chem 259:12241–12245
7. von Bergen M, Friedhoff P, Biernat J, Heberle J, Mandelkow EM, Mandelkow E (2000) Assembly of tau protein into Alzheimer paired helical filaments depends on a local sequence motif (^{306}VQIVYK^{311}) forming b structure. Proc Natl Acad Sci U S A 97:5129–5134
8. Congdon EE, Kim S, Bonchak J, Songrug T, Matzavinos A, Kuret J (2008) Nucleation-dependent tau filament formation: the importance of dimerization and an estimation of elementary rate constants. J Biol Chem 283:13806–13816
9. Zhong Q, Congdon EE, Nagaraja HN, Kuret J (2012) Tau isoform composition influences rate and extent of filament formation. J Biol Chem 287:20711–20719
10. George RC, Lew J, Graves DJ (2013) Interaction of cinnamaldehyde and epicatechin with tau: implications of beneficial effects in modulating Alzheimer's disease pathogenesis. J Alzheimers Dis 36:21–40
11. Alonso AC, Grundke-Iqbal I, Iqbal K (1996) Alzheimer's disease hyperphosphorylated tau sequesters normal tau into tangles of filaments and disassembles microtubules. Nat Med 2: 783–787
12. Kuret J, Congdon EE, Li G, Yin H, Yu X, Zhong Q (2005) Evaluating triggers and enhancers of tau fibrillization. Microsc Res Tech 67:141–155
13. Tomoo K, Yao TM, Minoura K, Hiraoka S, Sumida M, Taniguchi T, Ishida T (2005) Possible role of each repeat structure of the microtubule-binding domain of the tau protein in in vitro aggregation. J Biochem 138:413–423
14. Chirita CN, Necula M, Kuret J (2003) Anionic micelles and vesicles induce tau fibrillization in vitro. J Biol Chem 278:25644–25650
15. Chirita CN, Kuret J (2004) Evidence for an intermediate in tau filament formation. Biochemistry 43:1704–1714
16. Chirita CN, Congdon EE, Yin H, Kuret J (2005) Triggers of full-length tau aggregation: a role for partially folded intermediates. Biochemistry 44:5862–5872
17. LeVine H 3rd (1999) Quantification of beta-sheet amyloid fibril structures with thioflavin T. Methods Enzymol 309:274–284
18. Naiki H, Higuchi K, Hosokawa M, Takeda T (1989) Fluorometric determination of amyloid fibrils in vitro using the fluorescent dye, thioflavin T1. Anal Biochem 177:244–249
19. Kelenyi G (1967) On the histochemistry of azo group-free thiazole dyes. J Histochem Cytochem 15:172–180
20. King ME, Ahuja V, Binder LI, Kuret J (1999) Ligand-dependent tau filament formation: implications for Alzheimer's disease progression. Biochemistry 38:14851–14859
21. Friedhoff P, Schneider A, Mandelkow EM, Mandelkow E (1998) Rapid assembly of Alzheimer-like paired helical filaments from microtubule-associated protein tau monitored by fluorescence in solution. Biochemistry 37:10223–10230
22. Honson NS, Johnson RL, Huang W, Inglese J, Austin CP, Kuret J (2007) Differentiating Alzheimer disease-associated aggregates with small molecules. Neurobiol Dis 28:251–260
23. Crowe A, Huang W, Ballatore C, Johnson RL, Hogan AM, Huang R, Wichterman J, McCoy J, Huryn D, Auld DS, Smith AB 3rd, Inglese J, Trojanowski JQ, Austin CP, Brunden KR, Lee VM (2009) Identification of aminothienopyridazine inhibitors of tau assembly by quantitative high-throughput screening. Biochemistry 48:7732–7745
24. Pickhardt M, von Bergen M, Gazova Z, Hascher A, Biernat J, Mandelkow EM, Mandelkow E (2005) Screening for inhibitors of tau polymerization. Curr Alzheimer Res 2:219–226
25. Wanker EE, Scherzinger E, Heiser V, Sittler A, Eickhoff H, Lehrach H (1999) Membrane filter assay for detection of amyloid-like polyglutamine-containing protein aggregates. Methods Enzymol 309:375–386
26. Chang E, Kuret J (2008) Detection and quantification of tau aggregation using a membrane filter assay. Anal Biochem 373:330–336
27. Necula M, Kuret J (2004) Electron microscopy as a quantitative method for investigating tau fibrillization. Anal Biochem 329: 238–246
28. Huseby CJ, Kuret J (2016) Analyzing tau aggregation with electron microscopy. Methods Mol Biol 1345:101–112

29. Hatters DM, Griffin MD (2011) Diagnostics for amyloid fibril formation: where to begin? Methods Mol Biol 752:121–136
30. Carmel G, Mager EM, Binder LI, Kuret J (1996) The structural basis of monoclonal antibody Alz50's selectivity for Alzheimer's disease pathology. J Biol Chem 271: 32789–32795
31. Carmel G, Leichus B, Cheng X, Patterson SD, Mirza U, Chait BT, Kuret J (1994) Expression, purification, crystallization, and preliminary x-ray analysis of casein kinase-1 from Schizosaccharomyces pombe. J Biol Chem 269:7304–7309
32. Jarrett JT, Lansbury PT Jr (1993) Seeding "one-dimensional crystallization" of amyloid: a pathogenic mechanism in Alzheimer's disease and scrapie? Cell 73:1055–1058
33. Chang E, Congdon EE, Honson NS, Duff KE, Kuret J (2009) Structure-activity relationship of cyanine tau aggregation inhibitors. J Med Chem 52:3539–3547
34. Binder LI, Frankfurter A, Rebhun LI (1985) The distribution of tau in the mammalian central nervous system. J Cell Biol 101:1371–1378
35. LoPresti P, Szuchet S, Papasozomenos SC, Zinkowski RP, Binder LI (1995) Functional implications for the microtubule-associated protein tau: localization in oligodendrocytes. Proc Natl Acad Sci U S A 92:10369–10373
36. Wolozin BL, Pruchnicki A, Dickson DW, Davies P (1986) A neuronal antigen in the brains of Alzheimer patients. Science 232:648–650
37. Zhao D, Moore JS (2003) Nucleation-elongation: a mechanism for cooperative supramolecular polymerization. Org Biomol Chem 1:3471–3491
38. Goedert M, Spillantini MG, Potier MC, Ulrich J, Crowther RA (1989) Cloning and sequencing of the cDNA encoding an isoform of microtubule-associated protein tau containing four tandem repeats: differential expression of tau protein mRNAs in human brain. EMBO J 8:393–399

Chapter 7

Fourier Transform Infrared (FTIR) Spectroscopy, Ultraviolet Resonance Raman (UVRR) Spectroscopy, and Atomic Force Microscopy (AFM) for Study of the Kinetics of Formation and Structural Characterization of Tau Fibrils

Gayathri Ramachandran

Abstract

Kinetic studies of tau fibril formation in vitro most commonly employ spectroscopic probes such as thioflavinT fluorescence and laser light scattering or negative stain transmission electron microscopy. Here, I describe the use of Fourier transform infrared (FTIR) spectroscopy, ultraviolet resonance Raman (UVRR) spectroscopy, and atomic force microscopy (AFM) as complementary probes for studies of tau aggregation. The sensitivity of vibrational spectroscopic techniques (FTIR and UVRR) to secondary structure content allows for measurement of conformational changes that occur when the intrinsically disordered protein tau transforms into cross-β-core containing fibrils. AFM imaging serves as a gentle probe of structures populated over the time course of tau fibrillization. Together, these assays help further elucidate the structural and mechanistic complexity inherent in tau fibril formation.

Key words Tau, FTIR spectroscopy, UVRR spectroscopy, AFM, Amyloid fibril, Kinetics, Heparin, Neurodegeneration

1 Introduction

The aggregation of the intrinsically disordered protein tau into ordered amyloid fibrils is causally associated with Alzheimer's disease and the neurodegenerative tauopathies [1]. Biophysical studies of the mechanism of tau fibrillization in vitro benefit from the usage of multiple spectroscopic probes since different probes illuminate different aspects of the aggregation reaction. The kinetics of tau fibrillization in vitro in the presence of polyanionic inducers such as heparin or fatty acid micelles, has been most extensively studied using probes such as thioflavin T (ThT) fluorescence and light scattering, and the aggregation mechanism has been described as ligand-induced nucleation dependent polymerization [2–4]. These probes, although robust and essential for studying tau

Caroline Smet-Nocca (ed.), *Tau Protein: Methods and Protocols*, Methods in Molecular Biology, vol. 1523,
DOI 10.1007/978-1-4939-6598-4_7,

fibrillization, report on the coupled processes of conformational conversion and fibril growth that occur when monomeric, disordered tau transforms into polymeric, β-sheet-enriched fibrils. Vibrational spectroscopic techniques such as Fourier transform infrared (FTIR) spectroscopy and ultraviolet resonance Raman (UVRR) spectroscopy have the potential to complement these standard techniques since they can discriminate finely between different classes of secondary structure (disordered, α-helix, β-sheet) [5, 6]. Similarly, the atomic force microscopy (AFM) assay is a powerful complementary tool for morphological characterization of fibrillar structures that form over the time course of tau fibrillization [4, 7, 8].

FTIR spectroscopy is especially suited for distinguishing β-sheet from disordered/α-helical structures on the basis of the amide I band position [6]. Absorbance in the amide I region (1700–1600 cm^{-1}) arises from the stretching vibration of the amide carbonyl, which is sensitive to the difference in hydrogen bonding patterns found in different secondary structures. A common limitation of FTIR spectroscopy is the considerable absorbance of the water bending mode at ~1650 cm^{-1} in the middle of the crucial amide I region. Although this can be overcome by the use of thin-film attenuated total reflectance (TF-ATR) mode [6] for spectral collection as described in this chapter, the requirement of highly concentrated samples (>1 mM) limits the use of FTIR spectroscopy to equilibrium measurements of tau fibrillization, since kinetic studies of tau fibrillization are generally performed using protein in the more physiologically relevant μM concentration range. Fortunately, however, UVRR spectroscopy can overcome these limitations associated with FTIR spectroscopy and can hence be used to interrogate the kinetics of tau fibrillization [9]. The use of UV radiation enables the selective resonance enhancement of amide (amide I, II, III and Cα-H bending) and aromatic (Trp, Tyr, and Phe) vibrational modes; band positions of the amide vibrational modes can be used to distinguish between different secondary structures (Table 1). Although UVRR spectroscopic studies of protein aggregation have predominantly used deep UV excitation wavelengths (197 nm) to resonance enhance amide vibrations, the absence of tryptophan residues in tau allows for good quality data to be obtained with 220 nm excitation, as described here.

Imaging of tau fibrils using AFM, a form of scanning probe microscopy, is performed in intermittent contact/tapping mode on native, dried samples that have neither been stained nor fixed [11]. The oscillation of the AFM cantilever at frequencies close to its fundamental resonant frequency allows for *x-y* scans to be carried out with greatly reduced tip–sample interaction forces and increased sensitivity, when compared with contact mode imaging. AFM is an especially powerful tool for the imaging of sparsely populated structures formed during the time course of tau fibrillization and for the mapping of structural polymorphism inherent in tau fibrils [4, 7–9].

Table 1
Amide bands associated with secondary structures and their assignments

Amide band	Band position (cm^{-1})	Secondary Structure	Mode assignment
Amide I (Am I)	1660–1682 1670–1675	Disordered β-sheet	C=O stretching [5, 12–18]
Amide II (Am II)	1548–1561 1550–1564	Disordered β-sheet	Out-of-phase combination of N–H in-plane bending (NH ib) and C–N stretching (CN s) [5, 12–14, 17, 19]
Amide III (Am III)	1240–1279 1220–1230	Disordered β-sheet	In-phase combination of NH ib and CN s [5, 12–14, 16–18, 20]
C_α–H bending (C_α–H/Amide S)	1380–1390 1395–1406	Disordered β-sheet	C_α–H bending coupled to N–H bending [5, 12–14, 17, 21, 22]

Table is reprinted from ref. [9] with permission of the American Chemical Society, Copyright (2014)

In this chapter, I provide detailed protocols for the use of FTIR, UVRR spectroscopy, and AFM for the study of tau fibrillization kinetics. Protocols detailed here were standardized and used for the study of fibrillization kinetics of the functional, microtubule binding, four repeat domain of tau (tau4RD) in the presence of heparin in near-physiological buffer conditions. The tau repeat domain serves as a useful proxy for in vitro studies since it is the primary constituent of the tau fibril core and fibrillization kinetics are robust and reproducible [10]. The addition of these probes to the repertoire of biophysical probes used in studies of tau fibril formation helps unveil the structural and mechanistic complexity inherent in the aggregation pathway.

2 Materials

2.1 Buffers, Solutions, and Experimental Conditions

1. Prepare all buffers and solutions fresh using ultrapure water from a Milli-Q system (18.2 MΩ-cm at 25 °C) and filter through 0.22 μm syringe filters (Millipore) before use. Use chemical reagents of the highest purity grade.
2. Aggregation conditions: 25 mM Tris buffer, 50 mM NaCl, 1 mM DTT, pH 7 with heparin as inducer, stationary, 37 °C.
3. Aliquots of tau protein at a concentration of ≤250 μM in 25 mM Tris buffer, pH 7.
4. 2 M NaCl stock solution.
5. 1 M DTT stock solution.
6. Heparin stock is made by dissolving high molecular weight heparin (12,000 Da, HiMedia Laboratories) in 25 mM Tris buffer, 50 mM NaCl, pH 7 to obtain 7.5 mg/ml (625 μM) solution. Aliquots are stored at 4 °C and warmed up to 37 °C, before inducing aggregation reactions (*see* **Note 1**).

2.2 FTIR Spectroscopy

1. FT-IR spectrometer equipped with liquid N2-cooled MCT detector (Thermo-Nicolet-6700, Thermo Scientific).
2. Smart Orbit ATR accessory equipped with diamond crystal (Thermo Scientific).

2.3 UVRR Spectroscopy

1. UVRR spectrometer: Spectra are recorded an excitation wavelength of 220 nm, produced as the output of the fourth harmonic of a Ti-S laser, pumped by a Nd-YLF laser (Photonics Industries, NY). Raman scattering is dispersed using a monochromator with 3600 grooves/mm grating and recorded by means of a 1024 × 256 pixels, back-illuminated CCD camera (Jobin Yvon).
2. Suprasil NMR tubes, 5 mm (Wilmad-LabGlass).
3. Magnetic beads, PTFE, 3 × 3 mm (Bel-Art Products).
4. Solvents for spectral calibration: cyclohexane, acetonitrile and isopropyl alcohol.
5. 2 mM L-phenyl-alanine and 1 mM *N*-acetyl-L-tyrosinamide.
6. 0.5 M sodium perchlorate.
7. SynerJY™ software (Jobin Yvon).

2.4 AFM

1. Muscovite (mica) squares: 0.15 mm thickness, Grade V1 (SPI supplies).
2. Scotch tape.
3. Precision wipes (Kimtech Science).
4. AFM cantilevers (Nanoworld AG): Arrow™ Silicon AFM probes with Aluminum reflex coating on cantilever backside, spring constant of 2.8 N/m, 75 kHz resonance frequency, tip height 10–15 μm, rounding tip radius <10 nm.
5. Dumont #5 forceps (Fine Science Tools).
6. Molecular Imaging PicoPlus AFM equipped with small (maximum *x*-*y* scan size of ~10 μm) and large (maximum *x*-*y* scan size of ~100 μm) piezo scanners (Agilent Technologies).
7. WSxM software (WSxM solutions) [11].

3 Methods

3.1 Study of Tau Fibrils Using FTIR Spectroscopy

3.1.1 Sample Preparation

1. Thaw required aliquots of tau repeat domain protein (*see* **Note 2**) on ice for 1.5 h, followed by complete thawing at room temperature.
2. Spin protein aliquots at 20,000 g for 10 min and then dilute to the desired protein concentration in aggregation buffer (*see* **Note 3**).

3. The desired salt concentration (50 mM NaCl) is obtained by the addition of the appropriate volume of solution from a 2 M NaCl stock solution. The desired DTT concentration (1 mM) is obtained by addition of the appropriate volume of solution from a freshly thawed 1 M DTT frozen stock solution.
4. Incubate the protein for 2 h at 37 °C for the reduction of covalent, intramolecular tau disulfide bond.
5. Initiate fibril formation by the addition of heparin solution to obtain the desired inducer concentration. Withdraw aliquots for measurement using different spectroscopic probes at defined times (*see* **Note 4**).
6. At the designated time, mix the aggregation reaction thrice using a P200 pipette, withdraw 0.5 ml of the fibril sample and spin at 25,000 g for 10 min at 4 °C to pellet fibrils.
7. Resuspend the pellet in 20 μl of supernatant (*see* **Note 5**) and use 3 μl of suspension to prepare a thin hydrated film.

3.1.2 Data Acquisition and Analysis

1. FTIR spectrometers come with different default parameters based on the manufacturer. For the Thermo-Nicolet-6700, before the start of the experiment, ensure that the MCT detector is fully cooled in liquid N_2 (after fresh liquid N_2 is added to the dewar, the cooling may take 0.5–1 h) and that the instrument is purged with ultrapure N_2 gas for at least 1 h to remove residual traces of atmospheric CO_2 and water vapor. The decrease in the CO_2 IR band at ~2300 cm^{-1} and the water IR band at ~1600 cm^{-1} as a function of purging time, can be viewed in default transmission mode to monitor the efficiency of purging, but this will not be apparent in ATR mode.
2. Set the following acquisition parameters: wavenumber range: 2000–1000 cm^{-1}; resolution: 4 cm^{-1}; interferograms averaged: 1024; Smart Orbit ATR mode. Leave the ultrapure N_2 gas purge on during data acquisition.
3. Acquire a background spectrum using the selected acquisition parameters (when there is no sample on the diamond crystal). This background spectrum can be stored and applied to relevant sample spectra. A sample spectrum is recorded by the instrument as a ratio of the sample to the background spectrum, in order to correct for background.
4. Make a thin film of the aggregation buffer by applying 3 μl of sample onto the diamond crystal and drying it gently under a stream of ultrapure N_2 gas (*see* **Note 6**). Acquire buffer spectrum. If the hydrated film was dried evenly and fully, the spectrum will be mostly flat, with no significant signal in the region of interest (1700–1500 cm^{-1}). If signal is visible in this region, clean the diamond crystal using MilliQ water and

acetone and make a thin film once again to obtain a clean buffer spectrum.

5. Once a clean buffer spectrum has been obtained, acquire the buffer spectrum again for use as a 'background' spectrum. Clean the diamond crystal using MilliQ water and acetone.
6. Make a thin film of the concentrated fibril sample by gently drying 3 μl of sample onto the diamond crystal. Acquire a fibril 'sample' spectrum. Tau fibrils display a characteristic amide I band at 1626 cm^{-1} with a shoulder at 1655 cm^{-1} (Fig. 1).
7. Switch of the ultrapure N_2 gas purge and allow atmospheric water vapor to flood the sample compartment for 15–20 min. Acquire a water vapor spectrum (for background spectrum, use the spectrum acquired of diamond crystal alone). If required, this can be used during data analysis to remove water vapor spikes that appear prominently in the amide I band region.
8. Data analysis: Basic data analysis includes baseline correction and water vapor subtraction (*see* **Note 7**). Advanced methods such as deconvolution may be applied and have been described elsewhere [6]. However, for studies of tau fibrillization, FTIR spectroscopy is used more for equilibrium rather than kinetic measurements and hence, we have not carried out quantitative deconvolution of the spectra.

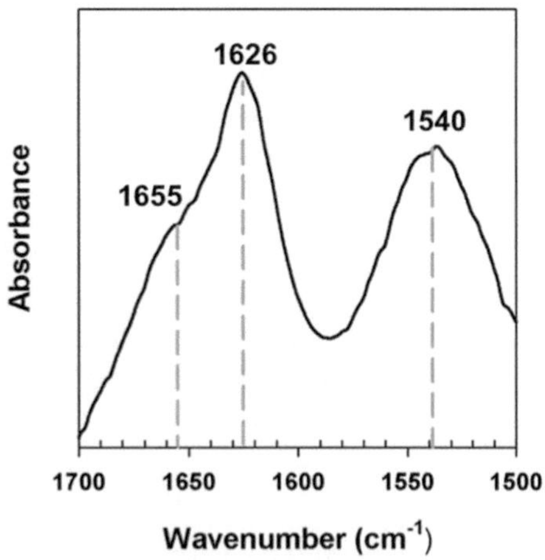

Fig. 1 FTIR spectra. FTIR spectra were acquired at the end-point of fibril formation (24 h) of 100 μM tau4RD in the presence of 37.5 μM heparin, at pH 7 and at 37 °C, using the protocol described in the text. The *dashed lines* through the spectra indicate important bands. The 1626 cm^{-1} band position corresponds to parallel β-sheet (characteristic of tau fibril core) and the 1655 cm^{-1} band position to random coil/α-helical configuration. Figure is reprinted from ref. [9] with permission of the American Chemical Society

3.2 Study of Tau Fibrillization Using UVRR Spectroscopy

3.2.1 Sample Preparation

1. Prepare tau fibrils as described in Subheading 3.1.1, **step 5**.
2. At designated time-points, mix the aggregating protein mixture thrice using a P200 pipette, withdraw a 400 μl aliquot of the sample and add to a Suprasil NMR tube containing a PTFE magnetic bead (*see* **Note 8**). Record spectra immediately.
3. Monomer spectra are recorded using 400 μl aliquots per acquisition without magnetic bead in tube and without heparin since the addition of heparin instantaneously initiates aggregation (*see* **Note 9**).

3.2.2 Data Acquisition

1. Optimize alignment of lasers and other components of the Raman spectrometer, at least a day in advance, in order to maximize signal output from a tau monomer sample.
2. Check laser alignment once again before the start of the experiment. Ensure that the laser power at the sample is ≤1 mW. Record all sample spectra using identical acquisition settings: slit width, 300 μm; slit height, 2 mm; total exposure time, 10 min (20 s, 1 accumulation, 30 cycles), from tubes spun about their axes to minimize photo-damage. For all spectra, check for sample degradation in the presence of laser light by comparing the first and last cycles and confirming that there is no significant difference (*see* **Note 10**).
3. Record spectra of solvents—cyclohexane, acetonitrile and isopropyl alcohol, for spectral calibration.
4. Record spectra of water, aggregation buffer (with and without heparin) and at least five different empty Suprasil tubes.
5. Record spectra of L-phenylalanine and *N*-acetyl-L-tyrosinamide.
6. Record a spectrum of the external standard, sodium perchlorate.
7. Record a spectrum of tau4RD monomer (100 μM). Repeat the acquisition with three additional, fresh samples of monomer (*see* **Note 11**).
8. Record a spectrum of the external standard, sodium perchlorate once again. The spectrum of the external standard recorded before and after every set of four sample spectra, is used to monitor laser stability. For data shown in Fig. 2, recorded over a ~20 h period, the change in intensity of the 933 cm^{-1} band of sodium perchlorate was found to be <4%, when averaged over the entire time period (*see* **Note 12**).
9. Record spectra of tau4RD fibril samples at designated time points. Make sure to record at least four spectra per sample time point, in addition to the spectrum of external standard before and after each sample set (*see* **Note 13**).
10. Record spectra of calibration solvents following the last test sample acquisition.

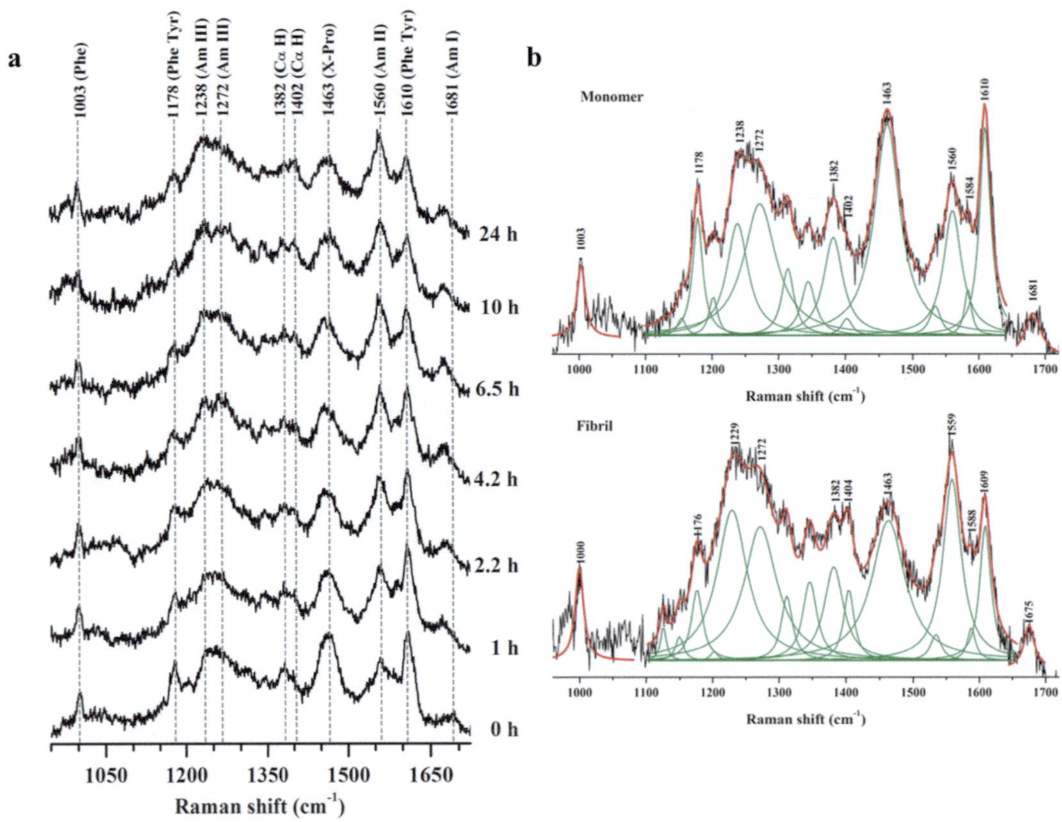

Fig. 2 UVRR spectra. (**a**) UVRR spectra acquired with Raman excitation at 220 nm at different times of aggregation of 100 μM tau4RD in the presence of 37.5 μM heparin, at pH 7 and at 37 °C, using the protocol described in the text. The *dashed lines* through the spectra indicate important bands. The band position labels correspond to the 0 h data. (**b**) UVRR spectra of monomer (0 h) and fibril (24 h), were deconvoluted by fitting to a sum of Lorentzian functions. Lorentzian decompositions are shown as *green lines*, and the sum of Lorentzian bands is shown as the *red solid overlay*. Note changes in band positions and band intensities (area under each band) that occur as monomeric tau4RD is converted into fibrils. Figure is reprinted from ref. [9] with permission of the American Chemical Society

3.2.3 Data Analysis

1. All data processing is carried out using SynerJY™. The first step involves the manual removal of cosmic rays (anomalous sharp spikes that appear in the spectra and are unrelated to the sample). This is aided by the availability of four spectra per time point; by comparing the different spectra, remove cosmic rays while ensuring that spectral shape and signal are not altered during this process.
2. Spectral calibration: fit Lorentzian functions to the distinct bands seen in the solvent spectra of cyclohexane, acetonitrile and isopropyl alcohol. Use the known band positions for each of these solvent bands to generate a calibration curve and apply this to all other sample spectra.

3. Subtract tube and buffer spectra from all samples (*see* **Note 14**). At the end of this subtraction, the baseline for all spectra should appear largely flat instead of downward sloping over the region of interest (between 950 and 1750 cm−1) (Fig. 2).
4. Generate an averaged spectrum from the four spectra acquired for each of the aggregation time points under investigation (Fig. 2a).
5. Fitting aggregation spectra: band positions and intensities are calculated by fitting Lorentzian functions to the bands of the averaged spectra in an iterative deconvolution process. As a first step, examine the entire data set to determine the minimum number of bands that can be adequately resolved throughout (*see* **Note 15**). Perform a free (unconstrained) fit over the entire data set with this identified number of bands. Each band will be characterized by its position, width and area. Save these values for future reference.
6. The spectra of L-phenylalanine (L-Phe) and *N*-acetyl-L-tyrosinamide (NAYA) are fitted independently to determine positions of Phe and Tyr bands, respectively. The 1003 cm^{-1} Phe band is especially useful as a guide to width expansion that occurs when Phe is present in the amide backbone of tau4RD as opposed to being free in solution (in L-Phe). The width of the 1003 cm^{-1} band in tau4RD monomer vs. L-Phe is thus compared and the extent of increase in the width is used to define the allowed window of width expansion for other aromatic bands. Perform a second round of fitting where the widths of aromatic bands alone are constrained such that they conform to the window described above and are within ±2 cm^{-1} of each other, when examined across time points. In the example shown in Fig. 2b, these constraints did not alter fit quality.
7. Width constraint for each amide band is identified from the spectrum that best resolves that particular amide band, across all time points. Perform a third round of fitting for all of the spectra with the width of each amide band constrained to the identified value such that the widths, across time points, are within ±2 cm^{-1} of each other. In the example shown in Fig. 2b, these constraints did not alter fit quality and this is important to confirm.
8. Once the fitting exercise is complete, band positions and area under each band (equivalent to intensity) can be examined as a function of aggregation time to determine the structural changes that occur during tau fibrillization [9]. Band intensities are normalized to the mean intensity of the 933 cm^{-1} band of sodium perchlorate spectra, acquired before and after each time point of aggregation, to account for changes in laser stability. Band positions are assigned to various amide and aromatic modes based on comparison with literature [9].

3.3 Study of Tau Fibrils Using AFM

3.3.1 Sample Preparation

1. Prepare tau fibrils as described in Subheading 3.1.1, **step 5**. Freshly cleave an upper layer off the top of a mica square using scotch tape (*see* **Note 16**). If the purpose is to image time-points of tau fibrillization, this process should be carried out such that a freshly cleaved mica square is available ~30 s before the time-point of interest.
2. At the designated time point, mix the aggregating protein mixture using a P200 pipette, withdraw a 100 μl aliquot and apply onto the freshly cleaved mica square. Incubate for 2 min.
3. Gently grasp a corner of the mica square using forceps and wash excess fibril sample off the surface by using ~25 drops of freshly filtered MilliQ water. Wick the last of the water off the mica square by gently touching an edge to a Precision wipe.
4. Tape the mica squares to a plastic petri dish using scotch tape (such that the sample side faces upwards) and make sure that the samples are labeled correctly. Dry the samples using a desiccator, under vacuum for 1 h. The samples can be imaged immediately or left overnight in the desiccator.

3.3.2 Data Acquisition and Analysis

1. Set up the atomic force microscope using standard operating procedures: make sure that the laser spot is aligned on the tip of the cantilever and at the origin position of the quadrant detector. Ensure that the relevant calibration file has been selected for the piezo scanner under use (*see* **Note 17**). Imaging will be carried out in intermittent contact (acoustic AC), constant amplitude mode.
2. Perform an amplitude-frequency sweep between 50 and 100 kHz in order to detect the fundamental resonance peak of the cantilever, expected to be close to 75 kHz. Adjust the drive such that the peak amplitude of cantilever oscillation is ~5 V. Demarcate the operating frequency, offset slightly towards the left of the resonance peak (0.1–0.2 kHz offset is standard), such that the value obtained is ~95% of the peak amplitude of cantilever oscillation.
3. Tape the mica square using small pieces of scotch tape onto the sample plate. Place the sample plate in the instrument and begin approach, using default approach speeds. Set the approach to stop at an amplitude set point that is equivalent to ~90% of the free amplitude. Before starting approach, ensure that other scan parameters are also at their default values: servo range at maximum value, I and P gain at default values, *x-y* scan size set to 10% of maximum scan size and *x-y* scan offset set to 0.
4. Start scanning the sample with the following parameters to obtain an image overview: square scan size of 9 μm (small scanner) or 90 μm (large scanner); pixel intensity of 256 data points/line and scan rate of 2–4 lines/s; other scan parameters can remain at default values at this stage.

5. Once an interesting feature has been identified, use the zoom tool to move to the region of interest and acquire a low-resolution image (256 data points/line, 2–4 lines/s). This is recommended since scanner hysteresis may result in some inaccuracies in zooming to the area of interest (sequential zooming is helpful in this regard). Once the *x*-*y* scan size has been established, parameters that can be manually optimized for each sample to obtain the highest resolution image include the amplitude set point, I and P gain, servo range and scan settings. Images shown in Fig. 3 were acquired with 1024 data points/line and 1–2 Hz. scan speed.

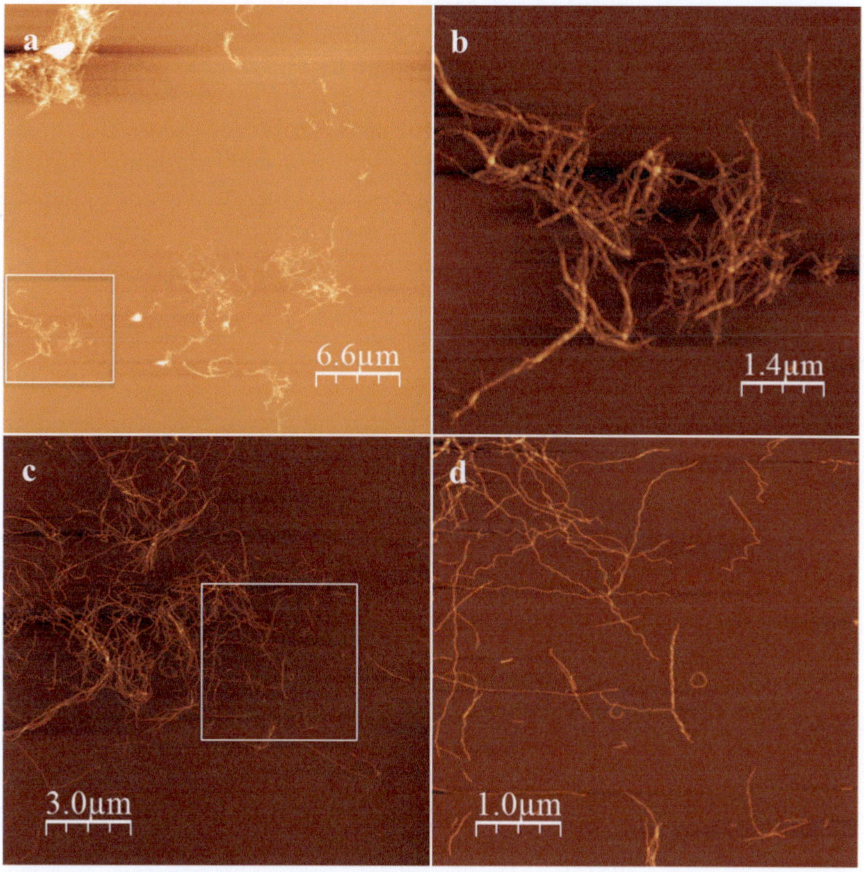

Fig. 3 AFM images acquired in tapping mode using a large piezo scanner and the protocol described in the text, shown in topography format after second order flattening. (**a**) AFM image acquired at end-point (24 h) of aggregation of 100 μM tau4RD (residues Q244-F378) in the presence of 37.5 μM heparin, at pH 7 and at 37 °C. Image was acquired at a square scan size of 33 μm, with 1024 data points/line and scan rate of 1.21 Hz. The *white square* indicates the zoom area for the image shown in (**b**). After zooming, the image was acquired at a square scan size of 7 μm, with 1024 data points/line and scan rate of 1.21 Hz. (**c**) AFM image acquired at end-point (4 h) of aggregation of 25 μM tauK18 (residues Q244-E372) in the presence of 37.5 μM heparin, at pH 7 and at 37 °C. Image was acquired at a square scan size of 15 μm, with 1024 data points/line and scan rate of 1.51 Hz. The *white square* indicates the zoom area for the image shown in (**d**). After zooming, the image was acquired at a square scan size of 5 μm, with 1024 data points/line and scan rate of 2 Hz. Note the increase in resolution obtained on zooming and the different fibril morphologies visible. Panel (**b**) is reprinted from [9] with permission of the American Chemical Society. Panel (**d**) is reprinted from ref. [8] with permission of Elsevier

6. Every scan will produce topography, amplitude and phase channel images. Ensure that both trace and retrace scans are recorded for the first two channels since this will alert one to the presence of imaging artifacts (*see* **Note 18**).
7. Data analysis: image analysis can be carried out using the freeware WSxM [11]. All images are flattened (first or second order flattening should suffice). Heights (diameters) of fibrils are measured in the direction perpendicular to the structures using the profile tool. Tau fibrils formed in certain aggregation conditions have a tendency to form clumps; in such circumstances, height measurements are necessarily carried out on individual fibrils at the edges of clumps [4]. Periodicities are measured using the profile tool along the length of the fibril. While making quantitative height measurements across different conditions, it is important to acquire images at similar scan sizes and if possible, using the same cantilever and on the same day (provided the cantilever performs with the same efficiency for the complete imaging session). Height histograms generated from a replicate imaging experiment can be compared to gauge the extent of reproducibility in fibril height measurements.

4 Notes

1. Different lots of heparin from the same supplier have been checked and found to have no effects on reproducibility of aggregation kinetics. Varying heparin concentration alters tau aggregation kinetics [4]. Protocols described here correspond to an aggregation condition that uses 37.5 μM heparin for inducing tau aggregation.
2. Tau repeat domain protein is purified as described elsewhere [4] using cation exchange chromatography and size-exclusion chromatography, and stored as aliquots of defined concentration (≤250 μM) in 25 mM Tris buffer, pH 7 at −80 °C. Boiling of lysate is not employed during protein purification because this results in protein with anomalous absorption at ~320 nm although the same protein preparation appears pure by SDS-PAGE analysis, and monomeric by size-exclusion chromatography and DLS measurements.
3. The protocol for FTIR spectroscopy of end-point tau fibrils concentrates fibril samples and hence can be implemented at lower protein concentrations as well. The protocol for UVRR spectroscopic measurements described here was standardized for an aggregation condition that used 100 μM tau4RD and 37.5 μM heparin [9]. AFM measurements use undiluted samples and hence the same protocol can be used over a full range of protein concentrations (10–200 μM).

4. Reproducible and consistent mixing of the aggregation reaction is critical for reproducibility of aggregation kinetics. In addition, reaction volumes are always 1 ml in a 1.5 ml tube or 1.5 ml in a 2 ml tube. While withdrawing aliquots for measurement using different spectroscopic probes, the last ~300 μl in the tube is not used.
5. TF-ATR FTIR measurements of tau fibril samples require around 2.5 mM of concentrated fibril sample for good quality spectra.
6. Making even thin films requires practice. If helpful, a P200 micropipette tip can be used to spread the sample evenly on the diamond crystal while drying under a stream of ultrapure N_2 gas. The angle and the pressure of the gas stream should be empirically adjusted for even and gentle drying.
7. Water vapor subtraction: a certain fraction of the water vapor spectrum is subtracted from the sample spectrum such that (a) random spikes in the amide I region disappear but (b) shape and intensity of the sample spectrum remains unaltered. This requires careful observation of spectral features prior to subtraction.
8. The PTFE magnetic bead, while required to keep aggregates in suspension has been observed to fragment fibrils. Important controls to be performed include measurement of ThT fluorescence and scattering intensity of fibril samples before and after UVRR spectral acquisition to ensure that the effects of fibril fragmentation are insignificant (signal change ≤15 %). Sample volumes and acquisition conditions detailed here have been optimized to this end.
9. The magnetic bead is not added during acquisition of monomer spectra to ensure that samples remain monomeric throughout. Tau monomer has been observed to aggregate following agitation, even in the absence of inducer. Since samples are spun in Suprasil tubes during data acquisition, an important control to be performed is acquiring fluorescence scans of tau monomer before and after UVRR data acquisition to ensure that the acquisition process does not alter the monomer.
10. These acquisition conditions have been optimized to ensure that there is no sample degradation but this should always be verified for every spectrum acquired.
11. The average of four spectra provided data of sufficiently good quality for quantitative analyses.
12. The best strategy is the use of a salt such as sodium perchlorate or sodium nitrate as an internal standard during data acquisition. However, since tau aggregation is exquisitely sensitive to

the presence of salt, this cannot be done unless aggregation conditions have been optimized ahead of time to include an internal standard. The next best strategy in the absence of an internal standard is to use an external standard as done here.

13. Since every fibrillization time point requires four fresh samples for data acquisition and spectra take 10 min to record, remember to stagger aggregation reactions at the time of induction by heparin by 15 min at least.
14. Tube subtraction is done empirically by subtracting a fraction of the tube spectrum from the sample spectrum, such that the features contributed by the tube are removed without the introduction of new features. It is often useful to pick the tube spectrum whose slope parallels the slope seen in the sample spectrum. Buffer subtraction removes the contribution of the weak and broad, water Raman band at ~1650 cm^{-1} and any other contributions from buffer components.
15. The entire data set is examined carefully in this global fitting routine, prior to the start of fitting exercises.This is required since some bands are more sharply defined at some time points than at others. This helps one identify the minimum number of bands that can be consistently resolved.
16. This requires some practice. While cleaving, smooth down the scotch tape on the mica square and then peel the topmost mica layer off, by pulling the tape upwards in one fluid motion. One way to check the efficiency of the cleaving is to examine the mica layer on the scotch tape—if it appears smooth, that is a good cleaving.
17. Images can be acquired using a large piezo scanner (maximum scan size ~100 μm) or a small piezo scanner (maximum scan size ~10 μm). The former is useful when imaging early time points of fibrillization since shorter fibrils or sparsely populated fibril clumps can be detected more easily. The latter is useful when the objective is to acquire higher resolution images at small scan sizes ($\leq 1 \times 1$ μm^2).
18. Common scanner artifacts include hysteresis and creep. *x*-*y* hysteresis can be identified by looking to see if there's any displacement of structures across the trace and retrace channels. Creep results in a drift in structures being imaged after zooming to a new scan location, for the initial few lines. Once the scanner relaxes to its equilibrium position, this problem will disappear and at this stage, a fresh scan can be acquired. Tip imaging is another artifact that can occur which can be identified by rotating the sample and imaging again. If the image orientation does not change, tip convolution effects are dominating the image and one should then choose a different cantilever.

Acknowledgments

The protocols described here were standardized during graduate studies in the laboratory of Jayant B. Udgaonkar at the National Centre for Biological Sciences, Tata Institute of Fundamental Research (NCBS-TIFR), India. FTIR spectra and AFM images were acquired at the core facilities of NCBS-TIFR; UVRR spectra were collected in the laboratory of Mrinalini Puranik (then at NCBS-TIFR; now at IISER Pune, India). I'd like to acknowledge colleagues in the Puranik laboratory for teaching me the nuances of UVRR spectroscopy.

References

1. Ballatore C, Lee VM, Trojanowski JQ (2007) Tau-mediated neurodegeneration in Alzheimer's disease and related disorders. Nat Rev Neurosci 8:663–672
2. Friedhoff P, von Bergen M, Mandelkow EM et al (1998) A nucleated assembly mechanism of Alzheimer paired helical filaments. Proc Natl Acad Sci U S A 95:15712–15717
3. King ME, Ahuja V, Binder LI et al (1999) Ligand-dependent tau filament formation: implications for Alzheimer's disease progression. Biochemistry 38:14851–14859
4. Ramachandran G, Udgaonkar JB (2011) Understanding the kinetic roles of the inducer heparin and of rod-like protofibrils during amyloid fibril formation by Tau protein. J Biol Chem 286:38948–38959
5. Chi Z, Chen XG, Holtz JS et al (1998) UV resonance Raman-selective amide vibrational enhancement: quantitative methodology for determining protein secondary structure. Biochemistry 37:2854–2864
6. Seshadri S, Khurana R, Fink AL (1999) Fourier transform infrared spectroscopy in analysis of protein deposits. Methods Enzymol 309: 559–576
7. Wegmann S, Jung YJ, Chinnathambi S et al (2010) Human Tau isoforms assemble into ribbon-like fibrils that display polymorphic structure and stability. J Biol Chem 285:27302–27313
8. Ramachandran G, Udgaonkar JB (2012) Evidence for the existence of a secondary pathway for fibril growth during the aggregation of tau. J Mol Biol 421:296–314
9. Ramachandran G, Milan-Garces EA, Udgaonkar JB et al (2014) Resonance Raman spectroscopic measurements delineate the structural changes that occur during tau fibril formation. Biochemistry 53:6550–6565
10. Ramachandran G, Udgaonkar JB (2013) Mechanistic studies unravel the complexity inherent in tau aggregation leading to Alzheimer's disease and the tauopathies. Biochemistry 52:4107–4126
11. Horcas I, Fernandez R, Gomez-Rodriguez JM et al (2007) WSXM: a software for scanning probe microscopy and a tool for nanotechnology. Rev Sci Instrum 78:013705
12. JiJi RD, Balakrishnan G, Hu Y et al (2006) Intermediacy of poly(L-proline) II and beta-strand conformations in poly(L-lysine) beta-sheet formation probed by temperature-jump/UV resonance Raman spectroscopy. Biochemistry 45:34–41
13. Huang CY, Balakrishnan G, Spiro TG (2006) Protein secondary structure from deep-UV resonance Raman spectroscopy. J Raman Spectrosc 37:277–282
14. Roach CA, Simpson JV, JiJi RD (2012) Evolution of quantitative methods in protein secondary structure determination via deep-ultraviolet resonance Raman spectroscopy. Analyst 137:555–562
15. Maiti NC, Apetri MM, Zagorski MG et al (2004) Raman spectroscopic characterization of secondary structure in natively unfolded proteins: alpha-synuclein. J Am Chem Soc 126:2399–2408
16. Shashilov VA, Lednev IK (2008) 2D correlation deep UV resonance raman spectroscopy of early events of lysozyme fibrillation: kinetic mechanism and potential interpretation pitfalls. J Am Chem Soc 130:309–317
17. Xu M, Shashilov V, Lednev IK (2007) Probing the cross-beta core structure of amyloid fibrils by hydrogen-deuterium exchange deep ultraviolet resonance Raman spectroscopy. J Am Chem Soc 129:11002–11003

18. Popova LA, Kodali R, Wetzel R et al (2010) Structural variations in the cross-beta core of amyloid beta fibrils revealed by deep UV resonance Raman spectroscopy. J Am Chem Soc 132:6324–6328
19. Sikirzhytski V, Topilina NI, Higashiya S et al (2008) Genetic engineering combined with deep UV resonance Raman spectroscopy for structural characterization of amyloid-like fibrils. J Am Chem Soc 130:5852–5853
20. Mikhonin AV, Ahmed Z, Ianoul A et al (2004) Assignments and conformational dependencies of the amide III peptide backbone UV resonance Raman bands. J Phys Chem B 108:19020–19028
21. Wang Y, Purrello R, Jordan T et al (1991) Uvrr spectroscopy of the peptide-bond. 1. Amide-S, a nonhelical structure marker, is a C-alpha-H bending mode. J Am Chem Soc 113:6359–6368
22. Austin JC, Jordan T, Spiro TG (1993) Ultraviolet resonance Raman studies of proteins and related model compounds. Wiley & Sons Ltd., New York

Chapter 8

Assays for the Screening and Characterization of Tau Aggregation Inhibitors

Janet E. Rickard, David Horsley, Claude M. Wischik, and Charles R. Harrington

Abstract

Aggregation of tau protein is a pathological hallmark of Alzheimer's disease and other neurodegenerative tauopathies. Inhibition of tau aggregation could provide a method for the treatment of these disorders. Methods to identify tau aggregation inhibitors (TAIs) in vitro are useful and here we describe assays for TAIs using purified recombinant tau protein fragments in a cell-free immunoassay format and a stably transfected cell model to create a more physiological environment.

Key words Alzheimer's disease, ELISA, Immunoassay, Tau protein, Tau aggregation inhibitor

1 Introduction

Tau comprises a family of microtubule-associated proteins that are involved in the stabilization and assembly of the microtubular network in neurons [1]. Alzheimer's disease is a neurodegenerative disorder characterized by the presence of neurofibrillary tangles, which are made up of paired helical filaments (PHFs) composed predominantly of a fragment of tau protein truncated to nearly 95 amino acids in length [2, 3]. In Alzheimer's disease, there is a correlation between neurofibrillary pathology and the presence of PHF-core tau [4] and the extent of dementia, whether assessed clinically [5, 6] or by molecular imaging [7, 8]. Methylthioninium chloride (MTC), the first compound to be identified as a tau aggregation inhibitor (TAI) [9], has shown potential for the treatment of Alzheimer's disease [10]. Several other TAIs have since been identified [11] and their activity has been demonstrated in a number of cell-free [9, 11, 12] and cellular models [11, 13]. We have developed assays that allow screening of compounds to identify TAIs prior to their testing in animal models of tauopathy and in the clinic with Alzheimer's disease and other tauopathies [10, 13,

Caroline Smet-Nocca (ed.), *Tau Protein: Methods and Protocols*, Methods in Molecular Biology, vol. 1523,
DOI 10.1007/978-1-4939-6598-4_8, © Springer Science+Business Media New York 2017

14]. The assays are based upon interactions between truncated tau fragments that constitute PHFs. The PHF-core tau is a ca. 95-amino acid fragment that, for these assays, corresponds to tau297-391 (or tau295-391 for the cellular assay) (Fig. 1). Truncated tau serves as a substrate for the binding of tau that, after proteolytic processing, can act as a complex able to bind further tau in an autocatalytic manner [9].

In the first assay, purified, bacterially expressed fragments of tau protein comprising the core domain found in the PHF core are used in an ELISA [9]. This assay has been used to screen large numbers of compounds relatively quickly. Typically, one person could screen 450 compounds per week, assayed at a single concentration, using 96-well format assays and 10 compounds per day to obtain B_{50} values. This could be scaled up with the aid of automated dispensers and 384- or 1536-well format assays. The second assay is based on a cell culture model, in which TAIs can be tested in a more physiological environment [13], where one can simultaneously assess the cytotoxicity of compounds and their ability to enter cells. The cellular assay takes longer to perform than the first assay and thus has a lower screening capacity. Typically, 48 compounds per week can be tested using replicates at a single

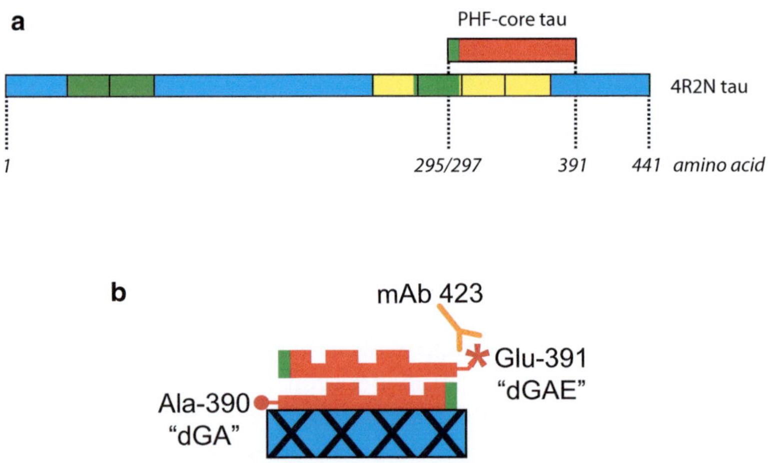

Fig. 1 (**a**) Schematic representation of the truncated PHF-core tau fragment in relation to the largest CNS tau isoform, generally referred to as 4RN2. This isoform contains an insert (*green*) within the tandem 4-repeat domain (*yellow*) and two inserts in the N-terminus (*green*) [1]. This full-length tau isoform is 441 amino acids in length (htau40), with the PHF-core tau corresponding to residues 295/297-391. (**b**) In the cell-free tau-tau binding assay, truncated tau terminating at Ala-390 (dGA) is adsorbed to the solid phase and allowed to bind tau terminating at Glu-391 (dGAE). Binding of the second protein is detected using mAb 423, an antibody whose reactivity is dependent on C-terminal truncation of tau at Glu-391; it does not recognize the dGA protein bound in the solid-phase. TAIs inhibit the tau-tau binding interaction in this assay

concentration, with a quarter of the compounds showing activity being followed up in greater detail the following week to obtain EC_{50} values.

2 Materials

2.1 Cell-Free Immunoassay

1. Poly(vinyl chloride) (PVC), 96-well round bottom plates (Corning) are used for the assay. A variety of alternative plates can be used, such as polycarbonate, polypropylene or polystyrene, or plates with proprietary enhanced binding properties. It is important to note that the solid phase can be critical and the effect of different surfaces should be evaluated for specific proteins.
2. Carbonate buffer: carbonate/bicarbonate (50 mM, pH 9.6). For 1 L of buffer, add 1.59 g Na_2CO_3 and 2.93 g $NaHCO_3$ to deionised water; the pH does not need to be adjusted. Store at 4 °C.
3. PBS: 10 mM phosphate, 0.8 % NaCl, 0.02 % KCl (pH 7.4). For 1 L of buffer, add 8.0 g NaCl, 0.2 g KCl, 0.2 g KH_2PO_4, and 1.46 g $Na_2HPO_4.2H_2O$ to deionised water; the pH does not need to be adjusted. Store at room temperature.
4. Blocking buffer: 2 % (w/v) dried milk powder in PBS (*see* **Note 1**). This solution is made fresh on the day of use.
5. PBST: PBS containing 0.05 % Tween® 20. Store at room temperature.
6. Plate washing buffer: 0.05 % Tween® 20 in deionised water. Store at room temperature.
7. Binding buffer: 50 mM MES, 25 mM NaCl, 10 mM PIPES, 0.05 % Tween® 20, 1 % (w/v) fish skin gelatine (pH 6.0). For 1 L of twofold concentrated binding buffer, add 19.52 g MES, 2.92 g NaCl, and 6.05 g PIPES to ~800 mL deionised water and adjust the pH to 6.0 (*see* **Note 2**). After the pH has been adjusted, add 1 mL of Tween® 20 and make the volume to 1 L. Store this solution at 4 °C. On the day of use, add fish skin gelatine to the required volume at a final concentration of 2 %.
8. Substrate solution: 0.1 mg/mL 3,3,5′,5′-tetramethylbenzidine (TMB; Sigma Aldrich) in 50 mM sodium acetate (pH 5.0) containing 0.003 % H_2O_2. A stock 0.5 M sodium acetate solution can be prepared in advance and stored at room temperature. For 1 L, add 41.01 g sodium acetate to ~900 mL deionised water, adjust the pH to 5.0 with glacial acetic acid, and make the volume to 1 L. Prepare the final substrate solution immediately before use. For 50 mL of substrate solution, add 5 mL of 0.5 M sodium acetate to 45 mL water. Make a 10 mg/mL solution of TMB in DMSO (*see* **Note 3**) and

slowly add 0.5 mL of this solution to the 50 mM sodium acetate, while gently swirling the solution. Finally, add 5 μL of 30 % H_2O_2. Inadequate swirling of the solution can result in the appearance of white precipitate. Caution should be taken to avoid contamination of the solution; traces of peroxidase activity will rapidly convert the solution to blue. *Drosophila* and other flies can be very effective at converting the color of the substrate solution.

9. Tau proteins: recombinant tau fragments dGA (tau297-390) and dGAE (tau297-391) [2] are purified from bacteria using standard tau purification protocols [15] (*see* **Note 4**).
10. Test compounds (TAIs): compounds should be dissolved at twice the final assay concentration in water. Compounds prepared as concentrated stock solutions in DMSO should be diluted to the working concentration in water just prior to addition to the assay.
11. Primary detection antibody: monoclonal antibody mAb 423 (hybridoma supernatant) [2].
12. Secondary detection antibody: goat anti-mouse IgG-HRP conjugate (BioRad).
13. Microplate spectrophotometer such as SpectraMax190 spectrophotometer (Molecular Devices).

2.2 Cell-Based Assay

1. Cell line: mouse 3T6 fibroblast cells with stable, inducible expression of full-length 4R2N tau under the control of the *lac* repressor protein and stable, constitutive expression of tau295-391 [13] (*see* Fig. 1a).
2. Culture medium: Dulbecco's modified Eagle's medium (DMEM) with Glutamax I and pyruvate (Gibco, Life Technologies), supplemented with glucose (4.5 g/L), fetal calf serum (10 %), penicillin (50 Units/mL), and streptomycin (50 μg/mL).
3. Tissue culture: 10-cm diameter Petri tissue culture dishes and 24-well tissue culture plates (Nunc). Incubator set at 37 °C with a 5 % CO_2 atmosphere.
4. Test compounds (TAIs): prepare TAIs either as a 1 mM stock solution in water or at up to 100 mM in DMSO, to be diluted in PBS to a suitable concentration for adding to cells immediately prior to use. TAIs that are water-insoluble can be prepared as dilutions in 10 % DMSO or in hydroxypropyl β-cyclodextrin as an aid to solubility. If used, TAIs are added at a concentration of 1 mM to a 40 % (w/v) aqueous solution of cyclodextrin.
5. Isopropyl thio-β-d-galactoside (IPTG) stock solution: 1 M solution in water stored frozen in aliquots and diluted to a final concentration of 5 mM in PBS immediately prior to use.

6. Polyacrylamide gel electrophoresis (PAGE) and immunoblotting apparatus.
7. Laemmli sodium dodecyl sulfate (SDS)-sample buffer.
8. SDS-PAGE and immunoblotting buffers: 15 % acrylamide gels, Tris/glycine buffer system; transferred, by blotting, in 3-(cyclohexylamino)-1-propane sulfonic acid (CAPS; 10 mM, pH 11; Sigma) to polyvinylidene difluoride (PVDF) membrane (BioRad) [16].
9. Blocking and wash buffers: blocking buffer is 5 % dried milk in PBS containing 0.1 % Tween® 20 and wash buffer is PBS containing 0.1 % Tween® 20.
10. Primary antibody: monoclonal antibody 7/51 [17] diluted 1:5 in blocking buffer.
11. Secondary antibody: HRP-coupled goat anti-mouse antibody (from BioRad used at a 1:2000 dilution in blocking buffer).
12. Enhanced chemiluminescence detection system: Clarity™ ECL reagent (BioRad).
13. Charge-coupled device (CCD) camera imaging system (e.g., BioSpectrum 810 Image System, UVP).

2.3 Cytotoxicity Measurement

Cellular toxicity is determined using the Cytotox 96-well kit (Promega). The assay measures lactate dehydrogenase (LDH), a stable cytosolic enzyme released upon cell lysis. LDH released from live cells is measured with a coupled enzymatic assay that results in the conversion of a tetrazolium salt into a red formazan product. The amount of color formed is proportional to the number of remaining cells.

3 Methods

3.1 Cell-Free Tau-Tau Binding Immunoassay

1. Dilute dGA to a concentration of 1 μM in carbonate buffer (*see* **Note 5**). Add 50 μL to wells in 96-well plates. Samples are included in triplicate for each assay condition; and carbonate buffer alone is included for measurement of background non-specific binding.
2. Incubate the plates for 1 h at 37 °C.
3. Rinse the plates with plate-washing buffer using an automated plate washer (*see* **Note 6**).
4. Add blocking buffer (200 μL) to each well.
5. Incubate the plates for 1 h at 37 °C.
6. While the plates are in blocking buffer, prepare dilutions of TAIs in the presence of dGAE. Dilute dGAE into binding buffer at two times the final required concentration (*see* **Note 7**). Aliquot 100 μL into microcentrifuge tubes and add water with

or without TAI to the required final concentration and in a final volume of 200 μL. Wells are included without TAI to serve as a control for 100 % binding.

7. Wash the plates (*see* **Note 8**).
8. Add TAI samples (50 μL), in triplicate, to the microplate wells.
9. Incubate the plates for 1 h at 37 °C.
10. Wash the plates.
11. Dilute the primary antibody, mAb 423, with blocking buffer (*see* **Note 9**) and add 50 μL to each well.
12. Incubate the plates for 1 h at 37 °C.
13. Wash the plates (*see* **Note 8**).
14. Dilute the secondary antibody, HRP-conjugated anti-mouse IgG, 1:1000 with PBST and add 50 μL to each well.
15. Incubate the plates for 1 h at 37 °C.
16. Wash the plates. After washing the plates, manually rinse them with deionised water to remove residual Tween® 20, tap plates to remove residual water, and dry the upper and lower surfaces of the plates with tissue.
17. Prepare TMB substrate solution. Add substrate (50 μL per well) to plates one at a time and measure rate of change in absorbance in a spectrophotometer, with shaking, at 650 nm over 2 min ([18]; *see* **Note 10**).
18. For calculation of inhibition curves, data are corrected for background by subtraction of the binding value measured in the absence of solid-phase protein (dGA). These values are then normalized to the binding value measured at the lowest concentration of TAI. Values are plotted as a function of TAI concentration. The TAI concentration at which the binding value is 0.5 is defined as the B_{50} value (Fig. 2).

3.2 Cell-Based Tau Aggregation Assay

1. Culture cells to 80 % confluency in 10-cm dishes in medium containing appropriate antibiotics (*see* **Note 11**).
2. Split the cultured cells into two 24-well plates (2 plates from one 10-cm dish) and culture for 24 h. This needs to be optimized so that the cells approach confluence by the end of the experiment.
3. Prepare TAIs as described in Subheading 2.2, **item 4**. Add TAIs, in triplicate, to cells and incubate for 24 h. Add IPTG (50 μM) at the end of day 2.
4. Incubate cells overnight, then remove medium and wash cells once with PBS.
5. Harvest cells by solubilization in Laemmli SDS-sample buffer (100 μL) and store samples at −20 °C.

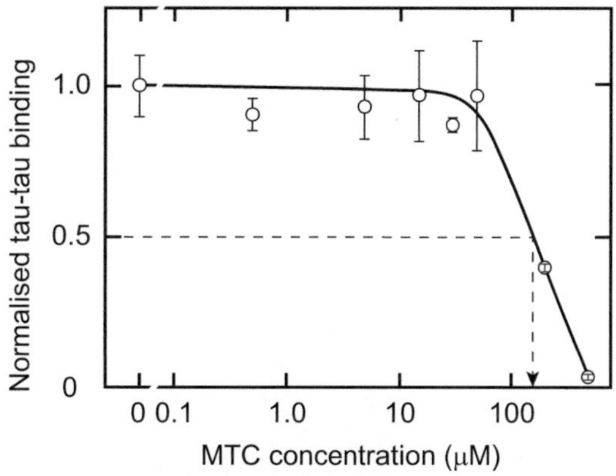

Fig. 2 Measurement of B_{50} value for methylthioninium chloride (MTC). Normalized tau-tau binding measured in the presence of increasing concentrations of MTC. The calculated B_{50} for MTC is 173.5 μM (indicated by the *dashed line*)

6. Separate proteins in samples using 15 % SDS-PAGE gels. Run all samples (20 μL per lane) from each TAI on a single gel. After electrophoresis (200 V constant voltage for 30–35 min or 35 mA constant current per gel for 55 min), transfer proteins to PVDF membrane.
7. Incubate blot with mAb 7/51 (hybridoma supernatant diluted 1:5 with blocking buffer) for 1 h at room temperature.
8. Incubate with HRP-conjugated goat anti-mouse IgG for 1 h at room temperature and develop immunoblot by enzyme chemiluminescence.
9. Image immunoblots with a CCD camera imaging system and quantify the full-length htau40 and truncated tau295-391 bands densitometrically (Fig. 3).
10. Calculate the ratio of the quantity of the two protein bands, to correct for the level of expression of full-length tau and for cell density, particularly at high concentration of TAIs. The EC_{50} value for the TAI is defined as the concentration at which the tau295-391:htau40 ratio is 50 % of that ratio measured in the absence of TAI (Fig. 4).

3.3 Cellular Toxicity

1. The toxicity of compounds is evaluated in the same 3T6 fibroblast cells used to determine TAI activity as described in Subheading 3.2 (*see* **Note 12**).
2. Dispense fibroblasts into 96-well plates (50 μL medium/well) at a density sufficient for the wells to be approximately 80 % confluent after 3 days of incubation. Allow cells to grow overnight before testing their exposure to compounds for 2 days.
3. Leave eight wells without cells to provide a blank control. Prepare test compounds as a dilution series in culture medium

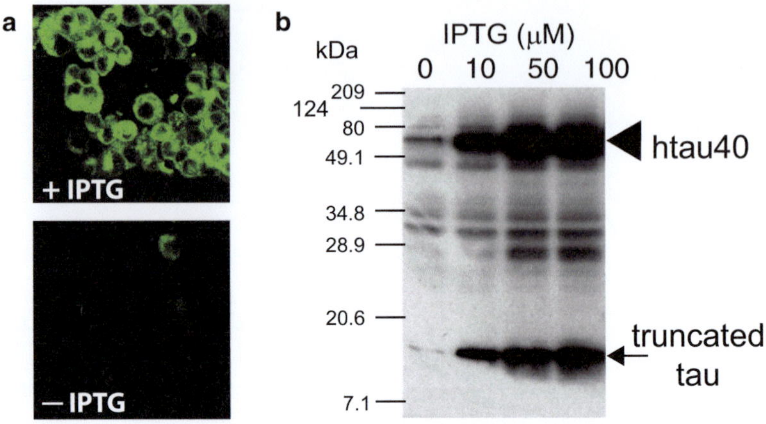

Fig. 3 Cellular model of tau aggregation. (**a**) Fibroblasts, stably transfected with both constitutive truncated tau295–391 and uninduced full-length tau (htau40), express both proteins at low levels. Following induction of expression of the full-length tau with IPTG, cells exhibit aggregated tau that binds a fluorescent PHF ligand. (**b**) SDS-PAGE and immunoblot analysis of cells shows that the induction of full-length tau with increasing concentrations of IPTG (*arrowhead*) is associated with the accumulation of a 12-kDa truncated tau neofragment (*arrow*). [From Harrington, C.R. et al. 2015, J. Biol. Chem., 290, 10862-10875]

in a separate 96-well plate and then transfer samples (50 μL per well) to cultured cells. Two compounds are tested per 96-well plate, including cells treated in the absence of compound.

4. Grow cells in the presence of compound for 48 h (to match the activity assay). After this period, remove culture medium and wash cells briefly with PBS.
5. Remaining live cells are lysed by the addition of lysis buffer (50 μL/well), prepared by dilution in water from a 10× stock solution, and incubated at room temperature for 30 min.
6. Add assay reagent (50 μL/well) to the plates and incubate at room temperature in the dark for 30 min.
7. Terminate the reaction by the addition of stop solution (50 μL) and determine the absorption at 490 nm spectrophotometrically.
8. Cytotoxicity is expressed as the LD_{50} value, the concentration at which there is 50 % of maximum LDH activity found for cells treated without compound (A_{490}), as determined graphically, and having first subtracted the blank control values.

3.4 Therapeutic Index

The therapeutic index (RxI) is expressed as $RxI = LD_{50}/EC_{50}$. In terms of screening compounds, a greater value for the therapeutic index would be consistent with a more effective TAI. Representative data for three related TAIs are shown in Table 1.

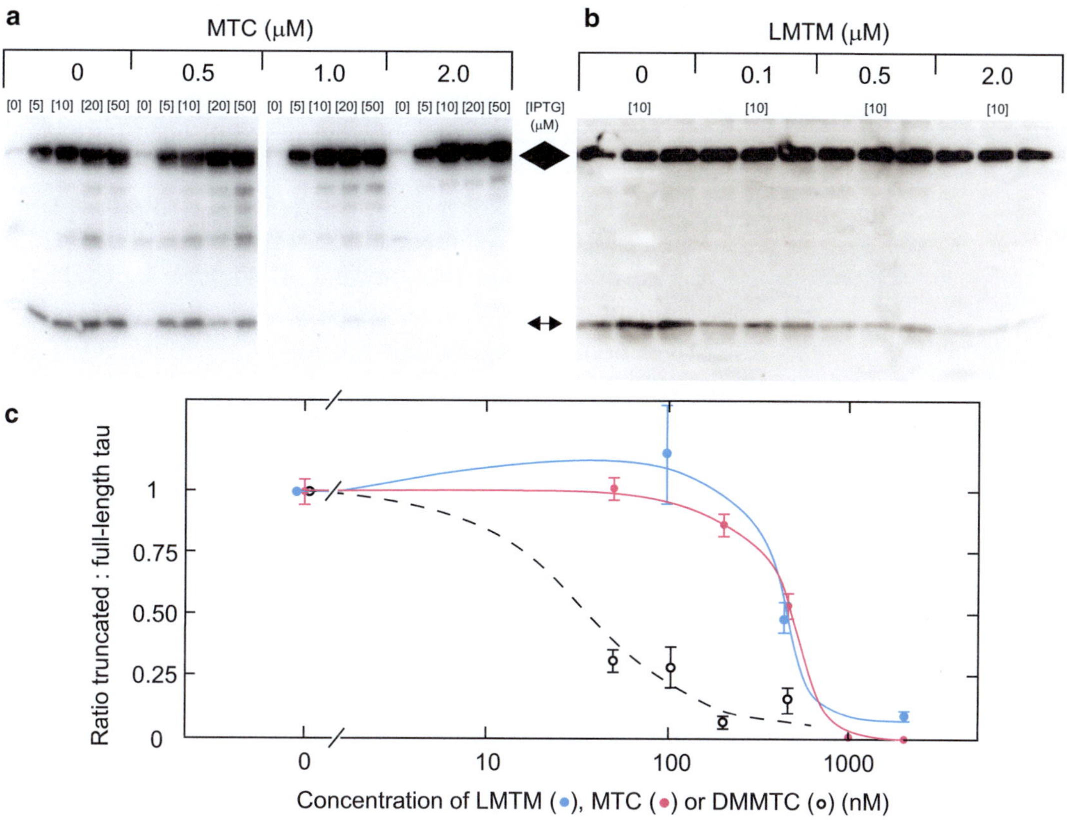

Fig. 4 Inhibition of tau aggregation in a cell model by MTC, leucomethylthioninium dihydromesylate (LMTM), and 1,9-dimethyl methylthioninium chloride (DMMTC) shown by representative immunoblots for MTC, tested with varied inducing concentrations of IPTG (**a**), and for LMTM using 10 μM IPTG (**b**). The positions of full-length and truncated 12-kDa tau are indicated by *arrowhead* and *arrow*, respectively. (**c**) Inhibition curves for three compounds following induction with 10 μM IPTG are shown from replicate experiments (mean ± S.E.). [From Harrington, C.R. et al. 2015, J. Biol. Chem., 290, 10862-10875]

Table 1

Inhibition of cell-free tau-tau binding and tau aggregation in cells[a]

Compound	B_{50} (μM)	EC_{50} (μM)	LD_{50} (μM)	RxI
MTC	195.6 ± 16.1 (10)	0.59 ± 0.04 (73)	65 ± 5 (38)	110
DMMTC	4.3 ± 0.5 (19)	0.041 ± 0.004 (22)	3.5 ± 0.97 (6)	85
LMTM	238.2 ± 74.2 (3)	0.19 ± 0.04 (9)	34 ± 4 (8)	179

[a]B_{50}, concentration at which tau-tau binding in the cell-free assay is inhibited by 50 %; EC_{50}, effective concentration at which aggregation-dependent production of truncated tau is inhibited by 50 % in the cell model; LD_{50}, concentration at which 50 % of 3T6 mouse cells are killed; and therapeutic index, RxI = LD_{50}/EC_{50}. Data expressed as the mean (μM) ± S.E., with number of replicate experiments indicated in *parentheses*

4 Notes

1. Marvel dried milk powder (Premier International Foods) is used. Other brands of milk powder have not been tested and may affect the results.
2. PIPES does not dissolve until the pH is close to 6.0. A large amount of KOH is required; add KOH pellets until the PIPES starts to dissolve and then use a 1 M solution of KOH for adjustment to pH 6.0.
3. TMB is noncarcinogenic but full toxicological testing is not available; protection to skin, eyes, and lungs should be used and waste disposed of appropriately.
4. Tau proteins are thermostable; purification protocols with or without heat treatment have been used. The behavior in the assay of the aqueous-phase protein (dGAE) is more sensitive than the solid-phase protein (dGA) to batch variation and so pH and ionic strength of the binding buffer may need to be modified for individual batches of protein to optimize the signal-to-noise ratio in the assay.
5. The concentration of dGA selected to coat the solid phase should be determined from a tau-tau binding curve, and is selected as the minimum concentration at which the binding of tau in the aqueous phase is saturating. This is usually 1 μM for dGA in our assay conditions.
6. We use an automatic plate washer (SkanWash) in which there is a program consisting of: 2-s aspiration; three cycles of a 1.5-s wash and 2-s aspiration; followed by a final 4-s aspiration. Hand washing plates leads to greater variability in results and is not recommended.
7. The concentration of dGAE used should be determined from a tau-tau binding curve, and is selected as the minimum concentration at which binding saturates. The concentration of dGAE is more sensitive to assay conditions than for dGA, but is used in the range of 10–100 nM.
8. Following blocking, it is advisable to rinse plates manually once with deionised water before using the plate washer. This helps to avoid particulate matter blocking tubes in the washer.
9. A 1:10 dilution of hybridoma supernatant fluid is usually suitable for mAb 423 (ca. 1 μg IgG/mL); the concentration should be optimized for other antibodies.
10. Development of substrate conversion is followed for 2 min with shaking at room temperature Nine measurements of absorption are taken at intervals of 14 s and SoftMax v.5.0 (Molecular Devices) used to calculate the initial rate of change of absorbance [18]. There is gradual, but slight color develop-

ment in the substrate solution as it stands during the measurement of sequential plates. This change is circumvented by kinetically measuring the rate of change in absorbance.

11. Antibiotics used for plasmid maintenance are hygromycin (100 μg/mL) for the lac repressor protein-encoding plasmid, G418 (200 μg/mL) for the htau40 plasmid, and zeocin (200 μg/mL) for the tau295-391 plasmid. Culture medium also contains penicillin (50 units/mL) and streptomycin (50 μg/mL).
12. The toxicity reported is for mouse fibroblasts. It can be applied readily to other cell lines, but the protocol is designed to measure cytotoxicity within the same assay system in which tau aggregation inhibitors are examined.

References

1. Buée L, Bussière T, Buée-Scherrer V et al (2000) Tau protein isoforms, phosphorylation and role in neurodegenerative disorders. Brain Res Rev 33:95–130
2. Novak M, Kabat J, Wischik CM (1993) Molecular characterization of the minimal protease resistant tau unit of the Alzheimer's disease paired helical filament. EMBO J 12:365–370
3. Wischik CM, Novak M, Thøgersen HC et al (1988) Isolation of a fragment of tau derived from the core of the paired helical filament of Alzheimer's disease. Proc Natl Acad Sci U S A 85:4506–4510
4. Mukaetova-Ladinska EB, Garcia-Sierra F, Hurt J et al (2000) Staging of cytoskeletal and b-amyloid changes in human isocortex reveals biphasic synaptic protein response during progression of Alzheimer's disease. Am J Pathol 157:623–636
5. Arriagada PW, Growdon JH, Hedley-White ET, Hyman BT (1992) Neurofibrillary tangles but not senile plaques parallel duration and severity of Alzheimer's disease. Neurology 42:631–639
6. Wilcock GK, Esiri MM (1982) Plaques, tangles and dementia: a quantitative study. J Neurol Sci 56:343–356
7. Maruyama M, Shimada H, Suhara T et al (2013) Imaging of tau pathology in a tauopathy mouse model and in Alzheimer patients compared to normal controls. Neuron 79:1094–1108
8. Okamura N, Furumoto S, Fodero-Tavoletti MT et al (2014) Non-invasive assessment of Alzheimer's disease neurofibrillary pathology using F-18-THK5105 PET. Brain 137:1762–1771
9. Wischik CM, Edwards PC, Lai RYK et al (1996) Selective inhibition of Alzheimer disease-like tau aggregation by phenothiazines. Proc Natl Acad Sci U S A 93:11213–11218
10. Wischik CM, Staff RT, Wischik DJ et al (2015) Tau aggregation inhibitor therapy: an exploratory phase 2 study in mild or moderate Alzheimer's disease. J Alzheimers Dis 44:705–720
11. Bulic B, Pickhardt M, Mandelkow E (2013) Progress and developments in tau aggregation inhibitors for Alzheimer disease. J Med Chem 56:4135–4155
12. Taniguchi S, Suzuki N, Masuda M et al (2005) Inhibition of heparin-induced tau filament formation by phenothiazines, polyphenols, and porphyrins. J Biol Chem 280:7614–7623
13. Harrington CR, Storey JMD, Clunas S et al (2015) Cellular models of aggregation-dependent template-directed proteolysis to characterize tau aggregation inhibitors for treatment of Alzheimer's disease. J Biol Chem 290:10862–10875
14. Melis V, Magbagbeolu M, Rickard JE et al (2015) Effects of oxidized and reduced forms of methylthioninium in two transgenic mouse tauopathy models. Behav Pharmacol 26:353–368
15. Brandt R, Kempf M, Lee G (1997) Expression and purification of tau for in vitro studies. In: Avila J, Brandt R, Kosik KS (eds) Brain microtubule associated proteins: modifications in disease. Harwood Academic Publishers, Amsterdam, pp 245–257

16. Matsudaira PT, Burgess DR (1978) SDS microslab linear gradient polyacrylamide gel electrophoresis. Anal Biochem 87:386–396
17. Novak M, Jakes R, Edwards PC et al (1991) Difference between the tau protein of Alzheimer paired helical filament core and normal tau revealed by epitope analysis of mAbs 423 and 7.51. Proc Natl Acad Sci U S A 88:5837–5841
18. Harrington CR, Edwards PC, Wischik CM (1990) Competitive ELISA for measurement of tau proteins in Alzheimer's disease. J Immunol Methods 134:261–271

Chapter 9

Tau Oligomers as Pathogenic Seeds: Preparation and Propagation In Vitro and In Vivo

Julia E. Gerson, Urmi Sengupta, and Rakez Kayed

Abstract

Tau oligomers have been shown to be the main toxic tau species in a number of neurodegenerative disorders. In order to study tau oligomers both in vitro and in vivo, we have established methods for the reliable preparation, isolation, and detection of tau oligomers. Methods for the seeding of tau oligomers, isolation of tau oligomers from tissue, and detection of tau oligomers using tau oligomer-specific antibodies by biochemical and immunohistochemical methods are detailed below.

Key words Tau oligomers, Amyloid seeding, Tau oligomer isolation, Tauopathies

1 Introduction

A large body of well-supported evidence suggests that toxicity in neurodegenerative diseases presenting with tau pathology is dependent upon tau aggregation intermediates known as tau oligomers on route to the formation of neurofibrillary tangles [1–7]. Tau is believed to exist primarily as an unfolded monomer bound to microtubules in its native state. When tau is misfolded or released from the microtubules, tau is believed to aggregate to form stable, lower energy structures with increased β-sheet structure due to the presence of exposed hydrophobic patches [8, 9]. However, the initiation of tau misfolding and aggregation in vivo is complex and does not translate to straightforward formation of aggregates in vitro. Native, un-mutated tau protein does not aggregate spontaneously in the laboratory and requires pro-aggregatory conditions such as fragmentation, introduction of mutations, or treatment with anionic compounds [8–15]. Therefore, in order to attain biologically relevant tau oligomers for the study of tau aggregation and toxicity, we have designed protocols for the preparation and characterization of tau oligomers in vitro using other amyloid oligomeric seeds, as well as for the isolation of tau oligomers from biological samples [1, 16–18]. We have also

Caroline Smet-Nocca (ed.), *Tau Protein: Methods and Protocols*, Methods in Molecular Biology, vol. 1523,
DOI 10.1007/978-1-4939-6598-4_9,

created novel antibodies and optimized techniques for the detection of tau oligomers using common biochemical techniques including ELISA, dot blot, Western blot, as well as immunohistochemistry that have been used by our laboratory and others to characterize oligomeric tau in Alzheimer's disease, progressive supranuclear palsy, Parkinson's disease, dementia with Lewy bodies, Huntington's disease, and transgenic mouse models [18–25].

2 Materials

2.1 Purification of Aβ42 and α-Synuclein

1. Human Aβ42 > 95 % purity purchased (Keck Laboratory, Yale University) (*see* **Note 1**) or pET30a wild-type α-synuclein plasmid.
2. *Escherichia coli* cells.
3. 0.5 mM isopropyl-1-thio-D-galactopyranoside (IPTG).
4. Lysis buffer: 500 mM NaCl, 100 mM Tris–HCl, pH 8.0, 1 mM EDTA, and 1 mM phenylmethylsulfonyl fluoride.
5. Equilibration/dialysis buffer: 20 mM Tris–HCl, pH 8.0, 1 mM dithiothreitol, 1 mM EDTA.
6. HiTrap ANX column (Amersham Biosciences).
7. Elution buffer: 20 mM Tris–HCl, pH 8.0, 1 mM dithiothreitol, 1 mM EDTA, 1 M NaCl gradient.
8. HiTrap Q XL column (Amersham Biosciences).

2.2 Preparation of Amyloid Oligomeric Seeds (HFIP)

1. Purified Aβ42 or α-synuclein protein.
2. 50 % acetonitrile in water.
3. Hexafluoroisopropanol (HFIP).

2.3 Preparation of Amyloid Oligomeric Seeds (NaOH/ PBS)

1. Purified Aβ42 or α-synuclein protein.
2. 50 % acetonitrile in water.
3. 100 mM NaOH (for Aβ42).
4. 0.02 % sodium azide in 1× Phosphate-Buffered Saline (PBS) pH 7.4.

2.4 Preparation of Tau Oligomers

1. Purified recombinant tau protein tau-441 (2N4R; M.Wt 45.9 kDa).
2. 8 M urea.
3. 1× PBS, pH 7.4.
4. Bicinchoninic acid (BCA) protein assay kit (Pierce).
5. Aβ42 or α-synuclein oligomers (0.3 mg/mL).
6. Microcon centrifugal filter device with a cutoff of 25 kDa (Millipore).

7. LC-6 AD Shimadzu high-performance liquid chromatography (HPLC) system.
8. TSK-GEL G3000 SWXL (30 cm × 7.8 mm) column (Supelco).

2.5 Preparation of Tissue Extracts

1. Fresh frozen tissue.
2. 1× PBS, pH 7.4.
3. Protease inhibitor cocktail.
4. Tissue lyser (Qiagen).

2.6 Isolation of Tau Oligomers from Biological Samples

1. Tosyl-activated magnetic Dynabeads (Dynal Biotech).
2. T22 antibody.
3. 0.1 M borate, pH 9.5.
4. 0.1 % Bovine serum albumin in 0.2 M Tris–Cl pH 8.5.
5. 1× PBS, pH 7.4.
6. 0.1 M glycine, pH 2.8.
7. 1 M Tris–Cl pH 8.0.
8. Microcon centrifugal filter device, 25 kDa molecular weight cutoff (Millipore).
9. BCA protein assay kit (Pierce).

2.7 ELISA for Tau Oligomers

1. 96 well ELISA plates (Nunc polysorp).
2. 0.05 M sodium bicarbonate, pH 9.6.
3. 1× TBS low tween: 1× Tris buffered saline (19.98 mM Tris, 136 mM NaCl, pH 7.4) 0.01 % Tween-20.
4. 5 % and 10 % milk solution: 5 % and 10 % nonfat dry milk in 1× TBS low tween.
5. Goat α-mouse IgG Horseradish-peroxidase (HRP) (BioRad).
6. Goat α-rabbit IgG HRP (BioRad).
7. TMB (3,3′-5,5′ tetramethylbenzidine) One-Step Substrate System (DAKO).
8. 2 M hydrochloric acid (HCl).

2.8 Dot Blot for Tau Oligomers

1. Nitrocellulose membrane 0.45 μm pore size.
2. 1× TBS low tween: 1× Tris-Buffered Saline (19.98 mM Tris–Cl, 136 mM NaCl, pH 7.4), 0.01 % Tween-20.
3. 5 % and 10 % milk solution: 5 % and 10 % nonfat dry milk in 1× TBS low tween.
4. Goat α-mouse IgG HRP (BioRad).
5. Goat α-rabbit IgG HRP (BioRad).
6. ECL 2 Western Blotting Substrate (Pierce).

2.9 Western Blot for Tau Oligomers

1. 4× loading buffer (Invitrogen).
2. Nupage Novex 4–12% Bis-Tris Gels 12well*1 mm gels (Invitrogen).
3. SDS running buffer: 50 mM MES, 50 mM Tris base, 0.1% SDS, 1 mM EDTA, pH 7.3.
4. 1× Transfer buffer: 10% Transfer buffer stock (1.92 M Glycine, 250 mM Tris, pH 9), 10% methanol.
5. Transfer pads.
6. Filter paper.
7. Nitrocellulose membrane 0.45 μm pore size.
8. 1× TBS low tween: 1× TBS (19.98 mM Tris, 136 mM NaCl, pH 7.4), 0.01% Tween-20.
9. 5% and 10% milk solution: 5% and 10% nonfat dry milk in 1× TBS low tween.
10. Goat α-mouse IgG HRP (BioRad).
11. Goat α-rabbit IgG HRP (BioRad).
12. ECL 2 Western Blotting Substrate (Pierce).

2.10 Immunohistochemistry for Tau Oligomers

2.10.1 Immunohistochemistry in Frozen Sections

1. Humidity chamber.
2. Chilled methanol.
3. 3% H_2O_2.
4. 1× PBS.
5. Goat serum.
6. Antibody diluent (Dako).
7. Vectastain ABC kit (Vector).
8. 3,3-diaminobenzidine (DAB) (Vector).
9. Hematoxylin (Vector).
10. Ethanol (80%, 95%, and 100%).
11. Xylene.
12. Cytoseal 60 mounting medium (Richard-Allan Scientific).

2.10.2 Immunofluorescence in Frozen Sections

1. Humidity chamber.
2. 4% paraformaldehyde or chilled methanol.
3. 1× PBS.
4. 0.5% Triton-X 100 in 1× PBS.
5. 3% bovine serum albumin in 1× PBS or goat serum (Sigma-Aldrich).
6. Goat anti-mouse/rabbit IgG Alexa-568/488 (Invitrogen).
7. DAPI or TOPRO (if labeling cellular nuclei).
8. Fluoromount G™ (Southern Biotech).

9. Clear nail polish.
10. Autofluorescence eliminator reagent (Millipore) or other autofluorescence inhibitor (for human tissue sections).

2.10.3 Immunohistochemistry in Paraffin-Embedded Sections

1. Xylene.
2. Ethanol (80 %, 95 %, and 100 %).
3. Target retrieval solution (Dako).
4. 1× PBS.
5. 3 % H_2O_2.
6. Goat serum (Sigma-Aldrich).
7. Antibody diluent (Dako).
8. Vectastain ABC kit (Vector).
9. DAB (Vector).
10. Hematoxylin (Vector).
11. Cytoseal 60 mounting medium (Richard-Allan Scientific).

2.10.4 Immunofluorescence in Paraffin-Embedded Sections

1. Xylene.
2. Ethanol (80 %, 95 %, and 100 %).
3. Target retrieval solution (Dako).
4. 1× PBS.
5. Goat serum (Sigma-Aldrich).
6. Antibody diluent (Dako).
7. Goat anti-mouse/rabbit IgG Alexa-568/488 (Invitrogen).
8. DAPI or TOPRO (if labeling cellular nuclei).
9. Fluoromount G™ (Southern Biotech).
10. Clear nail polish.
11. Autofluorescence eliminator reagent (Millipore) or other autofluorescence inhibitor (for human tissue sections).

2.11 Tau Oligomer-Specific Antibodies

A description of T22 and TOMA-1 antibodies with their specificity, reactivity, and applications is provided in Table 1.

3 Methods

3.1 Preparation of Tau Monomeric Samples

1. Express and purify recombinant tau protein tau-441 (2N4R; M.Wt 45.9 kDa) as described [26, 27].
2. Treat 0.6 mg of recombinant tau with 200 μl of 8 M urea to obtain monomeric tau.
3. Dialyze overnight against 1× PBS, pH 7.4.

Table 1
Description of T22 and TOMA-1 antibodies. Isotype and immunogen used for antibody generation are described as well as their species reactivity and the recommended dilution ranges depending on application

Antibody	Antibody 1 T22 [24]		Antibody 2 Tau oligomer-specific monoclonal antibody (TOMA-1) [22]		
Description	Rabbit polyclonal to tau oligomers		Mouse monoclonal to tau oligomers		
Isotype	IgG		IgG		
Specificity	Specifically recognizes tau oligomers, but not monomeric or fibrillar tau protein on Western and dot blots, ELISA, and immunohistochemistry. Detects bands at ~90 kDa corresponding to tau dimers, as well as higher molecular weight oligomeric bands		Specifically recognizes tau oligomers, but not monomeric or fibrillar tau protein on Western and dot blots, ELISA, and immunohistochemistry. Detects bands at ~90 kDa corresponding to tau dimers, as well as higher molecular weight oligomeric bands.		
Tested applications	ELISA, WB, IHC-Fr, ICC/IF, IHC-P		ELISA, WB, IHC-Fr, ICC/IF, IHC-P		
Species reactivity	*Reacts with*: Mouse, Rat, Human, Drosophila *Predicted to work with*: All Mammals		*Reacts with*: Mouse, Rat, Human *Predicted to work with*: All Mammals		
Immunogen	The initial immunization was performed with a preparation of full-length human recombinant tau protein (Tau-441, 2N4R isoform) in Freund's complete adjuvant, followed by two boosts at 4 week intervals		The initial immunization was performed with a preparation of full-length human recombinant tau protein (Tau-441, 2N4R isoform) in an equal volume of saline and Freund's complete adjuvant, followed by a second injection 2 weeks later of antigen in Freund's incomplete adjuvant and boosts given after 28, 47, 60, 80, and 103 days. Mice were given booster injections for 4 consecutive days prior to fusion		
Purity	Whole antiserum				
Application	*WB*	*IHC-Fr, ICC/IF, IHC-P*	*ELISA*	*WB*	*IHC-Fr, ICC/IF, IHC-P*
Dilution	1:250–1:500	1:200–1:300	1:200	1:200–1:500	1:50–1:200
Purity	Size exclusion chromatography purified				
Application	ELISA	WB			
Dilution	1:250–1:500	1:500–1:700			

4. Measure concentration of tau using BCA protein assay and normalize to 1 mg/mL using PBS (*see* **Note 2**).

3.2 Preparation of Amyloid Seeds

The preparation of amyloid oligomers to seed tau can be effectively completed using either Aβ42 or α-synuclein protein by the HFIP [24, 28, 29], NaOH, or PBS method (*see* **Note 3**) [30, 31].

3.2.1 Preparation of Aβ42 Protein

1. Dialyze Aβ42 overnight against water and lyophilize.
2. Resuspend 1 mg lyophilized Aβ42 protein in 1.5 mL 50% acetonitrile in water, divide into three aliquots of 0.3 mg in 500 μl and re-lyophilize.

3.2.2 Preparation of Aβ42 Oligomers (HFIP)

1. Dissolve 0.3 mg of protein in 200 μl of hexafluoroisopropanol and incubate for 10–20 min at room temperature.
2. Add the resulting solution to 700 μl dd H_2O in a siliconized eppendorf tube with holes placed on the top of the cap to allow the evaporation of HFIP.
3. Stir the samples at 500 r.p.m. using a Teflon-coated micro stir bar for 36 h in a fume hood at 22 °C.

3.2.3 Preparation of Aβ42 Oligomers (NaOH)

1. Dissolve re-lyophilized peptide (0.3 mg) in 30 μl of 100 mM NaOH.
2. Sonicate in a 22 °C water bath for 30 s.
3. Transfer solution to a tube containing 700 μl 0.02% sodium azide in 1× PBS.
4. Oligomerize the sample by shaking on a rocker for 2 days at room temperature.

3.3 Preparation of α-Synuclein Protein

1. Induce expression of pET30a wild-type α-synuclein with 0.5 mM isopropyl-1-thio-D-galactopyranoside at 25 °C overnight in *Escherichia coli*.
2. Harvest bacterial cells by centrifugation at 4000 ×*g* for 10 min.
3. Lyse cells in lysis buffer then boil cell lysate for 10 min.
4. Centrifuge lysate for 30 min at 13,000 ×*g*.
5. Precipitate resulting supernatant in acid at pH 3.5 and centrifuge for an additional 30 min at 15,000 ×*g*.
6. Dialyze the supernatant against dialysis buffer.
7. Load sample onto a HiTrap ANX column equilibrated in the same buffer.
8. Elute protein with elution buffer in a 0–1 M NaCl gradient.
9. Pool α-synuclein-containing fractions and apply to a HiTrap Q XL, equilibrated with 10 mL of equilibration buffer.
10. Elute α-synuclein with elution buffer over a linear gradient of 0–1 M NaCl.
11. Check the major peak eluted from the column by electrophoresis and pool fractions.

12. Dialyze α-synuclein overnight against water and lyophilize.
13. Resuspend 1 mg lyophilized α-synuclein protein in 1 mL 50% acetonitrile in water.
14. Divide into two aliquots of 0.5 mg in 500 μl and re-lyophilize.

3.3.1 Preparation of α-Synuclein Oligomers (HFIP)

1. Dissolve 0.5 mg of protein in 200 μl of hexafluoroisopropanol and incubate for 10–20 min at room temperature.
2. Add the resulting solution to 700 μl dd H_2O in a siliconized eppendorf tube with holes placed on the top of the cap to allow the evaporation of HFIP.
3. Stir the samples at 500 r.p.m. using a Teflon-coated micro stir bar for 36 h in a fume hood at 22 °C.

3.3.2 Preparation of α-Synuclein Oligomers (PBS)

1. Dissolve relyophilized protein (0.5 mg) in 700 μl 0.02% sodium azide in 1× PBS.
2. Stir the resulting solution in a closed tube at 500 r.p.m. using a Teflon-coated micro stir bar for 48 h at room temperature.
3. Dilute the samples as needed in 1× PBS before use.

3.4 Preparation of Tau Oligomers

1. Add 300 μl of the tau monomer stock (1 mg/mL) to 700 μl of 1× PBS to achieve a final concentration of 0.3 mg/mL.
2. Add 7 μl of Aβ42 or α-synuclein oligomers (0.3 mg/mL) prepared using methods previously described to seed the oligomerization of tau.
3. Mix the sample by pipetting for 1 min.
4. Incubate samples at room temperature for 1 h on an orbital shaker.
5. Purify resulting tau oligomers using a TSK-GEL G3000 SWXL column. Use 1× PBS, pH 7.4 as the mobile phase with a flow rate of 0.5 mL/min [16, 32].
6. Seed a new batch of tau monomer stock with purified tau oligomers and repeat seeding procedure three times (*see* **Notes 4** and **5**).

3.5 Preparation of Tissue Extracts

1. Prepare solution using one tablet of protease inhibitor cocktail in 50 mL 1× PBS.
2. Dilute frozen tissue 1/3 (w/v) in protease inhibitor solution.
3. Homogenize using tissue lyser.
4. Centrifuge samples at 9279.4 × *g* for 10 min at 4 °C.
5. Collect supernatant and aliquot (*see* **Note 6**).

3.6 Isolation of Tau Oligomers from Biological Samples

1. Coat tosyl-activated magnetic Dynabeads with 20 μg of T22 antibody (1.0 mg/mL) diluted in 0.1 M borate, pH 9.5, overnight at 37 °C.
2. Wash beads in 0.2 M Tris, 0.1 % bovine serum albumin, pH 8.5.
3. Incubate beads with PBS soluble fraction of tissue homogenate with rotation at room temperature for 1 h.
4. Wash beads three times with 1× PBS and elute using 0.1 M glycine, pH 2.8.
5. Adjust the pH of each eluted fraction to 7.0 using 1 M Tris–Cl pH 8.0.
6. Centrifuge fractions in a microcon centrifugal filter device with a molecular weight cutoff of 25 kDa (Millipore) at 14,000 × *g* for 25 min at 4 °C.
7. Resuspend oligomers in sterile PBS.
8. Determine total protein concentration using the BCA protein assay (*see* **Note 7**).
9. Centrifuge samples again in a microcon centrifugal filter device with a cutoff of 25 kDa at 14,000 × *g* for 25 min at 4 °C.
10. Resuspend oligomers in PBS in order to obtain the desired concentration.
11. Keep at 4 °C for 15–30 min.
12. Let rest at room temperature for 10 min prior to use (*see* **Note 5**).

3.7 ELISA for Tau Oligomers

This method is useful for the quantification of tau oligomers prepared in vitro as well as biological samples containing tau oligomers including serum, cerebrospinal fluid, and tissue homogenate.

1. Pipette 50 μl 0.05 M sodium bicarbonate pH 9.6 into each well of a 96-well ELISA plate (*see* **Note 8**).
2. Add 5 μl (15 μg) of coating protein to each well.
3. Incubate at 4 °C overnight.
4. Dump contents of plate into sink and wash 1× with 1× TBS low Tween.
5. Pipette 150 μl 10 % milk solution into each well and incubate 2 h at room temperature.
6. Dump contents of plate into sink and wash 1× with 1× TBS low Tween.
7. Pipette 100 μl of primary antibody diluted in 5 % milk solution into each well.
8. Incubate 1 h at RT
9. Let TMB warm to room temperature.

10. Dump contents of plate into sink and wash 4× with 1× TBS low Tween.
11. Pipette 100 μl of secondary antibody diluted 1:3000 in 5% milk solution into each well and incubate for 1 h at room temperature.
12. Dump contents of plate into sink and wash 4× with 1× TBS low Tween.
13. Pipette 100 μl TMB into wells. TMB should be room temperature before adding to wells.
14. Incubate plate in a location protected from light. Check every 15 min for a maximum of 1 h for blue color development.
15. To stop reaction: pipette 100 μl HCl into wells. The blue color should turn yellow.
16. Read in plate reader at 450 nm.

3.8 Dot Blot for Tau Oligomers

Analysis of a large number of samples can be completed quickly and easily using dot blot analysis for semi-quantitative evaluation of levels of tau oligomers.

1. Cut strips and demarcate squares for desired number of samples to make dot blot strips of appropriate size using nitrocellulose membrane.
2. Pipette 1.0–1.2 μl of each sample into each square on dot blot strips.
3. Let membranes dry for at least 30 min.
4. Block strips with 10% milk solution overnight at 4 °C.
5. Wash membrane strips once with 1× TBS low Tween for 5 min.
6. Incubate each strip with primary antibodies diluted in 5% milk solution for 1 h at room temperature.
7. Wash strips three times for 5 min in 1× TBS low Tween buffer.
8. Incubate strips with secondary antibodies diluted 1:3000 in 5% milk solution for 1 h at room temperature.
9. Wash strips three times for 5 min with 1× TBS low Tween buffer.
10. Apply ECL (Pierce) to strips following instructions and develop on X-ray autoradiography films.

3.9 Western Blot for Tau Oligomers

Analysis of samples containing tau oligomers by Western blot is critical for evaluating the presence of oligomeric species of different molecular weights.

1. Thaw samples on ice and bring 4× loading buffer to room temperature.
2. Dilute samples (8–10 μl) 4:1 in 4× loading buffer (2–2.5 μl) (*see* **Note 9**).

3. Load the samples in each well of the precast gel in a chamber filled with 1× SDS running buffer.
4. Run gel electrophoresis at 80 V for 10 min and then increase to 120 V for about an hour (until the dye approaches the bottom of the gel).
5. Following electrophoresis, remove gel from plates and place in a container filled with transfer buffer.
6. Immerse transfer pads, filter paper, and nitrocellulose membrane cut to gel size in transfer buffer and form sandwich with cassette in order from black to red: transfer pad, filter paper, gel, nitrocellulose membrane, filter, transfer pad.
7. Transfer proteins from the gel onto nitrocellulose membrane at 97 V for 1 h at 4 °C (*see* **Note 10**).
8. Remove membrane from cassette and block with 10% milk solution at 4 °C overnight or for 2 h at room temperature.
9. Wash the membrane once for 5 min in 1× TBS low Tween.
10. Incubate the membrane with primary antibody diluted in 5% milk solution for 1 h at room temperature.
11. Wash the membrane three times for 10 min in 1× TBS low Tween.
12. Incubate the membrane with secondary antibody diluted 1:3000 in 5% milk solution for 1 h at room temperature.
13. Wash the membrane three times for 10 min in 1× TBS low Tween.
14. Add detection substrate to the membrane.
15. Immediately develop the membrane on X-ray autoradiography films.

3.10 Immunohistochemistry for Tau Oligomers

Immunohistochemistry allows for the evaluation of the physical location of tau oligomers in tissue and cellular samples, while allowing for direct comparison between other proteins of interest.

3.10.1 Immunohistochemistry in Frozen Tissue

1. Fix sections in chilled methanol for 10 min at −20 °C.
2. Incubate for 15 min in 3% H_2O_2 to quench endogenous peroxidases.
3. Wash two times for 5 min each with 1× PBS.
4. Block nonspecific sites for 60 min in 5% goat serum diluted in 1× PBS.
5. Incubate sections overnight in a humidity chamber with primary antibody diluted in antibody diluent at 4 °C.
6. Wash three times for 10 min in 1× PBS.
7. Incubate in biotinylated goat anti-mouse IgG (Vectastain ABC kit) or biotinylated goat anti-rabbit IgG (Vectastain ABC Kit) prepared according to manufacturer's instructions for 1 h at room temperature.

8. Wash 4 times for 10 min in 1× PBS.
9. Incubate for 1 h in ABC solution (Vectastain ABC kit) made 30 prior to use according to manufacturer's instructions.
10. Wash three times for 10 min in 1× PBS.
11. Prepare and apply DAB according to manufacturer's instructions and monitor development (incubation time depends on antibody and tissue).
12. Wash three times for 10 min in 1× PBS.
13. Wash for 30 s in warm water.
14. Incubate for 1–3 min in Hematoxylin at room temperature.
15. Wash for 30 s in warm water.
16. Dehydrate: 1 min in 80% ethanol, 2 min in 80% ethanol, 1 min in 95% ethanol, 2 min in 95% ethanol, 2 min in 100% ethanol, 3 min in 100% ethanol, 3 min in xylene, and 4 min in xylene.
17. Coverslip using mounting medium (cytoseal 60).

3.10.2 Immunofluorescence in Frozen Tissue

1. Fix sections for 10–20 min in 4% paraformaldehyde (*see* **Note 11**).
2. Wash in 1× PBS for 5 min.
3. Permeabilize in 0.5% Triton-X 100 diluted in 1× PBS for 10 min.
4. Wash in 1× PBS for 10 min.
5. Block in 3% bovine serum albumin diluted in PBS for 30 min (*see* **Note 12**).
6. Incubate in primary antibodies diluted in prepared BSA block solution overnight at 4 °C in a humidity chamber.
7. Wash three times for 10 min in 1× PBS.
8. Incubate with goat anti-mouse IgG Alexa-568/488 and/or goat anti-rabbit IgG Alexa-568/488 diluted 1:500 in 1× PBS for 1 h at room temperature in a humidity chamber.
9. Wash in 1× PBS three times for 10 min.
10. Incubate in DAPI or TOPRO diluted 1:3000 in 1× PBS for 5 min to label cellular nuclei.
11. Wash three times for 5 min in 1× PBS.
12. Mount coverslips using Fluoromount G and seal with nail polish.

3.10.3 Immunohistochemistry in Paraffin-Embedded Tissue

1. Heat sections at 60 °C for 2 1/2 h.
2. Hydrate the tissue using a series of xylene and ethanol steps: Xylene (three times for 5 min), 100% ethanol (three times for 3 min), 95% ethanol (three times for 3 min), 80% ethanol (three times for 3 min), distilled water (30 s).

3. Heat via microwave in target retrieval solution diluted 10% in sterile water for 4 min at power level 2.
4. Let slides cool in the microwave for 1 min.
5. Wash two times for 5 min in 1× PBS.
6. Incubate for 15 min in 3% H_2O_2 to quench endogenous peroxidase.
7. Wash two times for 5 min each with 1× PBS.
8. Block nonspecific sites in 5% goat serum diluted in PBS for 60 min.
9. Incubate in primary antibody diluted in antibody diluent overnight at 4 °C in a humidity chamber.
10. Wash three times for 10 min in 1× PBS.
11. Incubate in biotinylated goat anti-mouse IgG (Vectastain ABC kit) or biotinylated goat anti-rabbit IgG (Vectastain ABC Kit) prepared according to manufacturer's instructions for 1 h at room temperature.
12. Wash four times for 10 min in 1× PBS.
13. Incubate for 1 h in ABC solution (Vectastain ABC kit) made 30 prior to use according to manufacturer's instructions.
14. Wash three times for 10 min in 1× PBS.
15. Prepare and apply DAB according to manufacturer's instructions and monitor development (incubation time depends on antibody and tissue).
16. Wash three times for 10 min in 1× PBS.
17. Wash for 30 s in warm water.
18. Incubate for 1–3 min in Hematoxylin at room temperature.
19. Wash for 30 s in warm water.
20. Dehydrate: 1 min in 80% ethanol, 2 min in 80% ethanol, 1 min in 95% ethanol, 2 min in 95% ethanol, 2 min in 100% ethanol, 3 min in 100% ethanol, 3 min in xylene, and 4 min in xylene.
21. Coverslip using mounting medium (cytoseal 60).

3.10.4 Immunofluorescence in Paraffin-Embedded Tissue

1. Heat sections at 60 °C for 21/2 h.
2. Hydrate the tissue using a series of xylene and ethanol steps: Xylene (three times for 5 min), 100% ethanol (three times for 3 min), 95% ethanol (three times for 3 min), 80% ethanol (three times for 3 min), distilled water (30 s).
3. Heat via microwave in target retrieval solution diluted 10% in sterile water for 4 min at power level 2.
4. Let slides cool in the microwave for 1 min.
5. Wash two times for 5 min in 1× PBS.

6. Block nonspecific sites in 5% goat serum diluted in PBS for 60 min.
7. Incubate in primary antibodies diluted in antibody diluent overnight at 4 °C in humidity chamber.
8. Wash three times for 10 min in 1× PBS.
9. Incubate with goat anti-mouse IgG Alexa-568/488 and/or goat anti-rabbit IgG Alexa-568/488 diluted 1:500 in antibody diluent for 1 h at room temperature.
10. Wash four times for 10 min in 1× PBS.
11. Incubate in DAPI or TOPRO diluted 1:3000 in 1× PBS for 5 min to label cellular nuclei.
12. Wash three times for 5 min in 1× PBS.
13. Mount coverslips using Fluoromount G and seal with nail polish.

4 Notes

1. Aβ 42 is synthesized by fluoren-9-yl-methoxycarbonyl chemistry, purified by reverse phase-high pressure liquid chromatography, and characterized by sequencing and electrospray mass spectrometry as described previously (any commercially available Aβ42 can be used).
2. Aliquots of tau monomer may be stored at −20 °C until use for 6 months.
3. To ensure that amyloid seeds are in oligomeric conformation, test for reactivity to A11, oligomer-specific antibody [29].
4. After three rounds of seeding with freshly prepared tau oligomers, there is estimated to be a ratio of less than 1:2,470,000 Aβ:Tau, which is below the detection limits of the methods available. The same protocol is also effective using HEPES buffer (50 mM) instead of 1× PBS. In order to stabilize tau oligomers after preparation for several months, NH_4OH can be added to a final concentration of 0.1% (final pH 9.5–10.5). Water sonication of the stock solutions for 30 s can also be completed to homogenize the samples before use. For toxicity or in vivo assays samples can be dialyzed (1–10 kDa cutoff) against 10 mM Na_2CO_3, pH 9–10 immediately before use to remove the azide or residual solvent [32].
5. Tau oligomers should be characterized after preparation using established methods including circular dichroism, fast protein liquid chromatography, Bis-ANS assay, atomic force microscopy, and electron microscopy [16].
6. Aliquots may be snap-frozen and stored at −80 °C until use.

7. Tau oligomers may be stored at −80 °C for up to 6 months. Prior to use, thaw on ice and sonicate two to three times for 30 s in water to homogenize samples.
8. As a second option, 1× PBS, pH 7.4 also works well as a coating buffer.
9. Samples are not boiled, but are mixed with sample buffer and loaded directly into the gel.
10. Polyvinylidene difluoride (PVDF) have not been effective for analysis of tau oligomers by Western blot; therefore, we suggest the use of nitrocellulose membranes.
11. Incubation in chilled methanol for 10 min at −20 C may also be used as for fixative step.
12. Goat serum may also be used for blocking step. Optimal conditions for each may differ between antibodies and should be tested for best results. For labeling human tissue, autofluorescence blocking is essential. There are many published methods to accomplish this, as well as commercial autofluorescence eliminator reagent (Millipore) that has been verified to be effective.

Acknowledgment

We thank the Kayed Lab members for their contributions to the optimization of the protocols and useful suggestions. This work was supported by the Mitchell Center for Neurodegenerative Diseases.

References

1. Lasagna-Reeves CA, Castillo-Carranza DL, Sengupta U, Clos AL, Jackson GR, Kayed R (2011) Tau oligomers impair memory and induce synaptic and mitochondrial dysfunction in wild-type mice. Mol Neurodegener 6:39
2. Spires TL, Orne JD, SantaCruz K, Pitstick R, Carlson GA, Ashe KH, Hyman BT (2006) Region-specific dissociation of neuronal loss and neurofibrillary pathology in a mouse model of tauopathy. Am J Pathol 168:1598–1607
3. Berger Z, Roder H, Hanna A, Carlson A, Rangachari V, Yue M, Wszolek Z, Ashe K, Knight J, Dickson D, Andorfer C, Rosenberry TL, Lewis J, Hutton M, Janus C (2007) Accumulation of pathological tau species and memory loss in a conditional model of tauopathy. J Neurosci 27:3650–3662
4. Cowan CM, Quraishe S, Mudher A (2012) What is the pathological significance of tau oligomers? Biochem Soc Trans 40:693–697
5. Kopeikina KJ, Carlson GA, Pitstick R, Ludvigson AE, Peters A, Luebke JI, Koffie RM, Frosch MP, Hyman BT, Spires-Jones TL (2011) Tau accumulation causes mitochondrial distribution deficits in neurons in a mouse model of tauopathy and in human Alzheimer's disease brain. Am J Pathol 179:2071–2082
6. Sahara N, DeTure M, Ren Y, Ebrahim AS, Kang D, Knight J, Volbracht C, Pedersen JT, Dickson DW, Yen SH, Lewis J (2013) Characteristics of TBS-extractable hyperphosphorylated Tau species: aggregation intermediates in rTg4510 mouse brain. J Alzheimers Dis 33:249–263
7. Maeda S, Sahara N, Saito Y, Murayama M, Yoshiike Y, Kim H, Miyasaka T, Murayama S, Ikai A, Takashima A (2007) Granular tau oligomers as intermediates of tau filaments. Biochemistry 46:3856–3861
8. Harada M, Isersky C, Cuatrecasas P, Page D, Bladen HA, Eanes ED, Keiser HR, Glenner

GG (1971) Human amyloid protein: chemical variability and homogeneity. J Histochem Cytochem 19:1–15
9. von Bergen M, Barghorn S, Li L, Marx A, Biernat J, Mandelkow E-M, Mandelkow E (2001) Mutations of tau protein in frontotemporal dementia promote aggregation of paired helical filaments by enhancing local β-structure. J Biol Chem 276:48165–48174
10. Wille H, Drewes G, Biernat J, Mandelkow EM, Mandelkow E (1992) Alzheimer-like paired helical filaments and antiparallel dimers formed from microtubule-associated protein tau in vitro. J Cell Biol 118:573–584
11. King ME, Ahuja V, Binder LI, Kuret J (1999) Ligand-dependent tau filament formation: implications for Alzheimer's disease progression. Biochemistry 38:14851–14859
12. Wilson DM, Binder LI (1997) Free fatty acids stimulate the polymerization of tau and amyloid beta peptides. In vitro evidence for a common effector of pathogenesis in Alzheimer's disease. Am J Pathol 150:2181–2195
13. King ME, Gamblin TC, Kuret J, Binder LI (2000) Differential assembly of human tau isoforms in the presence of arachidonic acid. J Neurochem 74:1749–1757
14. Alonso AD, Zaidi T, Novak M, Barra HS, Grundke-Iqbal I, Iqbal K (2001) Interaction of tau isoforms with Alzheimer's disease abnormally hyperphosphorylated tau and in vitrophosphorylation into the disease-like protein. J Biol Chem 276:37967–37973
15. Alonso A, Zaidi T, Novak M, Grundke-Iqbal I, Iqbal K (2001) Hyperphosphorylation induces self-assembly of τ into tangles of paired helical filaments/straight filaments. Proc Natl Acad Sci 98:6923–6928
16. Lasagna-Reeves CA, Castillo-Carranza DL, Guerrero-Muoz MJ, Jackson GR, Kayed R (2010) Preparation and characterization of neurotoxic tau oligomers. Biochemistry 49: 10039–10041
17. Lasagna-Reeves C, Castillo-Carranza DL, Sengupta U, Guerrero-Munoz MJ, Kiritoshi T, Neugebauer V, Jackson GR, Kayed R (2012) Alzheimer brain-derived tau oligomers propagate pathology from endogenous tau. Sci Rep 2:1–7
18. Castillo-Carranza DL, Gerson JE, Sengupta U, Guerrero-Muñoz MJ, Lasagna-Reeves CA, Kayed R (2014) Specific targeting of tau oligomers in Htau mice prevents cognitive impairment and tau toxicity following injection with brain-derived tau oligomeric seeds. J Alzheimers Dis 40:S97–S111
19. Vuono R, Winder-Rhodes S, de Silva R, Cisbani G, Drouin-Ouellet J, Spillantini MG, Cicchetti F, Barker RA (2015) The role of tau in the pathological process and clinical expression of Huntington's disease. Brain 138:1907–18
20. Blair LJ, Nordhues BA, Hill SE, Scaglione KM, O'Leary JC 3rd, Fontaine SN, Breydo L, Zhang B, Li P, Wang L, Cotman C, Paulson HL, Muschol M, Uversky VN, Klengel T, Binder EB, Kayed R, Golde TE, Berchtold N, Dickey CA (2013) Accelerated neurodegeneration through chaperone-mediated oligomerization of tau. J Clin Invest 123:4158–4169
21. Castillo-Carranza DL, Guerrero-Muñoz MJ, Sengupta U, Hernandez C, Barrett ADT, Dineley K, Kayed R (2015) Tau immunotherapy modulates both pathological tau and upstream amyloid pathology in an Alzheimer's disease mouse model. J Neurosci 35:4857–4868
22. Castillo-Carranza DL, Sengupta U, Guerrero-Munoz MJ, Lasagna-Reeves CA, Gerson JE, Singh G, Estes DM, Barrett AD, Dineley KT, Jackson GR, Kayed R (2014) Passive immunization with Tau oligomer monoclonal antibody reverses tauopathy phenotypes without affecting hyperphosphorylated neurofibrillary tangles. J Neurosci 34:4260–4272
23. Gerson J, Sengupta U, Lasagna-Reeves C, Guerrero-Munoz M, Troncoso J, Kayed R (2014) Characterization of tau oligomeric seeds in progressive supranuclear palsy. Acta Neuropathol Commun 2:73
24. Lasagna-Reeves CA, Castillo-Carranza DL, Sengupta U, Sarmiento J, Troncoso J, Jackson GR, Kayed R (2012) Identification of oligomers at early stages of tau aggregation in Alzheimer's disease. FASEB J 26:1946–1959
25. Sengupta U, Guerrero-Muñoz MJ, Castillo-Carranza DL, Lasagna-Reeves CA, Gerson JE, Paulucci-Holthauzen AA, Krishnamurthy S, Farhed M, Jackson GR, Kayed R (2015) Pathological interface between oligomeric alpha-synuclein and tau in synucleinopathies. Biol Psychiatry 78:672–683
26. Margittai M, Langen R (2004) Template-assisted filament growth by parallel stacking of tau. Proc Natl Acad Sci U S A 101:10278–10283
27. Margittai M, Langen R (2006) Side chain-dependent stacking modulates tau filament structure. J Biol Chem 281:37820–37827
28. Kayed R, Sokolov Y, Edmonds B, McIntire TM, Milton SC, Hall JE, Glabe CG (2004) Permeabilization of lipid bilayers is a common conformation-dependent activity of soluble amyloid oligomers in protein misfolding diseases. J Biol Chem 279:46363–46366
29. Kayed R, Head E, Thompson JL, McIntire TM, Milton SC, Cotman CW, Glabe CG (2003) Common structure of soluble amyloid oligomers implies common mechanism of pathogenesis. Science 300:486–489

30. Mina EW, Lasagna-Reeves C, Glabe CG, Kayed R (2009) Poloxamer 188 copolymer membrane sealant rescues toxicity of amyloid oligomers in vitro. J Mol Biol 391:577–585
31. Necula M, Kayed R, Milton S, Glabe CG (2007) Small molecule inhibitors of aggregation indicate that amyloid beta oligomerization and fibrillization pathways are independent and distinct. J Biol Chem 282:10311–10324
32. Lasagna-Reeves CA, Castillo-Carranza DL, Sengupta U, Guerrero-Munoz MJ, Kiritoshi T, Neugebauer V, Jackson GR, Kayed R (2012) Alzheimer brain-derived tau oligomers propagate pathology from endogenous tau. Sci Rep 2:700

Part III

Tau Post-Translational Modifications

Chapter 10

Mass Spectrometry Analysis of Lysine Posttranslational Modifications of Tau Protein from Alzheimer's Disease Brain

Stefani N. Thomas and Austin J. Yang

Abstract

Recent advances in mass spectrometry (MS)-based proteomics have greatly facilitated the robust identification and quantification of posttranslational modifications (PTMs), including those that are present at substoichiometric site occupancies. The abnormal posttranslational modification and accumulation of the microtubule-associated protein tau has been implicated in the pathogenesis of Alzheimer's disease (AD), and it is thought that the primary mode of regulation of tau occurs through PTMs. Several studies have been published regarding tau phosphorylation; however, other tau PTMs such as ubiquitylation, acetylation, methylation, oxidation, sumoylation, nitration, and glycosylation have not been analyzed as extensively. The comprehensive detection and delineation of these PTMs is critical for drug target discovery and validation. Lysine-directed PTMs including ubiquitylation, acetylation, and methylation play key regulatory roles with respect to the rates of tau turnover and aggregation. MS-based analytical approaches have been used to gain insight into the tau lysine-directed PTM signature that is most closely associated with neurofibrillary lesion formation. This chapter provides details pertaining to the liquid chromatography tandem mass spectrometry (LC-MS/MS)-based analysis of the lysine-directed posttranslational modification of tau.

Key words Alzheimer's disease (AD), Tau, Posttranslational modification, Methylation, Acetylation, Ubiquitylation, Mass spectrometry

1 Introduction

In Alzheimer's disease (AD), the microtubule-associated protein tau dissociates from the neuronal cytoskeleton and aggregates to form paired helical filaments (PHFs) that comprise neurofibrillary tangles (NFTs) that are one of the prominent neuropathological hallmarks of AD [1, 2]. The molecular and cellular mechanisms that regulate tau aggregation remain unclear; however, it has been postulated that posttranslational modifications (PTMs) influence tau stability, function, and aggregation propensity [3]. Phosphorylation is the most well-established tau PTM, and a

Caroline Smet-Nocca (ed.), *Tau Protein: Methods and Protocols*, Methods in Molecular Biology, vol. 1523,
DOI 10.1007/978-1-4939-6598-4_10, © Springer Science+Business Media New York 2017

distinct hierarchical pattern of tau phosphorylation has been shown to correlate with the progression of AD neuropathology [4, 5]. PHFs of hyperphosphorylated tau aggregates are known to form the degradation-resistant core of NFTs in AD [4, 6, 7]. In addition to phosphorylation, tau is also subjected to several other PTMs that confer pro- or anti-aggregation effects on tau [8]. This chapter will provide details for the analysis of lysine-directed PTMs of tau, with particular emphasis on ubiquitylation, methylation, and acetylation.

Lysine residues can participate in electrostatic and hydrophobic interactions [9, 10], and they are known to play critical roles in tau assembly and toxicity [11]. The lysine-directed ubiquitylation of tau has been demonstrated to modulate tau accumulation and aggregation [12, 13]. In non-pathological conditions, tau has been shown to be ubiquitylated and proteolytically degraded by the ubiquitin-proteasome system [14–16]. The accumulation of degradation-resistant PHF-tau is linked to the impaired function of the ubiquitin proteasome system, and although PHF-tau is ubiquitylated, it is not subsequently degraded by the proteasome system [17–19]. Several ubiquitylation sites within filamentous tau isolated from AD brain have been identified, and they are all localized to the microtubule-binding repeat region [20–22].

Lysine methylation has recently been shown to contribute to the regulation of tau metabolism by directly competing with ubiquitylation and acetylation [22]. Lysine methylation has also been identified on tau purified from normal human brain [23], wild-type mice, and a mouse model of AD [24]. These data indicate that lysine methylation is part of the modification signature that is associated with the preservation of normal tau function. Acetylation is a relatively novel posttranslational modification of tau that has been shown to inhibit its degradation and contribute to tauopathy [25, 26]. Studies have shown that inhibiting histone deacetylase (HDAC)6—a tau deacetylase—has neuroprotective and beneficial consequences including microtubule stabilization [27, 28]. Hence, the pharmacological modulation of the levels of tau acetylation has been proposed as a novel therapeutic strategy. Because the co-occurrence of lysine-directed PTMs on tau has the potential to affect diverse cellular events, it is imperative to differentiate the lysine-directed tau PTM signature that is most closely associated with neuropathology in AD.

Recent advances in mass spectrometry (MS)-based proteomics have enabled the identification and quantification of thousands of PTMs in a single experiment [29–35]. This is due in large part to the development of highly specific and selective PTM-specific antibodies, new high performance liquid chromatography, and mass spectrometry instrumentation and bioinformatic algorithms for peptide and protein identification and PTM site-localization scoring that enable unambiguous PTM site assignments. Because many regulatory PTMs, including ubiquitylation and methylation, have low stoichiometry site occupancies of their target proteins, the use of antibody-based enrichment strategies is essential to achieve the requisite sensitivity in MS analysis for the comprehensive detection of these PTMs in complex matrices such as cell lysate and tissue homogenate [29, 31, 34, 36]. In this context, the stoichiometry refers to the fraction of proteins that are modified. For the majority of these antibody-based enrichment strategies, the enrichment is conducted at the peptide level after the protein mixture has been enzymatically digested. The amount of peptide input is typically on the order of a few milligrams. Accordingly, these strategies are optimized for the large-scale analysis of PTMs. The analysis of PTMs on a single protein of interest requires a different analytical approach, similar to what will be discussed in this chapter.

An inherent challenge in the mass spectrometry-based analysis of PTMs is that modified peptides can be low abundance and more difficult to identify from their fragmentation spectra compared to unmodified peptides. Unlike global proteome analysis, PTM analysis relies on the identification of single peptide species. Although database search engines are effective at identifying PTM-containing peptides with a defined measure of confidence, their localization of specific PTM sites can be arbitrary and unreliable [37]. Thus, in addition to the unambiguous identification of the modified peptides, the localization of modified amino acids within the peptide(s) of interest requires the calculation of a "localization score" [38].

The quantification of PTMs is important because PTM-based regulation in a biological context suggests that an amino acid site could be functional. Conducting quantitative MS-based analysis is essential for the elucidation of the stoichiometry of the specific tau lysine-directed PTMs that correlate with AD neuropathology. Multiple Reaction Monitoring (MRM) is a targeted mass spectrometry (MS)-based technology that is becoming increasingly utilized for protein quantification. MRM-based approaches have been used to determine the relative abundance of tau polyubiquitylation in human AD brain [20] and global tau in human CSF [39, 40]. In contrast to MS-based discovery proteomics experiments, MRM entails the targeted, simultaneous measurements of peptides that serve as surrogates for the protein targets of interest. MRM-based assays are considered to be the "gold standard" for MS-based targeted protein quantification because they are highly specific, precise, and accurate, and they can be multiplexed (hundreds of peptides can be quantified in a single assay), standardized, and readily reproduced. A targeted proteomics method that is similar to MRM is parallel reaction monitoring (PRM) wherein an accurate mass and high resolution mass spectrometer is utilized to permit the parallel detection of all target product ions [41–46].

A common fit-for-purpose quantification strategy entails the use of stable isotope-labeled internal peptide standards for relative quantification based on the establishment of calibration curves of dilutions of the peptide mixtures [47–49]. Because quantitative MS-based assays are negatively affected by a lack of proper procedures for storing and handling peptides, careful attention should be paid to the generation, quantification, storage, and handling of these peptides. Guidelines have been recently established for these practices [50]. To achieve reliable relative quantification of peptides in complex matrices, targeted proteomic assays must be analytically characterized with respect to their specificity, lower limit of quantification (LLOQ), linear range, precision, and repeatability. Various open source software tools are available to facilitate quantitative proteomic assay data analysis and sharing [51–54]. A detailed overview of targeted proteomic assays such as MRM and PRM is provided in several comprehensive reviews [55–60].

2 Materials

2.1 SDS-PAGE and In-Gel Trypsin Digestion

1. Pre-cast NuPAGE Novex 4–20% Bis-Tris gels (ThermoFisher Scientific).
2. NuPAGE MES SDS running buffer (ThermoFisher Scientific).
3. NuPAGE LDS Sample buffer (ThermoFisher Scientific).
4. 500 mM Dithiothreitol (DTT) in water; aliquot and store at –20 °C.
5. SeeBlue Pre-stained standard (ThermoFisher Scientific) (*see* **Note 1**).
6. Scalpels (*see* **Note 2**).
7. Gel-loading pipette tips.
8. Microcentrifuge tubes rinsed with 50% methanol.
9. Coomassie blue stain or Silver stain (Silver Quest Silver Staining Kit; ThermoFisher Scientific) (*see* **Note 3**).
10. Optima LC/MS-grade Water (ThermoFisher Scientific).
11. 1 M Ammonium bicarbonate (Sigma) stock in water; check to ensure pH is between 7.5 and 9.0.
12. Optima LC/MS-grade Acetonitrile (ThermoFisher Scientific).
13. Optima LC/MS-grade Methanol (ThermoFisher Scientific).
14. 30% methanol.
15. 30% acetonitrile in 100 mM Ammonium bicarbonate.
16. 100 mM TCEP ("Bond-breaker neutral pH TCEP solution"; ThermoFisher Scientific): prepare 10 mM solution fresh in 100 mM Ammonium bicarbonate.
17. 55 mM Iodoacetamide (Sigma): prepared in 100 mM Ammonium bicarbonate; Prepare fresh and keep in the dark—iodoacetamide is light sensitive.

18. Digestion buffer: 50 mM Ammonium bicarbonate.
19. Trypsin solution: 1.5 μg sequencing-grade modified trypsin (Promega) in 50 mM Ammonium bicarbonate.
20. Acetonitrile/5 % formic acid.

2.2 C18 StageTip Desalting

1. Empore C18 extraction disks.
2. Optima LC/MS-grade Methanol.
3. Optima LC/MS-grade Acetonitrile.
4. Optima LC/MS-grade Water.
5. Wash solvent 1: 0.1 % Formic acid in Water.
6. Elution solvent: 50 % Acetonitrile/0.1 % Formic acid.

2.3 In-Solution Trypsin Digestion

1. 1 M Ammonium bicarbonate stock in water; ensure that the pH is between 7.5 and 9.0.
2. 0.1 % RapiGest SF™ (Waters) in 50 mM Ammonium Bicarbonate.
3. TCEP ("Bond-breaker neutral pH TCEP solution").
4. 1 M Iodoacetamide stock prepared in 50 mM Ammonium bicarbonate; prepare fresh.
5. 1 M DTT stock prepared in 50 mM Ammonium bicarbonate; 1 M DTT stock can be stored in aliquots at −20 °C.
6. Sequencing-grade modified trypsin.
7. Mass spectrometry-grade Trifluoroacetic acid (Sigma).

2.4 Liquid Chromatography

1. Mobile Phase "A" (aqueous solvent): 3 % Acetonitrile/0.1 % Formic acid in Water; Water and Acetonitrile: Optima, LC/MS-grade; Formic acid: mass spectrometry-grade (Sigma).
2. Mobile Phase "B" (organic solvent): 90 % Acetonitrile/0.1 % Formic acid.
3. C18 trap column: 5 mm length × 300 μm inner diameter, 5 μm Acclaim PepMap 100 C18 (ThermoFisher Scientific).
4. C18 analytical column: 50 cm length × 75 μm inner diameter, 2 μm Acclaim PepMap RSLC C18 EASY-Spray column (ThermoFisher Scientific).
5. Dionex UltiMate 3000 RSLCnano LC system (ThermoFisher Scientific).

2.5 Mass Spectrometry and Data Analysis

1. Q Exactive hybrid quadrupole-Orbitrap mass spectrometer equipped with an EASY-Spray nanospray ionization source (ThermoFisher Scientific) for peptide and protein identification and parallel reaction monitoring (PRM)-based targeted proteomics analysis (*see* **Note 4**). A triple quadrupole or a QqTOF mass spectrometer is required for targeted analyses such as selected/multiple reaction monitoring (SRM/MRM).

2. Trans-Proteomic Pipeline [61] (*see* **Note 5**).
3. MaxQuant quantitative proteomics software [62, 63].
4. MSConvert (freely available via ProteoWizard at http://proteowizard.sourceforge.net/tools.shtml).

2.6 Targeted MS Data Acquisition: Selected/Multiple Reaction Monitoring (SRM/MRM) or Parallel Reaction Monitoring (PRM)

1. Synthetic unlabeled and stable isotope-labeled peptides with sequences that are identical to the peptide(s) of interest to develop the targeted proteomics assay (*see* **Note 6**).
2. Skyline v. 3.5 targeted proteomics software [64].

3 Methods

The methods detailed in this chapter are written based on the analysis of PHF-tau purified from AD brain tissue. As there are several biological material-based sources of tau (e.g., tissue, cerebrospinal fluid, serum), a specific tau purification protocol is not discussed here. The recommended minimum starting amount of purified tau for the methods described in this chapter is 20 μg. The protein concentration of the starting material used for the methods described below should be accurately determined using a BCA assay or similar protein quantification assay. A workflow entailing the methods that are described in this chapter is presented in Fig. 1.

3.1 In-Gel Trypsin Digestion

1. Perform SDS-PAGE to electrophoretically separate the tau-containing sample. Following electrophoresis, wash the gel three times for 10 min per wash with deionized water to remove any residual SDS.
2. Conduct Coomassie blue staining or Silver staining. After staining is complete, wash the gel for 10 min with deionized water.
3. Capture an image of the gel. Use a clean scalpel to excise a gel region corresponding to 50–75 kDa, divide the region into ~2–3 sections and cut into 1 × 1 mm cubes. Transfer gel pieces to a microcentrifuge tube washed with 50% methanol (*see* **Note 7**). If using Silver stain, de-stain the gel pieces before proceeding to the next step.
4. Add 100% methanol for 5 min to dehydrate the gel pieces. Add a sufficient volume of methanol to cover the gel pieces.
5. Remove 100% methanol and add 30% methanol for 5 min to rehydrate the gel pieces (*see* **Note 8**).
6. Remove 30% methanol and wash the gel pieces twice for 10 min per wash with water.

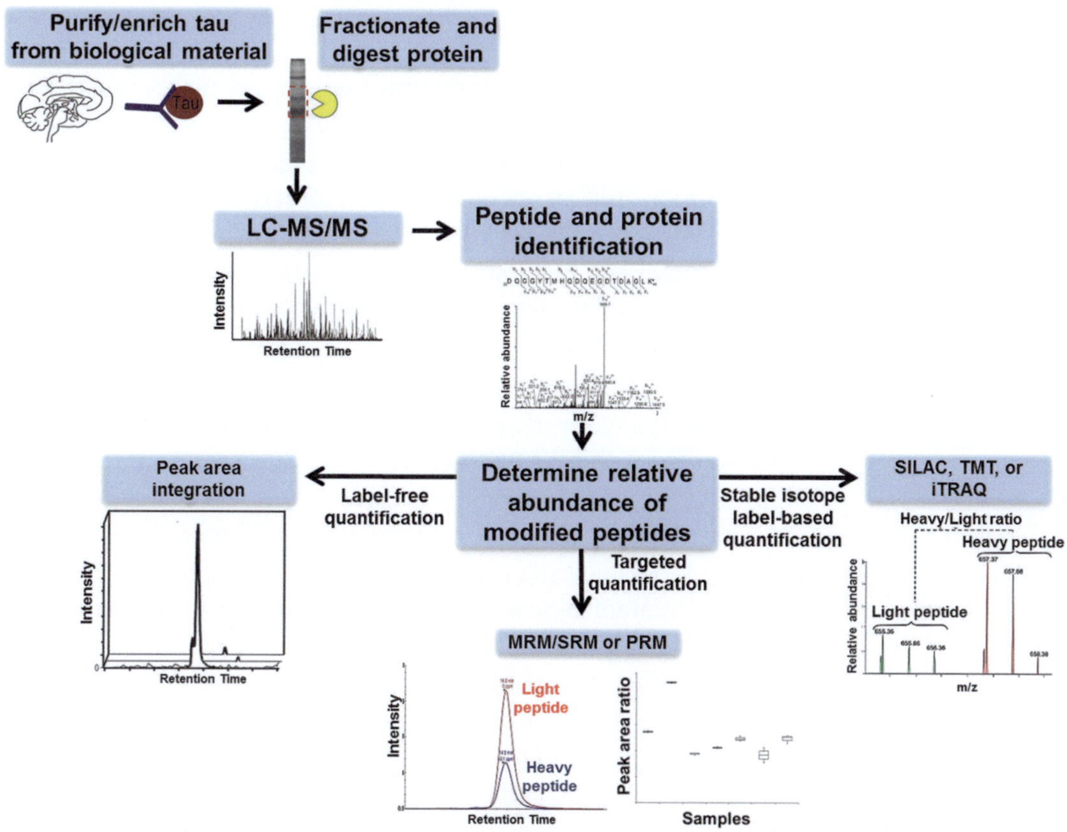

Fig. 1 Workflow for the mass spectrometry-based analysis of lysine-directed PTMs of tau. *MRM/SRM* multiple reaction monitoring/selected reaction monitoring, *PRM* parallel reaction monitoring, *TMT* tandem mass tags, *iTRAQ* isobaric tags for relative and absolute quantification, *SILAC* stable isotope labeling by amino acids in cell culture

7. Wash the gel pieces three times for 10 min per wash with 30% acetonitrile in 100 mM Ammonium bicarbonate.
8. Dry gel pieces in a vacuum centrifuge (speed vacuum) for 15 min. Gel pieces will become opaque when dry.
9. Add 10 mM TCEP (enough volume to cover the gel pieces) and let incubate for 1 h at 60 °C to reduce protein disulfide bonds.
10. Briefly centrifuge the gel pieces and discard the liquid.
11. Add 55 mM iodoacetamide (enough volume to cover the gel pieces) and let incubate for 45 min at room temperature in the dark to alkylate cysteine residues.
12. Briefly centrifuge the gel pieces and discard the liquid.
13. Wash the gel pieces for 15 min with 100 mM Ammonium bicarbonate.
14. Discard the liquid. Shrink the gel pieces with 100% acetonitrile.

15. Discard the liquid. Dry the gel pieces in a speed vacuum for 15 min. Ensure that the gel pieces are completely dry to facilitate the absorption of trypsin into the gel pieces in the next step.
16. Rehydrate the gel pieces in trypsin solution at 4 °C (on ice) for 45 min. Add an adequate volume of trypsin solution to completely cover the gel pieces (*see* **Note 9**).
17. Remove any remaining trypsin solution and add a sufficient volume of digestion buffer to completely cover the gel pieces.
18. Let incubate at 37 °C overnight with gentle shaking.
19. Briefly centrifuge the gel pieces and transfer the supernatant (containing tryptic peptides) to another microcentrifuge tube washed with 50 % methanol.
20. Add 50 mM Ammonium bicarbonate to cover the gel pieces and incubate at 37 °C with shaking for 15 min.
21. Add a volume of acetonitrile equal to the volume of 50 mM Ammonium bicarbonate utilized in **step 20** and incubate at 37 °C with shaking for 15 min.
22. Centrifuge the gel pieces and add the supernatant to the supernatant collected in **step 19**.
23. Extract the peptides: add acetonitrile/5 % formic acid (50:50) to the gel pieces and incubate at 37 °C with shaking for 15 min.
24. Remove the supernatant and combine with the supernatant collected in **steps 19** and **22**.
25. Repeat **step 23**.
26. Remove the supernatant and combine with the supernatant collected in **steps 19**, **22**, and **24**.
27. Dry the combined supernatant in a speed vac.
28. Desalt the sample using C18 StageTips.

3.2 In-Solution Trypsin Digestion

1. Use at least 20 μg protein at a concentration of 0.2–1.0 μg/μL for in-solution digestion. If protein is in a pellet form, add 0.1 % RapiGest in 50 mM Ammonium Bicarbonate. If protein is in solution, the ideal buffer is 50 mM Ammonium bicarbonate. Conduct buffer exchange if necessary and add 1:1 (v/v) 0.2 % RapiGest:protein solution.
2. Denature protein by incubating at 90 °C for 3 min. Cool to room temperature.
3. Add TCEP to 5 mM final concentration to reduce protein disulfide bonds. Vortex and incubate for 30 min at 60 °C. Let cool to room temperature.
4. Alkylate cysteine residues by adding iodoacetamide to 10 mM final concentration. Vortex and incubate at room temperature in the dark for 30 min.

5. Add trypsin to sample at a trypsin:protein ratio of 1:50 (w/w). Let incubate overnight at 37 °C with gentle shaking (*see* **Note 9**).
6. Stop enzymatic digestion and hydrolyze RapiGest by adding trifluoroacetic acid to 1% final concentration. Incubate for 45 min at 37 °C.
7. Centrifuge at 14,000 × *g* for 10 min at 4 °C and transfer the supernatant to a new microcentrifuge tube (*see* **Note 10**).
8. Desalt the sample using a C18-based spin column, cartridge, or tip (*see* **Note 11**).

3.3 C18 StageTip Desalting

1. Prepare C18 StageTips using two plugs of C18 material in each StageTip (*see* **Notes 12** and **13**).
2. Reconstitute the dried protein digest sample in 50 μL of Wash Solvent 1.
3. Wash the StageTip once with 50 μL of 100% Methanol.
4. Equilibrate the StageTip twice with 50 μL of Wash Solvent 1.
5. Load the sample onto the equilibrated StageTip.
6. Wash and desalt the sample twice with 50 μL of Wash Solvent 1.
7. Elute the sample into a clean microcentrifuge tube using 50 μL of Elution Solvent.
8. Dry the desalted sample in a speed vac. Reconstitute in liquid chromatography Mobile Phase A. For in-gel digests, use ~1/3 of the sample for LC-MS/MS analysis. For in-solution digests, load ~1 μg of peptides.

3.4 Liquid Chromatography

Utilize an LC method with the following parameters:

1. Gradient pump flow rate = 0.3 μL/min; loading pump flow rate = 5.0 μL/min.
2. Column oven temperature = 35 °C.
3. Autosampler temperature = 6.0 °C.
4. Sample loading time = 2.5× sample loop volume.
5. Separation gradient: 60 min linear gradient from 3 to 35% Mobile Phase B, followed by 35–90% B in 5 min, 90% B for 5 min, and column re-equilibration with 3%B for 10 min.

3.5 Mass Spectrometry

1. Set spray voltage to 1.8 kV and heated capillary temperature to 180 °C.
2. Create a data-dependent MS/MS instrument method with the following parameters:
   (a) Full scan range: 400–1800 *m/z*; positive polarity
   (b) Ten most abundant ions selected for MS/MS fragmentation

(c) Resolution = 70,000
(d) Maximum injection time = 100 ms
(e) Dynamic exclusion enabled—exclude ions for 20 s after being detected once within 20 s (see Note 14).
(f) Automatic gain control target = 5e4
(g) Activation time = 30 ms
(h) Isolation width = 2.0 m/z
(i) Normalized collision energy = 28 %
(j) Charge state exclusion: z = unassigned, 1 and ≥6

3.6 Mass Spectrometry Data Analysis of Lysine PTMs

1. Use MSConvert to convert the acquired .raw files to .mzML format.
2. Conduct database search and peptide-spectrum match (PSM) validation using tools available in the Trans-Proteomic Pipeline [61]. Use the following database search parameters:
   (a) Database: UniProtKB human protein database with reviewed sequences. The data should also be searched against a targeted library of tau isoform sequences with the following UniProt accession numbers: P10637, P10637-2, P10637-3, P10637-4, P-10637-5, and P10637-6.
   (b) Fixed modification: Cys +57.02146 (carbamidomethylation).
   (c) Variable Lys-directed modifications: Lys +14.01565 (monomethylation), Lys +28.03130 (dimethylation), Lys +42.04695 (trimethylation), Lys +42.01056 (acetylation), and Lys +114.042927 (di-Gly; tryptic ubiquitylation remnant). Other non-Lys-directed variable modifications to which tau is known to be subjected include the following: Met +15.99491 (oxidation); Ser, Thr, Tyr +79.96633 (phosphorylation); and Protein N-terminus +42.01056 (acetylation).
   (d) Precursor ion mass tolerance = 20 ppm.
   (e) Fragment ion (MS/MS) mass tolerance = 0.05 Da.
   (f) Maximum missed cleavages = 2.
   (g) Maximum modifications per peptide = 6.
   (h) Enzyme cleavage specificity = trypsin (or enzyme that was used for the in solution or in gel digestion).
3. Use PeptideProphet (available as part of the Trans-Proteomic Pipeline) to validate the peptide search results and set the false discovery rate (FDR) at 1 %.
4. Infer proteins from the identified peptides using ProteinProphet (available as part of the Trans-Proteomic Pipeline).

5. Manually inspect spectra of peptides passing the 1% FDR filter to verify that the majority of the major fragment ions are assigned. Peptides with C-terminal ubiquitylation sites should be removed and considered false positive identifications because ubiquitylation commonly results in a missed trypsin cleavage site at the modified Lys residue on the ubiquitylation substrate protein [65].
6. Determine the PTM site localization probability (PTM score) using MaxQuant [62, 63]. When evaluating the quality of the data, take into consideration the delta score, which measures the difference between the best spectrum-match and the next best match with a different amino acid sequence. Also consider the score difference between the best and second-best positioning of a PTM within the same peptide sequence.
7. If the sample was isotopically labeled (e.g., chemical labeling with iTRAQ or TMT tags or metabolically labeled using SILAC), use a software package such as MaxQuant [62, 63] or QuantiMORE [66] to calculate the peptide and protein ratios (*see* **Notes 15** and **16**).
8. Researchers are encouraged to deposit their raw mass spectrometry data to repositories such as the ProteomeXchange Consortium via the PRIDE partner repository [67].

3.7 Label-Free Quantification of Mass Spectrometry Data Using Intensity-Based Peak Area Integration

Peptide features are quantified by measuring their corresponding spectral intensities or areas under the peaks in MS or MS/MS spectra (*see* **Note 17**).

Search the mass spectrometry data using MaxQuant with the default settings for label-free quantification wherein peptide intensity values are aggregated by charge state and modification status [68]. The peptide feature intensities can then be $\log_2$-transformed followed by conducting pairwise comparisons between the samples using an appropriate statistical test based on how the data are distributed. Intensity-based quantification can also be performed using other software such as SIEVE.

3.8 Targeted Mass Spectrometry-Based Quantitative Measurement of Peptides Using SRM or PRM

1. The main steps of targeted mass spectrometry method development are as follows: (1) generate a hypothesis that can be tested by quantitative protein measurements; (2) develop the SRM or PRM method using synthetic peptide standards; (3) refine the method; and (4) validate the method. These steps are outlined in further detail in [58].
2. Import tau protein sequence into Skyline to conduct in silico digestion, select peptides of interest, select all *y*-ion transitions (m/z > precursor m/z), select optimal collision energies predicted via mass spectrometer-specific linear equations, and export the instrument method that will be used for MS analysis.

If the peptide of interest contains a posttranslational modification, ensure that at least one transition contains the modified residue.

3. Conduct transition refinement by manual inspection of data analyzed using Skyline.
4. Establish a scheduled instrument method using retention time prediction in Skyline.
5. Use the refined scheduled method to develop assay including response curve generation and determination of repeatability. An example of guidelines for assay development can be found at https://assays.cancer.gov/guidance-document/ [69].

4 Notes

1. Other pre-stained protein standards are suitable; however, the standard should include proteins with molecular weights close to those of tau isoforms (~45–65 kDa).
2. The use of scalpels as opposed to razor blades is preferred so as to minimize the chances of gloves making contact with the gel when excising gel bands and cutting the bands into smaller pieces. Avoid using latex gloves as these are a common source of keratin contamination.
3. There are alternative commercially available mass spectrometry compatible silver staining kits. Mass spectrometry compatible silver staining methods are those that do not involve the use of glutaraldehyde to fix the proteins in the gel.
4. Other types of mass spectrometers can be used to analyze the samples; however, for the best results, a mass spectrometer with high resolution and mass accuracy should be used. This is particularly important for differentiating among PTMs with similar masses such as trimethylation (42.046950 Da) and acetylation (42.010565 Da); Δ mass = 0.036385 Da.
5. Several other open source and freely available database search engines, algorithms, and software tools such as Mascot [70], MaxQuant [62, 63], Comet [71], MSGF+ [72], and ID Picker (MyriMatch) [73] can be used; however, be certain to adhere to the parameters that are outlined in Subheading 3.6.
6. Detailed guidelines for peptide procurement, characterization, storage, and handling, as well as approaches to the interpretation of the data generated by targeted MS-based assays have been developed [50].
7. All steps for in-gel digestion should be performed in a clean laminar flow hood to minimize keratin contamination.
8. For all steps of the in-gel digestion protocol that entail removing solutions from gel pieces, use a gel-loading pipette tip.

9. Enzymes other than trypsin such as LysC, ArgC, AspN, and GluC can be used to digest tau. The use of multiple proteases has been shown to improve the protein sequence coverage obtained from mass spectrometry-based analysis [74].
10. A cloudy pellet should appear after centrifugation. When pipetting the supernatant (containing peptides) after completion of the centrifugation, take caution to not disturb the pellet.
11. Consider the amount of peptides (measure using UV-based absorbance at 280 nm) and the binding capacity of the column, tip, or cartridge when selecting the format of the C18-based material used for desalting.
12. The original publication of the StageTip procedure provides detailed instructions and illustrations for how to prepare StageTips [75].
13. Use 2 mL microcentrifuge tubes and StageTip adapters so that the procedure can be completed using a microcentrifuge format. Conduct all centrifugation steps for the StageTip desalting at 3000–4000 × *g* for 3–5 min at room temperature or until all of the top-loaded solution has passed through the extraction disks in the StageTip.
14. The optimum duration for dynamic exclusion should be empirically determined based on the average chromatographic peak width. Using a dynamic exclusion duration of 20 s with an LC system that generates peak widths of ~40 s permits the acquisition of MS/MS data for a peptide as it begins to elute as a peak, at the apex of the peak and at the peak tail. The dynamic exclusion settings also affect the performance of spectral counting methods wherein low abundance proteins are under-sampled and high abundance proteins can be sampled to a degree such that near-maximum sequence coverage is achieved [76]. Optimization of the dynamic exclusion settings can increase the reproducibility of spectral counting and the quantification of low or high abundant proteins.
15. QuantiMORE (formerly named IsoQuant) is freely available for download at http://www.proteomeumb.org/MZw.html.
16. MaxQuant consists of several independent modules that enable the complete processing of raw mass spectrometry data files. It supports protein identification, quantification, recalibration, and the quality control of the raw and annotated mass spectra. A thorough step-by-step protocol explaining how MaxQuant can be used to analyze proteomics datasets was published in a recent edition of *Methods in Molecular Biology* [77].
17. Among the label-free methods for the quantification of mass spectrometry data, intensity-based quantification methods have been demonstrated to be more sensitive than spectral counting [78]. Although it is possible to conduct intensity-based

quantification by summing the MS/MS-level feature intensities, quantification on the MS-level has been demonstrated to be more accurate [79].

Acknowledgments

This work was supported by National Institutes of Health grants MH59786 and AG25323 (A.Y.) and the Maryland Cigarette Restitution Fund.

References

1. Braak H, Braak E, Strothjohann M (1994) Abnormally phosphorylated tau protein related to the formation of neurofibrillary tangles and neuropil threads in the cerebral cortex of sheep and goat. Neurosci Lett 171:1–4
2. Goedert M (1993) Tau protein and the neurofibrillary pathology of Alzheimer's disease. Trends Neurosci 16:460–465
3. Martin L, Latypova X, Terro F (2011) Post-translational modifications of tau protein: implications for Alzheimer's disease. Neurochem Int 58:458–471
4. Augustinack JC, Schneider A, Mandelkow EM et al (2002) Specific tau phosphorylation sites correlate with severity of neuronal cytopathology in Alzheimer's disease. Acta Neuropathol 103:26–35
5. Hampel H, Burger K, Pruessner JC et al (2005) Correlation of cerebrospinal fluid levels of tau protein phosphorylated at threonine 231 with rates of hippocampal atrophy in Alzheimer disease. Arch Neurol 62:770–773
6. Buee L, Bussiere T, Buee-Scherrer V et al (2000) Tau protein isoforms, phosphorylation and role in neurodegenerative disorders. Brain Res Brain Res Rev 33:95–130
7. Holzer M, Holzapfel HP, Zedlick D et al (1994) Abnormally phosphorylated tau protein in Alzheimer's disease: heterogeneity of individual regional distribution and relationship to clinical severity. Neuroscience 63:499–516
8. Gong CX, Liu F, Grundke-Iqbal I et al (2005) Post-translational modifications of tau protein in Alzheimer's disease. J Neural Transm (Vienna) 112:813–838
9. Marshall KE, Morris KL, Charlton D et al (2011) Hydrophobic, aromatic, and electrostatic interactions play a central role in amyloid fibril formation and stability. Biochemistry 50:2061–2071
10. Sinha S, Lopes DH, Du Z et al (2011) Lysine-specific molecular tweezers are broad-spectrum inhibitors of assembly and toxicity of amyloid proteins. J Am Chem Soc 133:16958–16969
11. Norlund MA, Lee JM, Zainelli GM et al (1999) Elevated transglutaminase-induced bonds in PHF tau in Alzheimer's disease. Brain Res 851:154–163
12. Petrucelli L, Dickson D, Kehoe K et al (2004) CHIP and Hsp70 regulate tau ubiquitination, degradation and aggregation. Hum Mol Genet 13:703–714
13. Shimura H, Schwartz D, Gygi SP et al (2004) CHIP-Hsc70 complex ubiquitinates phosphorylated tau and enhances cell survival. J Biol Chem 279:4869–4876
14. Arnaud LT, Myeku N, Figueiredo-Pereira ME (2009) Proteasome-caspase-cathepsin sequence leading to tau pathology induced by prostaglandin J2 in neuronal cells. J Neurochem 110:328–342
15. David DC, Layfield R, Serpell L et al (2002) Proteasomal degradation of tau protein. J Neurochem 83:176–185
16. Liu YH, Wei W, Yin J et al (2009) Proteasome inhibition increases tau accumulation independent of phosphorylation. Neurobiol Aging 30:1949–1961
17. de Vrij FM, Fischer DF, van Leeuwen FW et al (2004) Protein quality control in Alzheimer's disease by the ubiquitin proteasome system. Prog Neurobiol 74:249–270
18. Zhang JY, Liu SJ, Li HL et al (2005) Microtubule-associated protein tau is a substrate of ATP/Mg(2+)-dependent proteasome protease system. J Neural Transm (Vienna) 112:547–555
19. Riederer IM, Schiffrin M, Kovari E et al (2009) Ubiquitination and cysteine nitrosylation during aging and Alzheimer's disease. Brain Res Bull 80:233–241

20. Cripps D, Thomas SN, Jeng Y et al (2006) Alzheimer disease-specific conformation of hyperphosphorylated paired helical filament-Tau is polyubiquitinated through Lys-48, Lys-11, and Lys-6 ubiquitin conjugation. J Biol Chem 281:10825–10838
21. Morishima-Kawashima M, Hasegawa M, Takio K et al (1993) Ubiquitin is conjugated with amino-terminally processed tau in paired helical filaments. Neuron 10:1151–1160
22. Thomas SN, Funk KE, Wan Y et al (2012) Dual modification of Alzheimer's disease PHF-tau protein by lysine methylation and ubiquitylation: a mass spectrometry approach. Acta Neuropathol 123:105–117
23. Funk KE, Thomas SN, Schafer KN et al (2014) Lysine methylation is an endogenous post-translational modification of tau protein in human brain and a modulator of aggregation propensity. Biochem J 462:77–88
24. Morris M, Knudsen GM, Maeda S et al (2015) Tau post-translational modifications in wild-type and human amyloid precursor protein transgenic mice. Nat Neurosci 18:1183–1189
25. Cohen TJ, Friedmann D, Hwang AW et al (2013) The microtubule-associated tau protein has intrinsic acetyltransferase activity. Nat Struct Mol Biol 20:756–762
26. Min SW, Cho SH, Zhou Y et al (2010) Acetylation of tau inhibits its degradation and contributes to tauopathy. Neuron 67:953–966
27. Cook C, Stankowski JN, Carlomagno Y et al (2014) Acetylation: a new key to unlock tau's role in neurodegeneration. Alzheimers Res Ther 6:29
28. Irwin DJ, Cohen TJ, Grossman M et al (2012) Acetylated tau, a novel pathological signature in Alzheimer's disease and other tauopathies. Brain 135:807–818
29. Choudhary C, Kumar C, Gnad F et al (2009) Lysine acetylation targets protein complexes and co-regulates major cellular functions. Science 325:834–840
30. Imamura H, Sugiyama N, Wakabayashi M et al (2014) Large-scale identification of phosphorylation sites for profiling protein kinase selectivity. J Proteome Res 13:3410–3419
31. Kim W, Bennett EJ, Huttlin EL et al (2011) Systematic and quantitative assessment of the ubiquitin-modified proteome. Mol Cell 44:325–340
32. Olsen JV, Vermeulen M, Santamaria A et al (2010) Quantitative phosphoproteomics reveals widespread full phosphorylation site occupancy during mitosis. Sci Signal 3:ra3
33. Trinidad JC, Barkan DT, Gulledge BF et al (2012) Global identification and characterization of both O-GlcNAcylation and phosphorylation at the murine synapse. Mol Cell Proteomics 11:215–229
34. Udeshi ND, Svinkina T, Mertins P et al (2013) Refined preparation and use of anti-diglycine remnant (K-epsilon-GG) antibody enables routine quantification of 10,000s of ubiquitination sites in single proteomics experiments. Mol Cell Proteomics 12:825–831
35. Olsen JV, Mann M (2013) Status of large-scale analysis of post-translational modifications by mass spectrometry. Mol Cell Proteomics 12: 3444–3452
36. Thomas SN, Zhang H, Cotter RJ (2015) Application of quantitative proteomics to the integrated analysis of the ubiquitylated and global proteomes of xenograft tumor tissues. Clin Proteomics 12:14
37. Chalkley RJ, Clauser KR (2012) Modification site localization scoring: strategies and performance. Mol Cell Proteomics 11:3–14
38. Rigbolt KT, Prokhorova TA, Akimov V et al (2011) System-wide temporal characterization of the proteome and phosphoproteome of human embryonic stem cell differentiation. Sci Signal 4:rs3
39. Bros P, Vialaret J, Barthelemy N et al (2015) Antibody-free quantification of seven tau peptides in human CSF using targeted mass spectrometry. Front Neurosci 9:302
40. McAvoy T, Lassman ME, Spellman DS et al (2014) Quantification of tau in cerebrospinal fluid by immunoaffinity enrichment and tandem mass spectrometry. Clin Chem 60:683–689
41. Peterson AC, Russell JD, Bailey DJ et al (2012) Parallel reaction monitoring for high resolution and high mass accuracy quantitative, targeted proteomics. Mol Cell Proteomics 11:1475–1488
42. Tsuchiya H, Tanaka K, Saeki Y (2013) The parallel reaction monitoring method contributes to a highly sensitive polyubiquitin chain quantification. Biochem Biophys Res Commun 436:223–229
43. Tang H, Fang H, Yin E et al (2014) Multiplexed parallel reaction monitoring targeting histone modifications on the QExactive mass spectrometer. Anal Chem 86:5526–5534
44. Yu Q, Liu B, Ruan D et al (2014) A novel targeted proteomics method for identification and relative quantitation of difference in nitration degree of OGDH between healthy and diabetic mouse. Proteomics 14:2417–2426
45. Thomas SN, Harlan R, Chen J et al (2015) Multiplexed targeted mass spectrometry-based assays for the quantification of N-linked glycosite-containing peptides in serum. Anal Chem 87:10830–10838

46. Kim YJ, Gallien S, El-Khoury V et al (2015) Quantification of SAA1 and SAA2 in lung cancer plasma using the isotype-specific PRM assays. Proteomics 15:3116–3125
47. Abbatiello SE, Schilling B, Mani DR et al (2015) Large-scale interlaboratory study to develop, analytically validate and apply highly multiplexed, quantitative peptide assays to measure cancer-relevant proteins in plasma. Mol Cell Proteomics 14:2357–2374
48. Carr SA, Abbatiello SE, Ackermann BL et al (2014) Targeted peptide measurements in biology and medicine: best practices for mass spectrometry-based assay development using a fit-for-purpose approach. Mol Cell Proteomics 13:907–917
49. Abbatiello SE, Mani DR, Schilling B et al (2013) Design, implementation and multisite evaluation of a system suitability protocol for the quantitative assessment of instrument performance in liquid chromatography-multiple reaction monitoring-MS (LC-MRM-MS). Mol Cell Proteomics 12:2623–2639
50. Hoofnagle AN, Whiteaker JR, Carr SA et al (2016) Recommendations for the generation, quantification, storage, and handling of peptides used for mass spectrometry-based assays. Clin Chem 62:48–69
51. Sharma V, Eckels J, Taylor GK et al (2014) Panorama: a targeted proteomics knowledge base. J Proteome Res 13:4205–4210
52. Reiter L, Rinner O, Picotti P et al (2011) mProphet: automated data processing and statistical validation for large-scale SRM experiments. Nat Methods 8:430–435
53. Abbatiello SE, Mani DR, Keshishian H et al (2010) Automated detection of inaccurate and imprecise transitions in peptide quantification by multiple reaction monitoring mass spectrometry. Clin Chem 56:291–305
54. Aiyetan P, Thomas SN, Zhang Z et al (2015) MRMPlus: an open source quality control and assessment tool for SRM/MRM assay development. BMC Bioinformatics 16:411
55. Ebhardt HA, Root A, Sander C et al (2015) Applications of targeted proteomics in systems biology and translational medicine. Proteomics 15:3193–3208
56. Domon B (2012) Considerations on selected reaction monitoring experiments: implications for the selectivity and accuracy of measurements. Proteomics Clin Appl 6:609–614
57. Picotti P, Aebersold R (2012) Selected reaction monitoring-based proteomics: workflows, potential, pitfalls and future directions. Nat Methods 9:555–566
58. Bereman MS, MacLean B, Tomazela DM et al (2012) The development of selected reaction monitoring methods for targeted proteomics via empirical refinement. Proteomics 12: 1134–1141
59. Gallien S, Duriez E, Domon B (2011) Selected reaction monitoring applied to proteomics. J Mass Spectrom 46:298–312
60. Khristenko N, Domon B (2015) Quantification of proteins in urine samples using targeted mass spectrometry methods. Methods Mol Biol 1243:207–220
61. Deutsch EW, Mendoza L, Shteynberg D et al (2015) Trans-Proteomic Pipeline, a standardized data processing pipeline for large-scale reproducible proteomics informatics. Proteomics Clin Appl 9:745–754
62. Cox J, Mann M (2008) MaxQuant enables high peptide identification rates, individualized p.p.b.-range mass accuracies and proteome-wide protein quantification. Nat Biotechnol 26:1367–1372
63. Cox J, Neuhauser N, Michalski A et al (2011) Andromeda: a peptide search engine integrated into the MaxQuant environment. J Proteome Res 10:1794–1805
64. MacLean B, Tomazela DM, Shulman N et al (2010) Skyline: an open source document editor for creating and analyzing targeted proteomics experiments. Bioinformatics 26:966–968
65. Peng J, Schwartz D, Elias JE et al (2003) A proteomics approach to understanding protein ubiquitination. Nat Biotechnol 21:921–926
66. Liao Z, Wan Y, Thomas SN et al (2012) IsoQuant: a software tool for stable isotope labeling by amino acids in cell culture-based mass spectrometry quantitation. Anal Chem 84:4535–4543
67. Vizcaino JA, Deutsch EW, Wang R et al (2014) ProteomeXchange provides globally coordinated proteomics data submission and dissemination. Nat Biotechnol 32:223–226
68. Goeminne LJ, Argentini A, Martens L et al (2015) Summarization vs peptide-based models in label-free quantitative proteomics: performance, pitfalls, and data analysis guidelines. J Proteome Res 14:2457–2465
69. Whiteaker JR, Halusa GN, Hoofnagle AN et al (2014) CPTAC assay portal: a repository of targeted proteomic assays. Nat Methods 11:703–704
70. Perkins DN, Pappin DJ, Creasy DM et al (1999) Probability-based protein identification by searching sequence databases using mass spectrometry data. Electrophoresis 20: 3551–3567

71. Eng JK, Jahan TA, Hoopmann MR (2013) Comet: an open-source MS/MS sequence database search tool. Proteomics 13:22–24
72. Kim S, Pevzner PA (2014) MS-GF+ makes progress towards a universal database search tool for proteomics. Nat Commun 5:5277
73. Tabb DL, Fernando CG, Chambers MC (2007) MyriMatch: highly accurate tandem mass spectral peptide identification by multivariate hypergeometric analysis. J Proteome Res 6:654–661
74. Swaney DL, Wenger CD, Coon JJ (2010) Value of using multiple proteases for large-scale mass spectrometry-based proteomics. J Proteome Res 9:1323–1329
75. Rappsilber J, Mann M, Ishihama Y (2007) Protocol for micro-purification, enrichment, pre-fractionation and storage of peptides for proteomics using StageTips. Nat Protoc 2: 1896–1906
76. Hoehenwarter W, Wienkoop S (2010) Spectral counting robust on high mass accuracy mass spectrometers. Rapid Commun Mass Spectrom 24:3609–3614
77. Tyanova S, Mann M, Cox J (2014) MaxQuant for in-depth analysis of large SILAC datasets. Methods Mol Biol 1188: 351–364
78. Milac TI, Randolph TW, Wang P (2012) Analyzing LC-MS/MS data by spectral count and ion abundance: two case studies. Stat Interface 5:75–87
79. Krey JF, Wilmarth PA, Shin JB et al (2014) Accurate label-free protein quantitation with high- and low-resolution mass spectrometers. J Proteome Res 13:1034–1044

Chapter 11

The Study of Posttranslational Modifications of Tau Protein by Nuclear Magnetic Resonance Spectroscopy: Phosphorylation of Tau Protein by ERK2 Recombinant Kinase and Rat Brain Extract, and Acetylation by Recombinant Creb-Binding Protein

Haoling Qi, Clément Despres, Sudhakaran Prabakaran, François-Xavier Cantrelle, Béatrice Chambraud, Jeremy Gunawardena, Guy Lippens, Caroline Smet-Nocca, and Isabelle Landrieu

Abstract

Nuclear magnetic resonance (NMR) spectroscopy can be used as an analytical tool to investigate posttranslational modifications of protein. NMR is a valuable tool to map the interaction regions of protein partners. Here, we present protocols that have been developed in the course of our studies of the neuronal Tau protein. Tau is found aggregated in the neurons of Alzheimer's disease patients. Development of the disease is accompanied by increased, abnormal phosphorylation and acetylation of Tau. We have used NMR to investigate how these posttranslational modifications of Tau affect the interactions with its partners. We present here detailed protocols of *in vitro* phosphorylation of Tau by recombinant kinase, ERK2, or kinase activity of rat brain extracts, and acetylation by recombinant Creb-binding protein (CBP) acetyltransferase. The analytical characterization of the modified Tau by NMR spectroscopy is additionally described.

Key words Phosphorylation, Acetylation, ERK kinase, Creb-binding protein, Acetyltransferase, NMR spectroscopy, Recombinant proteins

1 Introduction

Tau protein is hyperphosphorylated in the paired helical filaments (PHF) found in neurons in brain affected by Alzheimer's disease. The longest Tau protein isoform (441-residue) has 80 Threonine (T) and Serine (S) residues exposed since Tau is an intrinsically disordered protein with all these sites prone to modification by numerous kinases [1]. In the PHF-Tau, up to about 45

Caroline Smet-Nocca (ed.), *Tau Protein: Methods and Protocols*, Methods in Molecular Biology, vol. 1523,
DOI 10.1007/978-1-4939-6598-4_11, © Springer Science+Business Media New York 2017

phosphorylation sites can be modified [2, 3] while 15–30 Tau phosphorylation sites were identified by mass spectrometry in Tau extracted from both normal mice and a mouse AD disease model [4] or cognitively normal human brain. Most of them correspond to proline-directed sites located mainly in the proline-rich region of Tau. Given the number and the proximity of these phosphorylation sites, to identify the phosphorylation sites in a phosphorylated Tau sample and to evaluate the level of modification for each of these sites are difficult [4]. Recently, acetylation of Tau protein was shown to be associated with Tauopathies [5, 6]. In this case, the lysine residues are modified by an acetyl group on the NH3 moiety of the side chain, neutralizing the positive charge of the lysine and modifying the steric characteristics of the side chain. Mass spectrometry analysis revealed 14 lysine residues were acetylated in Tau samples purified from mice, and these were mainly located in the microtubule binding region of Tau [4].

In addition, these posttranslational modifications affect the binding of Tau to several molecular partners, an aspect of interest to study by NMR spectroscopy [7]. The need for *in vitro* modified Tau for these NMR experiments has led us to develop protocols to generate the required amount of modified protein. In addition, we have used NMR spectroscopy to characterize enzymatically modified Tau samples with success [8–12]. Given the improvement of the probe sensitivity, we can now work with samples in the 10 μM range, in volumes that can be limited to 200 μl. A typical good sample of Tau would be 1 mg in 200 μl or 100 μM of Tau protein. This still is a lot of material compared to most of the applications of biochemistry, such as the immunodetection for the phosphorylation characterization or the GST pull-down for the interaction investigations. Yet compared to these methods we can obtain information at the molecular level and define the interaction sites up to the residue. Hence, the drawback of the material quantity is compensated by the resolution of the results.

We will here first describe the preparation of isotopically labeled Tau for NMR investigation. The in vitro phosphorylated samples of Tau by rat brain extract have been used for years as a model of hyperphosphorylated Tau [13–15]. Phosphorylation by rat brain extract has been described in Goedert et al. 1993 [14]. The protocol here presented is an adaptation to the need of biophysical analysis requiring the preparation of sufficient quantity of protein. In the course of our investigation, we also observed that a Tau protein phosphorylated by ERK2 is modified on numerous sites described as pathological sites of phosphorylation, and thus represents an interesting model of hyperphosphorylated Tau. ERK2 is activated by phosphorylation by MEK [16–18]. The acetylation of Tau is obtained in vitro by a fragment of the CBP protein, a histone acetyltransferase [12]. In addition to the modified, isotopically labeled Tau protein preparation, we describe the NMR strategy used for the identification of the PTMs.

2 Materials

Prepare all solutions using ultrapure water and analytical grade reagents. Media and stock solutions for bacterial culture, and all buffers are sterilized.

2.1 Preparation of Recombinant ^{15}N-Tau, ^{15}N, ^{13}C-Tau and $^{15}N_2$-Lysine/^{13}C-Tau

1. TaupET15b recombinant T7 expression plasmid (Novagen).
2. BL21(DE3) transformation competent *E. coli* bacteria (New England Biolabs).
3. Autoclaved LB Broth, Lennox (DIFCO).
4. M9 buffer: 6 g Na_2HPO_4, 3 g KH_2PO_4, 0.5 g NaCl per liter; dissolve in water and autoclave.
5. 1 M $MgSO_4$: prepare in ultrapure water and autoclave.
6. 100 mM $CaCl_2$: prepare in ultrapure water and autoclave.
7. MEM vitamin complements 100× (Sigma; *see* **Note 1**).
8. 100 mg/ml stock solution of ampicillin antibiotics: prepare in ultrapure water and store 1 ml aliquots at −20 °C.
9. 1 M stock solution of IPTG (Isopropyl β-D-1-thiogalactopyranoside): prepare in ultrapure water, 0.2 μM filter-sterilize and store 1 ml-aliquots at −20 °C.
10. ^{15}N-NH_4Cl (Isotec) or ^{14}N-NH_4Cl.
11. $^{13}C_6$-Glucose (Isotec).
12. $^{12}C_6$-Glucose.
13. $^{15}N_2$/$^{13}C_6$-L-Lysine hydrochloride (Isotec).
14. ^{15}N-ISOGRO complete medium powder (Isotec) (*see* **Note 1**).
15. ^{15}N, ^{13}C-ISOGRO complete medium powder (Isotec) (*see* **Note 1**).
16. The M9 medium is reconstituted by addition of 1 ml 1 M $MgSO_4$, 1 ml 100 mM $CaCl_2$ (*see* **Note 2**), 10 ml 100× MEM vitamin complement (Sigma), 1 ml 100 mg/ml Ampicillin to 1 L of M9 buffer. In addition, depending on the isotope labeling scheme, the components 17, 18, and 19 are dissolved in 10 ml of M9 buffer and directly 0.2 μM filter-sterilized into the M9 medium (*see* **Note 3**).
17. For the ^{15}N uniform labeling: 300 mg ^{15}N-ISOGRO complete medium, 1 g ^{15}N-NH_4Cl, and 4 g $^{12}C_6$-glucose.
18. For the ^{15}N, ^{13}C uniform labeling: 300 mg ^{15}N, ^{13}C-ISOGRO complete medium, 1 g ^{15}N-NH_4Cl, and 2 g $^{13}C_6$-glucose (*see* **Note 4**).
19. For the selective ^{15}N-Lysine and uniform ^{13}C-labeling: 1 g ^{14}N-NH_4Cl, 2 g $^{13}C_6$-glucose and 20 min before induction 150 mg of $^{15}N_2$/$^{13}C_6$-L-Lysine hydrochloride.

20. Cation-exchange chromatography (CEX) A buffer (*see* **Note 5**): 50 mM phosphate buffer pH6.5, 1 mM EDTA; autoclave.
21. Extraction buffer: CEX A buffer supplemented with Protease inhibitors 1× (1 pellet for 50 ml, Roche) and 2000 U DNaseI for 50 ml (20,000 U/ml, EUROMEDEX).
22. Cation-exchange chromatography CEX B buffer: CEX A buffer with 1 M NaCl; autoclave.
23. SP sepharose FF 5 ml column (GE Healthcare).
24. HiPrep 26/10 desalting column (GE healthcare).
25. Lyophilization buffer: 50 mM Ammonium Bicarbonate.
26. Emulsiflex-C3 homogenizer (Avestin).
27. Lyophilizer.

2.2 Preparation of Recombinant ERK2

1. His-ERK recombinant T7 expression plasmid.
2. BL21(DE3) transformation competent *E. coli* bacteria (New England Biolabs).
3. Autoclaved LB broth Lennox (Difco).
4. 100 mg/ml stock solution of ampicillin antibiotics.
5. 1 M stock solution of IPTG.
6. PBS: phosphate buffer saline.
7. Extraction buffer: 50 mM phosphate buffer pH7.6, 300 mM NaCl, 10 mM Imidazole, Protease inhibitor 1× (1 tablet for 50 ml, EDTA-free; Roche), 2000 U DNaseI for 50 ml (20,000 U/ml, EUROMEDEX).
8. Ni-NTA chromatography cartridge 1 ml (Thermo Science).
9. Wash buffer: 50 mM phosphate buffer pH7.6, 300 mM NaCl, 10 mM Imidazole; autoclave.
10. Elution buffer: 50 mM phosphate buffer pH7.6, 300 mM NaCl, 250 mM Imidazole (*see* **Note 6**).
11. Dialysis bags (12–14 kDa cutoff, 0.32 ml/cm; Spectrum Laboratories).
12. Centrifugal concentration devices (10 kDa cutoff, 2 ml; Sartorius).
13. 12% SDS Polyacrylamide precast gels (Pierce).
14. Activation buffer: 10 mM Hepes.KOH, pH7.3, 1 mM DTT, 5 mM $MgCl_2$, 100 mM NaCl, and 10% Glycerol.
15. Conservation buffer: 50 mM Tris–HCl, pH7.5, 150 mM NaCl, 0.1 mM EGTA, 20% Glycerol.
16. Emulsiflex-C3 homogenizer (Avestin).

2.3 Preparation of Recombinant GST-MEK1 R4F

1. GST-MEK1 R4F recombinant pGEX expression plasmid [19].
2. BL21(DE3) transformation competent *E. coli* bacteria (New England Biolabs).

3. Autoclaved LB broth Lennox (Difco).
4. 100 mg/ml stock solution of ampicillin antibiotics.
5. 1 M stock solution of IPTG.
6. PBS: phosphate buffer saline.
7. Extraction buffer: 50 mM Tris–HCl, pH7.5, 1 mM EDTA, 300 mM NaCl, Protease inhibitor 1× (1 tablet for 50 ml, EDTA-free; Roche), 2000 U DNaseI for 50 ml (20,000 U/ml, EUROMEDEX).
8. Glutathione Agarose 4B chromatography cartridge 5 ml (Protino, Macherey Nagel).
9. Wash buffer: 50 mM Tris–HCl, pH7.5, 1 mM EDTA, 300 mM NaCl.
10. Elution buffer: 50 mM Tris–HCl, 10 mM Glutathione reduced, pH7.5 (*see* **Note 6**).
11. Dialysis bags (12–14 kDa cutoff, 0.32 ml/cm; Spectrum Laboratories).
12. Centrifugal concentration devices (10 kDa cutoff, 2 ml; Sartorius).
13. Activation buffer: 10 mM Hepes.KOH, pH7.3, 1 mM DTT, 5 mM $MgCl_2$, 100 mM NaCl and 10% Glycerol.
14. Conservation buffer: 50 mM Tris–HCl, pH7.5, 150 mM NaCl, 0.1 mM EGTA, 20% Glycerol.
15. Emulsiflex-C3 homogenizer (Avestin).

2.4 Recombinant ERK2 Activation

1. GST-MEK1 R4F recombinant protein (Subheadings 3.5 and 3.6) or commercial activated-MEK1 (Millipore 14-429).
2. Recombinant His-ERK2 protein in activation buffer or conservation buffer (Subheadings 3.3 and 3.4).
3. Activation buffer: 10 mM Hepes.KOH, pH7.3, 1 mM DTT, 5 mM $MgCl_2$, 100 mM NaCl.
4. PD-10 G25 resin desalting column (GE-Healthcare).
5. Centrifugal concentration devices (10 kDa cutoff, 2 ml, Sartorius).
6. Glutathione Agarose 4B beads (Protino).

2.5 Phosphorylation of Tau by Activated ERK2

1. 1 mg of ^{15}N and 5 mg of ^{15}N, ^{13}C lyophilized recombinant full-length Tau (Subheadings 3.1 and 3.2).
2. Activated His-ERK2 in activation buffer (Subheading 3.7).
3. Phosphorylation buffer 5×: 250 mM Hepes.KOH, pH8.0, 62.5 mM $MgCl_2$, 5 mM EDTA, 250 mM NaCl.
4. Protease inhibitor 40× (1 tablet in 1 ml Phosphorylation buffer, EDTA-free; Roche).
5. PD MidiTrap G25 resin desalting column (GE-Healthcare).

6. Lyophilization buffer: 50 mM Ammonium Bicarbonate.
7. NMR buffer: 50 mM deuterated *d11*-Tris-HCl (Isotec), pH6.5, 30 mM NaCl, 2.5 mM EDTA.
8. 12% SDS Polyacrylamide precast gels (Pierce).
9. Lyophilizer.

2.6 Phosphorylation of Tau by Rat Brain Extract

1. 1 mg of ^{15}N-labeled recombinant full-length Tau per phosphorylation assay (Subheadings 3.1 and 3.2).
2. 100 μM okadaic acid (Sigma).
3. Homogenizing buffer H: 10 mM Tris–HCl pH7.4, 5 mM EGTA, 2 mM DTT, 1 μM okadaic acid (Sigma) supplemented with 20 μg/ml Leupeptin and 40 mM Pefabloc.
4. HMED buffer 5×: 200 mM Hepes.KOH, pH7.3, 10 mM $MgCl_2$, 25 mM EGTA, 10 mM DTT.
5. PD-10 G25 resin desalting column (GE-Healthcare).
6. Lyophilization buffer: 50 mM Ammonium Bicarbonate.
7. 12% SDS Polyacrylamide precast gels (Pierce).
8. NMR buffer: 50 mM deuterated *d11*-Tris–HCl (Isotec), pH6.5, 30 mM NaCl, 2.5 mM EDTA, 1 mM DTT, 5% D_2O, and 1 mM TMSP.
9. Lyophilizer.
10. Ultracentrifuge and type 50 Ti rotor (Beckman Coulter).

2.7 Preparation of Recombinant CBP

1. GST-CBP[1202-1848] recombinant T7 expression plasmid (pGEX-6P-1).
2. BL21(DE3) transformation competent *E. coli* bacteria (New England Biolabs).
3. Autoclaved LB Broth, Lennox (DIFCO).
4. 100 mg/ml stock solution of ampicillin antibiotics.
5. 1 M stock solution of IPTG.
6. PBS 10×: phosphate buffer saline.
7. Extraction buffer: PBS 1× pH7.6, 10% glycerol, 1% Triton X-100, 10 mM EDTA, 2 mM DTT, Protease inhibitor 1× (1 tablet for 50 ml; Roche), 2000 U DNaseI for 50 ml (20,000 U/ml, EUROMEDEX).
8. Glutathione sepharose resin beads (GE Healthcare): 20 μl of resin beads/mL of soluble extract.
9. Wash buffer: PBS 1× pH7.6, 10% glycerol, 1% Triton X100, 10 mM EDTA, 2 mM DTT.
10. Conservation buffer: 25 mM Hepes.KOH pH7.8, 0.1 mM EDTA, 1 mM THP (Tris (3-hydroxypropyl) phosphine), 50% glycerol (i.e., acetylation buffer with 50% glycerol).

11. 10% SDS Polyacrylamide gels.
12. Acetylation buffer: 25 mM Hepes.KOH pH7.8, 0.1 mM EDTA, 1 mM THP.
13. Stock solution of acetyl-coenzyme A (AcCoA) at 20 mM: 25 mg of AcCoA trisodium salt (FW 875; Sigma) are dissolved in 1.423 ml acetylation buffer and stored at −20 °C.
14. Stock solution of CBP peptide substrate (Ac-QPVEPKKPVES KKSGKSAKSKEKQ-NH_2, 8 TFA) at 10 mM: 18 mg of peptide was dissolved in 0.5 ml of deionized water. The pH was adjusted to 7.0 with a solution of NaOH 10 M.
15. C18 Zorbax 300SB analytical column 4.6 × 150 mm (Agilent).
16. 10% TFA (v/v).
17. Ultrasonic bath.
18. Buffers for reverse-phase chromatography: Buffer A (equilibration): 0.1% TFA (10 ml of the 10% TFA solution per liter), 2% acetonitrile in water; Buffer B (elution): 0.1% TFA: 80% acetonitrile in water; both buffers are degassed in an ultrasonic bath for 15 min at room temperature.
19. Emulsiflex-C3 homogenizer (Avestin).

2.8 Acetylation of Tau by CBP

1. Acetylation buffer: 25 mM Hepes.KOH pH7.8, 0.1 mM EDTA, 1 mM THP.
2. Stock solution of acetyl-coenzyme A (AcCoA) at 20 mM: 25 mg of AcCoA trisodium salt (Sigma) is dissolved in 1.423 ml acetylation buffer and stored at −20 °C.
3. Recombinant GST-CBP[1202-1848] fragment (Subheadings 3.11 and 3.12).
4. ^{15}N-labeled or ^{15}N-Lysine-labeled recombinant Tau protein (Subheadings 3.1 and 3.2).
5. C8 Zorbax 300SB semi-preparative column 9.4 × 250 mm (Agilent).
6. 10% TFA (v/v).
7. Ultrasonic bath.
8. Buffers for reverse-phase chromatography: Buffer A (equilibration): 0.1% TFA (10 ml of the 10% TFA solution per liter), 2% acetonitrile in water; Buffer B (elution): 0.1% TFA: 80% acetonitrile in water; both buffers are degassed in an ultrasonic bath for 15 min at room temperature.
9. HiPrep 26/10 desalting column (GE Healthcare).
10. Lyophilization buffer: 50 mM Ammonium Bicarbonate.
11. NMR buffer: 50 mM phosphate buffer pH6.4, 25 mM NaCl, 2.5 mM EDTA.
12. Lyophilizer.

3 Methods

3.1 Production of Recombinant ^{15}N-Tau, ^{15}N-Lysine,$^{13}C_u$-Tau or ^{15}N, ^{13}C-Tau

Bacterial fermentations have to be performed following the best practices of sterile manipulations.

3.1.1 Day 1: Bacterial Transformation

1. Transform Tau pET15b recombinant T7 expression plasmid (Novagen) into BL21(DE3) competent bacterial cells.
2. Mix 50 μl of competent BL21(DE3) bacteria with 100 ng of the plasmid DNA in a 1.5 ml plastic tube.
3. Incubate the mixture on ice for 30 min, followed by heat shock for 30 s at 42 °C. Place back the tube on ice for 5 min.
4. Add 1 ml of LB medium at room temperature and incubate the bacterial suspension at 37 °C for 30 min.
5. Spread 100 μl of the suspension at the surface of a LB plate containing 100 μg/ml of ampicillin antibiotics to select for the colonies having integrated the plasmid.
6. Incubate the plate overnight at 37 °C.

3.1.2 Day 2: Bacterial Growth

1. Start a small scale bacterial culture of 20 ml by using few colonies from the selection plate as inoculum.
2. Grow the Tau pET15b bacteria at 37 °C for 6 h in LB medium supplemented with 100 μg/ml of ampicillin, to reach saturation corresponding to an approximate OD600 of 4.0.
3. Meanwhile, reconstitute 1 L of M9 medium by the addition of the supplements and isotopes following one of the labeling schemes indicated in Subheading 2.1.
4. Start a large scale culture of 1 L by inoculating 20 ml of the LB saturated culture into 1 L of M9 medium supplemented with 100 μg/ml of ampicillin. Perform the culture in 2 L Erlenmeyer culture plastic baffled-flasks placed in an incubator programmed at 10 °C and 50 rpm for 14 h for the uniform ^{15}N or ^{15}N, ^{13}C-labeling or for 12 h for the ^{15}N-Lysine, ^{13}C-labeling considering that the growth phase will be slower due to the absence of complete medium for the later (*see* **Notes 7** and **8**).
5. Program the incubator to switch to 200 rpm and 37 °C after this period.

3.1.3 Day 3: Induction Phase

1. Check the bacterial growth by measuring the OD at 600 nm of an aliquot of 1 ml of the culture.
2. For the uniform ^{15}N or ^{15}N, ^{13}C-labeling, induce the protein production by the addition of 400 μM IPTG when bacterial culture arrives at an OD600 of 0.8–1.0. Alternatively, for the ^{15}N-Lysine, ^{13}C-labeling, add 150 mg of $^{15}N_2/^{13}C_6$-L-Lysine hydrochloride when bacterial culture reaches an OD600 of

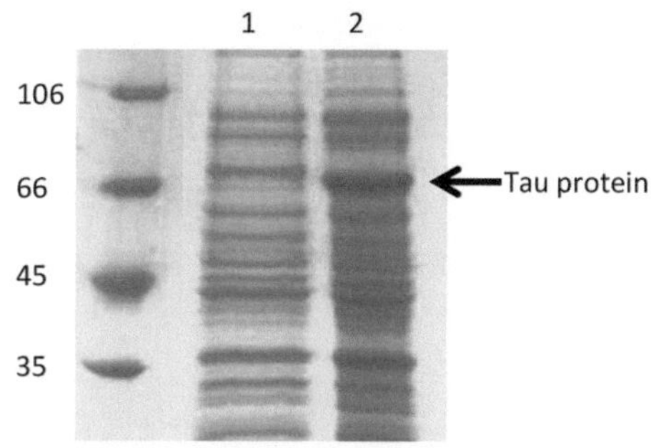

Fig. 1 Induction of recombinant Tau expression in BL21(DE3) bacteria. Lane 1: BL21(DE3) bacteria before the expression induction by IPTG. Lane 2: BL21(DE3) bacteria 4 h after induction. The band of recombinant Tau protein is shown by *black arrow* around 66 kDa

0.6–0.7, and continue the culture for 20 min. Then, induce the protein production by the addition of 400 μM IPTG.

3. Continue the bacterial culture for 4 h at 37 °C and 200 rpm.
4. Collect bacterial pellet by centrifugation for 20 min at 5500 × *g* in 1 L bottles.
5. Suspend the bacterial pellet in 50 ml of PBS buffer and centrifuge in 50 ml plastic tubes at 4000 × g.
6. Remove the supernatant solution and freeze the pellet at −20 °C until further use (Fig. 1).

3.2 Purification of ^{15}N-Tau, ^{15}N-Lysine,$^{13}C_u$-Tau or ^{15}N, ^{13}C-Tau

1. Unfreeze the bacterial pellet at room temperature and thoroughly suspend in 45 ml of extraction buffer.
2. Homogenize and lysate bacterial suspension with a high-pressure homogenizer.
3. Remove the insoluble material by centrifugation at 20,000 × *g* for 30 min.
4. Heat the bacterial protein extract for 15 min at 75 °C as a first purification step. The Tau protein is recovered in the soluble fraction after centrifugation at 15,000 × *g* for 30 min.
5. Perform purification of the ^{15}N-Tau, ^{15}N, ^{13}C-Tau, ^{15}N-Lysine,^{13}C-Tau protein by cation exchange chromatography on a 5 ml Hitrap SP sepharose FF column. After loading the sample on the column, wash the resin with CEX A buffer. Elute the protein by a three-step NaCl gradient with CEX B buffer. The first step corresponds to ten volumes of column to reach 250 mM NaCl, followed by a second step with five volumes of column to reach 500 mM NaCl, and the third step to two volumes of the column to reach 1 M NaCl (Fig. 2).

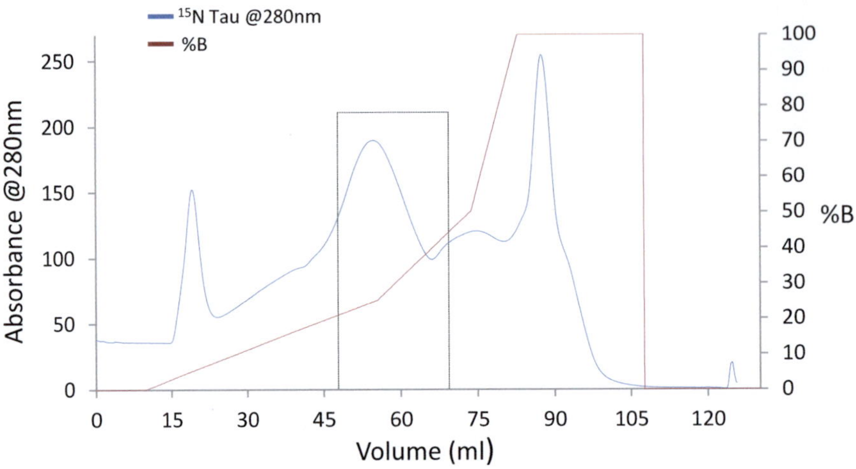

Fig. 2 Cation exchange chromatography purification step of the ^{15}N-Tau. The chromatogram shows the absorbance at 280 nm in milli-absorption units (graph in *blue*) and the percentage of CEX buffer B (graph in *marron*). The peak corresponding to ^{15}N-Tau is *boxed* in *black*

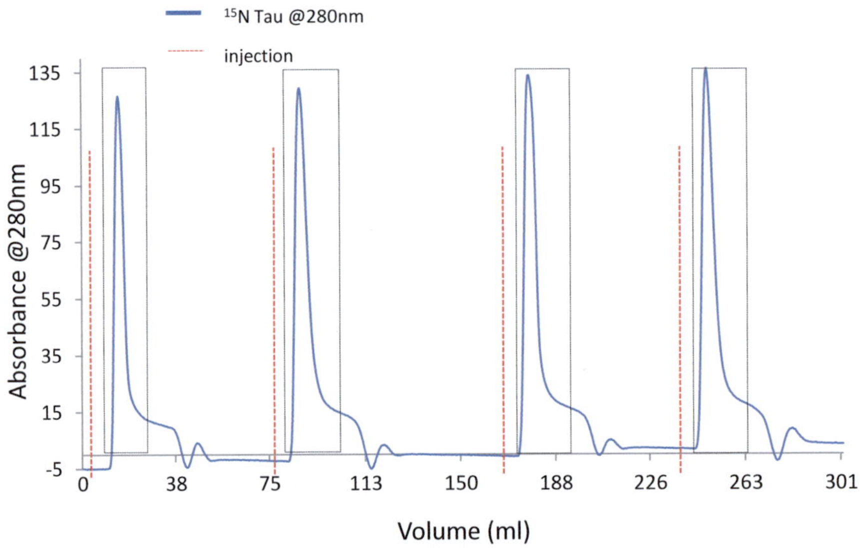

Fig. 3 Desalting of ^{15}N-Tau into ammonium bicarbonate buffer from the previous purification step of cation exchange chromatography (Fig. 2). The protein elution is monitored at 280 nm. Four injections of 5 ml ^{15}N-Tau are performed for one CEX chromatography. *Red lines* correspond to injections (5 ml loop). Collected ^{15}N-Tau fractions are shown in *black dotted box*

6. Transfer the pooled fractions from the chromatography purification step (corresponding to the box in Fig. 2, between 45 and 65 ml) on a 26/10 HiPrep Desalting column equilibrated in lyophilization buffer. Proteins elute in the void volume of about 12 ml. Collect the eluate and lyophilize the desalted fractions (*see* **Notes 9** and **10**). Injections of 5 ml are repeated four to five times, depending on the volume of the cation exchange pool (Fig. 3).

7. Suspend the lyophilized protein in a buffer suitable for further protein modification (phosphorylation or acetylation).

3.3 Production of Recombinant ERK2

Bacterial fermentations have to be performed following the best practices of sterile manipulations.

3.3.1 Day 1: Bacterial Transformation

1. Transform His6-tagged p42 MAP kinase from *Xenopus laevis* (His-ERK2) recombinant T7 expression plasmid into BL21(DE3) competent bacterial cells based on chemical transformation following the same procedure as described in Subheading 3.1.1.

3.3.2 Day 2: Bacterial Growth

1. Start a small scale bacterial culture of 10 ml with few colonies from the selection plate as inoculum.
2. Grow the His-ERK pET BL21(DE3) bacteria overnight in LB medium supplemented with 100 μg/ml of ampicillin to reach saturation corresponding to an approximate OD600 of 4.0.

3.3.3 Day 3: Induction Phase

1. Start a large scale culture of 1 L by inoculating 10 ml of the overnight culture into 1 L of LB medium supplemented with 100 μg/ml of ampicillin. Perform the culture in 2 L Erlenmeyer culture plastic baffled flasks placed in an incubator at 37 °C and shake at 200 rpm.
2. Check growth by measuring the OD at 600 nm of an aliquot of 1 ml of the culture.
3. When bacterial culture arrives at 0.8–1.0 of OD600, induce the protein production by the addition of 700 μM IPTG. Continue the bacterial culture during 4 h, at 30 °C and 200 rpm.
4. Collect bacterial pellet by centrifugation for 20 min at 5500 × *g* in 1 L bottles.
5. Suspend the bacterial pellet in 50 ml of PBS buffer and centrifuge in 50 ml plastic tubes at 4000 × *g*.
6. Remove the supernatant solution and freeze the pellet at −20 °C until further use.

3.4 Purification of Recombinant ERK2

1. Unfreeze the bacterial pellet at room temperature.
2. Suspend thoroughly in 45 ml extraction buffer.
3. Homogenize and lysate bacterial suspension with a high-pressure homogenizer.
4. Remove the insoluble material by centrifugation at 20,000 × *g* for 30 min.
5. Load the soluble bacterial lysate containing the His-ERK2 recombinant kinase (Fig. 4a) on a Ni-NTA resin equilibrated in wash buffer.
6. Wash the resin with wash buffer followed by 5 % elution buffer to remove residual lysate.

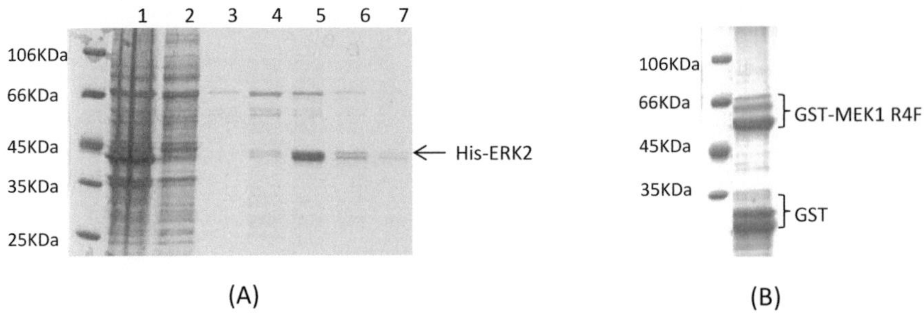

Fig. 4 (**a**) SDS-Polyacrylamide Gel Electrophoresis (SDS-PAGE) is run to monitor the preparation steps of His-ERK2. 5–10 μl of samples collected from the purification steps are run through 12 % SDS-Polyacrylamide Gel. Line 1: bacterial lysate; Line 2: soluble extract; Lines 3 and 4: washing fractions collected with 5 % elution buffer. Lines 5–7: purified sample fractions. Eluted His-ERK2 corresponds to the band appearing at 42 kDa in Line 5. A number of contaminating proteins are removed by washing the resin with 5 % elution buffer (*see* Lines 3 and 4). (**b**) Analysis of purified GST-MEK1 R4F by SDS-PAGE. 5 μl collected from the elution step is run through 12 % SDS-Polyacrylamide Gel. GST-MEK1 R4F fusion protein (*up*) and GST (*bottom*) are labeled. The GST tag is expected at 26 kDa and the GST-MEK1 R4F at 68 kDa. The multiple bands correspond to degradation products of the fusion protein

7. Elute His-ERK2 by equilibration of the resin into 100 % elution buffer. Collect the eluted fractions by 1 ml aliquots which are analyzed by SDS-PAGE before pooling homogeneous fractions. His-ERK2 is obtained in about 5 ml of pooled fractions at 1.0–1.2 mg/ml concentration. The yield is approximately 5 mg of His-ERK2 per liter of LB broth.
8. Dialyze His-ERK2 against the activation buffer or the conservation buffer.
9. Concentrate His-ERK2 in the activation buffer up to 3 mg/ml using centrifugal devices. At this step, His-ERK2 (*see* **Note 11**) is ready for the activation step (Subheading 3.7). Freeze His-ERK2 in conservation buffer in liquid nitrogen and keep at −80 °C until further use.

3.5 Production of Recombinant GST-MEK1 R4F

Bacterial fermentations have to be performed following the best practices of sterile manipulations.

3.5.1 Day 1: Bacterial Transformation

1. Transform the pGEX plasmid containing the cDNA sequence of recombinant GST-MEK1 R4F MAPK kinase into BL21(DE3) competent bacterial cells based on chemical transformation following the same procedure as described in Subheading 3.1.1.

3.5.2 Day 2: Bacterial Growth

1. Start a small scale bacterial culture of 20 ml with few colonies from the selection plate as inoculum in LB medium supplemented with 100 μg/ml of ampicillin to reach saturation corresponding to an approximate OD600 of 4.0.

3.5.3 Day 3: Induction Phase

1. Start 2 L of BL21(DE3) bacterial culture in LB medium supplemented by 100 μg/ml ampicillin antibiotics by adding the 20 ml overnight saturated bacterial culture to the fresh medium. Perform the culture in a 2 L Erlenmeyer culture plastic baffled flask placed in an incubator at 37 °C and shake at 200 rpm.
2. Check growth by measuring the OD at 600 nm of an aliquot of 1 ml of the culture.
3. When bacterial culture arrives at 0.8–1.0 of OD600, induce the protein production by the addition of 700 μM IPTG.
4. Continue the bacterial culture during 4 h at 30 °C and 200 rpm.
5. Collect bacterial pellet by centrifugation for 20 min at 5500 × *g* in 1 L bottles.
6. Suspend the bacterial pellet in 50 ml of PBS buffer and centrifuge in 50 ml plastic tubes at 4000 × *g*.
7. Remove the supernatant solution and freeze the pellet at −20 °C until further use.

3.6 Purification of Recombinant GST-MEK1 R4F

1. Unfreeze the bacterial pellets from 2 L bacterial culture at room temperature and then suspend bacteria thoroughly in extraction buffer.
2. Homogenize and lysate the suspension using a high-pressure homogenizer.
3. Remove the insoluble material by centrifugation at 20,000 × *g* for 30 min.
4. Load the soluble bacterial lysate containing the recombinant GST-MEK1 R4F kinase on a Glutathione Agarose 4B chromatography column equilibrated with wash buffer at a slow flow rate of 0.3–0.5 ml/min.
5. Rinse the loaded resin with wash buffer until the baseline at 280 nm is stabilized.
6. Collect the GST-MEK1 R4F protein into elution buffer by 1 ml aliquots.
7. Analyze aliquots by SDS-PAGE to choose the fractions to pool. The pooled fractions (Fig. 4b) correspond to about 6 ml at 1 mg/ml.
8. Dialyze GST-MEK1 R4F protein against the activation buffer or the conservation buffer.
9. Concentrate GST-MEK1 R4F protein in activation buffer up to 3 mg/ml using centrifugal devices to be used in His-ERK2 activation step (*see* Subheading 3.7).
10. Freeze dialyzed GST-MEK1 R4F in conservation buffer in liquid nitrogen and keep at −80 °C until further use.

3.7 Activation of Recombinant ERK2

1. Mix 1 ml of His-ERK (3 mg/ml, Subheading 3.4) with 1 ml of GST-MEK1 R4F (3 mg/ml, Subheading 3.6) in activation buffer in the presence of 5 mM ATP and protease inhibitors cocktail 1× in a 5 ml reaction volume, in 15 ml plastic tubes [20].
2. Perform the activation at 30 °C during 15 h.
3. If His-ERK or GST-MEK1 R4F is frozen in conservation buffer, dialyze the kinase solution against activation buffer. Concentrate the kinases with a centrifugal device for further reaction if needed.
4. Remove 5 μl from the reaction mixture at various time points for SDS-PAGE analysis to control the progress of the reaction. A slight delay of His-ERK2 migration is observed from 30 min of incubation. The amount of soluble His-ERK2 is decreasing during the incubation due to protein precipitation (Fig. 5a). Remove the insoluble material by centrifugation.
5. Remove GST-MEK1 R4F from the reaction mixture by using 500 μl Glutathione Agarose 4B beads equilibrated in activation buffer. Incubate the beads on a roller mixer with slow agitation (rotation of 20 rpm/min) for 1 h at 4 °C with the activation mix and finally eliminate resin beads by centrifugation at 1000 × *g* for 10 min. The final activated His-ERK is obtained at 0.1–0.15 mg/ml (2.5–4.0 μM) in 5 ml of reaction volume.
6. Add 10% Glycerol to the activated His-ERK2 in activation buffer.
7. Freeze aliquots of 100 μl in liquid nitrogen and conserve at −80 °C until further use.

3.8 Phosphorylation of Tau Protein by ERK2 Recombinant Kinase

1. Mix 100 μM recombinant ^{15}N-Tau protein (1 mg lyophilized protein, Subheading 3.2) with 50 μl activated ERK2 (4 μM in activation buffer, Subheading 3.7, final concentration 1 μM, *see* **Note 12**) in the presence of 2.5 mM ATP, 1 mM DTT, and

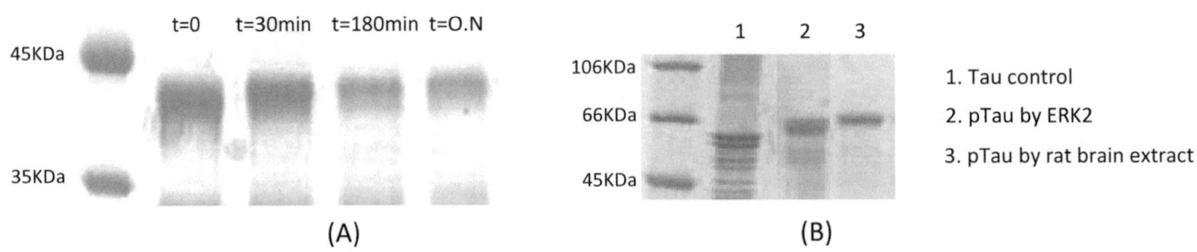

Fig. 5 (**a**) Time course of His-ERK2 *in vitro* activation by GST-MEK1R4F. 5 μl of samples is removed from the reaction mix at initial time, 30 min, 180 min, and at 16 h (*overnight*) incubation time. From 30 min of incubation, a slight migration delay of the band corresponding to His-ERK2 is observed compared to that at the initial time. The band upper shift increases until the end of reaction. The amount of soluble His-ERK2 decreases with the incubation time due to precipitation. (**b**) SDS-PAGE analysis of Tau phosphorylation. About 1.2 μg samples is loaded in 12% acrylamide gel. Compared to Tau control (*see* line 1), ERK2-phosphorylated Tau has a lower mobility resulting in a shift of the band from 52 kDa to 60 kDa (line 2) and phosphorylated Tau by rat brain extract a shift to 66 kDa (line 3)

1 mM EGTA complemented with protease inhibitors in a total sample volume of 200 μl in the phosphorylation buffer 1×.

2. Incubate the sample at 37 °C for 3 h. For the ^{15}N, ^{13}C-Tau protein, the reaction is scaled up to 5 mg lyophilized protein in 1 ml reaction mix.
3. After incubation, heat the samples at 75 °C for 15 min and centrifuge at 20,000 ×*g* for 15 min to remove the ERK kinase.
4. Desalt the phosphorylated ^{15}N-Tau or ^{15}N, ^{13}C-Tau (Fig. 5b) into NMR buffer and conserve at −20 °C.
5. For NMR spectroscopy analysis, complement the samples with 1 mM trimethylsilyl propionate (TMSP, which serves as internal reference), 2 mM DTT, and 10% D_2O for the purpose of NMR spectrometer field locking.

3.9 Preparation of Rat Brain Extract

1. Wash the rat brain (about 1.8 g) immediately after collection in an ice-cold homogenizing buffer.
2. Cut the brain crudely into pieces with dissection scissors and place in a homogenizing buffer (2.5 ml/g of brain).
3. Prepare the rat brain extract by homogenizing the brain using a Potter pestle fitted on a tissue grinder and a 10 ml glass tube.
4. Perform ultracentrifugation at 100,000 ×*g* for 1 h in a type 50 Ti rotor. The supernatant corresponds to the rat brain extract (*see* **Note 13**).

3.10 Phosphorylation of Tau Protein with Rat Brain Extract

1. Mix 4 μM recombinant ^{15}N-labeled Tau protein (around 1 mg protein in 5 ml of total volume) with 500 μl fresh rat brain extract in a final 5 ml reaction volume of HMED 1× buffer. Add 1 μM okadaic acid (phosphatase inhibitor), 2 mM ATP, and protease inhibitors 1×.
2. Incubate the phosphorylation mix for 24 h at 37 °C.
3. Inactivate enzymes in the sample by heating at 75 °C for 15 min.
4. Centrifuge the sample at 16,000 ×*g* for 15 min to remove out precipitated proteins.
5. Perform buffer exchange into the lyophilization buffer using a PD-10 column (G25 resin). Load 2.5 ml of the inactivated reaction mix on the column. Repeat operation to desalt the total reaction volume. The desalted sample is obtained in a final volume of 7 ml.
6. Lyophilize the phosphorylated Tau (Fig. 5b) and conserve at −20 °C.
7. To analyze by NMR spectroscopy, dissolve the lyophilized sample in 200 μl NMR buffer complemented with 2 mM DTT, 1 mM TMSP, and 10% D_2O.

3.11 Production of Recombinant GST-CBP Acetyltransferase

Bacterial fermentations have to be performed following the best practices of sterile manipulations.

3.11.1 Day 1: Bacterial Transformation

1. Transform the pGEX plasmid containing the cDNA sequence of recombinant CBP[1202-1848] fragment of mouse CBP acetyltransferase into BL21(DE3) competent bacterial cells based on chemical transformation following the same procedure as described in Subheading 3.1.1 except that plates used for colony selection are additionally supplemented with 1% glucose.

3.11.2 Day 2: Bacterial Growth

1. Start a small scale bacterial culture of 20 ml with few colonies from the selection plate as inoculum in LB medium supplemented with 1% glucose and 100 μg/ml of ampicillin to reach saturation corresponding to an approximate OD600 of 4.0.

3.11.3 Day 3: Induction Phase

1. Centrifuge the bacterial culture at 4000 × *g* for 15 min and discard supernatant.
2. Start 2 L of BL21(DE3) bacterial culture in LB medium supplemented by 1% glucose and 100 μg/ml ampicillin antibiotics by suspending the bacterial pellet from the 20 ml culture to the fresh medium. Perform the culture in 2 L Erlenmeyer culture plastic baffled flask placed in an incubator at 37 °C and shake at 200 rpm.
3. Check growth by measuring the OD at 600 nm of an aliquot of 1 ml of the culture.
4. When bacterial culture arrives at 0.8–1.0 of OD600, induce the protein production by the addition of 200 μM IPTG.
5. Continue the bacterial culture for 4 h at 18 °C and 200 rpm.
6. Collect bacterial pellet by centrifugation for 20 min at 6000 × *g* in 1 L-bottles.
7. Suspend the bacterial pellet in 100 ml of PBS buffer, aliquot into four 50 ml-plastic tubes, and centrifuge at 4000 × *g*.
8. Remove the supernatant and freeze the pellets at −20 °C until further use. For prolonged storage (superior to 6 months), freeze the pellet at −80 °C.

3.12 Purification of Recombinant GST-CBP Acetyltransferase

1. Unfreeze one bacterial pellet on ice (equivalent to a 0.5 L bacterial culture), then suspend thoroughly in a 20 ml extraction buffer.
2. Homogenize and lysate the suspension using a high-pressure homogenizer.
3. Remove the insoluble material by centrifugation at 25,000 × *g* for 30 min at 4 °C.

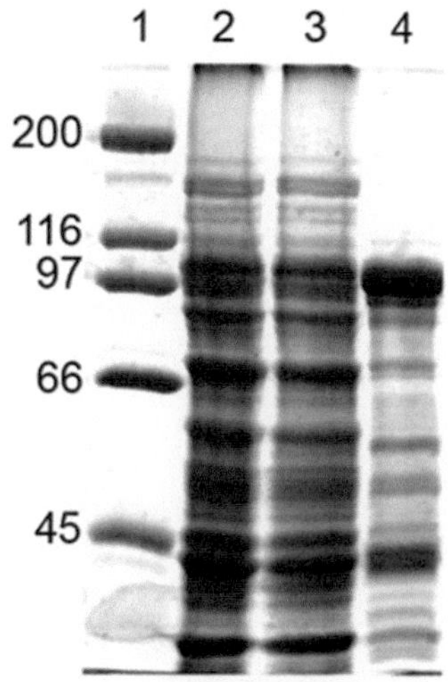

Fig. 6 SDS-10 % PAGE of the purification of GST-CBP[1202-1848] acetyltransferase on glutathione sepharose beads. Lane 1: molecular weight markers; lane 2: 5 μl of the soluble bacterial extract; lane 3: 7.5 μl of the pooled flow through and first wash supernatant; lane 4: 15 μl of the final glutathione beads slurry

4. Incubate the soluble bacterial lysate containing the recombinant GST-CBP[1202-1848] acetyltransferase (Fig. 6) for 3 h at 4 °C on a roller mixer with Glutathione Sepharose resin beads pre-equilibrated in the extraction buffer.
5. Rinse the loaded resin five times with 12 ml of wash buffer then five times with 12 ml of conservation buffer. The beads are kept as 50 % slurry in conservation buffer at −20 °C until further use (*see* **Note 14**). An aliquot of 15 μl is analyzed by SDS-PAGE.

3.13 Standard Analysis of CBP Acetyltransferase Activity on a Peptide Substrate

A peptide from the human Thymine DNA Glycosylase (TDG) is used as a substrate to check the activity of every new batch of CBP enzyme. The peptide sequence is Ac-QPVEPKKPVESKKSGKSA KSKEKQ-NH_2 (Ac and NH_2 mean that the N- and C-terminus are acetylated and amidated, respectively). The underlined residues in the peptide sequence correspond to three acetylation sites that have been previously identified [21].

1. For the acetylation reaction, mix 2.5 μl of the peptide solution at 10 mM (final concentration of 100 μM) with 25 μl of the stock solution of AcCoA at 20 mM (final concentration of 2 mM) or 25 μl of acetylation buffer as a control in 212.5 μl of acetylation buffer. Add 10 μl of the slurry of GST-CBP on glutathione beads (for a final volume of 250 μl).
2. Incubate the reactions overnight either at 20 °C or at 30 °C under rotative agitation.

3. Centrifuge the reaction mixtures at 16,000 × *g* for 10 min at 4 °C and analyze each supernatant by reverse-phase chromatography on a C18 column at room temperature. Equilibrate the column at 1 ml/min in buffer A until the absorbance and conductivity are stable. Inject the total reaction volume (250 μl) in a 1 ml injection loop for 4 ml. Then, wash the column at 1 ml/min with 2.5 ml of Buffer A. Elute peptides with a linear gradient of acetonitrile from 0% to 60% buffer B in 30 min at 1 ml/min (Fig. 7).
4. Collect fractions of 0.5 ml for MALDI-TOF mass spectrometry analysis. Mass increments of +42 Da ($[M+H]^+$ 2735.82 Da), +84 Da ($[M+H]^+$ 2777.88 Da), and +126 Da ($[M+H]^+$ 2819.81 Da) as compared to the mass of the unmodified peptide ($[M+H]^+$ 2693.53 Da) confirm the incorporation of one, two, or three acetyl moieties, respectively (Fig. 7, *see* **Note 15**).

3.14 Acetylation of Tau Protein by CBP

Perform acetylation of ^{15}N- or ^{15}N-Lysine,^{13}C-labeled Tau protein with minor modifications of the conditions described for the TDG peptide (Subheading 3.13).

1. Weigh 17.6 mg AcCoA and dissolve in 5 ml acetylation buffer to obtain a final concentration of 4 mM.
2. Filter the solution on a 0.22 μ-filter; this solution is sufficient for four acetylation reactions.
3. Dissolve 5 mg of Tau protein in 1.09 ml of AcCoA-containing acetylation buffer (final concentration of 100 μM). For the negative control, omit AcCoA and dissolve directly Tau protein in acetylation buffer. Add 100 μl of GST-CBP glutathione beads slurry per ml of reaction volume.
4. Incubate the acetylation reaction at 25 °C overnight (16 h) under rotative agitation.
5. Centrifuge the reaction mixture at 4000 × *g* for 5 min at 4 °C.

→

Fig. 7 Reverse-phase analysis on a C18 column of the acetylation reactions on a peptide substrate with recombinant GST-CBP[1202-1848]. Due to the presence of at least three acetylation sites, a complex pattern of elution was detected at 215 nm at both 20 °C and 30 °C. (**a**) Absorbance at 280 nm (*red*), 260 nm (*blue*), and 215 nm (*green*) of the acetylation reaction performed with 2 mM AcCoA at 30 °C overnight. (**b**) Comparison of the absorbance detected at 215 nm of the acetylation reactions performed with (*green*) or without AcCoA (*violet*) at 30 °C overnight or with AcCoA at 20 °C overnight (*blue*). The linear gradient of buffer B (%B) is indicated by *orange line.* (**c**) MALDI-TOF mass spectra of the fractions 2, 3, and 4 of the analytical C18 chromatography (*see* the fractions on panel **a**); *m*/*z* values of $[M+H]^+$ species are annotated. The "M" character indicates the mass of the unmodified peptide

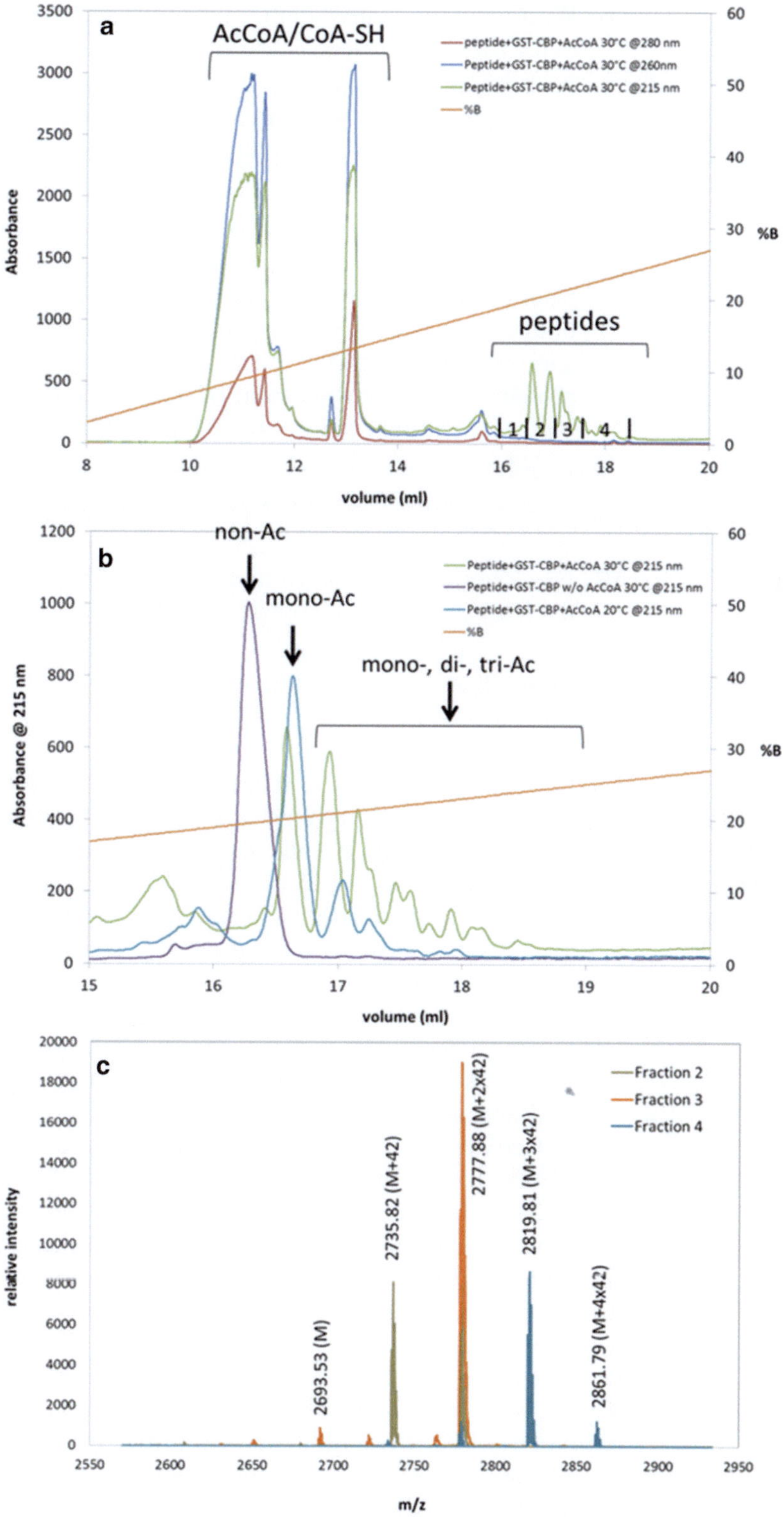
a
AcCoA/CoA-SH
peptide+GST-CBP+AcCoA 30°C @280 nm
Peptide+GST-CBP+AcCoA 30°C @260nm
Peptide+GST-CBP+AcCoA 30°C @215 nm
%B
peptides
1
2
3
4
Absorbance
volume (ml)
%B
b
non-Ac
mono-Ac
mono-, di-, tri-Ac
Peptide+GST-CBP+AcCoA 30°C @215 nm
Peptide+GST-CBP w/o AcCoA 30°C @215 nm
Peptide+GST-CBP+AcCoA 20°C @215 nm
%B
Absorbance @ 215 nm
volume (ml)
%B
c
Fraction 2
Fraction 3
Fraction 4
2693.53 (M)
2735.82 (M+42)
2777.88 (M+2x42)
2819.81 (M+3x42)
2861.79 (M+4x42)
relative intensity
m/z

6. Withdraw the supernatant cautiously and wash resin beads twice with 0.2 ml acetylation buffer.
7. Pool the supernatants and heat at 75 °C for 10 min followed by a centrifugation at 4000 × *g* for 15 min at 4 °C to remove potential contaminant proteins that might have leaked from glutathione beads.
8. Purify the acetylation and control reactions by reverse-phase chromatography on a C8 semi-preparative column at room temperature. Equilibrate the column at 4 ml/min in 20% buffer B until the absorbance and conductivity are stable. Inject the protein solution (1.5 ml) in a 2 ml injection loop for 20 ml, elute proteins with a linear gradient of acetonitrile from 20% to 60% buffer B in 24 min at 4 ml/min (Fig. 8). Collect 4 ml-fractions.
9. Lyophilize the collected fractions and dissolve in 200 μl of lyophilization buffer for SDS PAGE analysis on 10% acrylamide gels.
10. Desalt homogeneous fractions containing a high concentration of full-length protein in lyophilization buffer on a HiPrep 26/10 desalting column (Fig. 9), as described in Subheading 3.2, **step 6**.
11. Lyophilize the protein fraction and store at −20 °C until further use.

3.15 Assignment of Resonances of the Phosphorylated Residues: Identification of the Phosphorylation Pattern

A [^{1}H,^{15}N] HSQC 2D spectrum is recorded to detect the resonances of the phosphorylated S and T residues (*see* **Note 16**), because they are easily visualized, located around 8.5–9.5 ppm of amide proton and 117–125 ppm of ^{15}N-amide, outside the bulk of the ^{1}H, ^{15}N correlations (Fig. 10). As shown in Fig. 11, phospho-residue resonance assignment is based on [^{1}H,^{15}N,^{13}C] HNCACB 3D experiment that is recorded to link the resonances of the phosphorylated residues observed in the 2D [^{1}H,^{15}N] HSQC spectrum to a specific pS or pT residue in the Tau sequence. CA and CB chemical shift values of pS and pT are typical [22] and allow the identification of the nature of the phosphorylated i residue. pS and pT followed by a proline residue also in addition have a typical +2 ppm shift of the CA chemical shift value [23]. The nature of the residue at the i-1 position is defined by the chemical shift values of the CA and CB resonances corresponding to the i-1 residue (weaker set of signals compared to those of the i residue). A few identical phosphorylated patterns possess the same residue at position i-1 and cannot be uniquely identified from the HNCACB experiment alone, as is the case for K-pT175-P and K-pT181-P. Another 3D experiment lifts the redundancy by providing, in addition, the chemical shift value of ^{15}N of the i-1 residue. For example, the resonances corresponding to pT175 and pT181 can be successfully assigned due to the

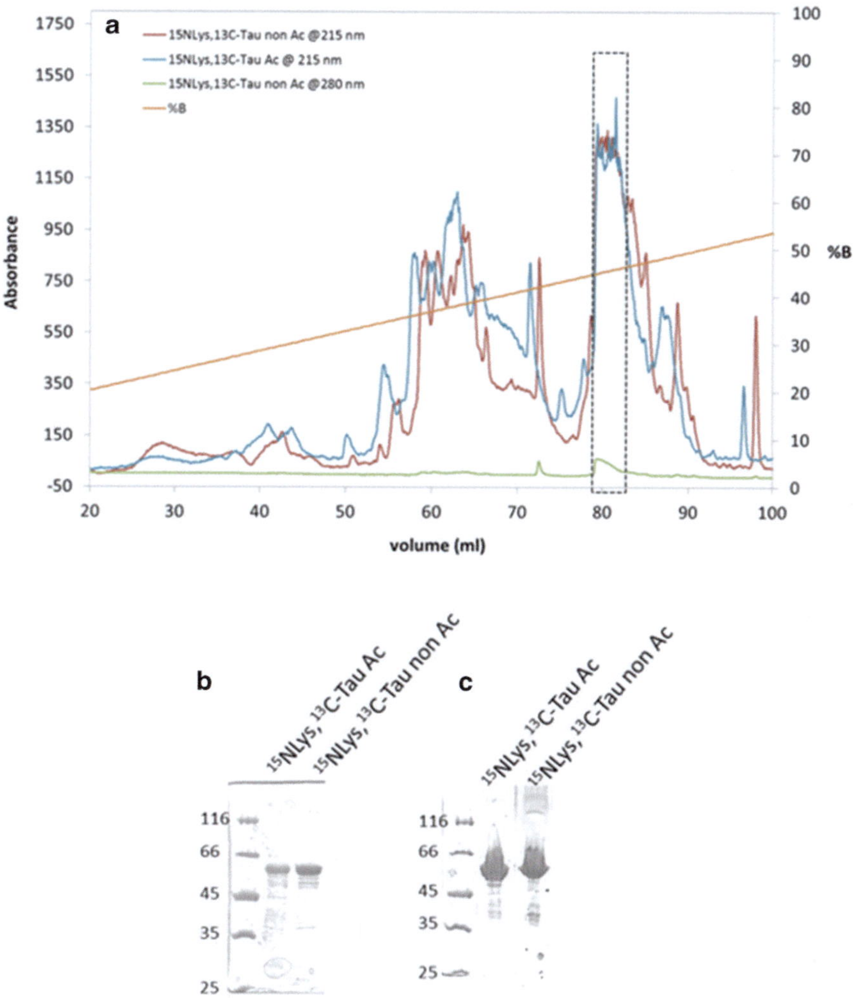

Fig. 8 Reverse-phase purification on a C8 column of the ^{15}N-Lys,^{13}C-labeled Tau protein before and after acetylation reaction with recombinant GST-CBP[1202-1848]. (**a**) Chromatogram of the purification of control and acetylated ^{15}N-Lys,^{13}C-labeled-Tau on a C8 column with a linear gradient of acetonitrile (*orange line*) as monitored by the absorbance at 215 nm for control ^{15}N-Lys,^{13}C-Tau (*red curve*) and acetylated ^{15}N-Lys,^{13}C-Tau (*blue curve*) and at 280 nm (*green curve*, control ^{15}N-Lys,^{13}C-Tau). (**b**, **c**) 12 % acrylamide SDS gel electrophoresis of the ^{15}N-Lys,^{13}C-Tau protein before and after the acetylation reaction (**b**) Fractions of non-acetylated and acetylated ^{15}N-Lys,^{13}C-Tau proteins purified by cation exchange chromatography. 2 μl of a 50 μM solution of protein are loaded on the gel. (**c**) Fractions of non-acetylated and acetylated ^{15}N-Lys,^{13}C-Tau proteins purified by reverse phase chromatography. Only the fractions corresponding to the boxed region on the chromatograms in (**a**) are shown. 2.5 μl of the lyophilized fractions dissolved in 200 μl water is loaded on the gel

different values of ^{15}N chemical shift observed for A-K174 and L-K180 (Figs. 11 and 12).

1. Dissolve 0.5–1 mg of ^{15}N-Tau in 200 μl NMR buffer (50–100 μM) to fill a 3 mm tube.

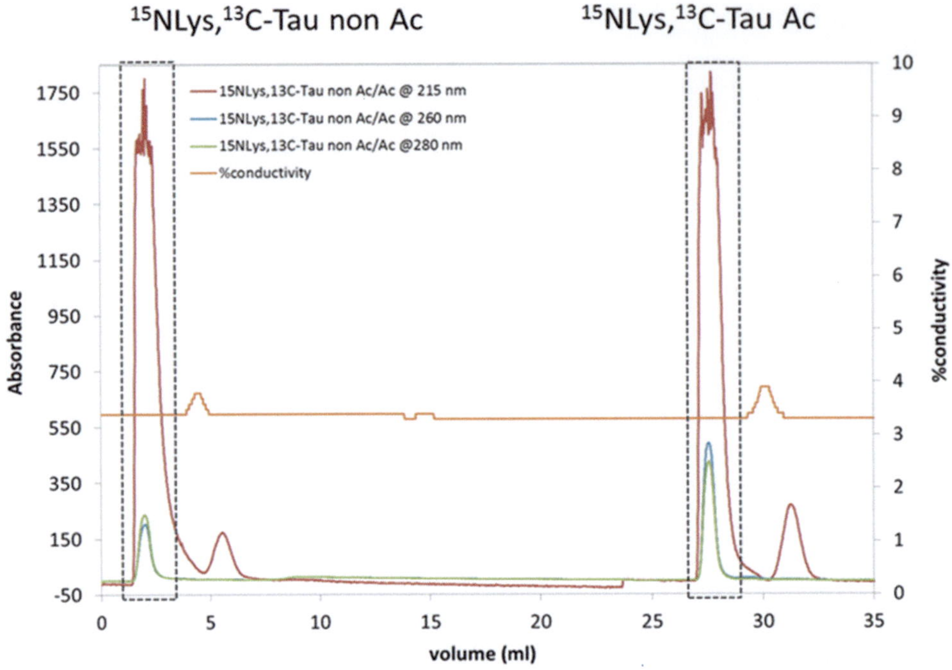

Fig. 9 Chromatogram of desalting in 50 mM ammonium bicarbonate of the ^{15}N-lysine,^{13}C-Tau protein in its non-acetylated and acetylated forms from fractions of the purification step by C8 reverse-phase chromatography

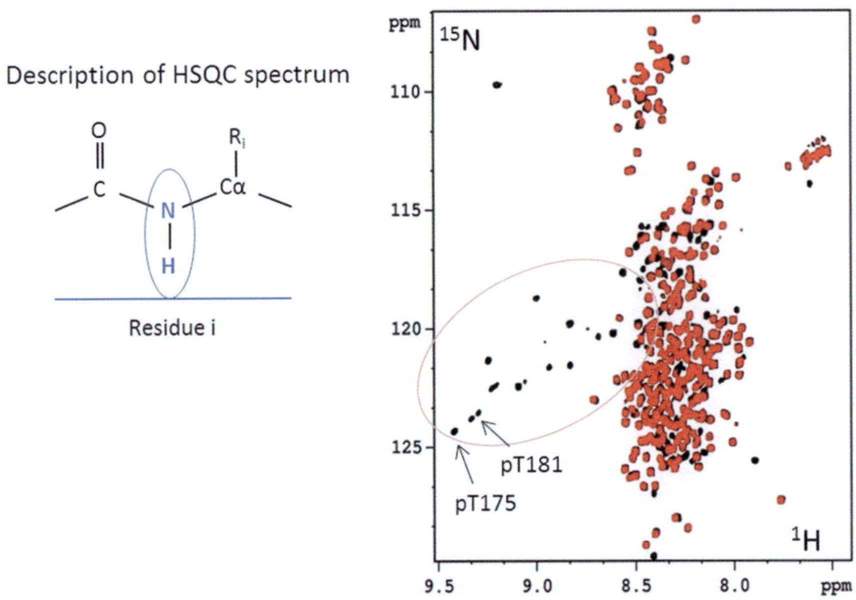

Fig. 10 Overlaid 2D [^{1}H,^{15}N] HSQC spectra of ^{15}N phosphorylated Tau in *black* with ^{15}N-Tau in *red*. HSQC spectrum records the J-coupling (through-bound magnetization transfer) correlation between ^{1}H and ^{15}N-nuclei. Each resonance thus corresponds to an amino acid of the protein backbone. The encircled multiple additional resonances in the black spectrum, compared to the red control spectrum of ^{15}N-Tau, correspond to phosphorylated residues of the Tau protein according to previous NMR studies of Tau protein [8] review in [11]

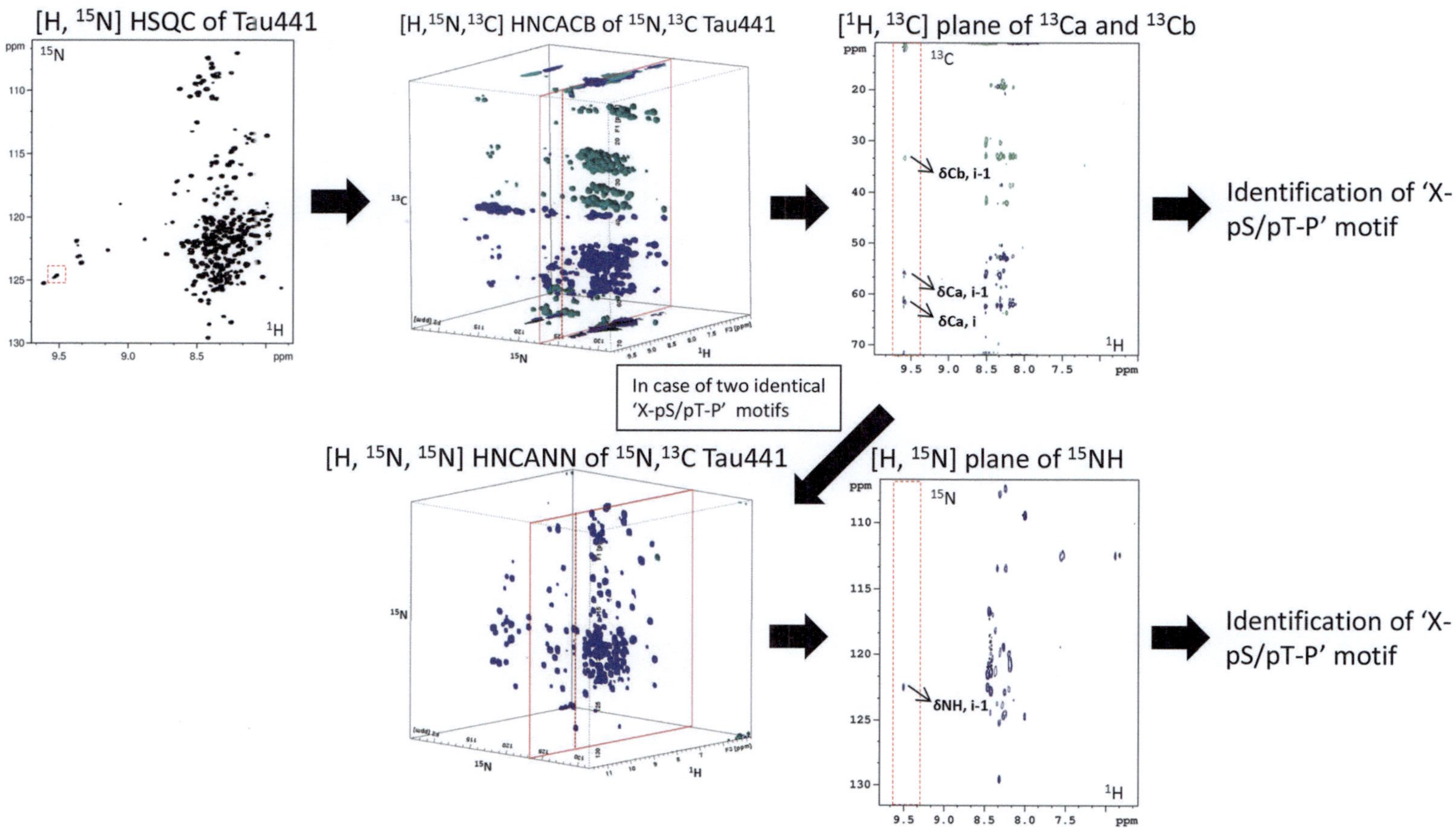

Fig. 11 Strategy of identification of X-pS/pT-P motifs in Tau protein. Chemical shifts of CA and CB of i and i-1 residues in [^{1}H,^{15}N,^{13}C] HNCACB 3D experiment and chemical shift of ^{15}N of the i-1 residues in [^{1}H,^{15}N,^{15}N] HNCANN 3D experiment are used to assign a resonance in the [^{1}H,^{15}N] 2D spectrum to a specific pS-P or pT-P motif in Tau sequence

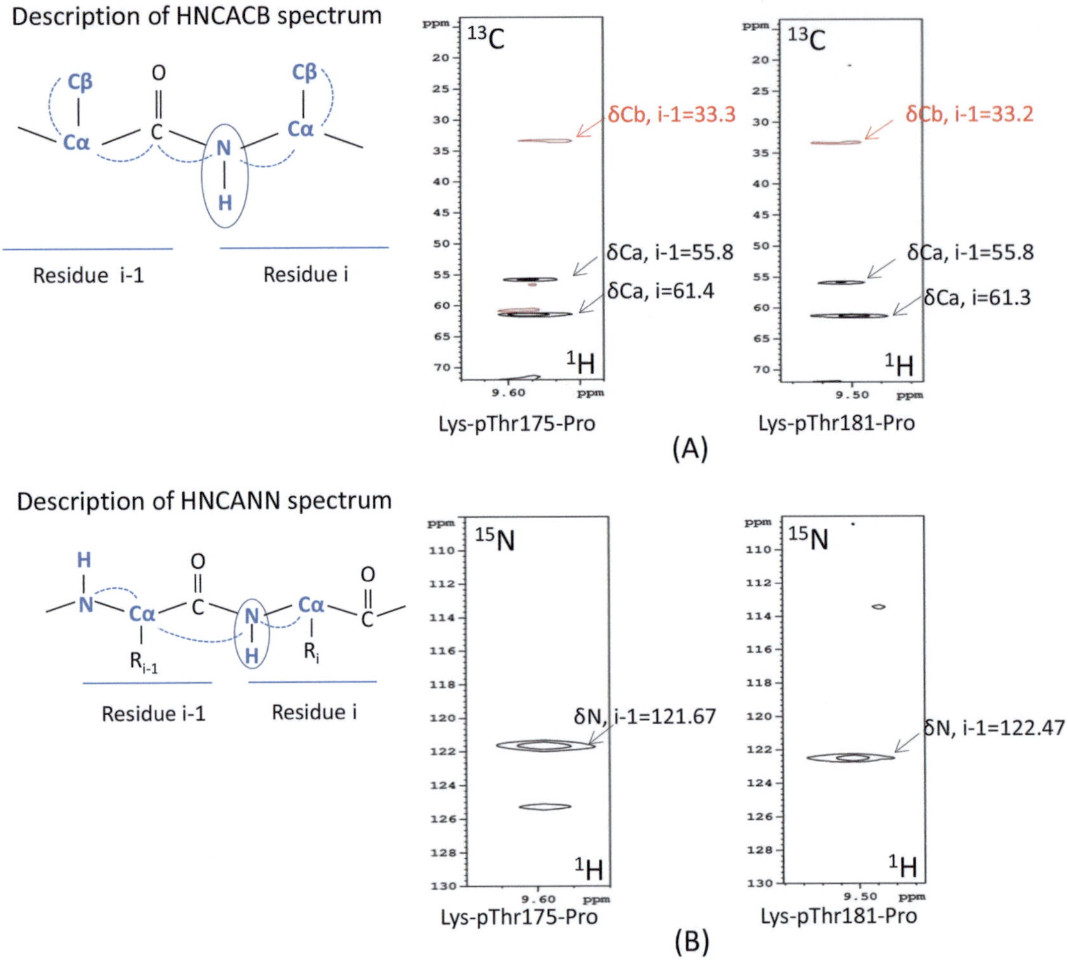

Fig. 12 (**a**) [^{1}H, ^{13}C] 2D planes extracted from the [^{1}H,^{15}N,^{13}C] HNCACB 3D spectrum, at ^{15}N-values of 125.24 ppm (*left*) and 124.76 ppm (*right*). The ^{13}C-CA resonances (in *black*) in side chain of pT175 (*left panel*) and pT181 (*right panel*) are indicated. ^{13}C-CB resonances for these residues are outside the spectrum window (74 ppm). This combination of CA and CB chemical shift values is typical of a pT-P dipeptide. The weaker set of signals (indicated by *arrows*) corresponds to ^{13}C-CA (in *black*) and ^{13}C-CB resonances (in *red*) of the prior residue. Because for both pT-P dipeptide the i-1 residue shows chemical shift values typical of a lysine residue (CA is 55.8 ppm, CB is 33.3 ppm), these resonances cannot be uniquely identified among the 9 pT-P motifs of Tau. (**b**) [^{1}H,^{15}N] 2D planes extracted from the [^{1}H,^{15}N,^{15}N] HNCANN 3D spectrum, at ^{15}N-values of 125.24 ppm (*left*) and 124.76 ppm (*right*). The ^{15}N-resonance in the third dimension represents the chemical shifts of the ^{15}N of the amide group of the K residue at position i-1. From the previously determined assignment of Tau [26–28], we could distinguish δN(i-1)K174 = 121.67 ppm and δN(i-1)K180 = 122.47 ppm, which allow discriminating the resonance of pT175 of pT181

2. Record a [^{1}H,^{15}N] HSQC 2D spectrum (hsqcetfpf3gpsi pulse sequence from Bruker, *see* **Note 17**) [24] (Fig. 10) at 293 K or 298 K on a Bruker900MHz Avance III NMR spectrometer equipped with a triple resonance cryogenic probehead (Bruker, Karlsruhe, Germany) or equivalent spectrometer,

with 3072 and 416 points for 14 and 25 ppm in the ^{1}H and ^{15}N dimensions, respectively. The relaxation delay is 1 s. The duration of the acquisition with 64 scans is 9 h.

3. Process the spectra with the Bruker TopSpin 3.1 software.
4. Dissolve 4 mg of ^{15}N, ^{13}C-Tau in 400 μl NMR buffer (200 μM) to fill a NMR Shigemi tube.
5. Record a [^{1}H,^{15}N,^{13}C] HNCACB 3D spectrum (pulse sequence hncacbgpwg3d from Bruker, *see* **Note 18**) at 298 K on a Bruker900MHz Avance III NMR spectrometer equipped with a triple resonance cryogenic probehead (Bruker, Karlsruhe, Germany) or equivalent spectrometer, with 2048, 256, 72 points for 14, 25, 61 ppm centered on 4.7, 119, 41 ppm in the ^{1}H, ^{15}N, and ^{13}C dimensions, respectively. The relaxation delay is 1 s. The duration of the acquisition with 16 scans is 4 days and 6 h. This experiment allows for recording the chemical shift values of CA and CB ^{13}C nuclei of the i and i-1 residues for each [^{1}H, ^{15}N] resonance.
6. Process the spectra with the Bruker TopSpin 3.1 software.
7. Extract [^{1}H,^{13}C] 2D planes from the [^{1}H,^{15}N,^{13}C] HNCACB 3D experiment (Fig. 12a) for each peak in the [^{1}H,^{15}N] HSQC. Make data analysis, peak picking and calculation of peak volumes with Sparky 3.114 software (T. D. Goddard and D. G. Kneller, SPARKY 3, University of California, San Francisco).
8. Record an additional [^{1}H,^{15}N,^{15}N] HNCANN 3D experiment (hncannhgpwg3d pulse sequence from Bruker, *see* **Note 19**) [25] to distinguish redundant phosphorylation patterns. Perform data acquisition on the same ^{15}N, ^{13}C-Tau sample in the same conditions as described in Subheading 3.15, **step 5** with 3072, 100, 100 points for 14, 21, 21 ppm centered on 4.7, 118.5, 118.5 ppm in the ^{1}H, ^{15}N, and ^{15}N dimensions, respectively. The duration of the acquisition with 16 scans is 1 day and 22 h.
9. Extract [^{1}H,^{15}N] 2D planes from the [^{1}H,^{15}N,^{15}N] HNCANN 3D experiment (Fig. 12b) for each ambiguous peak in the [^{1}H,^{15}N] HSQC (i.e., peaks that cannot be assigned with the HNCACB experiment only).^{15}N-chemical shift values in the third dimension of the recorded spectrum correspond to the ^{15}N chemical shift of the residue at the i, i+1, and i-1 positions.

3.16 Assignment of Resonances of Acetylated Residues: Identification of the Acetylation Pattern

The identification of acetylation sites involves the use of selective ^{15}N-lysine labeling of Tau samples to decrease the spectral complexity. The assignment of lysine residues in the non-acetylated Tau sample is made using the previously described [^{1}H,^{15}N] HSQC, HNCACB, and HNCANN experiments (*see* Subheading 3.15) recorded on uniformly ^{15}N-labeled Tau. These experiments are combined with a [^{1}H,^{15}N] HSQC recorded on a

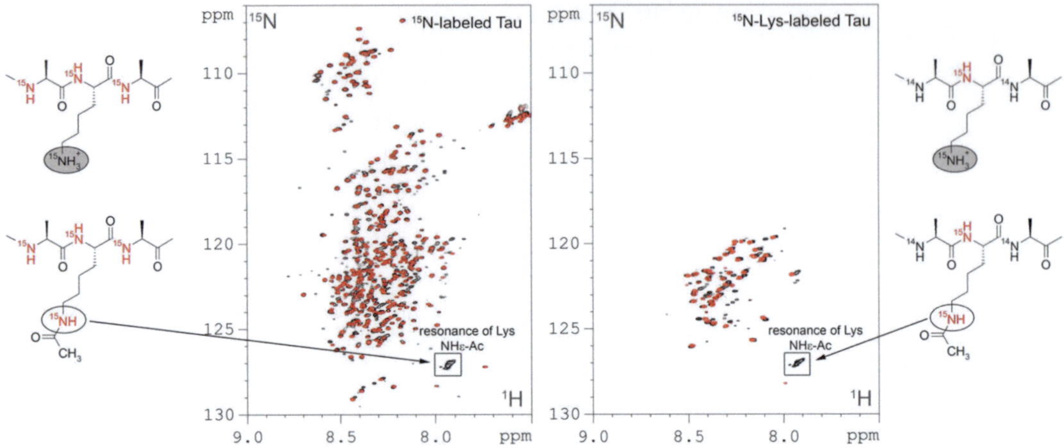

Fig. 13 Comparison of the [^{1}H,^{15}N] HSQC spectra of the non-acetylated (*red*) and acetylated (*black*) forms of uniformly ^{15}N-labeled Tau (*left spectra*) and selectively ^{15}N-lysine labeled Tau (*right spectra*)

selectively ^{15}N-lysine-labeled Tau sample for an easy identification of lysine resonances.

On the [^{1}H,^{15}N] HSQC of the ^{15}N-lysine labeled, acetylated Tau sample, one can detect some additional backbone amide resonances as compared to the control, non-acetylated Tau together with non-resolved peaks corresponding to the NHε acetamide functions of lysine side chains that are referred to as the indicator signal (Fig. 13, *see* **Note 20**) [11, 12]. Hence, identification of resonances corresponding to acetylated lysine residues is first required before their assignment. The resonances of acetylated lysine residues are identified in a selectively ^{15}N-lysine, uniformly ^{13}C labeled, acetylated Tau sample using the [^{1}H,^{15}N] HSQC-TOCSY experiment that enables the detection of proton resonances of the lysine side chain from either the backbone amide resonance or those of side chain acetamide (NHε-Ac, Fig. 14). Lysine assignment is then performed with the [^{1}H,^{15}N,^{13}C] HN(CO)CACB experiment for the identification of the i-1 residue (Fig. 15).

1. Dissolve 1–2 mg of either uniformly ^{15}N-labeled or selectively ^{15}N-lysine-labeled Tau in 200 μl NMR buffer (100–200 μM) to fill a 3 mm tube.
2. Record a [^{1}H,^{15}N] HSQC 2D spectrum (dipsihsqcf3gpsi3d pulse sequence from Bruker, *see* **Note 17**) [24] (Fig. 13) at 293 K on a Bruker900MHz Avance III NMR spectrometer equipped with a triple resonance cryogenic probehead (Bruker, Karlsruhe, Germany) or equivalent spectrometer.
3. For the uniformly labeled sample, acquire data with 2048 and 512 points for spectral widths of 14 and 30 ppm in the ^{1}H and ^{15}N-dimensions, respectively, centered on 4.7 ppm and

118.4 ppm. The duration of the acquisition with 32 scans is 5 h 15 min. For the selectively ^{15}N-lysine-labeled sample, acquire data with 2048 and 128 points for spectral widths of 14 and 9 ppm centered on 4.7 ppm and 123.5 ppm in the ^{1}H and ^{15}N-dimensions, respectively. The relaxation delay is 1 s. The duration of the acquisition with 32 scans is 80 min.

4. Process the spectra with the Bruker TopSpin 3.1 software.
5. Dissolve 11 mg of a ^{15}N-lysine selectively, ^{13}C uniformly labeled, acetylated Tau sample in 400 μl NMR buffer (600 μM) to fill a NMR Shigemi tube.
6. Acquire the three-dimensional spectra under a nonuniform sampling mode on a Bruker900MHz Avance III NMR spectrometer equipped with a triple resonance cryogenic probehead (Bruker, Karlsruhe, Germany) at 293 K.
7. Perform data processing, peak picking and calculation of peak intensities with the Bruker TopSpin 3.1 software.
8. Record a [^{1}H,^{15}N] HSQC-TOCSY spectrum (dipsihsqcf3gpsi3d pulse sequence from Bruker, *see* **Note 21**) with 32 scans per increment, a TOCSY mixing time of 120 ms and spectral widths of 16.0, 9.0, and 11.0 ppm in ^{1}H (F3), ^{15}N (F2), and ^{1}H (F1) dimensions that are sampled with 3072, 78, and 416 points, respectively. The proton and nitrogen dimensions are centered on 4.7 and 123.5 ppm, respectively. The relaxation delay is 1 s. The nonuniform sampling is done with 25 % sparse sampling, 2028 hypercomplex points in indirect dimension, T2 relaxation times of 0.3 s in each ^{1}H dimension, and 0.25 s in ^{15}N dimension. The duration of the acquisition is about 3 days and 21 h.
9. Extract [^{1}H, ^{1}H] 2D planes from the [^{1}H,^{15}N] HSQC-TOCSY 3days experiment for each peak in the [^{1}H,^{15}N] HSQC.
10. Measure the chemical shift values of ^{1}H nuclei of lysine side chain, especially the Hε chemical shifts. This latter value allows for discrimination of acetylated lysines from the non-acetylated ones (Fig. 14).
11. Extract [^{1}H,^{15}N] 2D planes from the [^{1}H,^{15}N] HSQC-TOCSY 3D experiment at 2.98 ppm and 3.14 ppm in the F1 dimension which corresponds to the Hε chemical shifts of a non-acetylated and acetylated lysine residue, respectively.
12. Superimpose each of this sub-spectrum to the [^{1}H,^{15}N] HSQC of acetylated Tau to discriminate between resonances of non-acetylated and acetylated lysine residues.
13. Record a [^{1}H,^{15}N,^{13}C] HN(CO)CACB 3D spectrum (hncocacbgpwg3d pulse sequence from Bruker, *see* **Note 22**) at 900 MHz and 293 K with 3072, 40, and 272 points for spectral widths of 14, 9, and 55 ppm in the ^{1}H, ^{15}N, and ^{13}C-dimensions, respectively, centered on 4.7, 123.5, and 39.0 ppm, respec-

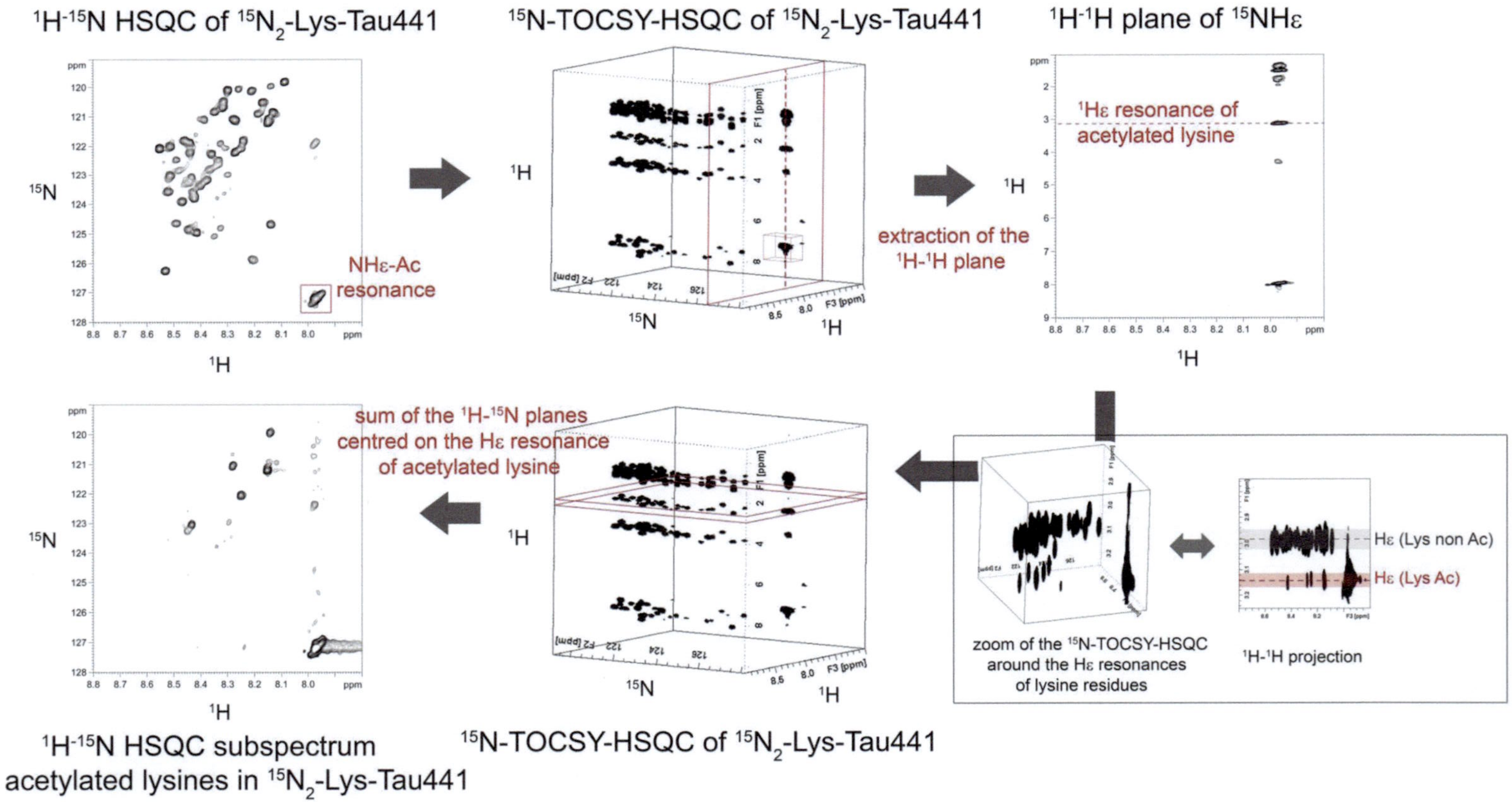

Fig. 14 Strategy of acetylated lysine identification based on the Hε chemical shift of lysine side chain from the [^{1}H,^{15}N] HSQC-TOCSY experiment acquired on ^{15}N-lysine,^{13}C-labeled Tau protein

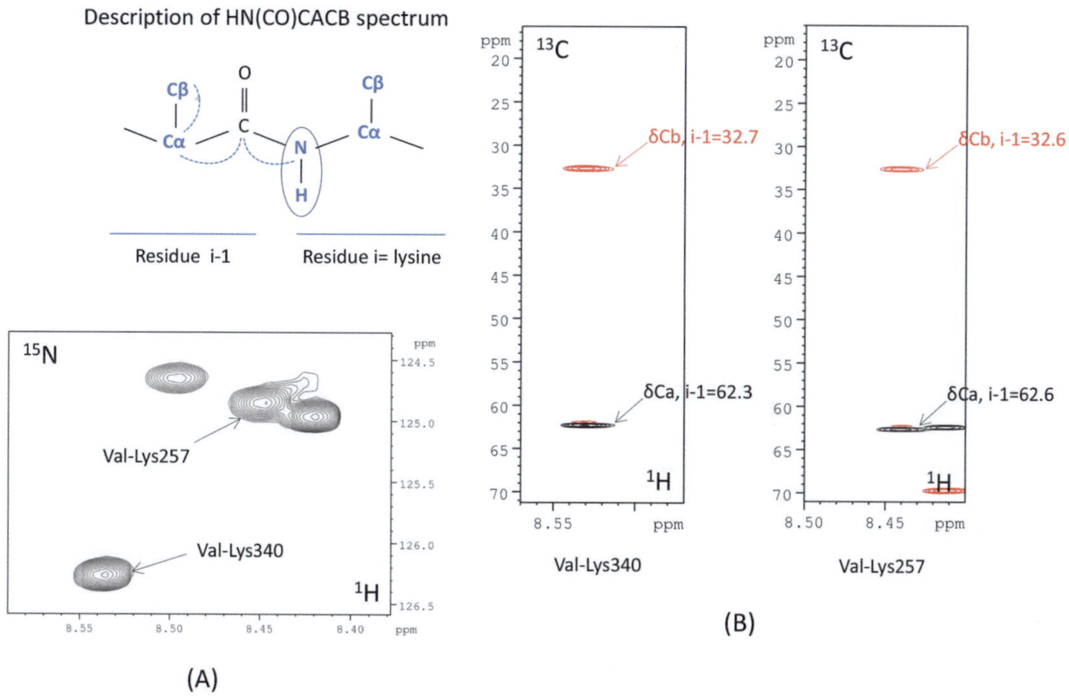

Fig. 15 Identification of lysine resonances using the [^{1}H,^{15}N,^{13}C] HN(CO)CACB experiment. (**a**) Zoom of the [^{1}H,^{15}N] HSQC experiment acquired on selectively ^{15}N-lysine-labeled, acetylated Tau. (**b**) [^{1}H, ^{13}C] 2D planes extracted from the [^{1}H,^{15}N,^{13}C] HN(CO)CACB 3D spectrum, at ^{15}N-values of 126.25 ppm (*left*) and 124.85 ppm (*right*). The ^{13}C-CA (*black*) and ^{13}C-CB resonances (*red*) of i-1 residue side chain of K340 (*left panel*) and K257 (*right panel*) are indicated. This combination of CA and CB chemical shift values is typical of a V residue. The unambiguous assignment of both V-K dipeptide has been made with a non-acetylated Tau sample and uniform ^{15}N-labeling using the previously described [^{1}H,^{15}N,^{13}C] HNCACB and [^{1}H,^{15}N,^{15}N] HNCANN three-dimensional experiments (Subheading 3.15)

tively. The relaxation delay is 1 s. The nonuniform sampling is done with 25% sparse sampling, 680 hypercomplex points in indirect dimension, T2 relaxation times of 0.3 s, 1 s, and 0.1 s in ^{1}H, ^{15}N, and ^{13}C dimensions, respectively. The duration of the acquisition with 48 scans is 1 day and 23 h.

14. Extract [^{1}H, ^{13}C] 2D planes from the [^{1}H,^{15}N,^{13}C] HN(CO) CACB 3D experiment for each peak in the [^{1}H,^{15}N] HSQC.
15. Measure the chemical shift values of "CA" and "CB" ^{13}C nuclei of the i-1 residues for each [^{1}H, ^{15}N] lysine resonance enabling residue identification (Figs. 15 and 16).
16. To avoid ambiguity related to redundant X-K dipeptides (where X is any amino acid) as illustrated in Fig. 15 with the V-K340/257 dipeptides, this experiment is combined with HNCACB and HNCANN experiments acquired on ^{15}N-Tau with uniform ^{15}N-labeling (*see* Subheading 3.15) for full assignment of lysine residues in the control, non-acetylated Tau sample. According to this latter strategy, a full sequential

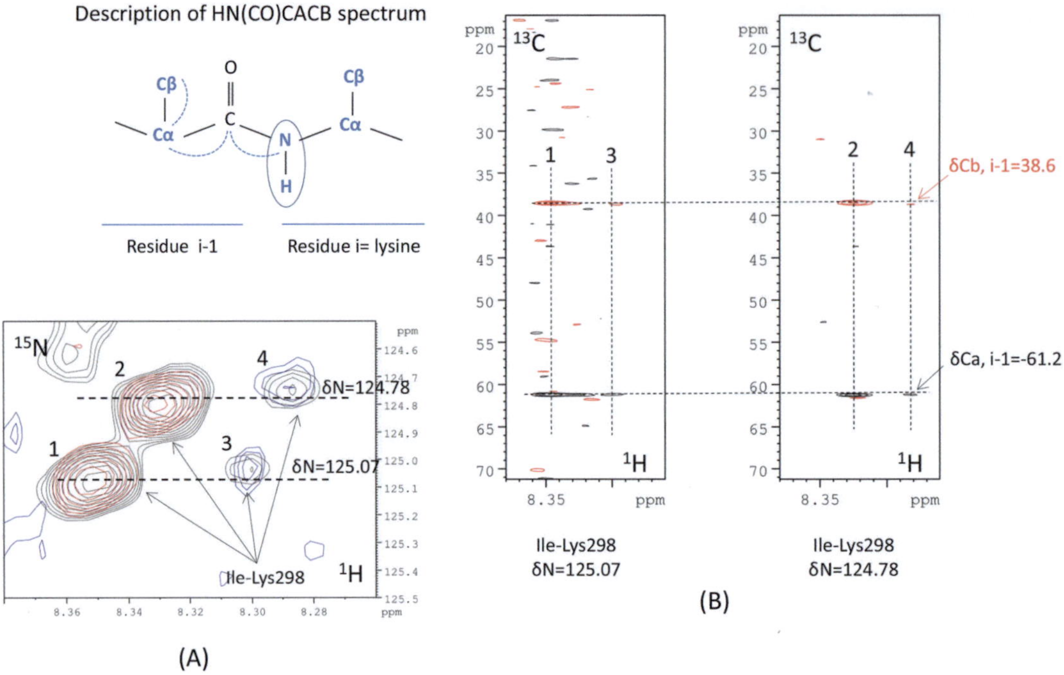

Fig. 16 Increasing spectral complexity in the case of two close acetylation sites showing four resonances for the single I-K298 dipeptide (numbered from 1 to 4). (**a**) Zoom of the [^{1}H,^{15}N] HSQC experiment acquired on selectively ^{15}N-lysine-labeled, acetylated Tau (*black*). Sub-spectra of non-acetylated (*red*) and acetylated (*blue*) lysine (Subheading 3.16) are overlaid and indicate that resonances 1 and 2 correspond to non-acetylated lysine while resonances 3 and 4 correspond to acetylated lysine. (**b**) [^{1}H,^{13}C] 2D planes extracted from the [^{1}H,^{15}N,^{13}C] HN(CO)CACB 3D spectrum at ^{15}N-values of 125.07 ppm (for resonances 1 and 3) and 124.78 ppm (for resonances 2 and 4). According to the ^{13}C-CA (*black*) and ^{13}C-CB values (*red*) of the i-1 residue, resonances numbered one to four are all identified as I-K dipeptide that is represented by the sole K298 along the Tau sequence. Hence, the splitting of each of non-acetylated and acetylated K298 resonance is due to the proximity of another acetylation site in the Tau sequence that has been identified as the i-4 residue according to the same assignment strategy

assignment enables the identification of the i-1 residue of X in each X-K dipeptide which is either E in the case of V-K340 or N in the case of V-K257.

17. Spectral complexity due to the presence of close acetylation sites is illustrated in Fig. 16 for the K298 residue. In this case, four resonances are detected for the single I-K dipeptide. Two of them come from a non-acetylated form of K298 and the two remaining from an acetylated K298 based on their respective Hε value (Fig. 16a), indicating the proximity of another acetylation site (identified as the K294 residue).

4 Notes

1. Complete medium and MEM vitamins addition are not compulsory but help to stimulate the growth and improve the

yield. MEM vitamins 100× (SIGMA) are aliquoted by 10 ml and aliquots are stored at −20 °C.

2. Upon addition of the 100 mM $CaCl_2$ stock a white precipitate will form.
3. The isotopes are added freshly at the time the bacterial culture is started. They are solubilized in M9 buffer and filtered sterilized at 0.2 μm.
4. To produce ^{15}N, ^{13}C-Tau, 300 mg of ^{15}N, ^{13}C-complete medium, 1 g of ^{15}N-NH_4Cl, and 2 g of ^{13}C-glucose are suspended in 10 ml of M9 buffer and are 0.2 μM filter-sterilized directly into the M9 medium. Glucose is the limiting factor for the bacterial growth and an OD600 of 1.4–1.6 is usually observed at the end of the fermentation.
5. Tau protein is a disordered protein sensitive to proteases. We find that sterilization (120 °C for 20 min) of buffers limits Tau degradation. The buffers are 0.22 μm-filtered or autoclaved and stored at 4 °C.
6. pH is adjusted after addition of 250 mM imidazole in phosphate buffer and 10 mM reduced Glutathione in Tris buffer, in the elution buffers.
7. We do not reconstitute a complete medium from the Isogro® powder, given the cost of the product, but use it as a culture supplement. Generally, bacterial cultures using M9 media supplemented with 0.05% Isogro® (w/v) reach the exponential growth phase (as measured by O.D. at 600 nm) after about 4 h of incubation at 37 °C under the conditions described to inoculate the culture. Without Isogro®, the exponential phase of growth is reached after 6 h at 37 °C.
8. The bacterial growth is slow in M9 medium. We use programmable incubators to start the culture at an early morning time, so that the protein production, collection of the bacterial pellet, and analytical control of the protein production can be conveniently scheduled during the work day hours.
9. The lyophilized Tau proteins are kept at −20 °C.
10. Tau protein concentration is difficult to define because its absorption coefficient at 280 nm is low (A280 nm = 0.14 for a solution at 1 mg/ml). To define Tau concentration, we use the surface of the peak of Tau protein absorption at 280 nm in the chromatogram in the desalting step in Ammonium Bicarbonate. This is the most reliable method we have found. This method was validated by analysis of one Tau sample by total amino acid quantification.
11. The His-ERK2 preparation is contaminated with a protein of 66 kDa (Fig. 4a).

12. The activated ERK kinase in activation buffer with 10% Glycerol is directly mixed with Tau protein in phosphorylation buffer, without prior buffer exchange. In an alternative protocol, His-ERK2, GST-MEK1 R4 and ^{15}N-Tau can be all mixed, to combine **steps 3.7** and **3.8**.
13. Rat brain extract can be stored at 4 °C for a few days only without loss of kinase activity.
14. Resin beads loaded with the GST-CBP[1202-1848] fragment can be stored without glycerol at 4 °C for few days. However, we have noticed a rapid decrease of acetyltransferase activity (within 2 or 3 days) in these conditions.
15. A peak with a mass increment of +168 Da ($[M+H]^+$ 2861.79 Da) corresponding to the incorporation of four acetyl moieties in the CBP standard substrate from TDG protein can be detected.
16. In contrast to S/T phosphorylation, Y phosphorylation does not induce a down field backbone-amide chemical shift of the modified Y residue resonance [22], probably because the phosphorylatable Y hydroxyl group is at a more distal position compared to the one of S and T residues.
17. [^{1}H,^{15}N] HSQC pulse sequence from Bruker is hsqcetfpf3gpsi avance-version: HSQC 2D H-1/X correlation via double INEPT (Insensitive Nuclei Enhanced by Polarization Transfer) using sensitivity improvement, phase sensitive using Echo/Antiecho-TPPI gradient selection with decoupling during acquisition using f3—channel and flip-back pulse.
18. [^{1}H,^{15}N,^{13}C] HNCACB pulse sequence from Bruker is hncacbgpwg3d avance-version: HNCACB 3D sequence with inverse correlation for triple resonance using multiple INEPT transfer steps (F1(H)→F3(N)→F2(Ca→Cb,t1)→F3(N,t2)→F1(H,t 3)), on/off resonance CA and CO pulses using shaped pulse, phase sensitive (t1), phase sensitive (t2), using constant time in t2 and water suppression using watergate sequence.
19. [^{1}H,^{15}N,^{15}N] HNCANN pulse sequence from Bruker is hncannhgpwg3d avance-version: (H)N(CA)NNH 3D sequence with inverse correlation for triple resonance using multiple INEPT transfer steps (F1(H)→F3(N,t1)→F2(Ca)→F3(N,t2)→F1(H,t3)), on/off resonance CA and CO pulses using shaped pulse, phase sensitive (t1), phase sensitive (t2), using semi-constant time in t1 and t2 and water suppression using watergate sequence.
20. In contrast to the phosphorylation sites that are easily identifiable due to the low field chemical shift of their backbone amide resonances (Fig. 10), the identification of acetylated lysine is not straightforward just by looking to the 2D [^{1}H,^{15}N] HSQC spectrum of acetylated ^{15}N-labeled Tau, or even in the

^{15}N-Lys-labeled Tau samples (Fig. 13). Furthermore, the CA and CB chemical shifts of lysine residues as extracted from the HNCACB experiment (Subheading 3.15, **step** 7) fail to inform about the lysine acetylation state. Only the extremity of the lysine side chain (from Cγ to Cε and their protons) is sensitive to the presence of an acetyl moiety [10].

21. [^{1}H,^{15}N] HSQC-TOCSY pulse sequence from Bruker is dipsihsqcf3gpsi3d avance-version: TOCSY-HSQC 3D sequence with homonuclear Hartman-Hahn transfer using DIPSI2 sequence for mixing, H-1/X correlation via double inept transfer using sensitivity improvement, phase sensitive (t1), phase sensitive using Echo/Antiecho-TPPI gradient selection (t2), using trim pulses in INEPT transfer, using f3—channel.
22. [^{1}H,^{15}N,^{13}C] HN(CO)CACB pulse sequence from Bruker is hncocacbgpwg3d avance-version: HNCOCACB 3D sequence with inverse correlation for triple resonance using multiple INEPT transfer steps (F1(H) → F3(N) → F2(C=O) → F2(Ca → Cb,t1) → F2(C=O) → F3(N,t2) → F1(H,t3)), on/off resonance CA and CO pulses using shaped pulse, phase sensitive (t1), phase sensitive (t2) using constant time in t2 and water suppression using watergate sequence.

Acknowledgment

The NMR facilities were funded by the Région Nord, CNRS, Pasteur Institute of Lille, European Community (FEDER), French Research Ministry and the University of Sciences and Technologies of Lille I. We acknowledge support from the TGE RMN THC (FR-3050, France) and the Research Federation FRABio (Univ. Lille, CNRS, FR 3688, FRABio, "Structural & Functional Biochemistry of Biomolecular Assemblies") for providing the scientific and technical environment. This study was supported by a grant from the LabEx (Laboratory of Excellence) DISTALZ (Development of Innovative Strategies for a Transdisciplinary approach to Alzheimer's disease), and in part by the French government funding agency Agence Nationale de la Recherche TAF. S.P. and J.G. were partially supported on NIH R01-GM08157.

References

1. Martin L, Latypova X, Wilson CM, Magnaudeix A, Perrin ML, Yardin C, Terro F (2013) Tau protein kinases: involvement in Alzheimer's disease. Ageing Res Rev 12:289–309. doi:10.1016/j.arr.2012.06.003, S1568-1637(12)00088-8 [pii]
2. Hasegawa M, Morishima-Kawashima M, Takio K, Suzuki M, Titani K, Ihara Y (1992) Protein sequence and mass spectrometric analyses of tau in the Alzheimer's disease brain. J Biol Chem 267:17047–17054
3. Morishima-Kawashima M, Hasegawa M, Takio K, Suzuki M, Yoshida H, Titani K, Ihara Y (1995) Proline-directed and non-proline-

directed phosphorylation of PHF-tau. J Biol Chem 270:823–829

4. Morris M, Knudsen GM, Maeda S, Trinidad JC, Ioanoviciu A, Burlingame AL, Mucke L (2015) Tau post-translational modifications in wild-type and human amyloid precursor protein transgenic mice. Nat Neurosci 18:1183–1189
5. Cohen TJ, Guo JL, Hurtado DE, Kwong LK, Mills IP, Trojanowski JQ, Lee VM (2011) The acetylation of tau inhibits its function and promotes pathological tau aggregation. Nat Commun 2:252
6. Min SW, Cho SH, Zhou Y, Schroeder S, Haroutunian V, Seeley WW, Huang EJ, Shen Y, Masliah E, Mukherjee C, Meyers D, Cole PA, Ott M, Gan L (2010) Acetylation of tau inhibits its degradation and contributes to tauopathy. Neuron 67:953–966. doi:10.1016/j.neuron.2010.08.044, S0896-6273(10)00687-2 [pii]
7. Qi H, Cantrelle F-X, Benhelli-Mokrani H, Smet-Nocca C, Buée L, Lippens G, Bonnefoy E, Galas M-C, Landrieu I (2015) Nuclear magnetic resonance spectroscopy characterization of interaction of Tau with DNA and its regulation by phosphorylation. Biochemistry (Mosc) 54:1525–1533. doi:10.1021/bi5014613
8. Landrieu I, Lacosse L, Leroy A, Wieruszeski JM, Trivelli X, Sillen A, Sibille N, Schwalbe H, Saxena K, Langer T, Lippens G (2006) NMR analysis of a Tau phosphorylation pattern. J Am Chem Soc 128:3575–3583. doi:10.1021/ja054656+
9. Amniai L, Barbier P, Sillen A, Wieruszeski J-M, Peyrot V, Lippens G, Landrieu I (2009) Alzheimer disease specific phosphoepitopes of Tau interfere with assembly of tubulin but not binding to microtubules. FASEB J 23:1146–1152
10. Smet-Nocca C, Wieruszeski JM, Melnyk O, Benecke A (2010) NMR-based detection of acetylation sites in peptides. J Pept Sci 16:414–423. doi:10.1002/psc.1257
11. Theillet F-X, Smet-Nocca C, Liokatis S, Thongwichian R, Kosten J, Yoon M-K, Kriwacki RW, Landrieu I, Lippens G, Selenko P (2012) Cell signaling, post-translational protein modifications and NMR spectroscopy. J Biomol NMR 54:217–236. doi:10.1007/s10858-012-9674-x
12. Kamah A, Huvent I, Cantrelle FX, Qi H, Lippens G, Landrieu I, Smet-Nocca C (2014) Nuclear magnetic resonance analysis of the acetylation pattern of the neuronal Tau protein. Biochemistry (Mosc) 53:3020–3032. doi:10.1021/bi500006v
13. Biernat J, Mandelkow EM, Schroter C, Lichtenberg-Kraag B, Steiner B, Berling B, Meyer H, Mercken M, Vandermeeren A, Goedert M et al (1992) The switch of tau protein to an Alzheimer-like state includes the phosphorylation of two serine-proline motifs upstream of the microtubule binding region. EMBO J 11:1593–1597
14. Goedert M, Jakes R, Crowther RA, Six J, Lubke U, Vandermeeren M, Cras P, Trojanowski JQ, Lee VM (1993) The abnormal phosphorylation of tau protein at Ser-202 in Alzheimer disease recapitulates phosphorylation during development. Proc Natl Acad Sci U S A 90:5066–5070
15. Alonso AC, Zaidi T, Grundke-Iqbal I, Iqbal K (1994) Role of abnormally phosphorylated tau in the breakdown of microtubules in Alzheimer disease. Proc Natl Acad Sci U S A 91:5562–5566
16. Boulton TG, Yancopoulos GD, Gregory JS, Slaughter C, Moomaw C, Hsu J, Cobb MH (1990) An insulin-stimulated protein kinase similar to yeast kinases involved in cell cycle control. Science 249:64–67
17. Anderson NG, Maller JL, Tonks NK, Sturgill TW (1990) Requirement for integration of signals from two distinct phosphorylation pathways for activation of MAP kinase. Nature 343:651–653. doi:10.1038/343651a0
18. Seger R, Ahn NG, Boulton TG, Yancopoulos GD, Panayotatos N, Radziejewska E, Ericsson L, Bratlien RL, Cobb MH, Krebs EG (1991) Microtubule-associated protein 2 kinases, ERK1 and ERK2, undergo autophosphorylation on both tyrosine and threonine residues: implications for their mechanism of activation. Proc Natl Acad Sci U S A 88:6142–6146
19. Mansour SJ, Matten WT, Hermann AS, Candia JM, Rong S, Fukasawa K, Vande Woude GF, Ahn NG (1994) Transformation of mammalian cells by constitutively active MAP kinase kinase. Science 265:966–970
20. Prabakaran S, Everley RA, Landrieu I, Wieruszeski JM, Lippens G, Steen H, Gunawardena J (2011) Comparative analysis of Erk phosphorylation suggests a mixed strategy for measuring phospho-form distributions. Mol Syst Biol 7:482. doi:10.1038/msb.2011.15, msb201115 [pii]
21. Tini M, Benecke A, Um SJ, Torchia J, Evans RM, Chambon P (2002) Association of CBP/p300 acetylase and thymine DNA glycosylase links DNA repair and transcription. Mol Cell 9:265–277, doi: S1097276502004537 [pii]
22. Bienkiewicz EA, Lumb KJ (1999) Random-coil chemical shifts of phosphorylated amino acids. J Biomol NMR 15:203–206
23. Wishart DS, Bigam CG, Holm A, Hodges RS, Sykes BD (1995) 1H, 13C and 15N random coil NMR chemical shifts of the common amino acids. I. Investigations of nearest-neighbor effects. J Biomol NMR 5:67–81

24. Schleucher J, Schwendinger M, Sattler M, Schmidt P, Schedletzky O, Glaser SJ, Sorensen OW, Griesinger C (1994) A general enhancement scheme in heteronuclear multidimensional NMR employing pulsed field gradients. J Biomol NMR 4:301–306
25. Weisemann R, Ruterjans H, Bermel W (1993) 3D triple-resonance NMR techniques for the sequential assignment of NH and 15N resonances in 15N- and 13C-labelled proteins. J Biomol NMR 3:113–120
26. Lippens G, Wieruszeski JM, Leroy A, Smet C, Sillen A, Buee L, Landrieu I (2004) Proline-directed random-coil chemical shift values as a tool for the NMR assignment of the tau phosphorylation sites. Chembiochem 5:73–78. doi:10.1002/cbic.200300763
27. Smet C, Leroy A, Sillen A, Wieruszeski JM, Landrieu I, Lippens G (2004) Accepting its random coil nature allows a partial NMR assignment of the neuronal Tau protein. Chembiochem 5:1639–1646. doi:10.1002/cbic.200400145
28. Mukrasch MD, Bibow S, Korukottu J, Jeganathan S, Biernat J, Griesinger C, Mandelkow E, Zweckstetter M (2009) Structural polymorphism of 441-residue tau at single residue resolution. PLoS Biol 7, e34

Chapter 12

Tag-Free Semi-Synthesis of the Tau Protein

Oliver Reimann, Caroline Smet-Nocca, and Christian P.R. Hackenberger

Abstract

Expressed protein ligation (EPL) is a valuable tool to study site-specific functionalities on proteins such as posttranslational modifications. The purification of such ligation products from EPL mixtures can be cumbersome due to a small size difference between the expressed protein portion and the desired ligated protein. Therefore, affinity tags are often required, which remain on the protein after purification. Herein, we present an efficient protocol to install a photocleavable biotin building block on synthetic C-terminal tau[390–441] and describe its use for purification of full-length semi-synthetic tau[1–441].

Key words Expressed protein ligation, Photocleavable biotin, Semi-synthesis, Affinity purification, Traceless linker

1 Introduction

The protein tau belongs to the family of microtubule-associated proteins and its abnormal accumulation is a feature of many neurodegenerative diseases referred to as tauopathies. The most-recognized malignancy associated with tau is Alzheimer's disease (AD), defined by the polymerization of tau into straight and paired helical filaments (SFs and PHFs). A large portion of tau in these filaments is abnormally phosphorylated [1, 2], which may be related to imbalanced kinase and phosphatase activity [3, 4]. It is assumed that abnormally phosphorylated tau species disrupt microtubule assembly [5], recruit normal tau into an aggregated state and are thus hallmarks of AD [6]. The C-terminus of tau is thought to constitute an inhibitory domain for tau polymerization [7]. This inhibitory effect is diminished if the C-terminal domain of tau becomes truncated or phosphorylated. Hence, deciphering the role of specific phosphorylation-sites at the C-terminus of tau that compromise its function and are involved in pathological features is of great importance. However, there are several obstacles in addressing the biological or pathological consequences of tau phosphorylation with the difficult access to homogeneous,

Caroline Smet-Nocca (ed.), *Tau Protein: Methods and Protocols*, Methods in Molecular Biology, vol. 1523,
DOI 10.1007/978-1-4939-6598-4_12, © Springer Science+Business Media New York 2017

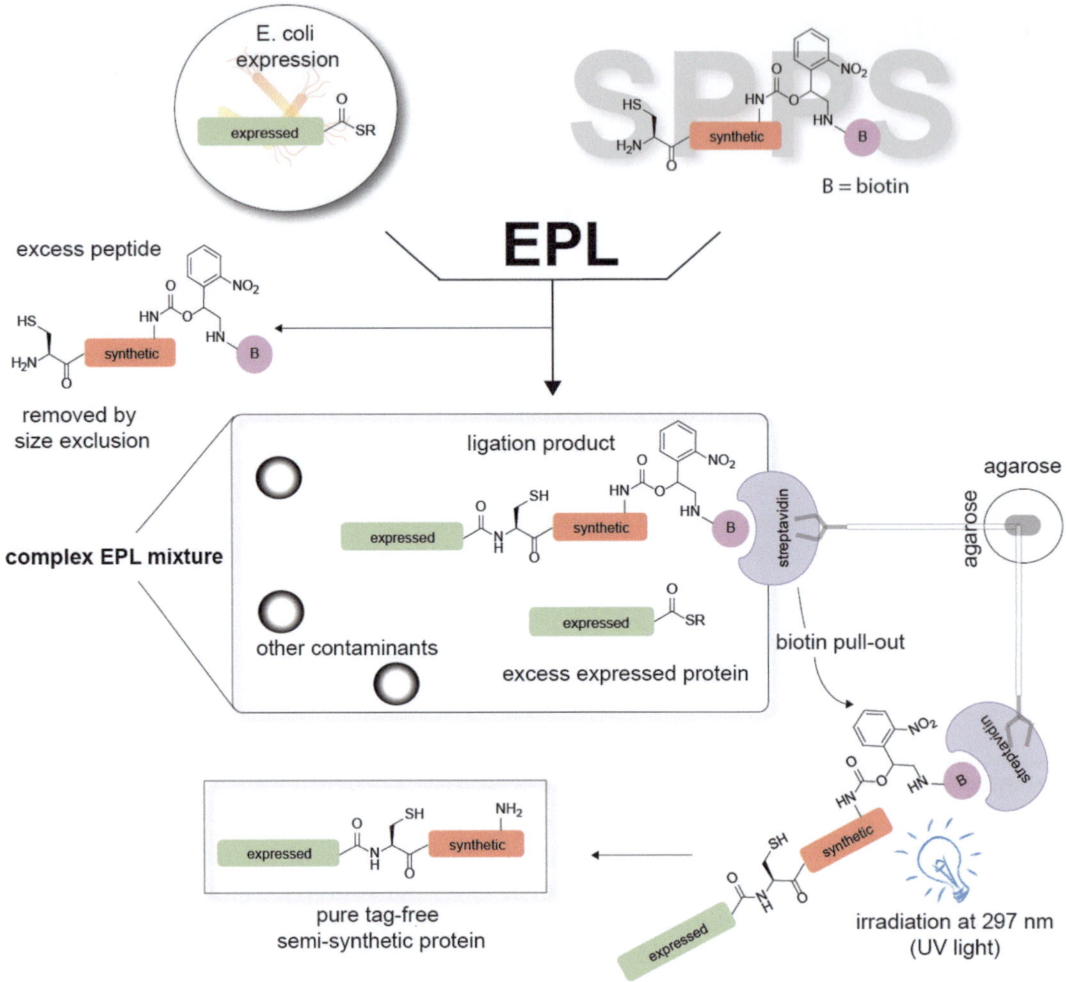

Scheme 1 Expressed protein ligation between an expressed protein fragment and a synthetic peptide bearing the photocleavable biotin, followed by the removal of excess peptide from the resulting complex mixture by size exclusion chromatography. The ligation product is then separated from the unreacted protein fragment and other contaminants by a biotin pull-out with streptavidin coated agarose beads. In a last step, the native protein is released from the streptavidin beads though UV light irradiation

site-specifically phosphorylated proteins for functional studies being one of the most challenging one. We have previously targeted these problems by chemical protein engineering, namely by expressed protein ligation (EPL), which allowed the site-specific introduction of phosphorylated serine at a biologically relevant position by solid phase peptide synthesis before the full-length tau protein is ligated to an expressed fragment by native chemical ligation [8–10]. Additionally, biotin is installed at a side-chain of an lysine residue within the synthetic peptide sequence as an affinity tag to ensure convenient purification (*see* Scheme 1) [11]. Affinity tags such as biotin are frequently used to increase the purity and

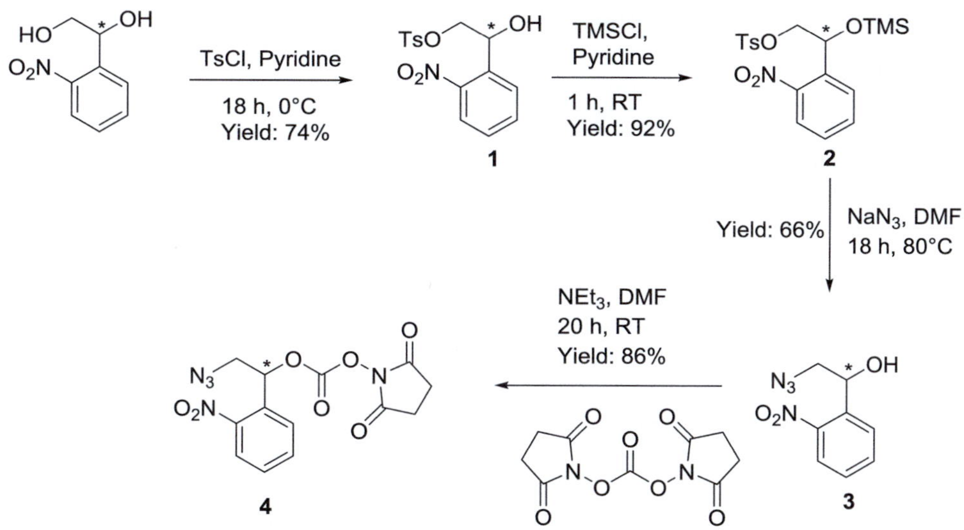

Scheme 2 Synthesis of a bifunctional photolinker according to [15]

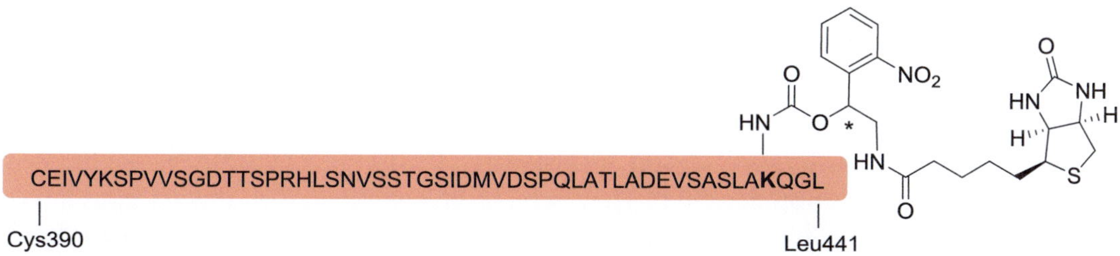

Fig. 1 Peptide sequence of C-terminal tau[390–441] (Ala390Cys) with Lys438 attached to photocleavable biotin building block

enhance the isolation of a recombinant protein; nevertheless, they remain covalently bound to the protein of interest and may have a structural or functional influence on the target protein. Therefore, we have recently implemented a new strategy for the traceless purification of peptides and proteins [12]. In this approach we have synthesized a photocleavable 2-nitrobenzyl group (*see* Scheme 2) and linked biotin to it at a Lys side chain of a synthetic peptide (*see* Fig. 1) that we further used for purification (*see* Scheme 3). This methodology allows for the rapid capture of ligation products onto streptavidin coated agarose beads, removal of excess substrates and by-products and the traceless release of ligation products in a buffer of choice in high yield. Thus, we are able to generate full-length tag-free semi-synthetic tau protein. Herein, we provide a very detailed and practical protocol for the synthesis of a C-terminal tau peptide bearing photocleavable biotin (PCB) and its use for the traceless purification of semi-synthetic full-length tau upon purification [12]. Recently, we have demonstrated the challenging

Synthesis of tau peptide with photocleavable biotin linker

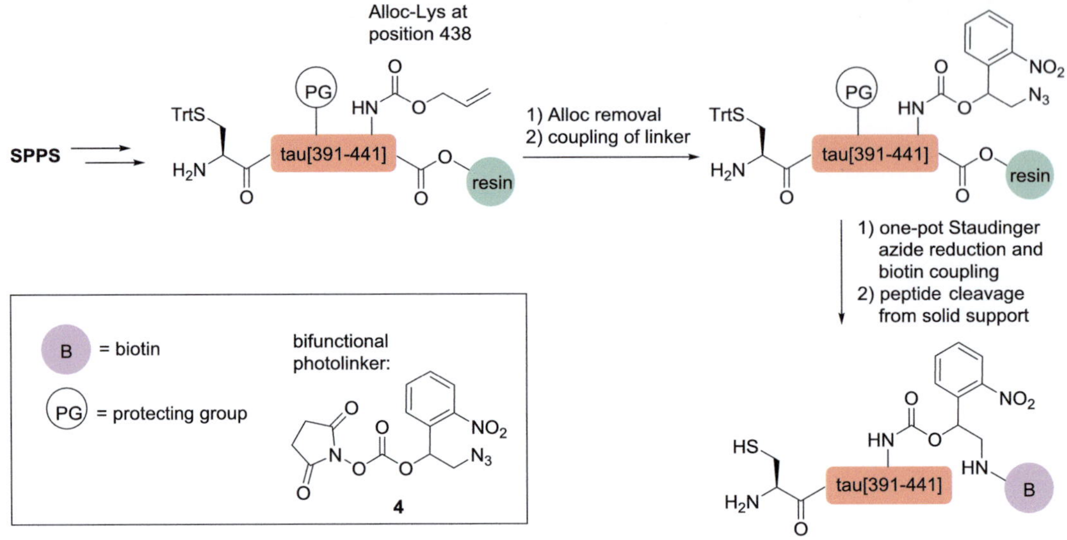

Scheme 3 Introduction of the photocleavable biotin unit to a peptide during SPPS

synthesis of C-terminal tau peptides phosphorylated at positions Ser396, Ser400, and Ser404 [11–13] or *O*-GlcNAcylated in position Ser400 [14], which can be combined with this traceless linker strategy to obtain tag-free full-length tau proteins, phosphorylated or O-GlcNAcylated at these specific sites. In addition, this methodology does not only allow for the investigation of their impact on the protein functions, but can also be adapted for the study of the protein conformation by NMR spectroscopy. In this latter case, the protein expression in *E. coli* allows for the isotopic labeling of the recombinant part of the Tau protein with ^{15}N or both ^{15}N and ^{13}C stable isotopes. As a result, EPL leads to segmental isotopic labeling of full-length tau allowing the detection of amide resonances of the ^{15}N-labeled region required for further conformational studies.

2 Materials

All solutions should be prepared with ultrapure water; solvents used are of analytical grade. Buffers are stored at 4 °C and do not contain sodium azide as additive.

2.1 Synthesis of a Photolinker

1. *p*-toluenesulfonyl chloride.
2. 1-(2-nitrophenyl)-1,2-ethanediol.
3. Round-bottom flask.

4. Nitrogen (N_2) and argon gas bottles.
5. Pyridine.
6. Ice-bath cooling.
7. Diethyl ether (Et_2O).
8. 1 M $KHSO_4$ solution.
9. Saturated aqueous solution of $NaHCO_3$.
10. Saturated aqueous solution of NaCl (Brine).
11. $MgSO_4$.
12. Rotary evaporator.
13. SiO_2 flash column chromatography.
14. Ethyl acetate (EtOAc)/Hexane (Hex) 1:1 and 1:4.
15. Trimethylsilyl chloride (TMSCl).
16. Sodium azide.
17. *N,N*-dimethylformamide (DMF).
18. *N*-hydroxysuccinimidyl carbonate.
19. Triethyl amine (NEt_3).

2.2 Solid Phase Peptide Synthesis (SPPS)

1. H-Leu-HMPB NovaPEG resin (Novabiochem) with a substitution grade of 0.49 mmol/g.
2. Commercially available fluorenylmethyloxycarbonyl (Fmoc)-protected amino acids (Table 1), Fmoc-L-Lys(Alloc)-OH, Boc-Cys(Trt)-OH, reagents and solvents (Table 2) used for peptide synthesis and purchased from Bachem, Iris Biotech and Novabiochem.
3. Activo-P11 automated peptide synthesizer.
4. Allyloxycarbonyl (Alloc)-deprotection solution (for 0.1 mmol scaled peptide synthesis): $Pd(PPh_3)_4$ (12 mg, 0.01 mmol, 0.1 equivalent (eq)), phenylsilane (300 μL, 2.4 mmol, 24 eq), CH_2Cl_2 (4 mL). Prepare freshly and use up.
5. Photolinker coupling solution (for 0.1 mmol scaled synthesis): Photocleavable linker (**4**) (80 mg, 0.22 mmol, 2.2 eq), diisopropylethylamine (DIPEA) (120 μL, 0.68 mmol, 6.8 eq), DMF (4 mL). Prepare freshly and use up.
6. Azide reduction and biotin coupling solution (for a 0.1 mmol scaled reaction): Biotin (106 mg, 0.4 mmol, 4 eq), diisopropylcarbodiimide (DIC) (62 μL, 0.4 mmol, 4 eq), tri-*n*-butylphosphine (120 μL, 0.5 mmol, 5 eq), hydroxybenzotriazole (HOBt) (54 mg, 0.4 mmol, 4 eq), N-methyl-2-pyrrolidone (NMP) (4 mL) (*see* **Note 1**). Prepare freshly and use up.
7. Peptide cleavage cocktail: trifluoroacetic acid (TFA)–dithiothreitol (DTT)–triisoprolylsilane (TIS)–thioanisol mixed in a ratio of 95:2:2:1 (*see* **Note 2**). Prepare freshly and use up.

Table 1
Building blocks used for peptide synthesis

L-amino acids used for SPPS	Provider
Fmoc-Gly-OH	Novabiochem
Fmoc-Gln(Trt)-OH	Iris Biotech
Fmoc-Lys(Alloc)-OH	Novabiochem
Fmoc-Ala-OH	Iris Biotech
Fmoc-Leu-OH	Iris Biotech
Fmoc-Ser(*t*Bu)-OH	Novabiochem
Fmoc-Val-OH	Novabiochem
Fmoc-Glu(O*t*Bu)-OH	Novabiochem
Fmoc-Asp(O*t*Bu)-OH	Iris Biotech
Fmoc-Thr(*t*Bu)-OH	Iris Biotech
Fmoc-Pro-OH	Iris Biotech
Fmoc-Met-OH	Novabiochem
Fmoc-Ile-OH	Novabiochem
Fmoc-Asn(Trt)-OH	Iris Biotech
Fmoc-His(Trt)-OH	Iris Biotech
Fmoc-Arg(Pbf)-OH	Iris Biotech
Fmoc-Lys(Boc)-OH	Iris Biotech
Fmoc-Tyr(*t*Bu)-OH	Novabiochem
Boc-Cys(Trt)-OH	Iris Biotech

8. Diethyl ether.
9. Buffer A for HPLC: 0.1 % TFA in water.
10. Buffer B for HPLC: 0.1 % TFA in acetonitrile.
11. Preparative high performance liquid chromatography (HPLC).
12. Ultra Performance Liquid Chromatography-Mass Spectrometry (UPLC-MS).

2.3 Expression of Proteins

1. T7 express cells (New England Biolab).
2. Super optimal broth with catabolite repression (SOC).
3. Lysogeny broth (LB) agar plates containing ampicillin (100 μg/mL).
4. Lysogeny broth (LB) medium.

Table 2
Settings on the Activo-P11 automated peptide synthesizer during the synthesis of the tau peptide

Swelling	Quantity	10 mL/g of resin
	Number of swelling steps	2
	Duration of first swelling	15 min (DMF)
	Duration of second swelling	15 min (DMF)
Washes	Quantity	10 mL/g of resin (DMF)
	Duration of each wash step	1 min
Deprotection	Reagent	25 % piperidine in DMF
	Quantity	10 mL/g of resin
	Number of deprotections	2
	Duration of first deprotection	3 min
	Duration of second deprotection	12 min
	Number of washes after deprotection (solvent)	5 (DMF)
Coupling	Number of couplings	1
	Equivalents of amino acid	10 (powder)
	Concentration of amino acid when applied to resin	0.5 M
	Coupling reagents	0.48 M HBTU (10 eq), 0.48 M HOBt (10 eq), 1 M DIPEA (10 eq) (DMF)
	Pre-activation time of coupling reagents with amino acids	3 min
	Coupling time	20 min
	Number of washes after coupling (solvent)	4 (DMF)

5. M9 minimal medium for protein labeling (for 1 L of medium): 6 g Na_2HPO_4 (anhydrous), 3 g KH_2PO_4, 0.5 g NaCl. Add 1 L of deionized water, adjust pH to 7.4 with NaOH if required and autoclave at 120 °C for 20 min.
6. Prepare a 1 M stock solution of $MgSO_4$ and autoclave.
7. Prepare a stock solution of ampicillin at 100 mg/mL, filtrate on a 0.22 μm filter, aliquot per 1 mL and store at −20 °C.
8. Sterilized MEM vitamin cocktail 100× (Sigma).
9. Prepare a 1 M solution of isopropyl β-D-1-thiogalactopyranoside (IPTG), filtrate on a 0.22 μm filter, aliquot per 1 mL and store at −20 °C.

10. Reconstitution of the M9 medium for ^{15}N labeling (for 1 L of medium): 1 g $^{15}NH_4Cl$, 4 g glucose, 0.5 g ^{15}N-ISOGRO® (Sigma Isotec) are dissolved in 20 mL M9 medium and filtrate on a 0.22 μm filter. Add 1 mL of the stock solution of $MgSO_4$ 1 M (1 mM final concentration), 10 mL of a sterilized MEM vitamin cocktail 100× and 1 mL of the stock solution of ampicillin (100 mg/L final concentration).
11. Reconstitution of the M9 medium for $^{15}N/^{13}C$ labeling (for 1 L of medium): 1 g $^{15}NH_4Cl$, 2 g $^{13}C_6$-glucose, 0.5 g $^{15}N/^{13}C$-ISOGRO® (Sigma Isotec) are dissolved in 20 mL M9 medium and filtrate on a 0.22 μm filter. Add 1 mL of the stock solution of $MgSO_4$ 1 M (1 mM final concentration), 10 mL of a sterilized MEM vitamin cocktail 100× (Sigma) and 1 mL of the stock solution of ampicillin (100 mg/L final concentration).
12. Lysis-buffer: 20 mM Tris–HCl, 500 mM NaCl, 1 mM ethylenediaminetetraacetic acid (EDTA), 0.1% Triton X-100, 0.1 mM tris(2-carboxyethyl)phosphine (TCEP), protease inhibitor (cOmplete, Roche) (*see* **Note 3**).
13. Chitin resin (New England Biolabs).
14. Equilibration-buffer: 20 mM Tris–HCl, pH 8.0, 500 mM NaCl, 1 mM EDTA, 0.1% Triton X-100, 1 mM TCEP (*see* **Note 3**).
15. Wash buffer: 20 mM Tris–HCl, pH 8.0, 500 mM NaCl, 1 mM EDTA, 1 mM TCEP (*see* **Note 3**).

2.4 Expressed Protein Ligation (EPL)

1. Cleavage-buffer: 20 mM Tris–HCl, pH 8.0, 500 mM NaCl, 1 mM EDTA, 0.1 mM TCEP, 200 mM sodium 2-mercaptoethane sulfonate (MESNa) (*see* **Note 3**).
2. Spin filter (Amicon Ultra-4, Millipore) with a 30 kDa cutoff.
3. Streptavidin-agarose resin (Novagen).
4. Centrifuge column (Pierce centrifuge column, 5 mL).
5. Streptavidin binding-buffer: 20 mM Tris–HCl, pH 7.0, 150 mM NaCl, 1 mM DTT (*see* **Note 4**).

2.5 Sodium Dodecyl Sulfate Polyacrylamide-Gel Electrophoresis (SDS-PAGE)

1. Stacking gel-buffer: 0.5 M Tris–HCl, pH 6.8.
2. Resolving gel-buffer: 1.5 M Tris–HCl, pH 8.8.
3. Acrylamide–bisacrylamide (Rotiphorese, Roth), 30%, 37.5:1.
4. SDS: 10% w/v (SDS, ultra pure, Roth).
5. Ammonium persulfate (APS): 10% w/v.
6. *N,N,N,N'*-tetramethyl-ethylenediamine (TEMED) (Roth).
7. Precast gradient gels (4–20% Mini-Protein TGX precast gel, Bio-Rad).

8. SDS-PAGE running buffer: 1:10 dilution in water from a commercially available 10× stock (Rotiphorese 10× SDS-PAGE, Roth).
9. 1 L of Coomassie blue staining solution: Coomassie brilliant blue R250 (2.5 g), 100 mL acetic acid, 300 mL ethanol, 600 mL water.
10. SDS Loading buffer: 4× Laemmli sample buffer (900 μL), 100 μL β-mercaptoethanol (*see* **Note 5**).
11. 1 L of destaining solution: 500 mL methanol, 100 mL acetic acid, 400 mL water.
12. Mini-Protean Tetra System (Bio-Rad).

2.6 Western Blots

1. 0.2 μm Nitrocellulose membrane.
2. PBS: Phosphate Buffered-saline.
3. PBS-T: Phosphate Buffered-saline with 0.1 % Tween.
4. Tau 46 antibody (mouse, monoclonal IgG1, Santa Cruz Biotechnology).
5. Secondary antibody mouse horseradish peroxidase (HRP)-conjugate (Thermo Scientific).
6. Streptavidin peroxidase conjugate (Merck Millipore).
7. Blocking solution: Roti-block (Roth) working solution is prepared by mixing Roti-block with phosphate buffered saline-Tween (PBS-T) by 1:10 and used for blocking of nitrocellulose membranes.
8. WesternBright chemiluminescence solution (WesternBright ECL, advansta).
9. Fluorescence imager (Lumi-Imager F1, Boehringer Mannheim).

2.7 Mass Spectrometry of Tau Proteins

1. 0.1 % TFA in water for peptide dissolution.
2. Super-2,5-dihydroxybenzoic acid (Super-DHB) matrix: Super-DHB (Fluka) is dissolved in a mixture of acetonitrile and water containing 0.1 % TFA in a 30:70 ratio. The final concentration of Super-DHB is 50 mg/mL.

3 Methods

3.1 Synthesis of a Photolinker [15] (Scheme 1)

1. Pour *p*-toluenesulfonyl chloride (5.84 g, 30.56 mmol, 1.4 eq) and 1-(2-nitrophenyl)-1,2-ethanediol (4 g, 21.84 mmol, 1 eq) into a 250 mL round-bottom flask and then sparge the reaction vessel directly with N_2 for 5 min. Subsequently, add pyridine (100 mL) and stir the reaction under ice-bath cooling for 20 h. Allow the mixture to warm up to room tempera-

ture and then quench the reaction with 300 mL of water. Extract the desired crude product with Et_2O (4×100 mL). The combined organic phases are washed with 1 M $KHSO_4$ (3×100 mL), saturated aqueous $NaHCO_3$ (3×100 mL) and brine (3×100 mL). Dry the organic layer over $MgSO_4$, concentrate by rotary evaporation and then purify by SiO_2 flash column chromatography (eluted with EtOAc/Hex (1:1); R_f=0.5). The product derivative **1** is obtained as orange oil (5.50 g, 16.24 mmol, 72 %).

2. Place a solution of *p*-toluenesulfonyl derivative **1** (5.50 g, 16.24 mmol, 1 eq) into a 250 mL round-bottomed flask. After the flask is purged for 5 min with N_2, add dry pyridine (100 mL, 99.7 % dry). Then add TMSCl (7.66 g, 9 mL, 70 mmol, 4 eq) dropwise to the solution under Argon counterflow and leave the reaction stirred for 90 min at ambient temperature. Subsequently, quench the reaction with 280 mL water. Extract the crude product twice with Et_2O and wash the combined organic layers with 1 M $KHSO_4$ (3×100 mL), saturated $NaHCO_3$ (3×100 mL) and Brine (2×100 mL). Afterwards, dry the organic layer over $MgSO_4$ and concentrate in vacuo, which yields a dark yellow solid **2** (6.12 g, 14.95 mmol, 92 %).
3. Place sodium azide (6.11 g, 29.6 mmol, 2 eq) in a 250 mL round-bottom flask and purge the reaction vessel with N_2 for 5 min. Dissolve compound **2** (6.12 g, 14.95 mmol, 1 eq) in 100 mL DMF and add it to that flask while stirring. Heat the reaction to 80 °C and stir for 16 h. Quench the reaction with 200 mL of water and extract the crude product with Et_2O (4×100 mL). Wash the combined organic layers with Brine (2×100 mL), dry over $MgSO_4$ and concentrate in vacuo. Purify by SiO_2 flash column chromatography (product elutes with hexanes/EtOAc (4:1); R_f=0.3) and yields compound **3** as a yellow oil (2.04 g, 9.8 mmol, 66 %).
4. Place compound **3** (2.0 g, 9.8 mmol, 1 eq) in a 100 mL round-bottomed flask and dissolve it in DMF (50 mL). Hereafter, add *N*-hydroxysuccinimidyl carbonate (4.05 g, 15.72 mmol, 1.65 eq) to the solution, followed by NEt_3 (4 mL, 28.57 mmol, 3 eq). Let the reaction stir at room temperature for 16 h. Then, remove the solvent by evaporation in vacuo and purify the crude material by SiO_2 flash column chromatography (product elutes with hexanes/EtOAc 1:1; R_f=0.5) and yields the desired linker **4** as a colorless sticky solid (2.8 g, 8.04 mmol, 82 %).

3.2 Synthesis of the C-terminal Tau Peptide Tau[390–441] Bearing PCB on Lys438 [12]

1. The synthesis of the C-terminal tau peptide is carried out on H-Leu-HMPB NovaPEG (*see* **Note 6**) resin with a substitution grade of 0.49 mmol/g (*see* Scheme 3) (*see* **Note 7**). The scale of the reaction is 0.1 mmol (*see* **Note 8**). The entire synthesis is carried out on an Activo-P11 automated peptide syn-

thesizer (*see* **Note 9**). Orthogonally protected Fmoc-L-Lys(Alloc)-OH is used instead of otherwise standard Fmoc-L-Lys(Boc)-OH (Boc, *tert*-butyloxycarbonyl) in position 438 (*see* **Note 10**). Furthermore, Boc-Cys(Trt)-OH (Trt, trityl) is introduced as last amino acid in position 390 (*see* **Notes 11** and **12**). The other building blocks are standard Fmoc-protected amino acids and are summarized in Table 1. Automated coupling is achieved according to the settings on the Activo-P11 automated peptide synthesizer denoted in Table 2.

2. The resin-bound peptide is transferred from the Activo-P11 vessel to a 8 mL reactor (Extract Clean 8 mL reservoir, Grace) with a frit, washed with dichloromethane (CH_2Cl_2) (5 × 4 mL) and subsequently lyophilized for 2 h (*see* **Note 13**). Hereafter, Alloc-deprotection solution (4 mL) is poured on the resin. The reactor is closed with an adequate cap and sealed with several stripes of Parafilm (*see* **Note 14**). The resin is shook at 300 rpm for 10 min at room temperature (RT). Then, the liquid is filtrated out, the resin is washed 5 times with CH_2Cl_2 and the procedure is repeated (*see* **Note 12**). The resin is extensively washed with CH_2Cl_2 (5 × 4 mL) and DMF (5 × 4 mL).
3. Then, photolinker coupling solution is added to the resin. The coupling is allowed to proceed overnight (o.n.) during which the reactor is shaking at 300 rpm at RT (*see* **Note 12**). Then, the resin is thoroughly washed with DMF (3 × 4 mL), CH_2Cl_2 (3 × 4 mL) and NMP (3 × 4 mL).
4. In a next step, the azide reduction and biotin coupling solution is added to the resin and the reaction vessel cautiously sealed with Parafilm (*see* **Note 15**). The reactor is rocked at 300 rpm for 2 h at RT (*see* **Note 12**). The resin is washed with NMP (3 × 4 mL), DMF (3 × 4 mL) and CH_2Cl_2 (3 × 4 mL).
5. The peptide is cleaved from the resin by the addition of 4 mL peptide cleavage cocktail while rocking for 4 h at 300 rpm (*see* **Note 16**). The liquid phase containing the peptides is separated from the resin by filtration through a frit into the reactor and collected in a 50 mL canonical tube. The filtrated resin is quickly flushed with 2 mL peptide cleavage cocktail and the liquid is added to the same 50 mL canonical tube. Subsequently, the TFA cocktail containing the peptides is concentrated by strong nitrogen gas flow on its surface to a volume of about 200 μL (*see* **Note 17**).
6. Hereafter, 30 mL of ice-cold diethyl ether is poured into the canonical tube to precipitate the peptide, followed by quick vortexing and centrifugation at 4000 rpm for 5 min. The ether is decanted to another canonical tube and the pelleted peptide is allowed to dry on air for 10 min (*see* **Note 18**).

7. The peptide is resuspended in 2 mL acetonitrile (0.1 % TFA), intensively vortexed and ultrasonicated for a couple of minutes. Then, 3 mL of water (0.1 % TFA) are added to entirely dissolve the peptide (*see* **Note 19**).
8. The peptide is purified by preparative high performance liquid chromatography (HPLC) and obtained with a final yield of 3 % (18 mg, 3 μmol, *see* **Note 20**).
9. The purity and identity of the peptide is determined by analytical Ultra Performance Liquid Chromatography-Mass Spectrometry (UPLC-MS) analysis, but are not included in this description (*see* **Note 21**).
10. The synthesis of C-terminal tau[390–441] with phosphoserine in position Ser404 is described by Broncel *et al.* and may serve as a reference for the introduction of phosphate groups as relevant posttranslational modifications to this peptide (*see* **Note 22**) [11].

3.3 Protein Expression and Purification

We have previously reported the cloning of the tau gene into the pTXB1 plasmid (New England Biolabs), containing the Mxe GyrA intein-chitin binding domain (CBD) insert [11]. The successful expression of the EPL fragment tau[1–389] is demonstrated, which yields full-length human tau[1–441] upon ligation to the in Subheading 3.2 described peptide.

1. Take a 200 μL aliquot of T7 express cells and unthaw it on ice. Pipette 2 μL of plasmid to the unfrozen cells and let them incubate for 1 h on ice (*see* **Note 23**).
2. Hold the vial containing the cells for exactly 45 s into a 42 °C warm water bath. After that, put the cells back on ice for 30 additional minutes. Add 500 μL SOC medium to the vessel and incubate for 1 h at 37 °C at 180 rpm (*see* **Note 24**).
3. Centrifuge the mixture at 4000 ×*g* for 1 min, then discard 500 μL of the supernatant, resuspend the pellet in the remaining 200 μL and distribute them on LB agar plates containing ampicillin (100 μg/mL). Leave them o.n. at 37 °C in an incubator or for 2–3 days at ambient temperature outside to grow bacterial colonies.
4. Pre-warm 5 mL ampicillin containing LB medium (100 μg/mL) to 37 °C. Then, take an autoclaved inoculating loop, pick one colony from the bacterial plate and hold it into the LB medium (hereafter referred to as preculture). Incubate the inoculated medium at 37 °C and check occasionally the optical density (O_D).
5. When the O_D reaches a value of 0.5 (3 h in this case), pour the whole 5 mL culture into 500 mL ampicillin containing LB medium, pre-warmed to 37 °C. After 3 h the O_D had reached a value of 0.7, the temperature of the incubator is cooled down

to 20 °C and induction is initiated by the addition of IPTG (125 μL of a 1 M stock solution) to a final concentration of 0.25 mM. Continue the expression o.n. at 20 °C. For expression of ^{15}N or $^{15}N/^{13}C$ labeled proteins, 10 mL of a saturated preculture (typically incubated at 37 °C o.n.) is centrifuged for 10 min at 3000 × *g*. Then, discard cautiously the supernatant and inoculate 1 L of reconstituted M9 medium pre-warmed to 37 °C with the bacterial pellet. After 3–4 h of incubation at 37 °C, the O_D reached a value of 0.6–0.7, the temperature of the incubator is then cooled down to 20 °C and induction is initiated by the addition of IPTG to a final concentration of 0.25 mM. Then, the protein expression is performed at 20 °C overnight.

6. Pellet the cells by centrifugation at 6000 × *g* at 4 °C for 20 min. Then decant the supernatant and resuspend the pellet in 15–20 mL lysis-buffer per liter of culture.
7. Perform cell disruption by French press in three rounds and centrifuge at 25,000 × *g* for 30 min and 4 °C to pellet disrupted cell walls.
8. Prepare chitin beads to a final settled volume of 10 mL for the purification of proteins from a 1 L-culture (chitin resin, New England Biolabs) (*see* **Note 25**). Filtrate the liquid off and wash beads 10 times with 10 mL lysis-buffer (*see* **Note 26**).
9. Add the bacterial soluble extract and incubate slightly agitating on a rocker o.n. at 4 °C.
10. Remove the liquid by filtration and wash the beads extensively with equilibration-buffer (10 × 10 mL), followed by washing with wash buffer (10 × 10 mL)

3.4 Expressed Protein Ligation (EPL)

The amount of peptide that is used for the EPL reaction relies on very rough approximations and is related to the amount of protein that is expected to be immobilized on chitin beads. We expect around 0.3–0.5 mg tau[1–389]-intein/CBD fusion protein immobilized on 1 mL of the chitin resin and reason from our experiences that a tenfold excess of peptide is suited for our goal to obtain a high conversion to the ligation product.

1. Quickly flush 3 times the column that contains the chitin resin with 3 mL cleavage buffer (*see* **Note 27**).
2. Weigh 3 mg (472 nmol, approx. 10 eq) of peptide into a 1.5 mL microcentrifuge tube (*see* **Note 28**) and dissolve it in 1 mL cleavage buffer. Pipette the resulting mixture onto the chitin resin. Rinse the 1.5 mL microcentrifuge tube twice with 1 mL cleavage buffer and add them to the column with the chitin beads.
3. Additionally seal the column from top and bottom with Parafilm and wrap the column in aluminum foil (*see* **Note 29**).

Place the column on a rocker and let it gently rock for 1 day at RT. Check the efficacy of the EPL by SDS-PAGE, while noting that tau runs retarded on SDS-PAGE gels (*see* **Note 30**).

4. Elute the liquid phase containing the cleaved proteins including the ligation product into a 15 mL Falcon tube. Wash the beads twice with 1 mL cleavage buffer to entirely elute proteins from the column and combine the liquid phases in the Falcon tube.
5. Load the solution to a spin filter with a 30 kDa cutoff to remove unreacted peptide from the reaction mixture. Centrifuge at 4000 rpm for about 15 min until the solution is concentrated to a volume of about 150 μL. Subsequently add 4 mL of streptavidin binding-buffer on top of the membrane and repeat the centrifugation step (*see* **Note 31**).
6. Carefully pipette the residual 200 μL to a 1.5 mL microcentrifuge tube and wash the membrane with streptavidin binding buffer. Then add those portions to the 1.5 mL microcentrifuge tube (*see* **Note 32**). The final volume of buffer containing the protein of interest is now 1.5 mL.
7. Transfer 1 mL of streptavidin-agarose beads suspension to a centrifuge column (Pierce centrifuge column, 5 mL), which corresponds to 500 μL settled beads (*see* **Note 33**). Wash the beads 3 times with 2 mL streptavidin binding buffer to remove residual bead storage buffer containing sodium azide.

3.5 Purification of the EPL Product

The purification is carried out at normal daylight in the laboratory, since we previously did not observe any premature cleavage in such conditions. Prior to biotin purification save an aliquot for subsequent SDS-PAGE analysis.

1. Apply the sample described in Subheading 3.4, **step 6** onto streptavidin beads and incubate for 2 h gently rocking at RT (*see* **Note 34**). Then, the liquid is eluted from the column and kept for further analysis.
2. Wash the beads extensively with streptavidin binding-buffer (20 × 3 mL). Analyze the wash fractions by Ultraviolet (UV)-absorption (NanoDrop, ND-1000 spectrophotometer, Thermo Scientific). No residual unbound protein should be detected in the last wash fractions.
3. Add now 1 mL of streptavidin binding-buffer to the beads and a small stirring bar. Place the column in 20 cm distance to a UV-light source. Stir the emulsion so that the beads are homogeneously distributed in the solution and irradiate the sample for 10 min at 297 nm (*see* **Note 35**). Check the efficacy of the photolytic cleavage by UV-absorption. Repeat this cleavage step until no more protein is cleaved from the beads (*see* **Note 36**).

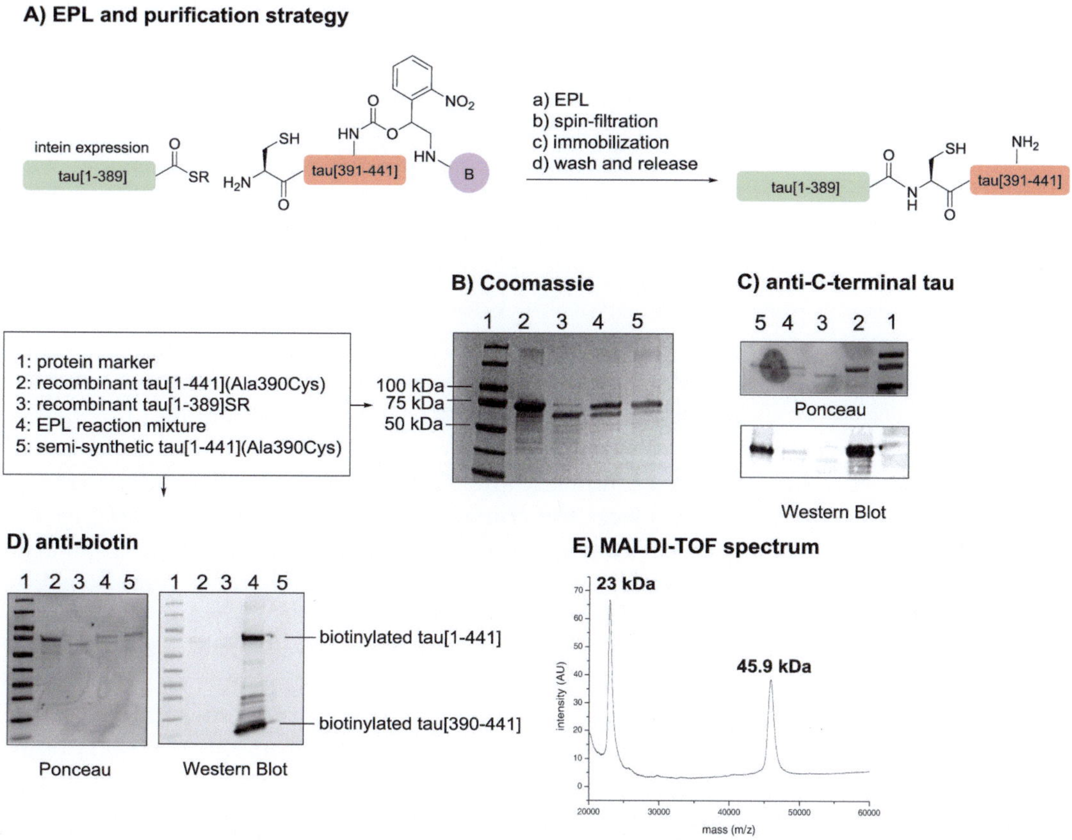

Fig. 2 (**a**) Tag-free purification of semi-synthetic tau: (a) EPL; (b) spin-filtration (30 kDa cut-off); (c) capture on streptavidin coated agarose beads; (d) wash and UV light irradiation (297 nm); (**b**) Coomassie Brilliant Blue stained SDS-PAGE gel; (**c**) anti-C-terminal tau western blot; (**d**) anti-biotin western blot; (**e**) MALDI-TOF spectrum of purified semi-synthetic tau[1–441](Ala390Cys). The calculated mass values for tau[1–441](Ala390Cys) are $[M+H]^+ = 45.9$ kDa and $[M+2H]^{2+} = 22.8$ kDa

4. Combine the elution fractions and store them for further use.

3.6 Sodium Dodecyl Sulfate Polyacrylamide Gel Electrophoresis (12%)

Comparing the sample before and after purification by SDS-PAGE helps to determine the efficacy of the isolation of pure tau protein (*see* Fig. 2).

1. To prepare two self-casted gels mix 4.1 mL H_2O, 4 mL acrylamide–bisacrylamide, 1.7 mL resolving gel-buffer, 100 μL SDS, 100 μL APS, and 10 μL TEMED in a 50 mL conical flask. Vortex quickly and cast the mixture into a 8.3 cm × 8.2 cm × 0.75 mm gel cassette (Bio-Rad). Add a layer of isopropanol on top and wait for 20 min until the polymerization is finished.
2. Then mix 3 mL H_2O, 650 μL acrylamide–bisacrylamide, 1.25 mL stacking gel-buffer, 50 μL SDS, 50 μL APS and 10 μL TEMED in a canonical flask, vortex quickly, remove the residual isopropanol from the casting apparatus and cast the mix-

ture on top. Immediately place a 10-well gel comb without generating air bubbles.

3. Place 15 μL of the solution containing the crude EPL mixture and a 15 μL aliquot of the purified EPL product in two separate 1.5 mL microcentrifuge tubes, mix each sample with 7.5 μL 4× loading buffer and then heat at 95 °C for 5 min. Then load 20 μL of each sample to a well. Also place 5 μL of protein standard to another well of the gel (*see* **Note 37**).
4. Run the gel at 250 V for 35 min.
5. After the electrophoresis open the gel with a spatula carefully and rinse the gel carefully with water to a microwave compatible container. Discard the water and add enough Coomassie staining-solution to ensure a liquid level in your container of about 1 cm. Close the container and place it in a microwave. Quickly heat the container until the liquid starts boiling and place the container on a rocker. After 5 min, pour the Coomassie blue staining-solution back to a container and wash the gel by pouring sink water on top. Discard the water and add destaining-solution again up to a 1 cm level. Heat again until the solution boils in the microwave and subsequently place the gel in the container in destaining solution on a rocker. Replace the destaining-solution after 10 min and from there on every hour.

3.7 Anti-tau Western Blot

The anti-tau western blot assures the identity of the protein (*see* Fig. 2).

1. Separate proteins by SDS-PAGE as described in Subheading 3.6 using a precast gradient gel (4–20% Mini-Protein TGX precast gel, Bio-Rad) (*see* **Note 38**).
2. Wet-blot the proteins to a nitrocellulose membrane (250 mA, 1 h).
3. Block the membrane o.n. at 4 °C with blocking-solution. Wash 5 × 10 min with PBS-T solution.
4. Incubate the blot for 1 h with antibody Tau 46 (1:200) in PBS-T at RT. Wash 5 × 10 min with PBS-T.
5. Incubate the blot with a secondary antibody mouse HRP-conjugate (1:2000) in PBS-T for 1 h at RT. Wash 3× with PBS-T and 3× with PBS for 10 min.
6. Pour 2–3 mL WesternBright chemiluminescence solution or any other chemiluminescence agent on top of the blot, wait for 2 min and view outcome on a fluorescence imager (*see* **Note 39**).

3.8 Anti-biotin Western Blot

After the purification the protein should not be detected by an anti-biotin western blot (*see* Fig. 2).

1. Separate proteins by SDS-PAGE as described in Subheading 3.6 using a precast gradient gel (4–20% Mini-Protein TGX precast gel).

2. Wet-blot the proteins to a nitrocellulose membrane (250 mA, 1 h).
3. Block the membrane o.n. at 4 °C with blocking-solution. Wash 5 × 10 min with PBS-T solution.
4. Incubate the blot for 1 h with streptavidin peroxidase conjugate (1:2000) at rt. Wash 3 × 10 min with PBS-T and then 3 × 10 min with PBS.
5. Rinse the blot with 2–3 mL WesternBright chemiluminescence solution or any other chemiluminescence agent, wait for 2 min and view outcome on a fluorescence imager (*see* **Note 39**) (Lumi-Imager F1, Boehringer Mannheim).

3.9 Matrix-Assisted Laser Desorption/Ionization (MALDI) Spectrometry

1. Pipette 1 μL of solution containing the purified peptide with 1 μL H_2O (0.1% TFA) and 1 μL of Super-DHB-matrix to a 0.5 mL reaction tube.
2. Mix them by pipetting up and down and scratching at the vessel walls with your pipette. Subsequently, spot 1 μL on a MALDI target and let the droplet dry on air.
3. Analyze the probe with a MALDI device (*see* **Note 40**).

4 Notes

1. Biotin is poorly soluble in DMF. NMP proved to work best, but required intensive vortexing and ultrasonification for approx. 5–10 min.
2. DTT and thioanisol help to avoid oxidation of methionine residues during acidic cleavage. Extreme caution, gloves, laboratory glasses and working under a fume hood are mandatory when working with highly corrosive TFA.
3. Prepare buffer without TCEP, autoclave at 120 °C for 20 min and store buffer at 4 °C. On demand add TCEP to the desired amount of buffer stock, set the pH, filtrate and use up. Be aware that the addition of TCEP strongly acidifies the solution.
4. Prepare buffer without DTT, autoclave at 120 °C for 20 min and store buffer at 4 °C. On demand add DTT to the desired amount of buffer stock, set the pH, filtrate and use up.
5. Probes are mixed with loading buffer in the probe–buffer ratio of 3:1.
6. The first coupling of an amino acid to a solid support is often hampered due to inaccessibility to functional groups of the resin. The use of preloaded resins is thus valuable in terms of time expenditure and overall synthesis efficacy.
7. The loading of the resin should be chosen wisely. Long peptide chains like our 52 amino acid long tau peptide become sterically

demanding and thus highly loaded resins (>1 mmol/g) are not favored.

8. If the peptide sequence is very long one often encounters the presence of many deletion sequences, which leads in the end to a lower yield. If further use of the target peptides for EPL reactions is planned one should stick to the 0.1 mmol scale, even if 0.05 mmol is the standard in many laboratories.
9. Neither automated reactor heating, nor UV-monitoring is applied during the synthesis.
10. Fmoc-L-Lys(Alloc)-OH is incorporated at the position where the photolinker is introduced.
11. By using a Boc-protected amino acid in the last position, possible side-reactions during later steps, in particular during the Staudinger reduction of the azide to an amine, can be encountered. Moreover, the last Fmoc-deprotection is circumvented, since the Boc-group is cleaved anyways during later TFA cleavage.
12. A test-cleavage is performed to confirm the successful synthesis up to this point. For this purpose, a spatula tip of resin is exposed to peptide cleavage cocktail (300 mL) and rocked at 300 rpm for 1 h at RT in a 1.5 mL reaction vessel equipped with a frit. The cleavage cocktail containing the peptides is then filtrated into a 1.5 mL microcentrifuge tube, before the liquid phase is removed by constant nitrogen flow on its surface (fume hood). When almost all of the TFA is evaporated, 1 mL of ice-cold diethyl ether is added to the vessel to precipitate the peptide and remove excess reagents cleavage products. Directly after the addition of the ether, the vessel is quickly vortexed and placed in a centrifuge and centrifuged at 14.000 rpm for 1 min afterwards the ether is decanted and the pellet left at air to dry for 5 min. Then, the pellet is dissolved in 30 μL acetonitrile, quickly vortexed and ultrasonificated and then 150 μL of water are added. The resulting mixture can be analyzed by analytical UPLC-MS.
13. The reactor vessel is opened at the bottom and the top, then a piece of paper is placed on top of it and fixed with a rubber band to avoid that the resin leaves the reactor vessel after the pressure drops and CH_2Cl_2 traces are evaporated immediately.
14. CO_2-gas is released during the Alloc-deprotection and causes overpressure in a sealed reactor. Caution is required when opening the reaction vessel. Instead of shaking the closed reactor vessel, one could also bubble nitrogen or argon through the opened solution with a needle. When doing so, one has to keep an eye on the liquid level, since CH_2Cl_2 is quite volatile and evaporates easily when left standing on air.

15. N_2-gas is released during this reaction. Sealing of the cap with Parafilm and careful opening post reduction is required.
16. Seal the reactor with Parafilm carefully open the reactor vessel after cleavage since the evaporation of residual CH_2Cl_2 can cause overpressure. If the resin is dried by lyophilization prior to cleavage, no extra cautions need to be considered.
17. A 50 mL Falcon tube is taken due to the presence of a larger surface that helps to quickly evaporate the TFA by nitrogen gas flow. Usually this takes then for 6 mL approx. 10 min.
18. The discarded ether phase is stored for 1–2 h at −20 °C to check for further precipitation of peptide.
19. If some undissolved particles are still present one can either filter the mixture through an adequate filter or quickly centrifuge the container to pellet residual particles.
20. The solvents used for HPLC are H_2O (A, 0.1 % TFA) and acetonitrile (B, 0.1 % TFA). The method is built up as follows: 0–5 min (B = 10 %); 5–70 min (10–70 % B). HPLC purification is carried on a Gilson PLC 2020 Personal Purification System using a Nucleodur VP 250/32 C18HTec 5 μm column from Macherey-Nagel at a flow rate of 32 mL/min equipped with a 5 mL injection loop.
21. UPLC-UV traces are collected using a Waters H-class instrument, equipped with a Quaternary Solvent Manager, a Waters autosampler, a Waters TUV detector connected to a 3100 mass detector using an Acquity UPLC-BEH C18 1.7 μM 2.1 × 50 mm RP column with a flow rate of 0.6 mL/min. The following method is applied: (A: 0.1 % TFA in H_2O, B: 0.1 % TFA in acetonitrile) 5 % B 0–5 min, 5–95 % 5–15 min, 95 % B 15–17 min. The UPLC-UV chromatogram is recorded at 220 nm.
22. Phosphoserine is introduced to C-terminal tau[390–441] peptide by double coupling with 5 eq Fmoc-Ser(PO(OBzl)OH)-OH, *N*,*N*,*N*′,*N*′-tetramethyl-O-(1H-benzotriazol-1-yl) uronium hexafluorophosphate, O-(benzotriazol-1-yl)-*N*,*N*,*N*′,*N*′-tetramethyluronium hexafluorophosphate (HBTU) and DIPEA for 3 h each coupling.
23. Aside from T7 express cells, expression worked also for ER2566 cells.
24. Pre-warm the SOC medium to 37 °C prior to addition to the cells.
25. Check the overall protein concentration of the cell lysate by for example UV absorption (NanoDrop, ND-1000 spectrophotometer, Thermo Scientific). Run a SDS-PAGE gel with your cell lysate and estimate the amount of target protein that you roughly expect from your expression. By this you can approximate how much chitin beads are required.

26. The supplier states that at least 2 mg of maltose-binding protein is immobilized on beads per mL of settled resin.
27. This distributes the thiol MESNa throughout the beads, but does not lead to cleavage of the protein from the beads due to very short exposure time.
28. Consider that lyophilized peptides are present as TFA salts after their purification by HPLC when TFA containing solvents are used. Add the molar mass of CF_3COOH (113.99) as many times as you have positively charged amino acids to the molar mass of your peptide to calculate the correct amount of employed peptide.
29. It is better to protect the reaction mixture from light, but we found the photolabile group to be quite robust to light exposure, which enabled us to handle all active steps at daylight or room light.
30. Take a 1 mL pipette and place a tip on it. Cut the very end of the tip, which facilitates the uptake of particles, take 10 μL out and place them in an extra tube. Mix this emulsion with 10 μL SDS loading buffer (4×). Heat tube at 95 °C for 5 min, centrifuge for 1 min at 14,000 rpm and load the supernatant to your gel. Avoid placing chitin bead particles into the gel pocket. Run a 12% gel and for 40 min at 250 V. Proceed as described in Subheading 3.6 and check for bands corresponding to the tau substrate (approx. 41 kDa, runs at approx. 60 kDa on gels) and the ligation product (approx. 46 kDa, runs at around 68 kDa).
31. The membrane of this spin-filter consists of low-binding degenerated cellulose. Prior to their use load 4 mL of water to the spin filter and let it run for 10 at 4000 rpm. Afterward remove the water from the collection tube and from the filter by pipetting. This procedure helps to remove azide that may stick to the membrane from some storage buffer when they are sold.
32. Carefully pipette up and down several times when additional portions of 200 μL buffer are added to the spin filter to wash out all the remaining proteins.
33. Biotin binds to streptavidin very quickly due to extremely high binding affinity ($K_d \sim 10^{-14}$ mol/L). This enables in principle a flow through binding to the beads. However, we observed slightly better binding when we incubated the samples for 1–2 h.
34. 1 mL settled streptavidin-agarose beads bind >85 nmol free biotin in solution according to the supplier. This corresponds to more than approx. 4 mg of protein in our case.
35. As light source, a LOT Hg(Xe) arc lamp (LOT-QuantumDesign GmbH) is used.
36. In our case, we did not observed protein cleavage from the streptavidin beads after the third irradiation for 10 min.

37. If available, also place a sample of the expressed full-length target protein (tau[1–441]) and of the expressed protein fragment (tau[1–389]) in the gel as references.
38. Separation on gradient gels proceeds quicker and thus, 20 min at 250 V is sufficient.
39. The blots can also be dried on air, stored and used for visualization at any given later time point.
40. Super-DHB matrix creates spots with a thicker corona at the exterior of the spot. Laser shots on this corona delivered the best spectra. MALDI spectra are recorded on a AB SCIEX 5800 tandem time-of-flight (TOF/TOF) system (Applied Biosystems) with nanoLC (Dionex) and robotic system for spotting (Probot, Dionex) and 4700 Proteomics Analyzer (Applied Biosystems).

References

1. Goedert M (1993) Tau-protein and the neurofibrillary pathology of Alzheimers-disease. Trends Neurosci 16:460–465
2. Buee L, Bussiere T, Buee-Scherrer V, Delacourte A, Hof PR (2000) Tau protein isoforms, phosphorylation and role in neurodegenerative disorders. Brain Res Rev 33:95–130
3. Trojanowski JQ, Lee VMY (1995) Phosphorylation of paired helical filament tau in Alzheimer's disease neurofibrillary lesions: focusing on phosphatases. FASEB J 9:1570–1576
4. Patrick GN, Zukerberg L, Nikolic M, de la Monte S, Dikkes P, Tsai LH (1999) Conversion of p35 to p25 deregulates Cdk5 activity and promotes neurodegeneration. Nature 402: 615–622
5. Lindwall G, Cole RD (1984) Phosphorylation affects the ability of tau-protein to promote microtubule assembly. J Biol Chem 259: 5301–5305
6. Alonso AD, Zaidi T, Grundkeiqbal I, Iqbal K (1994) Role of abnormally phosphorylated tan in the breakdown of microtubules in Alzheimer-disease. Proc Natl Acad Sci U S A 91:5562–5566
7. Abraha A, Ghoshal N, Gamblin TC, Cryns V, Berry RW, Kuret J, Binder LI (2000) C-terminal inhibition of tau assembly in vitro and in Alzheimer's disease. J Cell Sci 113: 3737–3745
8. Dawson PE, Muir TW, Clark-Lewis I, Kent SB (1994) Synthesis of proteins by native chemical ligation. Science 266:776–779
9. Muir TW, Sondhi D, Cole PA (1998) Expressed protein ligation: a general method for protein engineering. Proc Natl Acad Sci U S A 95:6705–6710
10. Hackenberger CP, Schwarzer D (2008) Chemoselective ligation and modification strategies for peptides and proteins. Angew Chem Int Ed Engl 47:10030–10074
11. Broncel M, Krause E, Schwarzer D, Hackenberger CP (2012) The Alzheimer's disease related tau protein as a new target for chemical protein engineering. Chem Eur J 18:2488–2492
12. Reimann O, Smet-Nocca C, Hackenberger CPR (2015) Traceless purification and desulfurization of tau protein ligation products. Angew Chem Int Ed 54:306–310
13. Reimann O, Glanz M, Hackenberger CP (2015) Native chemical ligation between asparagine and valine: application and limitations for the synthesis of tri-phosphorylated C-terminal tau. Bioorg Med Chem 23: 2890–2894
14. Schwagerus S, Reimann O, Despres C, Smet-Nocca C, Hackenberger CP (2016) Semisynthesis of a tag-free O-GlcNAcylated tau protein by sequential hemoselective ligation. J Pept Sci 22:327–333
15. Bindman N, Merkx R, Koehler R, Herrman N, van der Donk WA (2010) Photochemical cleavage of leader peptides. Chem Commun (Camb) 46:8935–8937

Chapter 13

Production of *O*-GlcNAc Modified Recombinant Tau in *E. coli* and Detection of Ser400 *O*-GlcNAc Tau In Vivo

Scott A. Yuzwa and David J. Vocadlo

Abstract

Assembly of the microtubule-associated protein tau (tau) into paired helical filaments that ultimately give rise to neurofibrillary tangles (NFTs) makes up one half of the two hallmark pathologies of Alzheimer's disease (AD). Tau has been shown to be modified with *O*-linked N-acetylglucosamine residues (*O*-GlcNAc), which is the modification of serine and threonine residues of nucleocytoplasmic proteins with *N*-acetyl-D-glucosamine (GlcNAc) moieties. Increasing *O*-GlcNAc in mouse models of tauopathy has been shown to hinder the progression of symptoms in these mice and impair the aggregation of tau into NFTs. In order to study how *O*-GlcNAc on tau may contribute to the protective effects observed in tauopathy mouse models, it is beneficial to study *O*-GlcNAc modified tau in vitro. Here we describe a method for producing, purifying and enriching recombinant tau that is *O*-GlcNAc modified. These methods have enabled the identification of *O*-GlcNAc modification sites on tau including Ser400. We further describe the detection of Ser400 *O*-GlcNAc on tau from brain lysates.

Key words *O*-GlcNAc, Microtubule-associated protein tau, Tau, Purification, Ser400 *O*-GlcNAc tau

1 Introduction

The aggregation of the microtubule-associated protein tau (tau) into paired-helical filaments (PHFs) that give rise to the larger neurofibrillary tangles is the defining characteristic of a group of disorders referred to as the tauopathies. The most common type of tauopathy is sporadic Alzheimer's disease (AD) where the aggregation of tau into NFTs is one of the two characteristic pathologies of AD. Understanding how tau transitions from a highly soluble and intrinsically unstructured protein into highly ordered paired helical filaments is critical in order to enable the development of disease modifying therapeutic strategies for AD and the associated tauopathies and has been the focus of much attention.

Caroline Smet-Nocca (ed.), *Tau Protein: Methods and Protocols*, Methods in Molecular Biology, vol. 1523,
DOI 10.1007/978-1-4939-6598-4_13, © Springer Science+Business Media New York 2017

The normal cellular function of tau within neurons is to bind to and stabilize microtubules [1]. Work carried out a number of years ago using tau antibodies observed tau immunoreactivity against paired helical filaments (PHFs) [2, 3]. Since this discovery that tau is the major component of PHFs from the AD brain, intensive research has been directed toward characterizing the molecular pathways that cause tau to transition from performing a role in catalyzing tubulin polymerization to aggregating into PHFs. One clear theme that has emerged through the work of a number of investigators is that post-translational modifications of tau play profound roles in this transition of tau into a central pathological driver of AD. The first post-translational modification identified on tau was the post-translational phosphorylation of serine and threonine residues. Tau from the healthy human brain has been shown to contain ~1.9 mol of phosphorylation per mole of tau protein whereas soluble tau from the AD brain contains ~2.6 mol of phosphorylation per mole of tau and tau from AD PHFs contains 6–8 mol of phosphorylation per mole of tau protein, indicating that tau is hyperphosphorylated in the AD brain [4]. The role of tau hyperphoshorylation in the AD brain remains somewhat controversial with some evidence suggesting that hyperphosphorylation drives tau aggregation [5] while other evidence suggests that hyperphosphorylation may occur subsequent to aggregation [6]. On balance, however, the general view is that tau phosphorylation drives its gain of toxicity.

A number of other post-translational modifications have been identified on tau including: nitration of tyrosine [7], ubiquitination [8, 9], acetylation of lysine residues [10–12] as well as *O*-GlcNAc [13–15]. The *O*-GlcNAc modification of nucleocytoplasmic proteins occurs via the enzyme catalyzed transfer of *N*-Acetyl-D-glucosamine (GlcNAc) residues from the donor sugar uridine diphosphate-*N*-acetyl-D-glucosamine (UDP-GlcNAc) to the hydroxyl groups of targeted serine and threonine residues [16]. The only enzyme catalyzing installation of this β-*O*-linked GlcNAc residue is the glycosyltransferase referred to as *O*-GlcNAc transferase (OGT) [17, 18]. The removal of *O*-GlcNAc from modified proteins is catalyzed by a single glycoside hydrolase referred to as *O*-GlcNAcase (OGA) [19, 20]. The action of these two enzymes allows reversible attachment of *O*-GlcNAc to modified proteins that can occur multiple times during the lifespan of a particular protein, making this modification conceptually similar to phosphorylation with respect to its dynamic nature and presence on serine/threonine [21]. A series of studies by different groups using various approaches have suggested that *O*-GlcNAc modification on tau can antagonize tau phosphorylation both in tissues as well as in vitro during acute perturbations in *O*-GlcNAc [14, 15, 22–24]. In these studies it was found that increasing tau *O*-GlcNAc leads to decreased tau phosphorylation while the converse is also true. Studies have mapped *O*-GlcNAc to several different sites on tau [23, 25], though tau samples from mammalian brain tissues having normal *O*-GlcNAc levels have only had *O*-GlcNAc mapped to Ser400 [25–27]. We, and others, have shown that Thiamet-G, a selective brain penetrant inhibitor of OGA, can increase global *O*-GlcNAc levels within brain and lead to a delay in the progression of motor impairments and tau aggregation in mouse models of tauopathy [28–30]. Notably, these studies show that long term inhibition of OGA over a period of weeks to months blocks the accumulation of the 64 kDa hyperphosphorylated tau species but does not affect soluble tau phosphorylation as observed when *O*-GlcNAc levels are increased acutely. The molecular mechanisms by which *O*-GlcNAc protects against tau toxicity are a topic of current interest. While the level of tau *O*-GlcNAcylation within brain is unknown, it has been suggested by some to be quite high [13] whereas others have suggested low modification levels only detectable at Ser400 [27]. These differences may arise from variation in the methods used to detect this modification. OGA inhibition has, in any event, been found to increase tau *O*-GlcNAc levels [29, 30], which has stimulated interest in the effects of tau *O*-GlcNAcylation on its aggregation.

In order to understand how increased *O*-GlcNAc might contribute to the beneficial effects in mouse models of tauopathy, in vitro models of tau aggregation and tau function have been explored. A key step, however, needed to study the aggregation and function of tau in vitro and the potential impacts of *O*-GlcNAc on these processes, is to be able to prepare recombinant *O*-GlcNAc modified tau. One way that *O*-GlcNAc modified tau can be prepared is by the co-transformation of expression plasmids for tau and OGT into *E. coli* so that OGT can catalyze the addition of *O*-GlcNAc onto tau using UDP-GlcNAc produced by the *E. coli*. This method, which we describe here, allows for the production of milligram quantities of *O*-GlcNAc modified recombinant tau. One caveat to this method, however, is that not all of the tau produced in co-transformed *E. coli* cells is *O*-GlcNAc modified with typically ~60% being *O*-GlcNAc modified. To overcome this limitation, we also describe a high-performance liquid chromatography (HPLC) method to enrich the sample in *O*-GlcNAc modified tau to obtain tau samples with near 100% *O*-GlcNAc modification. Using these methods we have shown that the *O*-GlcNAc modification of tau impairs its aggregation in vitro and has allowed us to implicate a particular serine residue on tau, Ser400, on which *O*-GlcNAc is particularly important [30, 31]. Finally, we describe briefly how *O*-GlcNAc at Ser400 on tau can be identified in brain lysates [25, 30].

While we have previously used these methods, described in detail below, to map sites of tau *O*-GlcNAcylation and to study the role of *O*-GlcNAc on tau aggregation and function in vitro [25, 30, 31], we anticipate that these methods will be of wide general use to study a number of interesting and outstanding questions pertaining to the role of *O*-GlcNAc on tau biology as well as other proteins where *O*-GlcNAc may have an influence on their propensity to aggregate.

2 Materials

2.1 O-GlcNAc Modified Recombinant Human Tau Expression, Purification and Enrichment

1. Expression plasmids: pMal-c2X vectors encoding wild-type human OGT (wtOGT-pMal-c2x) and H558A human OGT (mutOGT-pMal-c2x) as well as pET28a vector encoding wild-type human Tau441 (wtTau441-pET28a).
2. BL21(DE3) competent *E. coli* cells (Agilent).
3. Agar plates.
4. Luria broth (LB) growth media.
5. Expression flasks.
6. Shaker incubator.
7. High-speed centrifuge.
8. 1 M isopropyl β-D-1-thiogalactopyranoside (IPTG): dissolve IPTG at 1 M in water and store aliquots at −20 °C.
9. Probe sonicator (Fisher Scientific sonic dismembrator, model 500).
10. Peristaltic pump or FPLC apparatus.
11. HisTrap FF (5 mL, GE Healthcare) Ni-NTA columns.
12. Ni-NTA column binding buffer: 20 mM sodium phosphate, 500 mM NaCl, 5 mM imidazole, pH 7.4.
13. Ni-NTA column wash buffer: 20 mM sodium phosphate, 500 mM NaCl, 60 mM imidazole, pH 7.4.
14. Ni-NTA column elution buffer: 20 mM sodium phosphate, 500 mM NaCl, 250 mM imidazole, pH 7.4.
15. Dialysis tubing 3 kDa MWCO (Spectrum Laboratories).
16. Dialysis buffer: 20 mM sodium phosphate buffer, pH 6.7.
17. Ultra-15 centrifugal filter units (Amicon).
18. HPLC equipment.
19. HPLC grade water and acetonitrile, trifluoroacetic acid (TFA).
20. Semi-preparative HPLC column: Zorbax 300SB-C8 9.4 mm × 250 mm (Agilent) or similar.
21. HPLC Buffer A: Water, 0.1 % TFA.
22. HPLC Buffer B: Acetonitrile, 0.1 % TFA.
23. Lyophilization equipment.

24. Sodium dodecyl sulfate polyacrylamide gel electrophoresis (SDS-PAGE) equipment.
25. Thrombin (from human plasma, Roche).

2.2 Detection of O-GlcNAc on Tau in Brain Lysates

1. Mouse or rat brain tissue: whole brain from 6 to 8 week C57 mice/Sprague-Dawley rats.
2. Lysis buffer: phosphate-buffered saline (PBS), pH 7.4, 500 mM NaCl, 1 mM Thiamet-G, and complete-mini protease inhibitor tablet (Roche).
3. Tissue homogenizer (IKA).
4. 70 % perchloric acid (Fisher Chemical).
5. Calf intestinal alkaline phosphatase (Roche).
6. 3925 anti-Ser400 *O*-GlcNAc tau antibody (produced in rabbit).
7. PBS-T: PBS, pH 7.4, 0.1 % Tween 20.

3 Methods

3.1 O-GlcNAc Modified Recombinant Human Tau Expression and Purification

1. Prepare LB-agar plates with 16.6 μg/mL kanamycin and 33.3 μg/mL ampicillin (*see* **Note 1**).
2. Transform competent BL21(DE3) *E. coli* cells with wild-type OGT (wtOGT-pMal-c2x, ampicillin resistance) and wild-type Tau441 (wtTau441-pET28a, kanamycin resistance) or mutant OGT (mutOGT-pMal-c2x, ampicillin resistance) and wild-type Tau441(wtTau441-pET28a, kanamycin resistance). Here, mutant OGT co-transformants serve as a negative control because OGT is catalytically inactive and thus cannot *O*-GlcNAc modify Tau441 [32].
3. Pick a wtOGT/Tau441 colony and a mutOGT/Tau441 colony and inoculate overnight starter cultures using LB containing 50 μg/mL kanamycin and 100 μg/mL ampicillin.
4. Dilute the starter cultures 1:100 into LB growth media containing 50 μg/mL kanamycin and 100 μg/mL in 1 L shaker flasks (*see* **Note 2**) and grow at 37 °C until the optical density reaches 0.6–0.8.
5. Reduce the temperature of the shaker incubator to 22 °C and add isopropyl β-D-1-thiogalactopyranoside (IPTG) to a final concentration of 0.5 mM. Continue shaking for ~20 h (overnight) at 22 °C.
6. Collect *E. coli* cells by centrifugation.
7. Resuspend the bacterial pellets in Ni-NTA column binding buffer and add 2 mg/mL lysozyme and one Complete-mini

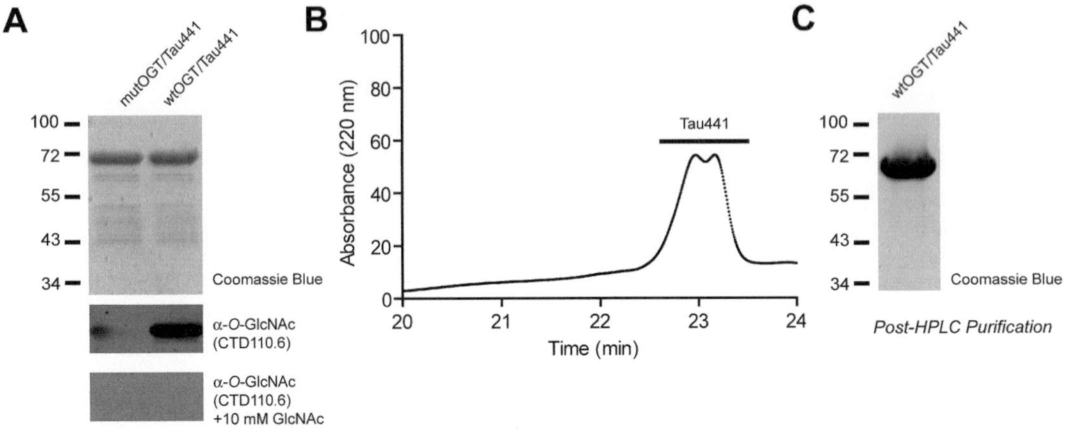

Fig. 1 Production and purification of recombinant *O*-GlcNAc modified human Tau441 in *E. coli*. (**a**) Only Tau441 co-expressed with wtOGT is *O*-GlcNAc modified as detected by the pan-*O*-GlcNAc antibody (CTD110.6). The *O*-GlcNAc immunoreactivity can be abolished by 10 mM GlcNAc indicating the specificity of the antibody biding to *O*-GlcNAc on Tau441. (**b**) HPLC trace of Tau441 indicates Tau441 elutes between 22 and 24 min. (**c**) *O*-GlcNAc modified Tau441 following HPLC purification. Note the bands below the Tau441 band in (**b**) (incompletely translated Tau441 and degradation products) have been removed by HPLC purification

protease inhibitor tablet (Roche) per 30 mL of resuspended bacterial pellet. Typically, we add 30 mL of Ni-NTA binding buffer to the combined pellet from 1 L of culture of wtOGT/Tau441 or mutOGT/Tau441. Incubate on ice for 30 min.

8. Sonicate the solution containing the resuspended bacterial pellet (on ice) using six cycles of 20 s of sonication at 30% power alternating with a 40 s cooling period. Pellet cellular debris by centrifugation.
9. Load the clarified supernatant onto a HisTrap FF Ni-NTA column which has been pre-equilibrated with Ni-NTA binding buffer using a peristaltic pump or FPLC apparatus.
10. Wash the HisTrap FF column with 80–100 mL of Ni-NTA column wash buffer and elute the column with 25 mL of Ni-NTA column elution buffer and confirm the elution of tau protein using a protein detection assay.
11. Dialyze the collected fractions containing Tau441 protein eluates against at least three changes of dialysis buffer (*see* **Note 3**). Analyze the resulting Tau441 protein by 12% SDS-PAGE followed by staining with Coomassie Blue. A typical Coomassie Blue stained gel of mutOGT/Tau441 and wtOGT/Tau441 is shown in Fig. 1a. Note that the Tau441 protein contains lower molecular weight bands attributable to degradation products and incompletely translated Tau441 products. Western blots using the pan-*O*-GlcNAc antibody (CTD110.6) can be used to verify the presence of *O*-GlcNAc on Tau441 while pre-incubating the CTD110.6 with 10 mM free GlcNAc can

confirm the specificity of this approach for *O*-GlcNAc (Fig. 1a lower panels).

12. Concentrate the Tau441 protein using centrifugal filter units to a volume of ~2 mL.
13. Acidify the Tau441 protein sample using 10% TFA to generate a final TFA concentration of 0.1% (v/v).
14. Equilibrate a semi-preparative HPLC column with 30% buffer B at a flow rate of 1 mL/min.
15. Inject 0.25 mg Tau441 per HPLC run.
16. Wash the column with 30% buffer B for 5 min at 1 mL/min.
17. Elute the column using a linear gradient of 30–75% buffer B over 45 min at a flow rate of 1 mL/min.
18. Collect 1 mL fractions over the entirety of the HPLC run.
19. Repeat **steps 14–18** until all of the Tau441 protein has been purified using the HPLC column.
20. Pool the Tau441 containing fractions and lyophilize to dryness. Tau441 typically elutes between 22 and 24 min (Fig. 1b).
21. Resuspend Tau441 in 20 mM sodium phosphate buffer, pH 7.
22. Determine the protein concentration (*see* **Note 4**) and estimate the yield of Tau441 protein by 12% SDS-PAGE followed by staining with Coomassie Blue. A typical Coomassie Blue stained gel is shown for wtOGT/Tau441 following HPLC purification and demonstrates that the lower molecular weight bands below the full length Tau441 band have been completely removed (Fig. 1c). The extent of tau *O*-GlcNAcylation may be estimated by mass spectrometric analysis of the isolated tau protein [33].

3.2 O-GlcNAc Modified Recombinant Human Tau HPLC Enrichment

1. Treat Tau441 samples with thrombin in a 1:8 (w/w) mass ratio of thrombin to tau for 20 min at room temperature (*see* **Note 5**).
2. Stop thrombin digestion by the addition of phenylmethanesulfonyl fluoride (PMSF) to a 1 mM final concentration.
3. Equilibrate a semi-preparative HPLC column with 30% buffer B at a flow rate of 1 mL/min.
4. Inject 0.25 mg Tau441 per HPLC run.
5. Elute the Tau441 protein using a linear gradient of 30–35% buffer B over 95 min with a flow rate of 1 mL/min. Note the much shallower gradient during the enrichment of the *O*-GlcNAc modified Tau441 protein. As shown in Fig. 2a, Tau441 typically elutes as two distinct peaks between 28 and 36 min. Using the CTD110.6 antibody, it can be demonstrated that the *O*-GlcNAc modified Tau441 species elute in the earlier eluting of these two peaks (Fig. 2a).

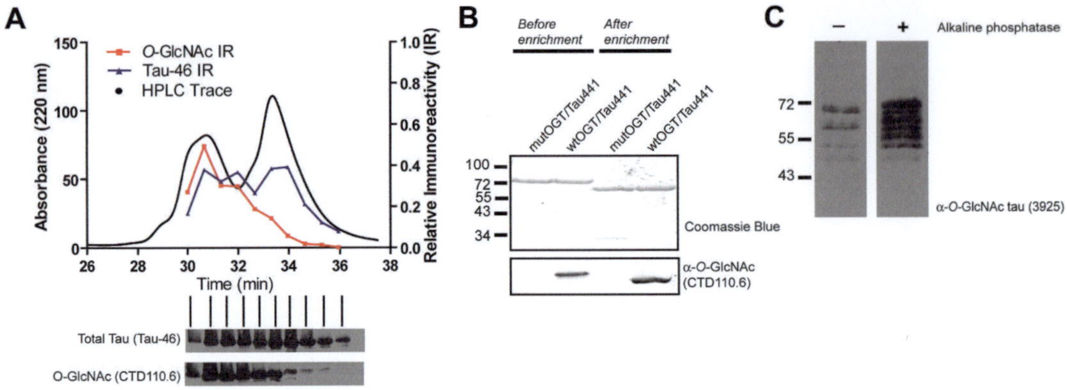

Fig. 2 HPLC enrichment of *O*-GlcNAc modified Tau441 and detection of Ser400 *O*-GlcNAc tau in vivo. (**a**) *O*-GlcNAc modified Tau441 elutes as two distinct peaks between 28 and 38 min. Both peaks are attributable to Tau441 as indicated by the total tau Western blot of fractions collected across these two peaks. Only the early eluting peak is attributable to *O*-GlcNAc modified Tau441 as identified by Western blot using a pan *O*-GlcNAc antibody (CTD110.6). *Lines* indicate the locations of the collected fractions used for Western blots. (**b**) Post-HPLC enrichment Tau441 contains roughly twofold more *O*-GlcNAc as detected by Western blot for total *O*-GlcNAc (CTD110.6). (**c**) Detection of Ser400 *O*-GlcNAc on Tau441 using the 3925 Ser400 *O*-GlcNAc tau antibody. Note the pronounced increase in immunoreactivity following dephosphorylation of the Western blot membranes

6. Collect 1 mL fractions over the entirety of the HPLC run.
7. Combine Tau441 fractions from the earliest eluting Tau441 peak (*see* Fig. 2a).
8. Lyophilize to dryness.
9. Resuspend Tau441 in 20 mM sodium phosphate buffer, pH 7.
10. Determine the protein concentration (*see* **Note 4**) and analyze Tau441 by 12 % SDS-PAGE followed by staining with Coomassie Blue. A typical Coomassie Blue stained gel is shown for Tau441 before and after HPLC enrichment of *O*-GlcNAc modified Tau441 (Fig. 2b). Using the *O*-GlcNAc antibody, CTD110.6, it can be shown that this HPLC enrichment method increases the proportion of *O*-GlcNAc on tau by ~2-fold as estimated by CTD110.6 immunoreactivity (Fig. 2b, lower panels).

3.3 Detection of Ser400 O-GlcNAc Modified Tau from Brain Lysates

1. Lyse mouse or rat brain tissue in 100 μl of lysis buffer/100 mg of frozen tissue weight using a 1 min pulse with an IKA tissue homogenizer.
2. Boil the lysate for 7 min.
3. Centrifuge the lysate at 17,900 × *g* for 20 min at 4 °C.
4. Add perchloric acid to 2.5 % (v/v) and incubate on ice for 15 min and centrifuge at 17,900 × *g* for 20 min at 4 °C.
5. Add 5× SDS-PAGE loading buffer to samples.

6. Add 1 M Tris–HCl pH 8 until the color of the solution turns from yellow to the characteristic blue color of bromophenol blue in SDS-PAGE loading buffer.
7. Carry out SDS-PAGE electrophoresis on samples and transfer to nitrocellulose membranes.
8. Block nitrocellulose membranes in 2% bovine serum albumin (BSA) in PBS-T for 1 h.
9. Incubate nitrocellulose membrane with 100 Units of calf intestinal alkaline phosphatase in 50 mM Tris–HCl, pH 8.5, 0.1 mM EDTA, 0.5 mM $MgCl_2$ for 4.5 h at 37 °C.
10. Wash briefly with PBS-T and incubate blots with 5–10 μg/mL of 3925 Ser400 *O*-GlcNAc tau antibody.
11. Wash three times with PBS-T, incubate with the appropriate secondary antibody, wash three times with PBS-T and develop blots using enhanced chemiluminescence. A typical Western blot result is shown in Fig. 2c, with and without the treatment of the blots with alkaline phosphatase. Alkaline phosphatase markedly increases the immunoreactivity for Ser400 *O*-GlcNAc on Tau441.

4 Notes

1. 16.6 μg/mL kanamycin and 33 μg/mL ampicillin represent one third of the traditional concentration of these antibiotics used on LB-agar plates for growth of *E. coli*. The combination of antibiotics increases the selection pressure on the bacteria and very few double transformants are obtained if the traditional 50 μg/mL and 100 μg/mL kanamycin and ampicillin concentrations are used on LB-agar plates. We have found the 16.6 μg/mL and 33 μg/mL concentrations of these antibiotics to be the most ideal to select for the presence of both plasmids but still yield a reasonable number of double transformants.
2. In our hands, the yield from 6 L of LB expression media of wtOGT/Tau441 or mutOGT/Tau441 is about 1–2 mg of control or *O*-GlcNAc modified Tau441 following HPLC purification (but prior to HPLC enrichment of *O*-GlcNAc modified Tau441).
3. Commonly, we observe precipitation of some contaminating proteins in the Ni-NTA column eluates during the dialysis into low salt buffer. These contaminating proteins are likely insoluble in the low-salt conditions used here.

4. Due to some of the unusual physical properties of tau protein, such as its high isoelectric point and the paucity of aromatic amino acids, careful attention to measuring the Tau441 protein concentration is a necessity. Due to the aforementioned physical properties of tau, the Bradford assay underestimates the concentration of tau. Modified Lowry assays, such as the Bio-Rad DC (a detergent compatible proprietary modification of the Lowry method) or the bicinchoninic acid (BCA) assays are superior for measuring Tau441 concentration colorimetrically. Concentrations of protein may also be determined by amino acid analysis or monitoring proteins absorption by the amide backbone at 214 nm.
5. Care should be taken to not let thrombin digestions of tau proceed longer than 20 min as thrombin may begin to degrade tau in addition to removing the 6×His-tag.

Acknowledgments

S.A.Y. would like to thank the National Science and Engineering Research Council of Canada (NSERC) and the Alzheimer Society of Canada for providing doctoral fellowship support while these methods were being established. D.J.V. would like to thank the Alzheimer Society of Canada and the Canadian Institutes for Health Research (MOP-123341) for support of this research and the Canada Research Chairs program for support as a Tier I Canada Research Chair in Chemical Biology.

References

1. Weingarten MD, Lockwood AH, Hwo SY, Kirschner MW (1975) A protein factor essential for microtubule assembly. Proc Natl Acad Sci U S A 72:1858–1862
2. Grundke-Iqbal I, Iqbal K, Quinlan M, Tung YC, Zaidi MS, Wisniewski HM (1986) Microtubule-associated protein tau. A component of Alzheimer paired helical filaments. J Biol Chem 261:6084–6089
3. Nukina N, Kosik KS, Selkoe DJ (1987) Recognition of Alzheimer paired helical filaments by monoclonal neurofilament antibodies is due to crossreaction with tau protein. Proc Natl Acad Sci U S A 84:3415–3419
4. Ksiezak-Reding H, Liu WK, Yen SH (1992) Phosphate analysis and dephosphorylation of modified tau associated with paired helical filaments. Brain Res 597:209–219
5. Alonso A, Zaidi T, Novak M, Grundke-Iqbal I, Iqbal K (2001) Hyperphosphorylation induces self-assembly of tau into tangles of paired helical filaments/straight filaments. Proc Natl Acad Sci U S A 98:6923–6928
6. Le Corre S, Klafki HW, Plesnila N, Hubinger G, Obermeier A, Sahagun H, Monse B, Seneci P, Lewis J, Eriksen J, Zehr C, Yue M, McGowan E, Dickson DW, Hutton M, Roder HM (2006) An inhibitor of tau hyperphosphorylation prevents severe motor impairments in tau transgenic mice. Proc Natl Acad Sci U S A 103:9673–9678
7. Reynolds MR, Reyes JF, Fu Y, Bigio EH, Guillozet-Bongaarts AL, Berry RW, Binder LI (2006) Tau nitration occurs at tyrosine 29 in the fibrillar lesions of Alzheimer's disease and other tauopathies. J Neurosci 26:10636–10645

8. Cripps D, Thomas SN, Jeng Y, Yang F, Davies P, Yang AJ (2006) Alzheimer disease-specific conformation of hyperphosphorylated paired helical filament-Tau is polyubiquitinated through Lys-48, Lys-11, and Lys-6 ubiquitin conjugation. J Biol Chem 281:10825–10838
9. Morishima-Kawashima M, Hasegawa M, Takio K, Suzuki M, Titani K, Ihara Y (1993) Ubiquitin is conjugated with amino-terminally processed tau in paired helical filaments. Neuron 10:1151–1160
10. Cohen TJ, Guo JL, Hurtado DE, Kwong LK, Mills IP, Trojanowski JQ, Lee VM (2011) The acetylation of tau inhibits its function and promotes pathological tau aggregation. Nat Commun 2:252
11. Cook C, Carlomagno Y, Gendron TF, Dunmore J, Scheffel K, Stetler C, Davis M, Dickson D, Jarpe M, DeTure M, Petrucelli L (2014) Acetylation of the KXGS motifs in tau is a critical determinant in modulation of tau aggregation and clearance. Hum Mol Genet 23:104–116
12. Min SW, Cho SH, Zhou Y, Schroeder S, Haroutunian V, Seeley WW, Huang EJ, Shen Y, Masliah E, Mukherjee C, Meyers D, Cole PA, Ott M, Gan L (2010) Acetylation of tau inhibits its degradation and contributes to tauopathy. Neuron 67:953–966
13. Arnold CS, Johnson GV, Cole RN, Dong DL, Lee M, Hart GW (1996) The microtubule-associated protein tau is extensively modified with O-linked N-acetylglucosamine. J Biol Chem 271:28741–28744
14. Lefebvre T, Ferreira S, Dupont-Wallois L, Bussiere T, Dupire MJ, Delacourte A, Michalski JC, Caillet-Boudin ML (2003) Evidence of a balance between phosphorylation and O-GlcNAc glycosylation of Tau proteins—a role in nuclear localization. Biochim Biophys Acta 1619:167–176
15. Liu F, Iqbal K, Grundke-Iqbal I, Hart GW, Gong CX (2004) O-GlcNAcylation regulates phosphorylation of tau: a mechanism involved in Alzheimer's disease. Proc Natl Acad Sci U S A 101:10804–10809
16. Torres CR, Hart GW (1984) Topography and polypeptide distribution of terminal N-acetylglucosamine residues on the surfaces of intact lymphocytes. Evidence for O-linked GlcNAc. J Biol Chem 259:3308–3317
17. Kreppel LK, Blomberg MA, Hart GW (1997) Dynamic glycosylation of nuclear and cytosolic proteins. Cloning and characterization of a unique O-GlcNAc transferase with multiple tetratricopeptide repeats. J Biol Chem 272: 9308–9315
18. Lubas WA, Frank DW, Krause M, Hanover JA (1997) O-Linked GlcNAc transferase is a conserved nucleocytoplasmic protein containing tetratricopeptide repeats. J Biol Chem 272: 9316–9324
19. Dong DL, Hart GW (1994) Purification and characterization of an O-GlcNAc selective N-acetyl-beta-D-glucosaminidase from rat spleen cytosol. J Biol Chem 269: 19321–19330
20. Gao Y, Wells L, Comer FI, Parker GJ, Hart GW (2001) Dynamic O-glycosylation of nuclear and cytosolic proteins: cloning and characterization of a neutral, cytosolic beta-N-acetylglucosaminidase from human brain. J Biol Chem 276:9838–9845
21. Roquemore EP, Chevrier MR, Cotter RJ, Hart GW (1996) Dynamic O-GlcNAcylation of the small heat shock protein alpha B-crystallin. Biochemistry 35:3578–3586
22. Li X, Lu F, Wang JZ, Gong CX (2006) Concurrent alterations of O-GlcNAcylation and phosphorylation of tau in mouse brains during fasting. Eur J Neurosci 23:2078–2086
23. Smet-Nocca C, Broncel M, Wieruszeski JM, Tokarski C, Hanoulle X, Leroy A, Landrieu I, Rolando C, Lippens G, Hackenberger CP (2011) Identification of O-GlcNAc sites within peptides of the Tau protein and their impact on phosphorylation. Mol Biosyst 7:1420–1429
24. Yuzwa SA, Macauley MS, Heinonen JE, Shan X, Dennis RJ, He Y, Whitworth GE, Stubbs KA, McEachern EJ, Davies GJ, Vocadlo DJ (2008) A potent mechanism-inspired O-GlcNAcase inhibitor that blocks phosphorylation of tau in vivo. Nat Chem Biol 4:483–490
25. Yuzwa SA, Yadav AK, Skorobogatko Y, Clark T, Vosseller K, Vocadlo DJ (2011) Mapping O-GlcNAc modification sites on tau and generation of a site-specific O-GlcNAc tau antibody. Amino Acids 40:857–868
26. Wang Z, Udeshi ND, O'Malley M, Shabanowitz J, Hunt DF, Hart GW (2010) Enrichment and site mapping of O-linked N-acetylglucosamine by a combination of chemical/enzymatic tagging, photochemical cleavage, and electron transfer dissociation mass spectrometry. Mol Cell Proteomics 9:153–160
27. Morris M, Knudsen GM, Maeda S, Trinidad JC, Ioanoviciu A, Burlingame AL, Mucke L (2015) Tau post-translational modifications in wild-type and human amyloid precursor protein transgenic mice. Nat Neurosci 18: 1183–1189

28. Borghgraef P, Menuet C, Theunis C, Louis JV, Devijver H, Maurin H, Smet-Nocca C, Lippens G, Hilaire G, Gijsen H, Moechars D, Van Leuven F (2013) Increasing brain protein O-GlcNAcylation mitigates breathing defects and mortality of Tau.P301L mice. PLoS One 8:84442
29. Graham DL, Gray AJ, Joyce JA, Yu D, O'Moore J, Carlson GA, Shearman MS, Dellovade TL, Hering H (2014) Increased O-GlcNAcylation reduces pathological tau without affecting its normal phosphorylation in a mouse model of tauopathy. Neuropharmacology 79C:307–313
30. Yuzwa SA, Shan X, Macauley MS, Clark T, Skorobogatko Y, Vosseller K, Vocadlo DJ (2012) Increasing O-GlcNAc slows neurodegeneration and stabilizes tau against aggregation. Nat Chem Biol 8:393–399
31. Yuzwa SA, Cheung AH, Okon M, McIntosh LP, Vocadlo DJ (2014) O-GlcNAc modification of tau directly inhibits its aggregation without perturbing the conformational properties of tau monomers. J Mol Biol 426: 1736–1752
32. Martinez-Fleites C, Macauley MS, He Y, Shen DL, Vocadlo DJ, Davies GJ (2008) Structure of an O-GlcNAc transferase homolog provides insight into intracellular glycosylation. Nat Struct Mol Biol 15:764–765
33. Shen DL, Gloster TM, Yuzwa SA, Vocadlo DJ (2012) Insights into O-linked N-acetylglucosamine (O-GlcNAc) processing and dynamics through kinetic analysis of O-GlcNAc transferase and O-GlcNAcase activity on protein substrates. J Biol Chem 287:15395–15408

Part IV

Detecting Tau Proteins, Their Modifications, Mutations and Interacting Partners

Chapter 14

Two-Dimensional Electrophoresis Protocols to Analyze the Microtubule-Associated Tau Proteins from Several Biological Sources

Nicolas Sergeant, Francisco-Jose Fernandez-Gomez, Helene Obriot, Sabiha Eddarkaoui, Valérie Buée-Scherrer, and Luc Buée

Abstract

Technology breakthrough in proteomics enables to gather qualitative and quantitative information about a protein or a complex mixture of proteins. Two-dimensional gel electrophoresis remains an interesting technique, which provides an overview of the complexity of isovariants from a single protein when coupled to western blotting. Here, we describe a detailed protocol for the two-dimensional analysis of microtubule-associated Tau isovariants from cell to human or mouse brain tissue. We provide protocol enabling to separate native and Tau proteins that are aggregated in neurodegenerative disorders such as Alzheimer's disease.

Key words Two-dimensional gel electrophoresis, Isoelecfocusing, Alzheimer's disease, Posttranslational modifications, Tau protein isovariants, Brain tissue

1 Introduction

Proteomics enable to gather quantitative and qualitative information of a complex mixture of polypeptides [1]. Recent advances in mass spectrometry also provide the opportunity to address posttranslational modifications and the 3D-structure of a single protein [2, 3]. Moreover, this technique is also very powerful to study the complexity of single protein and identify related posttranslational modifications and proteolytic fragments. Although mass spectrometry can provide an in depth analysis of a protein, it is also not easily accessible to the broader scientific community and not systematically applicable to the analysis of "one preferred protein". Mass spectrometry is at the top of the methods used to characterize a protein. Other methods are also useful to study the

Caroline Smet-Nocca (ed.), *Tau Protein: Methods and Protocols*, Methods in Molecular Biology, vol. 1523,
DOI 10.1007/978-1-4939-6598-4_14, © Springer Science+Business Media New York 2017

proteome of a complex mixture of proteins or even a single protein. These methods include chromatography, 1D or 2D-gel electrophoresis. The latest being of interest because proteins are resolved according to their net charge and apparent molecular weights.

Two-dimensional gel electrophoresis (2DE) was first introduced by O'Farrell in 1977 [4]. When coupled with protein staining or differential fluorescent labeling, 2DE is a simple method to provide an overview of a proteome. This technique shows its limits because of inter gel variability, difficulties to analyze protein in low abundance or transmembrane proteins or even insoluble proteins that are difficult to study using this method. By coupling 2DE to the use of antibodies and western blotting, some of these limitations have been stretched. In the context of Alzheimer's disease, 2DE coupled to western blot enabled the analysis of protein and peptides that are found aggregated in the neuropathological lesions; the so-called neurofibrillary tangles and amyloid deposits. Neurofibrillary tangles are made by intraneuronal accumulation of hyper- and abnormally phosphorylated tau proteins structured into paired helical filaments. Amyloid deposits are composed principally of a 1–42 amyloid-beta peptide, which derives from successive proteolytic cleavages, by secretases.

The microtubule-associated Tau protein is one such example for the use of 2DE coupled to western blotting to get a global overview of its proteome complexity. Tau protein belongs to the family of microtubule-associated proteins. A single gene located on chromosome 17q21 encodes several isoforms [5]. In human brain, six Tau isoforms differ by the alternative insertion or not of exon sequences numbered 2, 3, and 10. This complexity is further increased by posttranslational modifications such as phosphorylation, acetylation, O-GlcNAcylation and proteolysis (for review *see* refs. 6–8). There are 85 putative phospho-sites on the longest human brain tau isoform (441 aa) and recent MS/MS analysis brings the number of posttranslational modifications of the mouse brain tau to 100. Such a high number of posttranslational modifications when coupled to the expression of multiple isoforms renders the proteomic analysis of Tau very challenging [8, 9].

Alternatively, 2DE has been a powerful method to address such an heterogeneity [10, 11] and provide an overview of the heterogeneity of Tau proteome as well as the biochemical parameters that characterize Tau during neurofibrillary degeneration [12–14]. 2DE is also easy to investigate potential differences between two experimental conditions. To that purpose, we provide herein a simple 2DE method to study and analyze Tau proteome from different biological sources including human or mouse brain and cells.

The brain proteome remains difficult to achieve because the brain tissue is rich in transmembrane proteins and lipids. Previous work from our groups have shown that sodium-dodecyl sulfate is very useful to study the brain proteome [13, 14]. However, SDS could easily be used when performing capillary isoelectrofocalisation even at concentration above 1 %, whereas SDS should be avoided while using IPG strips. Other strategies should be used instead. Thus, the brain tissue is first homogenized in buffer containing SDS and proteins are precipitated in order to remove the SDS and change the detergent to CHAPS. An alternative approach such as liquid chromatography has been proposed to increase the identification of membrane proteins [15].

In neurodegenerative disorders, the challenge was to identify the constituents of the neuropathological lesions wherever it is neurofibrillary tangles, amyloid deposits, Lewy bodies, intranuclear aggregates or cytosolic ubiquitin inclusion. Isolation of the aggregates coupled to peptide sequencing, 1D gel electrophoresis, 2D gel electrophoresis, and mass spectrometry enabled the identification of the protein forming those proteinaceous inclusions. Figure 1 show the complexity of Tau proteome in human, mouse or cells as well as in the human brain of a patient who had suffered of Alzheimer's disease (Fig. 1b). 2DE when coupled with the use of Tau specific antibody reveals a complex pattern of isovariants which are poorly resolved in the first dimension. Thus, Tau proteins are revealed as a smear along a pH gradient between 5 and 9 (Fig. 1a). Note that basic are resolved as spots suggesting that posttranslational modifications are conferring a homogenous net charge to Tau isovariants. However, the number and position of posttranslational modifications cannot be deduced from this 2D western blots. Moreover, a partial dephosphorylation can occur during the preparation of the 2D samples which could contribute to the heterogeneity of Tau isovariant isofocalization. A heterogenous distribution of the charges between tau molecules may explain this smeary resolution.

When compared to Alzheimer's disease brain tissue (Fig. 1b), Tau isovariants are poorly resolved on both axes using the C-terminal specific Tau antiserum. On a sharp contrast, three main rockets of Tau isovariants are observed with the N-terminal specific Tau antiserum. When aggregated Tau proteins are protected from dephosphorylation that occurs with the postmortem delay [13, 16], Tau isovariants detected with the N-terminal Tau antibody are in more acidic pH range than that observed in a less affected AD patient or our Thy-Tau22 transgenic mouse model of neurofibrillary degeneration [17]. These Tau isovariants are solubilized using our TSU lysis buffer and are resolved by 2DE.

The 2DE protocol we provide can be useful to analyze the proteome of a single protein and can also be very useful to study protein found to be aggregated in neurodegenerative disease such as amyloid-beta peptides, synuclein, TDP-43, p62 ... This 2DE method can also be adapted for the use of isoelecfocusing concentrator in order to further study protein isovariants by mass spectrometry.

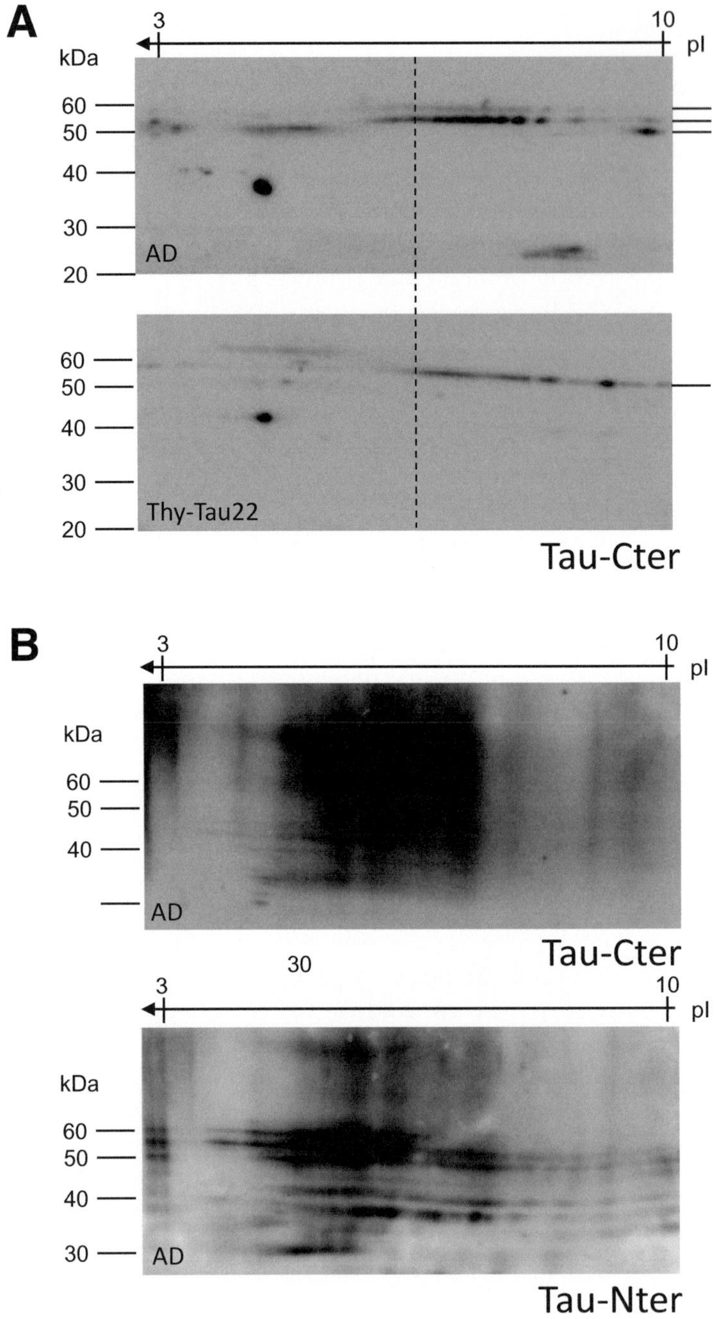

Fig. 1 Comparison of the 2DE profiles between human brain Tau from an Alzheimer's disease (AD) patient and the human Tau isoform 412 overexpressed in the Thy-Tau22 mouse model of hippocampal neurofibrillary degeneration. (**a**) In both condition 50 μg of brain protein were precipitated and resolved using IPG strip of pH 3–10 NL and 4–12 % Criterion XT MOPS precast gels. In the *upper panel* three pathological Tau proteins are resolved in a pH range comprised between 6.5 and 9. In the *lower panel*, the human Tau 412 mutated isoform is resolved between pI 7–9.5. Note that isovariants of Tau in the Thy-Tau22 are less acidic suggesting a lower level of phosphorylation. Isoelectric points are indicated on the *x*-axis *arrow*. Apparent molecular weights are

2 Materials

Solutions were prepared with pure water obtained with a Milli-Q Water Purification system providing deionized water with a resistivity below 18 MΩ. All chemicals are PlusOne GE Healthcare.

1. Half of a human brain hippocampus. Brain tissues are provided by the Lille Neurobank after agreement of the project by the scientific and ethic committee. All patients had an informed consent. The brain tissue is manipulated in a L2 secured cell laboratory under a fume hood.
2. TCU and TSU lysis buffer: for a total of 10 mL solution, weigh 5 g of Urea and 1.6 g of thiourea, complete to a final volume of 10 mL with pure water. This solution is incubated at room temperature with Amberlite and filtered before the addition of either 1 % SDS (TSU) or 4 % CHAPS (TCU) (w/v) (*see* **Note 1**). The TSU is used with tissues whereas TCU is used with cells and 2D-gel electrophoresis.
3. Tris-SDS buffer: 20 mM non-buffered Tris and 2 % (w/v) SDS.
4. Modified Laemmli sample buffer: 240 mg of non-buffered Tris (20 mM) is added to 20 ml of glycerol and completed to a final volume of 100 mL. This solution is gently mixed and 2 % (w/v) of SDS is added. Avoid making foam when mixing.
5. Tris-sucrose buffer: 20 mM non-buffered Tris and 0.32 M sucrose.
6. Pierce™ BCA protein assay kit (ThermoFisher).
7. Bradford reagent kit (BioRad).
8. 5 mg/mL BSA in PBS.
9. Serum-free cell culture medium.
10. Acetone.
11. Chloroform.
12. Methanol.
13. IEF Resuspension buffer: 8 M Urea, 2 M thiourea and 4 % CHAPS.
14. IEF buffer: add 1.1 μL of IPG Buffer and 2.2 μL of DeStreak (GE Healthcare), 0.005 % (w/vol) bromophenol blue or phenol red to 200 μL of IEF resuspension buffer.

Fig. 1 (continued) indicated on the *left side* of the 2DE. Human Tau protein isovariants were detected using a polysera against the C-terminus of Tau [18]. (**b**) Staining of pathological Tau proteins from AD brain tissue. 50 μg of human brain protein were resolved by 2DE and Tau proteins were revealed using a polyserum against the C-terminus (Tau-Cter) or a polyserum against the N-terminus of Tau (Tau-Nter). Note that Tau polymers are resolved as a smear in between pI 5–7. In sharp contrast, the N-terminal Tau antibody mainly stains the high molecular weight aggregates and three series of isovariants between 50 and 60 kDa and pI 4.5–7

15. DeStreak (GE Healthcare).
16. Second dimension loading buffer: 25 mM Tris–HCl pH 6.8, 6 M urea, 2% SDS, 30% glycerol, 20 mM DTT, 0.05% bromophenol blue.
17. Nitrogen gas bottle.
18. Ponceau Red.
19. TBS-T: 25 mM Tris–HCl pH 8.0, 150 mM NaCl, 0.1% Tween-20 (v/v).
20. TBS-M: 5% (w/v) of skimmed milk in TBS-T.
21. The primary Tau antibody used herein is an anti-Tau polyserum against the last 12 amino acids of human Tau. The antibody is diluted at 1:5000 in TBS-T.
22. Secondary antibody: peroxidase-labeled goat anti-rabbit IgG (H+L chains) (Vector Laboratories).
23. ECL™ western blotting kit.
24. IPG strips (GE Healthcare): Immobilin Dry-strips, 11 cm immobilized pH gradient strip of pH range 3–10 nonlinear (NL) or 7–11.
25. Nitrocellulose sheets of 0.45 μm of porosity.
26. Mineral oil.
27. A laminar-flow cabinet.
28. Glass potter.
29. Ultrasound generator (Bioblock Scientific).
30. Bench Top Centrifuge.
31. An orbital shaker.
32. Rehydration plates and rehydration cassettes.
33. IEF system: IPGPhor III isoelectrofocusing Unit (GE healthcare).
34. Criterion XT 4–12% polyacrylamide precast gradient gels (BioRad).
35. Criterion XT Blotter system (BioRad).
36. LAS-4000 Luminescence Image Analyzer (FujiFilm Lifesciences) or Amersham ECL™ Hyperfilm (GE Healthcare).

3 Methods

1. Homogenize the human brain tissue using a glass potter in either Tris-SDS or TSU buffer, although TSU buffered is preferred to recover proteins from aggregates. Add 10 volumes of lysis buffer to the tissue and force the tissue inside the conical glass by several go and return of the pestle until the tissue is

completely disrupted. Sonicate the resulting lysate with 30 consecutive 2 s-pulses at 60 Hz using an ultrasound generator. Remove tissue debris by centrifugation at 12,000 × *g* (*see* **Notes 2** and **3**). Determine protein concentration (*see* **Note 4**).

2. The mouse brain is often used for several biochemical methods. Homogenize first the mouse brain in an ice-cold Tris-sucrose buffer. Immerse half a hippocampus (±20 mg) in 200 μL of Tris-sucrose buffer in a 0.5 mL Eppendorf tube and sonicate on ice with 30 pulses of 0.5 s with an amplitude of 40 %. Achieve homogenization on an orbital shaker at 4 °C during 1 h. Determine protein concentration using the BCA protein assay kit. Keep the brain tissue lysate at −80 °C until use.
3. Add 1 volume of Tris-SDS buffer to the mouse brain Tris-sucrose lysate. Sonicate the resulting lysate on ice with 20 pulses of 0.5 s with an amplitude of 40 %.
4. For cell-derived proteins: rinse the cell layer with warm medium without serum. Discard the medium and add 200 μL of warm (60 °C) modified Laemmli lysis buffer per well of a 6-well plate. Remove the cell layer with a rubber policeman, recover cell lysate in a 1.5 mL Eppendorf tube and sonicate on ice with 20 pulses of 0.5 s with an amplitude of 40 %. Determine protein concentration using a BCA protein assay kit.
5. Determine protein concentrations with the Bradford reagent kit using BSA for the standard curve. Note that BSA dilution is always performed in the lysis buffer used.
6. Precipitate protein lysates from brain tissues or cells (50 μg) before changing the buffer for a 2D-gel buffer. Precipitate protein from the mouse brain tissue or from cells with 10 volumes of cold-acetone (−20 °C). Achieve precipitation on ice during 10 min and centrifuge at 12,000 × *g* during 10 min at 4 °C. Discard the supernatant and dry the pellet under nitrogen in a laminar-flow cabinet.
7. Precipitate protein lysates from human brain tissue (50 μg) with chloroform/methanol. Perform all steps on ice. 30 min before starting the precipitation, cool chloroform, methanol and water at 4 °C. Add three volumes of MeOH and one volume of chloroform to one volume of TSU lysate and gently mix. Add three volumes of cold water and resuspend the resulting solution using a vortex during 1 min. Recover the protein pellet after centrifugation at 12,000 × *g* during 30 min at 4 °C. Discard the supernatant using a Pasteur pipette. Add three volumes of cold MeOH and the resuspend protein pellet using a vortex. After centrifugation at 12,000 × *g* during 30 min at 4 °C, discard the supernatant, dry the protein pellet under nitrogen in a laminar-flow cabinet.

8. For the first dimension or isoelectrofocusing (IEF), solubilize between 20 to 100 μg (maximum) of protein pellet in 200 μL of IEF resuspension buffer. Achieve resuspension of the protein by applying 30 pulses of 0.5 s at 60 Hz. Add 1.1 μL of IPG Buffer as well as 2.2 μL of DeStreak. Add 0.005 % (w/v) of bromophenol blue or phenol red.
9. Layer the 2D-sample in a lane of a rehydration plate and recover with an 11 cm immobilized pH gradient strip of pH range 3–10 nonlinear (NL) or 7–11 (*see* **Note 5**). Achieve rehydration overnight at room temperature. Add 2D buffer if urea crystals appear (*see* **Note 6**).
10. For isoelectrofocusing, place IPG strips in the isoelectrofocusing tray with the plastic in contact with the tray. Wet the paper bridge with deionized water and place at each extremity of IPG strips. Fix electrodes and overlay IPG strips with mineral oil according to the manufacturer instruction.
11. Perform isoelectrofusing steps by applying 500 V with a maximum of 50 μA per strip during 1 h, 1000 V during 1H, 6000 V during 2 h30 and a final of 6000 V of 30 min. Apply this program for samples from the human brain. For the mouse brain or cell-derived samples, the 500 V step is not necessary and the 6000 V step is reduced to 2 h 00. Note that the same program is applied for IPG of 3–10 NL or 7–11 pH gradient (*see* **Note 7**).
12. Remove strips, eliminate the excess of mineral oil (*see* **Note 8**) and keep the 11 cm strips at −20 °C until the second dimension is performed.
13. Before the second dimension, equilibrate IPG strips with the second dimension loading buffer. equilibrate strips 3 times for 10 min each at room temperature (*see* **Note 9**).
14. Perform the second dimension using Criterion XT 4–12% polyacrylamide precast gradient gels. Criterion XT gels are well-adapted for the second dimension of 7 or 11 cm IPG strips. Carefully place the strip on the top of the gel (*see* **Note 10**). Perform the migration by applying 100 V during 1 h (*see* **Note 11**).
15. For western blotting, transfer the SDS-PAGE to a nitrocellulose sheet of 0.45 μm of porosity. Transfer proteins at a constant current intensity of 2.5 mA/cm^2.
16. After transfer, reversibly stain proteins with Ponceau Red to check the quality and efficacy of protein transfer. For western blot analysis, block the membranes in TBS-M for 30 min, then incubate overnight at 4 °C with primary anti-Tau antibodies. Rinse the nitrocellulose 3 times for 10 min each in TBS-T with gentle shaking. Incubate the membrane for 1 h at room

temperature with the secondary antibody (peroxidase-labeled goat anti-rabbit IgG). Reveal the immunoreactive complexes using the ECL™ western blotting kit and perform image acquisition with the LAS-4000 Luminescence Image Analyzer or by exposure to ECL™ Hyperfilm (*see* **Note 12**).

4 Notes

1. Uric acid is removed from solutions of urea or a mix of urea/thiourea using PlusOne Amberlite IRN-150 L. Solution is gently mixed and filtered using 0.22 μm seringe filter. Batch of solutions are made for a single project and kept at −80 °C. Solutions are defrosted at room temperature.
2. Centrifugation of the brain lysate obtained after sonication is not mandatory since the tissue debris do not enter the Immobilin Dry-strips during rehydration. It is however preferred to determine the protein concentration in each sample using Bradford reagent kit.
3. Extraction using the TSU buffer can be used for 1D SDS-PAGE. To that purpose, TSU-lysate can be diluted with one volume of a gel loading buffer as lithium dodecyl sulfate (LDS) gel loading buffer (Life Technologies).
4. If the human brain tissue is homogenized in Tris-SDS buffer, the protein concentration is determined using the BCA Protein Assay kit. The BCA kit is not recommended when urea or thiourea is in the lysis buffer.
5. GE Healthcare is proposing several IPG buffers depending on the range of pH in which the protein will be resolved. IPG buffers 3–10 NL or 7–11 are often used to resolve tau protein using IPG 3–10 NL or IPG 7–11, respectively.
6. During rehydration, after 2 h, overlay strips with mineral oil and place the rehydration plate in rehydration cassette according to the manufacturer instructions. However, it should be mentioned that the room temperature should not be too low at night (below 15 °C) since urea solution crystallizes below this temperature. The rehydration cassette can be placed in a hoven maintained at a constant temperature of 20 °C.
7. Isoelectrofocusing is performed using the IPGPhor III isoelectrofocusing Unit. No more than 6 IPG strips are loaded together. Traces of bromophenol blue or phenol red are used to ascertain that the isoelectric focalization of each strip is achieved. Thus, the bromophenol blue or phenol appears yellow at the plus end of the strip. If dyes are not focused at the plus end, a migration problem may have occurred.

8. After rehydration, remove strips from the rehydration plate and eliminate the excess of mineral oil on a Whatman filter paper sheet.
9. During the equilibration, in the last step of equilibration the reducing agent DTT can be replaced with 4.7 % iodoacetamide in the second dimension loading buffer to alkylate the cysteine residues.
10. When placing the IPG strip at the top of the second dimension gel, be careful to avoid bubbles in between both gels. Do not completely dry the gel surface and start to push from one extremity to the other.
11. The second dimension migration time can be optimized depending on the molecular weight range of interest (e.g., for low molecular weights, the bromophenol blue migration front should be at the bottom of the gel and not going out in the buffer tank).
12. When performing western blot of Tau in mouse brain tissue, secondary antibodies often react with heavy chains of mouse immunoglobulins resolved at the same molecular weight than Tau [1].

Acknowledgments

This work was supported by the Inserm, Lille 2 University, and CNRS. A grant was obtained from ANR, Laboratory of Excellence Labex DISTALZ, and France Alzheimer. The authors declare no conflict of interest.

References

1. Petry FR, Pelletier J, Bretteville A, Morin F, Calon F, Hébert SS, Whittington RA, Planel E (2014) Specificity of anti-tau antibodies when analyzing mice models of Alzheimer's disease: problems and solutions. PLoS One 9, e94251
2. Bizarro J, Charron C, Boulon S, Westman B, Pradet-Balade B, Vandermoere F, Chagot ME, Hallais M, Ahmad Y, Leonhardt H, Lamond A, Manival X, Branlant C, Charpentier B, Verheggen C, Bertrand E (2014) Proteomic and 3D structure analyses highlight the C/D box snoRNP assembly mechanism and its control. J Cell Biol 207(4):463–480
3. Dailing A, Luchini A, Liotta L (2015) Unlocking the secrets to protein-protein interface drug targets using structural mass spectrometry techniques. Expert Rev Proteomics 12:457–467
4. O'Farrell PZ, Goodman HM, O'Farrell PH (1977) High resolution two-dimensional electrophoresis of basic as well as acidic proteins. Cell 12:1133–1141
5. Caillet-Boudin ML, Buée L, Sergeant N, Lefebvre B (2015) Regulation of human MAPT gene expression. Mol Neurodegener 10:28
6. Sergeant N, Bretteville A, Hamdane M, Caillet-Boudin ML, Grognet P, Bombois S, Blum D, Delacourte A, Pasquier F, Vanmechelen E, Schraen-Maschke S, Buée L (2008) Biochemistry of Tau in Alzheimer's disease and related neurological disorders. Expert Rev Proteomics 5:207–224
7. Derisbourg M, Leghay C, Chiappetta G, Fernandez-Gomez FJ, Laurent C, Demeyer D, Carrier S, Buée-Scherrer V, Blum D, Vinh J,

Sergeant N, Verdier Y, Buée L, Hamdane M (2015) Role of the Tau N-terminal region in microtubule stabilization revealed by new endogenous truncated forms. Sci Rep 14:9659

8. Morris M, Knudsen GM, Maeda S, Trinidad JC, Ioanoviciu A, Burlingame AL, Mucke L (2015) Tau post-translational modifications in wild-type and human amyloid precursor protein transgenic mice. Nat Neurosci 18:1183–1189
9. Hanger DP, Byers HL, Wray S, Leung KY, Saxton MJ, Seereeram A, Reynolds CH, Ward MA, Anderton BH (2007) Novel phosphorylation sites in tau from Alzheimer brain support a role for casein kinase 1 in disease pathogenesis. J Biol Chem 282:23645–23654
10. Butler M, Shelanski ML (1986) Microheter ogeneity of microtubule-associated tau proteins is due to differences in phosphorylation. J Neurochem 47:1517–1522
11. Janke C, Beck M, Holzer M, Bigl V, Arendt T (2000) Analysis of the molecular heterogeneity of the microtubule-associated protein tau by two-dimensional electrophoresis and RT-PCR. Brain Res Brain Res Protoc 5:231–242
12. Greenberg SG, Davies P (1990) A preparation of Alzheimer paired helical filaments that displays distinct tau proteins by polyacrylamide gel electrophoresis. Proc Natl Acad Sci U S A 87:5827–5831
13. Sergeant N, Bussière T, Vermersch P, Lejeune JP, Delacourte A (1995) Isoelectric point differentiates PHF-tau from biopsy-derived human brain tau proteins. Neuroreport 6:2217–2220
14. Sergeant N, David JP, Goedert M, Jakes R, Vermersch P, Buée L, Lefranc D, Wattez A, Delacourte A (1997) Two-dimensional characterization of paired helical filament-tau from Alzheimer's disease: demonstration of an additional 74-kDa component and age-related biochemical modifications. J Neurochem 69: 834–844
15. Lohaus C, Nolte A, Blüggel M, Scheer C, Klose J, Gobom J, Schüler A, Wiebringhaus T, Meyer HE, Marcus K (2007) Multidimensional chromatography: a powerful tool for the analysis of membrane proteins in mouse brain. J Proteome Res 6:105–113
16. Buée L, Bussière T, Buée-Scherrer V, Delacourte A, Hof PR (2000) Tau protein isoforms, phosphorylation and role in neurodegenerative disorders. Brain Res Brain Res Rev 33:95–130
17. Laurent C, Burnouf S, Ferry B, Batalha VL, Coelho JE, Baqi Y, Malik E, Mariciniak E, Parrot S, Van der Jeugd A, Faivre E, Flaten V, Ledent C, D'Hooge R, Sergeant N, Hamdane M, Humez S, Müller CE, Lopes LV, Buée L, Blum D (2016) A2A adenosine receptor deletion is protective in a mouse model of Tauopathy. Mol Psychiatry 21:97–107. doi:10.1038/mp.2014.151
18. Bretteville A, Ando K, Ghestem A, Loyens A, Bégard S, Beauvillain JC, Sergeant N, Hamdane M, Buée L (2009) Two-dimensional electrophoresis of tau mutants reveals specific phosphorylation pattern likely linked to early tau conformational changes. PLoS One 4(3), e4843

Chapter 15

A Simple Method to Avoid Nonspecific Signal When Using Monoclonal Anti-Tau Antibodies in Western Blotting of Mouse Brain Proteins

Franck R. Petry, Samantha B. Nicholls, Sébastien S. Hébert, and Emmanuel Planel

Abstract

In Alzheimer's disease and other tauopathies, tau displays several abnormal post-translation modifications such as hyperphosphorylation, truncation, conformation, and oligomerization. Mouse monoclonal antibodies have been raised against such tau modifications for research, diagnostic, and therapeutic purposes. However, many of these primary antibodies are at risk of giving nonspecific signals in common Western blotting procedures. Not because they are unspecific, but because the secondary antibodies used to detect them will also detect the heavy chain of endogenous mouse immunoglobulins (Igs), and give a nonspecific signal at the same molecular weight than tau protein (around 50 kDa). Here, we propose the use of anti-light chain secondary antibodies as a simple and efficient technique to prevent nonspecific Igs signals at around 50 kDa. We demonstrate the efficacy of this method by removing artifactual signals when using monoclonal antibodies directed at tau phosphorylation (AT100, 12E8, AT270), tau truncation (TauC3), tau oligomerization (TOMA), or tau abnormal conformation (Alz50), in wild-type, 3×Tg-AD, and tau knockout mice.

Key words Tau phosphorylation, Antibody specificity, Alzheimer's disease, Mice models, Light chain antibody

1 Introduction

Alzheimer's disease (AD) is the most common neurodegenerative disease and the most prevalent tauopathy. AD is characterized by a progressive loss of cognitive function [1, 2] and is marked by the presence of extracellular senile plaques [2] and intracellular neurofibrillary tangles (NFTs) composed of abnormal aggregates of hyperphosphorylated tau [3, 4]. The main known tau function is to bind and stabilize the microtubule network in the axons of neurons. This capacity is negatively regulated by its phosphorylation [5]. In AD and others tauopathies, tau becomes hyperphosphorylated, resulting in reduced affinity for microtubules and self-aggregation

Caroline Smet-Nocca (ed.), *Tau Protein: Methods and Protocols*, Methods in Molecular Biology, vol. 1523,
DOI 10.1007/978-1-4939-6598-4_15,

into abnormal filaments, leading to formation of NFTs and cell death [6]. The mechanisms leading to tau hyperphosphorylation and which form tau is toxic are still a matter of debate [7, 8]. Thus, tau hyperphosphorylation is well studied in laboratories by using animal models, and especially transgenic mice expressing human tau, to better understand the link between phosphorylation and the events that occur in tauopathies.

The most commonly used method to analyze tau phosphorylation is Western blotting by immuno-detection. Antibodies are typically created in the mouse or the rabbit to detect specific epitopes of phosphorylated tau. Nevertheless, the use of monoclonal mouse antibodies in mice model can lead to nonspecific signals. Indeed, the secondary anti-mouse antibody used in Western blotting can bind to either the heavy chain (HC) (~50 kDa) or the light chain (LC) (~25 kDa) of endogenous Igs present within brain tissues [9], thus interfering with the tau signal at ~50 kDa. Our group has previously analyzed the specificity of the commonly used tau antibodies in various mouse models and reported some nonspecific signals from Igs [10].

In this chapter, we present our work in which we aimed at analyzing the specificity of others monoclonal antibodies directed against tau, and used secondary light chain (LC) antibodies to remove the nonspecific signals. We compared the anti-tau antibody immunoreactivity profiles in four different mouse models: (1) non-transgenic wild-type mice with low tau phosphorylation at basal levels [11], (2) tau knockout mice (TKO) used as negative control [12], (3) 3×Tg-AD mice [13] that express human mutated tau protein (P301L) as well as human mutated amyloid precursor protein (APPswe) on human mutated presenilin 1 (PS1) background, and (4) anesthetized wild-type mice were used as a positive control, as it has been shown that anesthesia-induced hypothermia leads to tau hyperphosphorylation [14]. For each primary anti-tau antibody, we used a conventional secondary anti-mouse antibody or the secondary LC antibody (*see* **Note 1**). The former is able to bind to both the light and the heavy chains of mouse antibodies whereas the latter only binds to the light chain of mouse antibodies. Thus, the secondary LC antibody does not interfere with the tau signal at 50 kDa. Instead, the LC secondary antibody will bind to the light chain of the mouse endogenous antibodies, giving a nonspecific signal at ~25 kDa.

Overall we found that use of LC antibody removed nonspecific signal around 50 kDa for AT100, Alz50, TauC3, TOMA, and AT270. Furthermore, we recommend always loading a TKO mouse sample as negative control, and a hypothermic mouse as positive control for tau hyperphosphorylation during Western blot analysis.

2 Materials

1. RIPA (Radioimmunoprecipitation Assay) Base Buffer (for 500 ml): 25 ml of 1 M Tris–HCl pH 7.4, 5 ml of IGEPAL CA-630, 4.5 g of NaCl, 1 ml of 0.5 M EDTA. Make up to 500 ml with distilled water. Store at 4 °C.
2. Modified RIPA buffer: 1/100 (v/v) of phosphatase inhibitor cocktail, 1/100 (v/v) of protease inhibitor cocktail, 1/100 (v/v) of 100 mM Phenylmethylsulfonyl fluoride (PMSF), 1/40 (v/v) of 10% Na-deoxycholate and complete to desired volume with RIPA Base Buffer. Make just before extraction and keep on ice to maintain inhibitors activity.
3. Western blot loading sample buffer: 1/4 (v/v) of lithium dodecyl sulfate (LDS) 4×, 1/100 (v/v) of phosphatase inhibitor cocktail, 1/100 (v/v) of protease inhibitor cocktail, 1/100 (v/v) of 100 mM PMSF, 5% of β-mercaptoethanol. Adjust samples for a final protein concentration of 2 μg/μl with modified RIPA buffer.
4. Pierce™ Bicinchoninic acid (BCA) Protein Assay Kit (Thermo Scientific).
5. PBS 10× (for 500 ml): 40 g of NaCl, 1 g of KCl, 7.2 g of Na_2PO_4, 1.2 g KH_2PO_4, adjust pH to 7.5. Make up to 500 ml with distilled water.
6. Blocking buffer: 5% dry milk in PBS-T 1×.
7. PBS-T (PBS-Tween) 1× (for 500 ml): 1/10 (v/v) of PBS 10× and 1/1000 (0.1%) of Tween 20. Make up to 500 ml with distilled water. Keep at room temperature.
8. Primary antibodies: Total tau (N-terminal tau, BD Biosciences), 12E8 (pSer262, Neotope Biosciences), AT100 (pSer212-pThr124, Fisher), Alz50 (Alzheimer tau, gift from Dr. Peter Davies), TOMA (tau oligomer-specific monoclonal antibody, gift from Dr. Rakez Kayed), TauC3 (tau cleaved at Asp421, Millipore) and AT270 (pThr181, Fisher). Primary antibodies are reported in Table 1.
9. All primary antibodies are diluted in SuperBlock Blocking Buffer in PBS (ThermoFisher Scientific) at a final concentration 1/1000. Add 1/1000 (v/v) of sodium azide 10%.
10. Secondary antibodies: Peroxidase-conjugated AffiniPure Goat Anti-Mouse IgG (H+L) (115-035-003, Jackson Immunoresearch Laboratories) and HRP-conjugated Goat Anti-mouse Light Chain (AP200P, Millipore).
11. Secondary antibodies are diluted in 5% dry milk in PBS-T 1× at a final concentration 1/5000. Secondary antibodies are reported in the Table. 1.

Table 1
List of all used antibodies. We incubated all the mouse primary antibodies overnight at 4 °C at a dilution of 1:1000 (final concentration). For each primary antibody, we have used two types of secondary anti-mouse: a conventional anti-mouse antibody and the anti-mouse LC antibody. The secondary antibodies were incubated at room temperature for 2 h at a dilution of 1:5000 (final concentration)

Antibody	Epitope	Supplier	Catalog number	Lot number	Dilution	Incubation period	Protein load
Monoclonal primary mouse antibodies							
Alz50	AD tau	Peter Davies	–	–	1/1000	Overnight	20
TauC3	Asp421	Millipore	36-017	27027	1/1000	Overnight	10
AT100	pThr212-pSer214	Fisher	MN1060	JK 1228858	1/1000	Overnight	20
TOMA	Oligomeric tau	Rakez Kayed	–	–	1/1000	Overnight	20
12E8	pSer262	BD Biosciences	RB90	956	1/1000	Overnight	10
AT270	pThr181	Fisher	MN1050	JK 128859	1/1000	Overnight	10
Total tau	N-Terminal 23-179	BD Biosciences	610672	21178	1/1000	Overnight	10
Secondary anti-mouse antibodies							
Anti-mouse IgG (H + L)	Anti-mouse	Jackson Immuno Research	115-035-003	97103	1/5000	2 h	–
Light Chain (LC)	Anti-mouse	Millipore	AP200P	2297170	1/5000	2 h	–

12. Lysis buffer: 20 mM MES pH 6.8, 1 mM EGTA, 0.2 mM $MgCl_2$, 5 mM DTT, protease inhibitor cocktail (Roche), phosphatase inhibitor cocktail (Sigma).
13. Cation exchange buffer A: 10 mM HEPES pH 7, 50 mM NaCl.
14. Cation exchange buffer B: 10 mM HEPES, 1 M NaCl.
15. TauC3 protein as described in Subheading 3.1.
16. Nitrocellulose membrane.
17. ECL detection kit.
18. ImageQuant™ LAS4000 imaging system (GE Healthcare).

3 Methods

3.1 Production of TauC3

1. TauC3 is expressed as described in [15]. A point mutation introducing a stop codon to the pET29b full-length (441 amino acids; 2N4R) Tau plasmid (Addgene plasmid # 16316) at residue 422 allows for expression of recombinant TauC3.

2. Protein is expressed in *E. coli* strain BL21(DE3) culture grown to an O.D. of 0.6 and induced with 1 mM IPTG for 4 h at 37 °C.
3. The pellet is resuspended in the lysis buffer.
4. Cells are disrupted using sonication and boiled for 20 min at a NaCl concentration of 500 mM.
5. Cells are then centrifuged at 127,000 × *g* for 45 min at 4 °C and the supernatant was dialyzed overnight into the cation exchange buffer A.
6. The dialysate is cleared by centrifugation at 127,000 × *g* for 45 min at 4 °C and loaded onto a cation exchange column.
7. TauC3 is eluted using a linear gradient over 6 column volumes to a 60% final concentration of cation exchange buffer B.

3.2 Sample Preparation

1. Collect mouse brain on dry ice (*see* **Note 2**) and isolate the whole cortex in a new 2 ml Eppendorf tube for further processing. Freeze on dry ice.
2. Weigh the cortex.
3. Homogenize the cortex with 5 volumes (v/w) of cold modified RIPA buffer (*see* **Note 3**).
4. Centrifuge at 20,000 × *g* for 20 min at 4 °C.
5. Harvest the supernatant and put in a new eppendorf tube. Discard the pellet.
6. Dose proteins with the BCA Protein Assay kit and mix them with the Western blot loading sample buffer (*see* **Note 4**).

3.3 Standard Western Blot Procedure (Fig. 1, Left)

1. Separate proteins by SDS-PAGE using Tris–Glycine gels containing 10% of acrylamide–bis-acrylamide and then blot onto nitrocellulose membranes.
2. Block the membrane with blocking buffer for 1 h at room temperature.
3. Incubate the membrane with the primary anti-tau antibody overnight at 4 °C.
4. The following day, wash the membrane 3 times for 10 min each with PBS 1×.
5. Incubate the secondary anti-mouse in 5% dry milk for 1 h at room temperature.
6. Wash the membrane 3 times for 10 min each with PBS 1×.
7. Peroxidase activity of the secondary antibody is revealed using ECL detection kit and visualized with an imaging system.

3.4 Western Blot Procedure with Secondary Anti-mouse Light Chain Antibodies (Fig. 1, Right)

1. Separate proteins by SDS-PAGE using Tris–Glycine gels containing 10% of acrylamide–bis-acrylamide and then blot onto nitrocellulose membranes.
2. Block the membrane with blocking buffer for 1 h at room temperature.
3. Incubate the membrane with the primary anti-tau antibody overnight at 4 °C.
4. The following day, wash the membrane 3 times for 10 min each with PBS 1×.
5. Incubate the secondary anti-mouse light chain antibody for 2 h at room temperature (*see* **Note 1**).
6. Wash the membrane 3 times for 10 min each with PBS 1×.
7. Peroxidase activity of the secondary antibody is revealed using ECL detection kit and visualized with an imaging system.

4 Notes

1. The secondary anti-mouse LC antibody is less stable and is light sensitive compared to conventional a secondary anti-mouse antibody. We recommend adding glycerol in the secondary LC antibody mixture for extended storage after reconstitution. We also recommend keeping the secondary LC antibody away from the light during the incubation period by covering the membrane with aluminum paper to avoid signal loss. The secondary LC antibody gives a strong nonspecific signal at 25 kDa (Fig. 1; Alz50 and TauC3). This signal is often stronger than the tau signal at 50 kDa. This must be kept in mind when exposing the blot. Alternatively, we recommend cutting the membrane at ~35 kDa or hiding the bottom of the membrane during the exposition, or taking incremental exposure times.
2. All mice were killed by decapitation without anesthesia. We have shown previously that anesthesia induces rapid tau phosphorylation [14]. Thus, this method is used to preserve the state of tau phosphorylation at the death of the mice.
3. Cortices were homogenized in cold modified RIPA buffer. Protease and phosphatases inhibitors were added to the RIPA buffer to avoid proteins degradation and tau dephosphorylation during the sample processing, respectively.
4. We prepare samples to a final concentration of 2 μg of proteins per μl. The amount of proteins loaded per primary antibody used is indicated in the Table 1. Typically, 10 μg of proteins are loaded for the Western blot, although some epitopes (Alz50, AT100, and TOMA) might need more total protein for optimal detection.

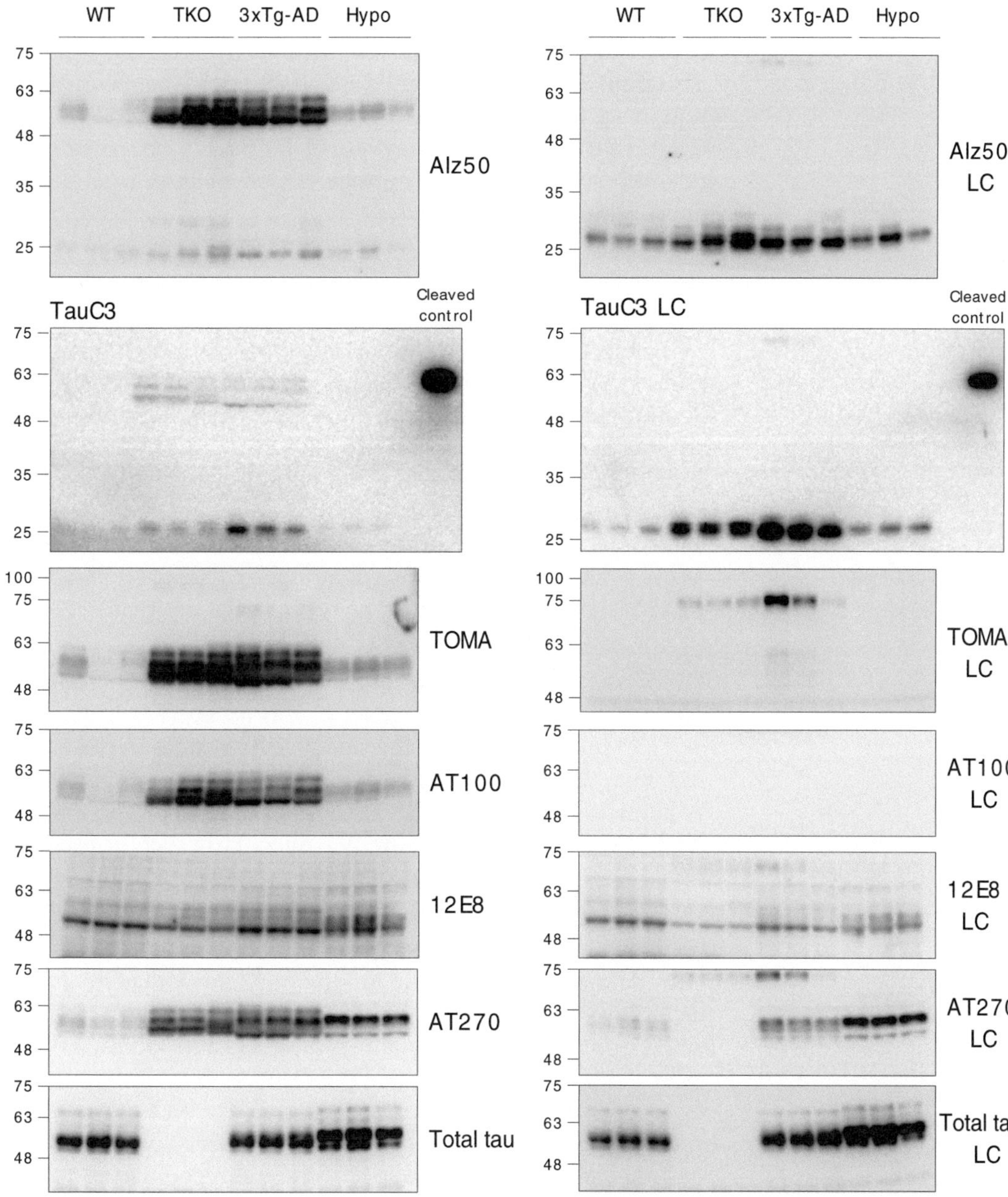

Fig. 1 Western blot analysis of tau using monoclonal anti-mouse antibodies without (*left*) or with (*right*) anti light chain secondary antibody. Total proteins were extracted from the cortex of three mouse lines at 18 months of age: control mice (WT and Hypothermic), Tau KO mice, and 3xTg-AD mice ($n = 3$ per group). Proteins were separated by SDS-PAGE and visualized using the following monoclonal anti-mouse antibodies directed to total-tau (*see* **Note** 5); phospho-tau 12E8 (*see* **Note** 6); AT270 (*see* **Note** 7); AT100; conformational tau Alz50, cleaved tau TauC3, and oligomeric tau TOMA (*see* **Note** 8 for these last three antibodies). Conventional anti-mouse (*left*) and anti-mouse light chain (*right*) secondary antibodies were used to detect the primary antibody. Full gels are shown for Alz50 and TauC3 antibodies to show a nonspecific band at 25 kDa due to the light chains of endogenous antibodies. Tau KO mice and 3xTg-AD mice displayed more Igs signal than wild-type mice (*see* Note 9)

5. Nonspecific signals are absent when total tau protein is detected using with the total tau antibody from BD Biosciences. This antibody was designed to recognize the N-terminal region of rat tau, but also bind to human and mouse tau. Contrarily to the other monoclonal antibodies examined here, this one is directed at an epitope abundant in normal mouse brain, so it gives a very strong signal and the unspecific signal from endogenous Igs becomes undetectable.
6. Despite the use of the anti-mouse LC antibodies, 12E8 (Ser262/Ser356) still gave nonspecific signal, likely due to the cross reactivity with other microtubule-associated proteins, such as MAP2 (low molecular weight) at the pS262 and Ser356 epitopes. Indeed tau and MAP2 share high homology around Ser262 and Ser356 [16]. In this case, the use of the anti-mouse light antibodies is not helpful. However, tau phosphorylation at the Ser262/Ser356 with the 12E8 antibody is detectable in adult mice, in absence of pathological condition or induction of tau phosphorylation [17]. The true phosphorylation signal of tau at this epitope might be quantified by subtracting the signal from the tau KO mice.
7. Here, the use of the secondary anti-mouse LC antibody revealed the true AT270 signal [11], as no signal was detected in TKO mice. Moreover, the signal was stronger in the 3xTg-AD mice model and in the hypothermic mice compared to WT mice. Unlike AT100 phosphorylation epitope of tau, AT270 signal is present in the WT mice. However, nonspecific signal is present with conventional secondary antibodies, so we recommend the use of the anti-mouse light chain antibodies to detect AT270 by Western blot.
8. Our data show that some of the tau antibodies analyzed, including Alz50, TauC3, and AT100 did not give a signal when using the secondary anti-mouse LC antibodies. TOMA gave a signal ~75 kDa in 3xTg-AD mice, but this is not tau since this band was also present in the Tau KO mice. Alz50 and TOMA are designed to recognize specific conformational and oligomeric epitopes, respectively, and recognize abnormal tau from AD brain on SDS-PAGE [18–20]. TauC3 was developed to recognize tau cleaved by the caspase 3 in AD brains [21]. Here, we added a positive control in the form of a recombinant tau (synthetized up to Asp421) [15], and our data show a strong signal in the positive control but not in the mouse brain samples when using the anti-mouse light chain antibody. Finally, AT100 is a phosphorylation epitope of tau only detectable in pathological conditions, and is used to investigate increased abnormal tau phosphorylation levels [22]. We failed to see signal in the 3xTg-AD and hypothermic mice with the anti-mouse LC antibody, suggesting that tau

phosphorylation at this epitope is absent even in conditions that were reported to increase the phosphorylation of tau, but are not pathological (hypothermic WT mice recover from tau hyperphosphorylation and do not display tau pathology). Taken together, all these results suggest that none of the mouse models used herein present a tau pathology extensive enough to allow detection by AT100, Alz50, TOMA, or TauC3 by conventional Western blotting.

9. Our results show that TKO and 3xTg-AD mice display a strong nonspecific signal, suggesting that the brain of these two transgenic mice contains more Igs than the non-transgenic WT mice. We do not know why this is, as the mice were of same age and sex and killed in the same conditions, but it is important to keep this result in mind since it can lead to artifactual results and erroneous conclusions. For example, in Fig. 1, if one uses normal secondary antibodies and compares the AT100 signal from the WT and 3xTg-AD mice, it will lead to the conclusion that tau is hyperphosphorylated at this epitope in 3xTg-AD mice. The result with the LC secondary clearly reveals that this is not the case. While we do not know why Igs signal was stronger in 3xTg-AD and TKO mice in this experiment, other conditions, such as inflammation and diabetes, are known to increase the amount Igs in the mouse brain [23], leading to potential artifactual results in the analysis of tau modifications. We thus recommend being careful when analyzing low tau phosphorylation levels in mouse model.

Acknowledgments

We would like to thank Dr. Peter Davies (Albert Einstein College of Medecine, Bronx, NY, USA) for the gift of the Alz50 antibody, Dr. Rakez Kayed (Mitchell Center for Neurodegenerative Diseases, Texas, USA) for the TOMA antibody, and Dr. Peter Seubert (Neotope Biosciences, CA, USA) for the gift of the 12E8 antibody.

References

1. Querfurth HW, LaFerla FM (2010) Alzheimer's disease. N Engl J Med 362:329–344. doi:10.1056/NEJMra0909142
2. Selkoe DJ (2001) Alzheimer's disease: genes, proteins, and therapy. Physiol Rev 81:741–766
3. Buée L, Bussière T, Buée-Scherrer V, Delacourte A, Hof PR (2000) Tau protein isoforms, phosphorylation and role in neurodegenerative disorders. Brain Res Brain Res Rev 33:95–130
4. Grundke-Iqbal I, Iqbal K, Tung YC, Quinlan M, Wisniewski HM, Binder LI (1986) Abnormal phosphorylation of the microtubule-associated protein tau (tau) in Alzheimer cytoskeletal pathology. Proc Natl Acad Sci U S A 83:4913–4917
5. Alonso AC, Zaidi T, Grundke-Iqbal I, Iqbal K (1994) Role of abnormally phosphorylated tau in the breakdown of microtubules in Alzheimer

disease. Proc Natl Acad Sci U S A 91: 5562–5566

6. Ballatore C, Lee VM-Y, Trojanowski JQ (2007) Tau-mediated neurodegeneration in Alzheimer's disease and related disorders. Nat Rev Neurosci 8:663–672. doi:10.1038/nrn2194
7. Bretteville A, Planel E (2008) Tau aggregates: toxic, inert, or protective species? J Alzheimers Dis 14:431–436
8. Duff K, Planel E (2005) Untangling memory deficits. Nat Med 11:826–827. doi:10.1038/nm0805-826
9. St-Amour I, Paré I, Alata W, Coulombe K, Ringuette-Goulet C, Drouin-Ouellet J, Vandal M, Soulet D, Bazin R, Calon F (2013) Brain bioavailability of human intravenous immunoglobulin and its transport through the murine blood-brain barrier. J Cereb Blood Flow Metab 33:1983–1992. doi:10.1038/jcbfm.2013.160
10. Petry FR, Pelletier J, Bretteville A, Morin F, Calon F, Hébert SS, Whittington RA, Planel E (2014) Specificity of anti-tau antibodies when analyzing mice models of Alzheimer's disease: problems and solutions. PLoS One 9, e94251. doi:10.1371/journal.pone.0094251
11. Goedert M, Jakes R, Crowther RA, Cohen P, Vanmechelen E, Vandermeeren M, Cras P (1994) Epitope mapping of monoclonal antibodies to the paired helical filaments of Alzheimer's disease: identification of phosphorylation sites in tau protein. Biochem J 301(Pt 3):871–877
12. Dawson HN, Ferreira A, Eyster MV, Ghoshal N, Binder LI, Vitek MP (2001) Inhibition of neuronal maturation in primary hippocampal neurons from tau deficient mice. J Cell Sci 114:1179–1187
13. Oddo S, Caccamo A, Shepherd JD, Murphy MP, Golde TE, Kayed R, Metherate R, Mattson MP, Akbari Y, LaFerla FM (2003) Triple-transgenic model of Alzheimer's disease with plaques and tangles: intracellular Abeta and synaptic dysfunction. Neuron 39:409–421
14. Planel E, Richter KEG, Nolan CE, Finley JE, Liu L, Wen Y, Krishnamurthy P, Herman M, Wang L, Schachter JB, Nelson RB, Lau L-F, Duff KE (2007) Anesthesia leads to tau hyperphosphorylation through inhibition of phosphatase activity by hypothermia. J Neurosci 27:3090–3097. doi:10.1523/JNEUROSCI.4854-06.2007
15. Barghorn S, Biernat J, Mandelkow E (2005) Purification of recombinant tau protein and preparation of Alzheimer-paired helical filaments in vitro. Methods Mol Biol 299:35–51
16. DeTure MA, Di Noto L, Purich DL (2002) In vitro assembly of Alzheimer-like filaments. How a small cluster of charged residues in Tau and MAP2 controls filament morphology. J Biol Chem 277:34755–34759. doi:10.1074/jbc.M201201200
17. Seubert P, Mawal-Dewan M, Barbour R, Jakes R, Goedert M, Johnson GV, Litersky JM, Schenk D, Lieberburg I, Trojanowski JQ (1995) Detection of phosphorylated Ser262 in fetal tau, adult tau, and paired helical filament tau. J Biol Chem 270:18917–18922
18. Carmel G, Mager EM, Binder LI, Kuret J (1996) The structural basis of monoclonal antibody Alz50's selectivity for Alzheimer's disease pathology. J Biol Chem 271:32789–32795
19. Castillo-Carranza DL, Sengupta U, Guerrero-Muñoz MJ, Lasagna-Reeves CA, Gerson JE, Singh G, Estes DM, Barrett ADT, Dineley KT, Jackson GR, Kayed R (2014) Passive immunization with Tau oligomer monoclonal antibody reverses tauopathy phenotypes without affecting hyperphosphorylated neurofibrillary tangles. J Neurosci 34:4260–4272. doi:10.1523/JNEUROSCI.3192-13.2014
20. Ksiezak-Reding H, Davies P, Yen SH (1988) Alz 50, a monoclonal antibody to Alzheimer's disease antigen, cross-reacts with tau proteins from bovine and normal human brain. J Biol Chem 263:7943–7947
21. Gamblin TC, Chen F, Zambrano A, Abraha A, Lagalwar S, Guillozet AL, Lu M, Fu Y, Garcia-Sierra F, LaPointe N, Miller R, Berry RW, Binder LI, Cryns VL (2003) Caspase cleavage of tau: linking amyloid and neurofibrillary tangles in Alzheimer's disease. Proc Natl Acad Sci U S A 100:10032–10037. doi:10.1073/pnas.1630428100
22. Mailliot C, Bussière T, Caillet-Boudin ML, Delacourte A, Buée L (1998) Alzheimer-specific epitope of AT100 in transfected cell lines with tau: toward an efficient cell model of tau abnormal phosphorylation. Neurosci Lett 255:13–16
23. Shuai H, Zhang J, Zhang J, Xie J, Zhang M, Ma J, Zhang L, Wang X (2012) Role of stereotaxically injected IgG from db/db mice in the phosphorylation of the microtubule-associated protein tau in hippocampus. Brain Res 1486: 14–26. doi:10.1016/j.brainres.2012.08.049

Chapter 16

Flow Cytometry Analysis and Quantitative Characterization of Tau in Synaptosomes from Alzheimer's Disease Brains

Karen Hoppens Gylys and Tina Bilousova

Abstract

Synaptosomes, resealed nerve terminals that form when tissue is homogenized in isotonic medium, are a model system that has been a key source of knowledge about neurotransmission. Synaptosomes contain mitochondria, cytoskeletal proteins, and release neurotransmitters; many have postsynaptic elements. Cryopreservation at the time of autopsy makes it possible to prepare synaptosomes from human samples. Flow cytometry is a powerful analytic technique that precisely measures fluorescence on a cell-by-cell basis, and also indicates particle size and complexity with a routine parameter that measures light scattering. We describe here a procedure for flow cytometry analysis of tau in synaptosomes, a procedure that enables (1) "purification" of synaptosomes from the P-2 fraction (crude synaptosomes) by gating on particle size, and (2) quantitative measure of tau immunofluorescence in individual terminals. Application of flow cytometry to study of synaptosomes has yielded important information, not possible with routine biochemistry, about synaptic pathology in Alzheimer's disease.

Key words Flow cytometry, HT7 antibody, Tau fragment, FACs, Cytoskeleton, Neurodegeneration

1 Introduction

The synapse is the functional unit of the brain, and synapse loss is thought to be an early correlate of cognitive deficits in Alzheimer's disease (AD) [1]. However, synaptic boutons are poorly visualized at the level of light microscopy, and immunodetection of proteins of interest by EM are primarily qualitative and face technical issues that include epitope masking and fixation artifacts. Therefore many researchers have used the biochemical synaptosome preparation to study synaptic proteins and function. Synaptosomes are resealed nerve terminals formed when fresh brain tissue is gently homogenized in isotonic sucrose, forming spherical particles that contain mitochondria and the organelles for neurotransmitter release (Fig. 1a). Synaptosomes are primarily presynaptic with some adherent postsynaptic structures, and the preparation has been used extensively in pharmacology since the 1960s to study in vitro

Caroline Smet-Nocca (ed.), *Tau Protein: Methods and Protocols*, Methods in Molecular Biology, vol. 1523,
DOI 10.1007/978-1-4939-6598-4_16, © Springer Science+Business Media New York 2017

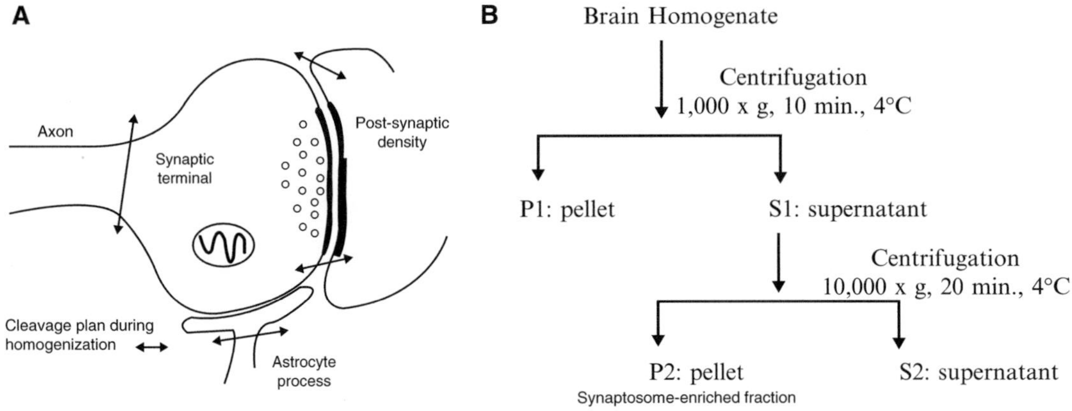

Fig. 1 Synaptosome preparation. (**a**) Homogenization of fresh brain tissue in isotonic sucrose solution induces formation of synaptosomes—avulsion and resealing of synaptic terminals, which may contain some postsynaptic elements and sometimes astrocyte end-feet. (**b**) Schematic illustration of synaptosome-enriched (P2) fraction preparation from brain homogenate

receptor binding and neurotransmitter release [2]. Classical protocols for purification include Ficoll discontinuous gradient purification of the initial crude synaptosomal pellet, also called the P2. Even with additional density gradient purification, synaptosome preparations are impure and contain free vesicles, mitochondria, myelin, and membrane fragments; quantified by flow cytometry, purified synaptosome fractions contain ~83 % synaptosomes labeled by a voltage-sensitive dye [3]. Gradient purification requires extra time and reduces viability, which has led to the use of the P2 fraction for study of functional processes not affected by the usual contaminants [2].

Flow cytometry, also referred to as fluorescence-activated cell sorting (FACS), is a widely used analytic instrument that provides rapid and simultaneous analysis of multiple characteristics of single cells, including cell size, complexity, and the fluorescence emitted by a wide range of dyes and fluorophore-conjugated antibodies. Flow cytometry offers the additional advantage of quantifying two parameters for fluorescence; both the positive fraction (% positive) and the brightness of fluorescence (relative fluorescence, RFU) are precisely measured for each particle. Several thousand cells/particles can be analyzed per second by suspension in a stream of sheath fluid that passes through multiple detectors [4]. Since synaptosomes are analogous to cells, flow cytometry analysis of synaptosomes represents a powerful approach to the study of synapses, and the earliest studies used Percoll gradient-purified synaptosomes [3, 5, 6]. However, reasoning that analysis of the P2 offered significant advantages in terms of faster preparation time and higher tissue yield, our lab developed procedures for flow cytometry analysis of P2 fraction beginning in 2000 [7]. Importantly, we have been able to demonstrate by a variety of methods that analysis of particles based on size and including only particles between 0.5 and 1.5 μm allows focus on a population of synaptosomes that is ~95 % pure [8–10], superior to the ~80 % purity obtained with gradient purification [3, 9].

A number of workers have studied synaptosomes prepared from human tissue obtained during surgery or at the time of autopsy [11–13], in which case it is cryopreserved in isotonic sucrose using protocols that have been previously optimized [14]. In some circumstances, including studies of tau release, we have expanded these cryopreservation strategies to include cryopreservation of P2 pellets after isolation. The structural and functional integrity of synaptosomes prepared from cryopreserved postmortem tissue has been repeatedly demonstrated by our group and more recently by others [15]; for example, 99 % of size gated synaptosomes label with Fluo-4NW, a calcium-sensitive dye that indicates intact membranes and esterase activity [16]. We have recently shown depolarization-dependent release of tau from human AD synaptosomes [17], and a number of other authors have also measured functional endpoints using cryopreserved postmortem human synaptosomes [12, 18, 19]. The combined techniques of flow cytometry and synaptosome preparation offer quantitative answers to questions it is not possible to ask with routine biochemistry or pathology techniques, and we report here a detailed protocol for flow cytometry analysis of immunolabeled synaptosomes from human AD samples.

2 Materials

Prepare all the solutions using deionized water.

2.1 Solutions for Brain Tissue Cryopreservation and Synaptosome-Enriched (P2) Fraction Preparation

For prolonged storage all water- based stock solutions need to be sterilized by filtration through 0.45 μm filters.

1. 0.2 M ethylene diamine tetraacetic acid (EDTA) stock solution: weigh 0.744 g of EDTA disodium salt (FW 372.2), add 8 ml of water, adjust the pH to 8.0 with constant stirring (*see* **Note 1**); when EDTA salt is completely dissolved adjust volume to 10 ml with water. Store the solution at room temperature (stable for more than 1 year).
2. 0.2 M ethylene glycol tetraacetic acid (EGTA) stock solution: weigh 0.761 g of EGTA (FW 380.4), add 8 ml of water, adjust the pH to 8.0 with constant stirring (*see* **Note 1**); when EGTA is completely dissolved adjust volume to 10 ml with water. Store the solution at room temperature (stable for more than 1 year).

3. 0.1 M phenylmethylsulfonyl fluoride (PMSF) stock solution: dissolve 0.174 g of PMSF (FW 174.2) in 10 ml of 100% ethanol (protect from light). Divide into 200 μl aliquots and store at −20 °C (stable for more than 1 year).
4. 0.1 M sodium pyrophosphate decahydrate (NaPP) stock solution: weigh 4.46 g of $Na_4P_2O_7{\cdot}10H_2O$ (FW 400.1) and dissolve in 100 ml of water. Store at 4 °C (stable for 3–4 weeks).
5. 1 M sodium fluoride (NaF) stock solution: dissolve 0.42 g of NaF (FW 41.99) in 10 ml of water. Store at 4 °C (stable for more than 1 year).
6. 1.5 mM pepstatin A stock solution: dissolve 5 mg of pepstatin A (FW 685.9) in 5 ml of 100% Methanol. Divide into 100 μl aliquots and store at −20 °C, avoid freeze–thaw cycles (stable for more than 1 year).
7. 20 mM leupeptin stock solution: dissolve 10 mg of leupeptin (FW 475.6) in 1 ml of water. Divide into 100 μl aliquots and store them at −20 °C, avoid freeze–thaw cycles (stable for more than 1 year).
8. 1.5 mM aprotinin stock solution: dissolve 10 mg of aprotinin in 1 ml of water. Divide into 100 μl aliquots and store them at −20 °C, avoid freeze–thaw cycles (stable for more than 1 year).
9. Trypsin inhibitor (from soybean): dissolve 100 mg of trypsin inhibitor in 10 ml of water. Divide into 200 μl aliquots and store them at −20 °C, avoid freeze–thaw cycles (stable for more than 1 year).
10. 1 M Tris–HCl, pH 8.0: dissolve 12.1 g of Tris base in 90 ml of water. Adjust pH to 8.0. Bring up the volume to 100 ml with water. Divide into 1 ml aliquots and store at −20 °C (stable for more than 1 year).
11. Cryopreservation buffer (0.32 M sucrose, 2 mM EDTA, 2 mM EGTA, 0.2 mM PMSF, 1 mM NaPP, 5 mM NaF, 10 mM Tris–HCl, pH 8.0): to make 100 ml of cryopreservation buffer, weigh 11 g sucrose and dilute it in 90 ml of deionized water (*see* **Note 2**), add 100 μl of 0.2 M EDTA, 100 μl of 0.2 M EGTA, 200 μl of 100 mM PMSF, 1 ml of 100 mM NaPP, 500 μl of 1 M NaF, and 1 ml of 1 M Tris–HCl, pH 8.0 stock solution. Adjust volume to 100 ml with water; keep the solution on ice until use.
12. P2 buffer (0.32 M sucrose, 1.5 μM pepstatin A, 0.75 μM aprotinin, 8 μM leupeptin, 1 μM trypsin inhibitor, 2 mM EDTA, 2 mM EGTA, 0.2 mM PMSF, 1 mM NaPP, 5 mM NaF, 10 mM Tris–HCl, pH 8.0): to make 100 ml of P2 buffer, weigh 11 g sucrose and dilute it in 90 ml of deionized water, add protease and phosphatase inhibitors using stock solutions—100 μl of 1.5 mM pepstatin A, 50 μl of 1.5 mM aprotinin, 40 μl of 20 mM leupeptin, 200 μl of 0.5 mM trypsin inhibitor, 100 μl

of 0.2 M EDTA, 100 μl of 0.2 M EGTA, 200 μl of 100 mM PMSF, 1 ml of 100 mM NaPP, 500 μl of 1 M NaF, and 1 ml of 1 M Tris–HCl, pH 8.0.

2.2 Solutions and Reagents for P2 Immunostaining and Flow Cytometry

1. Phosphate buffered saline (PBS): 0.01 M phosphate buffer, 0.0027 M potassium chloride, and 0.137 M sodium chloride, pH 7.4 (*see* **Note 3**).
2. 4% paraformaldehyde (PFA) stock solution: weigh 2 g of paraformaldehyde, move it into 50 ml centrifuge tube, add PBS to reach 50 ml total volume, close and seal the tube with Parafilm and incubate overnight into water bath (55 °C). Next day let the solution to cool down to room temperature, recheck pH and adjust it to 7.4 if required. Filter the solution through 0.45 μm filter and store at 4 °C (stable for 6 months).
3. Fixation buffer (0.25% PFA in PBS, pH 7.4): add 625 μl of PFA stock solution to 9 ml 375 μl of PBS, vortex. Prepare fixation buffer fresh right before fixation step.
4. 10% Tween 20 stock solution: add 1 ml or 100% Tween 20 (*see* **Note 4**) to 9 ml of water, invert tube several times to mix, store at room temperature, protect from light (the solution is stable for a month).
5. Permeabilization buffer (0.2% Tween 20 in PBS): add 1 ml of 10% Tween 20 stock solution to 49 ml of PBS. Invert tube or slightly vortex to mix.
6. Immunostaining/Blocking Buffer (2% fetal bovine serum (FBS) in PBS, pH 7.4): add 40 μl of FBS to 960 μl of PBS, pH 7.4. Invert tube or slightly vortex to mix.
7. Washing buffer (0.1% tween 20 in PBS): add 500 μl of 10% Tween 20 stock solution to 49 ml of PBS. Invert tube or slightly vortex to mix.
8. Tau/ptau-specific antibodies and isotype control antibodies.
9. Presynaptic marker-specific antibody (for example, SNAP25, or Synaptophysin, or VGluT1-specific antibody).
10. Isotype control antibodies.
11. Zenon kits (*see* **Note 5**). Particular Zenon kits need to be selected based on the primary antibody host, isotype and suitable fluorophore type.
12. Polystyrene beads with 0.75, 1.5 and 4.5 μm in diameter.

3 Methods

All the procedures were carried out on ice unless specified. Unfixed human tissue is considered biohazardous material; always follow corresponding biosafety protocols to protect yourself and environment (*see* **Note 6**).

3.1 Human Brain Tissue Cryopreservation

1. Make cryopreservation buffer and keep it chilled on ice (*see* Subheading 2.1, **step 11** and **Note 2**).
2. Weight fresh brain tissue (*see* **Note 7**) and calculate the volume of cryopreservation buffer based on 1:10 mass to volume ratio (10 ml of cryopreservation buffer for 1 g of fresh tissue).
3. Place brain tissue into a petri dish and add 2 ml of ice cold cryopreservation buffer. Mince the tissue into 1–3 mm fragments using a prechilled scalpel.
4. Transfer the minced tissue and buffer into a 15 ml or 50 ml caped centrifuge tube and add the rest of the cryopreservation buffer with inhibitors.
5. Placed the tube into Styrofoam boxes with insulating material for slow freezing to −80 °C (*see* **Note 8**).

3.2 Synaptosomal (P2)-Fraction Preparation and Cryopreservation

1. Remove a tube contained cryopreserved brain tissue from −80 °C freezer and place in warm (37 °C) water bath for 1–2 min to defrost perimeter.
2. Mix with 5 volumes of ice cold P2-buffer and shake vigorously until tissue is free.
3. Centrifuge the mix for 2 min at 1000 × *g*, 4 °C to pellet minced brain tissue.
4. Mix the pellet with 10 volumes of P2-buffer, based on the weight of fresh tissue before cryopreservation (1:10 mass to volume ratio; *see* Subheading 3.1, **step 2** and **Note 7**).
5. Homogenize tissue using a Teflon/glass homogenizer (clearance 0.1–0.15 mm) by eight gentle up and down strokes at 800 rpm. Save 200–500 μl of homogenate for P2-quality assay (*see* **Note 9**).
6. Centrifuge the rest of homogenate for 10 min at 1000 × *g*, 4 °C to isolate cell debris and nuclei (P1-fraction).
7. Collect supernatant (S1), place it into the clean pre-weighed centrifuge tube and centrifuge for 20 min at 10,000 × *g*, 4 °C to obtain crude synaptosomal P2 pellet (Fig. 1b).
8. Carefully discard supernatant (S2) and weight P2 pellet (*see* **Note 10**).
9. Resuspend P2 pellet in 10 (1:10 mass to volume ratio) volumes of P2-buffer.
10. Divide resuspended P2 fraction into 500 μl aliquots.
11. Placed the tube into Styrofoam boxes with insulating material for slow freezing to −80 °C (*see* **Note 11**).

3.3 Immunostaining of P2 Fraction

1. Remove a P2 fraction aliquot from −80 °C and thaw immediately in warm (37 °C) water bath for 1–2 min.

2. Transfer P2 fraction into a tube containing equivalent volume of ice-cold PBS and gently mix by pipetting up and down several times, avoid using vortex before fixation step. 10 cm
3. Centrifuge for 4 min at 4000 × *g* (4 °C) to pellet P2 fraction.
4. Decant the supernatant and resuspend the pellet in fixation solution, incubate for 1 h at 4 °C.
5. Centrifuge the sample for 4 min at 4000 × *g* (4 °C) to pellet P2 fraction.
6. Decant supernatant.
7. Resuspend fixed P2 fraction in 500 μl permeabilization buffer (*see* **Note 12**).
8. Place the tube into the water bath (37 °C) and incubate for 15 min, invert the tube every 5 min.
9. Add 500 μl of PBS and centrifuge for 4 min at 4000 × *g* (4 °C) to pellet P2-fraction. Decant supernatant.

Steps 10–12 *Subheading 3.3 (Antibody conjugation using Zenon kit) should be performed in parallel with* steps 7 *and* 8 *Subheading 3.3.*

10. Mix 18 μl of Immunostaining/Blocking Buffer with 1 μg of specific antibodies or isotype control antibodies for each immunostaining reaction (*see* **Note 13**).
11. Add 4.2 μl of corresponded Zenon kit Solution A (Fab fragments with conjugated Alexa dye) to the antibodies and incubate for 5 min at room temperature, protected from light.
12. Add 4.2 μl of Zenon kit Solution B (Stop solution) to the antibody/solution A mixture and incubate for additional 5 min at room temperature, protect from light.
13. Place 5 μl of P2 fraction (from **step 9**, Subheading 3.3) to a new tube
14. Add 25 μl of corresponded Zenon kit conjugated antibodies (from **step 12**, Subheading 3.3) or 25 μl of Immunostaining/Blocking Buffer (blank) to the tube, mix well and incubate for 30 min at room temperature, protect from light (*see* **Note 14**).
15. Add 1 ml of washing buffer to each reaction, mix well and centrifuge for 4 min at 4000 × *g* (4 °C).
16. Repeat the washing **step 2** times, protect from light.
17. Resuspend immunostained P2-fraction in washing buffer and transfer to FACS tubes, protected from light.

3.4 Flow Cytometry (FC) Analysis

Periodical checking and monitoring of cytometer performance (laser alignment, laser time delay etc.) is beyond the scope of this chapter.

1. Run size calibration standards, we use polystyrene beads (0.75, 1.5, and 4.5 μm) and adjust the forward scatter (FSC) and the

side scatter (SSC) detectors to place the populations of interest (synaptosome sizes are between 0.75 and 1.5 μm) on scale, adjust FSC threshold to exclude noise and apply FSC-based size gate (Fig. 2a).

2. Run controls: unstained samples (blanks), negative (isotype) controls, and positive controls (samples labeled with antibodies

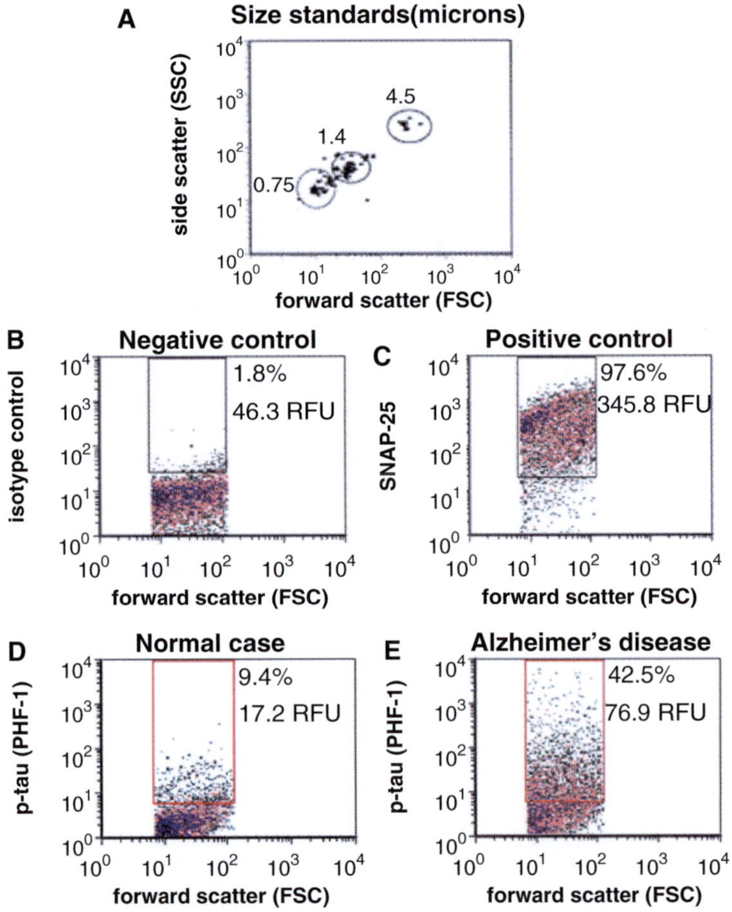

Fig. 2 Flow cytometry analysis of synaptosomal preparations. (**a**) An example of size standard acquisition for FSC-based size gate establishment. (**b**) Background labeling in presence of isotype control antibodies (*negative control*): only 1.8 % from the total 10,000 events collected within applied size gates show positive signal with average intensity 46.3 relative fluorescence units (RFU). (**c**) Presynaptic marker (SNAP25)-positive events within the size gate (*positive control*): 97.6 % positive events with average intensity 345.8 RFU. (**d**) An example of p-tau (Ser369/Ser404 phosphorylation-dependent epitopes accessed by PHF-1 antibody) signal in a synaptosomal fraction from normal human brain cortex: 9.4 % of p-tau positive events with average intensity 17.2 RFU. (**e**) An example of p-tau (PHF-1) signal in a synaptosomal preparation from brain of a subject with a late stage of Alzheimer's disease: 42.5 % of p-tau positive events with average intensity 76.9 RFU

against presynaptic markers; we routinely use SNAP25, synaptophysin or VGluT1 antibodies) and samples of interest to confirm and/or optimize the settings by adjusting voltage on corresponding detectors (Fig. 2b, c and **Note 15**).

3. Adjust fluorescent compensation by running single and multi-labeled samples, if multiple color data will be collected from each sample.
4. Collect from 5000 to 10,000 events within the size gate from each experimental sample.
5. Analyze the data (*see* Fig. 2d, e for an example).

4 Notes

1. EDTA and EGTA compounds, especially in the free acid form, are almost insoluble in water. Place magnet stirring bar to the EDTA (EGTA) water mixture and place the beaker on the magnet stirrer hot plate. Stir constantly, carefully titrate with sodium hydroxide (10 N NaOH) and continuously measure pH. Stop when pH reaches 8.0 and continue stirring until EDTA/EGTA is fully dissolved in water.
2. Sucrose solution without protease/phosphates inhibitors can be made in advance and either kept on ice or frozen at −20 °C for prolonged storage. Immediate cryopreservation of human brain autopsy samples is very important in order to minimize tissue degradation and obtain high quality synaptosomal fraction, thus all the possible advance preparation should be made to be able to start cryopreservation procedure as soon as the autopsy samples arrive to the laboratory.
3. To standardize PBS composition we routinely use commercially available PBS tablets.
4. Tween 20 (100%) is highly viscous liquid, in order to accurately measure 1 ml of the reagent cut the very tip portion of pipette tip (around 3–5 mm), fill the pipette slowly, after release tween 20 to water, wash pipette tip a couple of times by pipetting up and down to make sure that a whole 1 ml of tween 20 is added to the solution.
5. The ways to attach fluorescent tag to antibodies of interest is not limited to Zenon labeling technology; classical covalent primary antibody–fluorophore conjugates, secondary antibody conjugates etc. can be used. In our hands, Zenon kits perform extremely well: the labeling procedure is very fast (15 min), minimum amount of antibody is required for labeling reaction (starting from 1 μg), and Zenon reagents are isotype-specific which allows an additional flexibility for multicolor experiments. A limitation of Zenon kits, which needs to be taken into

account, is that Fab fragment/antibody interactions are not as strong as covalent bonds, therefore flow cytometry needs to be performed fairly soon after the immunostaining step.

6. Unfixed human tissue is considered hazardous material because it can potentially contain some infectious pathogens. Thus, always use personal protective equipment and work within a biosafety cabinet; it is especially important for the homogenization step when risk of aerosol formation is high. Aerosols can also be released during centrifugation steps, always use sealed centrifuge cups/rotors to protect yourself and environment.
7. Keeping good notes about tissue weight is important, because based on this information you will estimate amount of buffer for homogenization of tissue during P2 fraction preparation later on. P2 preparation may be performed in months, sometimes even years after original cryopreservation of brain tissue.
8. We use paper towels as insulation material, wrapping each tube into multiple layers of paper towels before placed in a Styrofoam box.
9. In addition to synaptosomal P2 fraction, we routinely save small portions of all other fractions (homogenate, P1, S1, and S2) to be able to monitor quality of P2 fraction by comparison of synaptic marker levels between those fraction by Western blotting. Moreover, the P1 fraction, with cell nucleus elements, can be used for genotyping.
10. Transferring of the P2 pellet to a new tube or a weigh boat to determine weight can lead to loss of material. To avoid this, we use set of centrifuge tubes with known mass and weigh P2 without transferring to another container.
11. The described double-cryopreservation protocol is not only adequate for maintenance of synaptosomal structures but also their functional integrity for functional assays, for example depolarization-induced tau release [17]. Based on our unpublished data, functional integrity of synaptosomes highly depends on postmortem interval (time between patient death and cryopreservation of brain tissue) but not so much on second cryopreservation of P2 fraction.
12. If P2 immunostaining and flow cytometry can't be performed on the same day, the fixed P2 fraction can be resuspended in 250 μl of PBS and kept at 4 °C overnight. Next day add 250 μl of 0.4 % Tween 20 in PBS to achieve the correct concentration of Tween 20 for permeabilization. Continue to follow the protocol from Subheading 3.3, **step 8**.
13. Usually more than one P2 sample needs to be labeled with antibodies in a single experiment. In this case the Antibody/Zenon kit conjugation reaction can be scaled up based on how

many individual staining reactions will be performed, and then the Antibody/Zenon kit conjugation mixture is divided between multiple immunostaining reactions (25 μl per one reaction).

14. In case of dual labeling experiments, the corresponding Zenon/Antibody conjugates can be added simultaneously (25 μl of each per reaction), but all volumes of blanks, isotype controls, and single labeled samples then need to be adjust to 50 μl.
15. If synaptosomal preparations are of a good quality, and immunostaining (Subheading 3.3) and FC adjustments (Subheading 3.4, **steps 1** and **2**) are performed correctly, more than 90 % of total event number within the size gate should be positive for the presynaptic markers, and less than 5 % should show nonspecific binding to isotype control antibodies.

References

1. Gouras GK, Tampellini D, Takahashi RH, Capetillo-Zarate E (2010) Intraneuronal beta-amyloid accumulation and synapse pathology in Alzheimer's disease. Acta Neuropathol 119:523–541. doi:10.1007/s00401-010-0679-9
2. Dunkley PR, Jarvie PE, Robinson PJ (2008) A rapid Percoll gradient procedure for preparation of synaptosomes. Nat Protoc 3:1718–1728. doi:10.1038/nprot.2008.171
3. Wolf ME, Kapatos G (1989) Flow cytometric analysis of rat striatal nerve terminals. J Neurosci 9:94–105
4. Woo J, Baumann A, Arguello V (2014) Recent advancements of flow cytometry: new applications in hematology and oncology. Expert Rev Mol Diagn 14:67–81. doi:10.1586/14737159.2014.862153
5. Wolf ME, Kapatos G (1989) Flow cytometric analysis and isolation of permeabilized dopamine nerve terminals from rat striatum. J Neurosci 9:106–114
6. Wolf ME, Kapatos G (1989) Stimulation of D2 dopamine receptors decreases intracellular calcium levels in rat anterior pituitary cells but not striatal synaptosomes: a flow cytometric study using indo-1. Synapse 4:353–370. doi:10.1002/syn.890040411
7. Gylys KH, Fein JA, Cole GM (2000) Quantitative characterization of crude synaptosomal fraction (P-2) components by flow cytometry. J Neurosci Res 61:186–192
8. Fein JA, Sokolow S, Miller CA, Vinters HV, Yang F, Cole GM, Gylys KH (2008) Co-localization of amyloid beta and tau pathology in Alzheimer's disease synaptosomes. Am J Pathol 172:1683–1692. doi:10.2353/ajpath.2008.070829
9. Gylys KH, Fein JA, Yang F, Cole GM (2004) Enrichment of presynaptic and postsynaptic markers by size-based gating analysis of synaptosome preparations from rat and human cortex. Cytometry A 60:90–96. doi:10.1002/cyto.a.20031
10. Gylys KH, Fein JA, Yang F, Wiley DJ, Miller CA, Cole GM (2004) Synaptic changes in Alzheimer's disease: increased amyloid-beta and gliosis in surviving terminals is accompanied by decreased PSD-95 fluorescence. Am J Pathol 165:1809–1817
11. Battista N, Bari M, Finazzi-Agrò A, Maccarrone M (2002) Anandamide uptake by synaptosomes from human, mouse and rat brain: inhibition by glutamine and glutamate. Lipids Health Dis 1:1
12. Kanekiyo T, Cirrito JR, Liu C-C, Shinohara M, Li J, Schuler DR, Shinohara M, Holtzman DM, Bu G (2013) Neuronal clearance of amyloid-β by endocytic receptor LRP1. J Neurosci 33:19276–19283. doi:10.1523/JNEUROSCI.3487-13.2013
13. Rausch JL, Johnson ME, Li J, Hutcheson J, Carr BM, Corley KM, Gowans AB, Smith J (2005) Serotonin transport kinetics correlated between human platelets and brain synaptosomes. Psychopharmacology (Berl) 180:391–398. doi:10.1007/s00213-005-2178-6
14. Dodd PR, Hardy JA, Baig EB, Kidd AM, Bird ED, Watson WE, Johnston GA (1986) Optimization of freezing, storage, and thawing conditions for the preparation of metabolically active synaptosomes from frozen rat and human brain. Neurochem Pathol 4:177–198
15. Jones PB, Adams KW, Rozkalne A, Spires-Jones TL, Hshieh TT, Hashimoto T, von

Armin CAF, Mielke M, Bacskai BJ, Hyman BT (2011) Apolipoprotein E: isoform specific differences in tertiary structure and interaction with amyloid-β in human Alzheimer brain. PLoS One 6, e14586. doi:10.1371/journal.pone.0014586

16. Sokolow S, Luu SH, Headley AJ, Hanson AY, Kim T, Miller CA, Vinters HV, Gylys KH (2011) High levels of synaptosomal Na(+)-Ca(2+) exchangers (NCX1, NCX2, NCX3) co-localized with amyloid-beta in human cerebral cortex affected by Alzheimer's disease. Cell Calcium 49:208–216. doi:10.1016/j.ceca.2010.12.008

17. Sokolow S, Henkins KM, Bilousova T, Gonzalez B, Vinters HV, Miller CA, Cornwell L, Poon WW, Gylys KH (2015) Pre-synaptic C-terminal truncated tau is released from cortical synapses in Alzheimer's disease. J Neurochem 133:368–379. doi:10.1111/jnc.12991

18. Begley JG, Butterfield DA, Keller JN, Koppal T, Drake J, Mattson MP (1998) Cryopreservation of rat cortical synaptosomes and analysis of glucose and glutamate transporter activities, and mitochondrial function. Brain Res Brain Res Protoc 3:76–82

19. Pankiewicz JE, Guridi M, Kim J, Asuni AA, Sanchez S, Sullivan PM, Holtzman DM, Sadowski MJ (2014) Blocking the apoE/Aβ interaction ameliorates Aβ-related pathology in APOE ε2 and ε4 targeted replacement Alzheimer model mice. Acta Neuropathol Commun 2:75. doi:10.1186/s40478-014-0075-0

Chapter 17

In Vivo Microdialysis of Brain Interstitial Fluid for the Determination of Extracellular Tau Levels

Kaoru Yamada

Abstract

Tau is a microtubule binding protein highly abundant in cytoplasm of neurons. However now there is compelling evidence that it is also physiologically released in the extracellular space. Tau is detectable in culture media of various cells, cerebrospinal fluid (CSF), and brain interstitial fluid (ISF). Being able to measure extracellular tau levels in vivo would deep our understanding on the underlying mechanisms of physiological/pathological tau release from neurons. To meet this need, this chapter describes an in vivo microdialysis protocol and materials for the measurement of extracellular tau in ISF from freely moving mice.

Key words Tau protein, Microdialysis, Extracellular tau, Alzheimer's disease, Tau secretion, ISF, CSF

1 Introduction

Although tau is predominantly present in the cytoplasm, it is also physiologically present in the extracellular space. The appreciable amount of tau is detectable in not only CSF of patients with Alzheimer's disease (AD) but also CSF of cognitively normal subjects without neurodegeneration. The several in vitro studies also detected tau in the media from immortalized cell lines overexpressing tau, primary neurons, and human iPS cell-derived cortical neurons [1–4]. The release of tau takes place without any signs of cell death, suggesting that the process occurs physiologically.

Microdialysis is a powerful in vivo technique to assess the extracellular levels of tau in ISF. Traditionally, it has been widely used to measure relatively small molecules such as neurotransmitters, amino acids, and compounds in neuroscience research for many years. However recent commercial availability of microdialysis probes with higher molecular cutoff membranes has now made it possible to detect various proteins related to neurodegeneration including tau in mice [5–9] and human as well [10].

Caroline Smet-Nocca (ed.), *Tau Protein: Methods and Protocols*, Methods in Molecular Biology, vol. 1523,
DOI 10.1007/978-1-4939-6598-4_17, © Springer Science+Business Media New York 2017

Microdialysis needs the stereotaxic implantation of probes in the brain. Once the probe is inserted, it is continuously perfused at relatively slow flow rates (0.1–5 μl/min) such that the perfusion buffer is in equilibrium with ISF. The molecules which are smaller enough to pass through microdialysis membranes diffuse from ISF into the perfusion buffer according to a concentration gradient and can be collected for further analysis. Using this technique, we have detected extracellular tau in ISF from both non-transgenic mice and human tau transgenic mice [5, 6, 11].

Microdialysis offers several advantages over other methods. First, while brain fractions obtained by tissue extraction will likely consist of tau from both intracellular and extracellular compartments in central nervous system (CNS), microdialysis specifically samples extracellular tau from ISF. Second, it does not involve the mechanical homogenization procedure, the in vivo structure/state of tau will be less compromised. Third, microdialysis enables a real-time measurement from a single living animal, allowing the comparison of tau levels before and after treatment. This reduces the number of animals used for the experiments as well as inter-animal variability in tau levels. Each animal has own intrinsic tau levels called "baseline." After baseline tau levels are established, pharmacological experiments can be started to assess the effect. Drugs can be administrated systemically or directly infused in the brain during microdialysis. Reverse microdialysis is one technique to deliver small compounds in brain. In this technique, small compounds in the perfusion buffer diffuse out through microdialysis membrane into ISF. Altered neuronal activity by picrotoxin or NMDA delivered by this method rapidly changes ISF tau [6], consistent with the observation in primary neuronal culture [2].

On the other hand, microdialysis is often limited by the recovery of molecules. Because the recovery of molecules into dialysate generally depends on multiple factors such as interaction with membrane materials or binding to other proteins in CNS, it should be taken into consideration that there might be tau species in ISF that are not efficiently recovered in microdialysis probes. Microdialysis samples will reflect local secretion and metabolism of tau surrounding probes at the particular brain region. This differs from another extracellular fluid, CSF, which will likely reflect changes of tau from entire CNS. Notably, there is no association of tau or Aβ levels between ISF and CSF [5, 12]. This may suggest that metabolism is different between ISF and CSF.

The basic lab setup for microdialysis includes microdialysis probes, free-moving system, pumps and refrigerated fraction collectors. This chapter chooses the hippocampus as a target brain region and describes the protocol to collect tau in ISF.

2 Materials

2.1 Chemicals and Solutions

1. Artificial cerebrospinal fluid (aCSF) is made from 20× aCSF. Make 20× aCSF for the storage: Dissolve 1.18 g of $MgSO_4$, 0.9 g of KCl, and 0.22 g of KH_2PO_4 in 200 ml of distilled water and stir the solution until completely dissolved. Dissolve 0.76 g of $CaCl_2$ and stir. The final solution for 20× aCSF will have the following concentration: 26 mM $CaCl_2$, 24 mM $MgSO_4$, 60 mM KCl, 8 mM KH_2PO_4. Filter the solution through 0.22 μm syringe filter (Millex). The solution can be stored for 1–2months at room temperature.
2. To make 1×aCSF, dissolve 1.05 g of $NaHCO_3$ and 3.56 g of NaCl in 475 ml of distilled water along with 25 ml of 20×aCSF and stir the solution. Adjust pH to 7.35 and filter the solution through 0.22 μm syringe filter (Millex). The solution can be stored at room temperature for 1–2 weeks. However the occasional check of pH will be encouraged. The final solution for 1× aCSF will have the following concentration: 1.3 mM $CaCl_2$, 1.2 mM $MgSO_4$, 3 mM KCl, 0.4 mM KH_2PO_4, 25 mM $NaHCO_3$, 122 mM NaCl.
3. Chlorate hydrate for anesthesia. Make 20 mg/ml solution by dissolving it in sterile saline and store it at room temperature. Alternative anesthetic drug such as isoflurane is also acceptable.
4. 30% Bovine serum albumin (BSA) solution (Sigma-Aldrich).
5. 70% ethanol.

2.2 Equipment and Accessories for Stereotaxic Surgery

1. Mice for stereotaxic surgery.
2. Disposable 1 ml syringe (sterile) and 27G needle for intraperitoneal injection of chloral hydrate.
3. Stereotaxic frame with manipulators (SAS-4120, ASI-instruments) equipped with a mouse adaptor.
4. Brain atlas for stereotaxic coordinates.
5. Guide cannula for microdialysis probe (PEG-8, shaft length 8 mm, Eicom).
6. Dummy cannula for temporal occupation for the guide cannula (PED-8: shaft length 8 mm, Eicom) (Reusable).
7. Stereotaxic adaptor (PESG-8: shaft length 8 mm, Eicom) (Reusable).
8. Cap nut (AC-5, Eicom) (Reusable).
9. Dental drill with clamp.
10. Dental cement.
11. Bone screws (BASi).

12. Screw driver for bone screws.
13. Electrode holder with clamp.
14. Scalpel or razor blade.
15. Surgical tweezers.
16. Cotton swabs.
17. Surgical clipper.

2.3 The Equipment and Accessories for Microdialysis

1. AtmosLM microdialysis probes with 1000 kDa cutoff membrane (PEP-8-02, shaft length 8 mm, membrane length (2) mm, Eicom).
2. Shaft length (X) mm of probe, guide, dummy and stereotaxic adaptor should be the same.
3. Disposable syringe filter unit, PVDF (pore size 0.1 μm).
4. Microdialysis syringe pump (KDS-101, KD scientific).
5. Microdialysis roller pump (ERP-10, Eicom).
6. Roller tubes for a roller pump (RT-5S, Eicom).
7. Refrigerated fraction collector (820 microsampler, Univentor).
8. Sample collection tubes for a refrigerated fraction collector.
9. Connection needle for roller tubes (RTJ, Eicom).
10. FEP tubing (JF-10-70, length 70 cm, Eicom).
11. Teflon tubing (JT-10-50, length 50 cm, Eicom), cut into half to generate two 25 cm tubings with byton tube on one end.
12. Cage for free moving (MD-1409, Ratturn microdialysis standalone system, BASi) (*see* **Note 1**).
13. Microdialysis dual species cage kit (CX-1600, BASi).
14. Mouse animal collar (BASi).
15. Intramedic luer-stub adaptor-23G (Beckton Dickinson).
16. Disposable 3 ml syringe (sterile).
17. Disposable 1 ml syringe (sterile).
18. Byton tube (Eicom).

3 Methods

3.1 Stereotaxic Surgery for Microdialysis Guide Implantation

1. Weigh the mouse and anesthetize by intraperitoneal injection of chloral hydrate at 400 mg/kg. Pay attention to whether the mouse breathes normally during the surgical procedure. If anesthesia is not deep enough, the additional dose of chloral hydrate up to 20% of the initial dose can be injected.
2. Shave the hair of mice head with a surgical clipper starting from in-between ears to in-between eyes.

3. Fix the mice head on the stereotaxic frame. Fix the mouse using nose/tooth clamp and then tighten the head with the ear bars. Make sure that the head is stable and no longer moves side by side.
4. Make an incision on the skin vertically using a scalpel.
5. Wipe off the connective tissue and blood on the skull using a cotton swab moistened with aCSF.
6. Determine the coordinate using Paxinos brain atlas of mouse. The coordinate (A/P: −3.1 mm, M/L: −2.5 mm, D/V: −1.2 mm, 12°) has been used to target the posterior hippocampus for microdialysis probe with 2 mm length [5, 6] (*see* **Note 2**). If one wants to choose a different brain region, determine the coordinate and probe length by brain atlas for mouse.
7. Mount a dental drill on the manipulator with clamp. Make sure that the drill is turned off. Before starting measurements, make sure the midline is straight by moving the dental drill vertically along the midline. If the midline is not straight, loose the ear bars and fix the mouse again.
8. Identify the lambda. Position the dental drill above the lambda and lower it until the drill bit gently touches the skull and record the ventral coordinate.
9. Move the drill vertically from the lambda and position it above the bregma. Lower the drill until the drill bit gently touches the skull and record the ventral coordinate. When the mouse head is leveled, the ventral coordinates of bregma and lambda are equal. If not, adjust the height of nose/tooth clamp accordingly. After the head gets leveled, record the anterior/posterior, lateral coordinate of the bregma.
10. Move the drill vertically 3.1 mm posterior from bregma (A/P: −3.1 mm, M/L: 0 mm). Then move the drill 2.0 mm lateral (right) from midline and lower the drill until the drill bit gently touches the skull and record the vertical coordinate. Move the drill 2.0 mm lateral from midline to the opposite side (left) and lower the drill until the drill bit gently touches the skull and record the vertical coordinate. If the skull is leveled, these vertical measurements of two equidistant points from midline will be equal. If not, adjust the height of ear bar on one side accordingly.
11. Position the drill above the desired coordinate (A/P: −3.1 mm, M/L: −2.5 mm). Turn on the drill and carefully drill a hole on the skull for the guide implantation. If a hole is not large enough for a guide cannula, make the second hole overlapping with the first one. Drill another hole on the right parietal bone and insert a bone screw using a screw driver.
12. Make a "crown" by cutting a circular locking piece from the lid of 1.5 ml centrifuge tube. Place it on the skull so that both

the hole for the guide and the bone screw are within the circle (*see* Fig. 1). This crown will prevent spreading of dental cement to the skin.

13. Set up an assembly for a guide implantation: Set a guide cannula on the shorter arm of a stereotaxic adaptor and fix it with a cap nut. Attach the long arm of the adaptor on the electrode holder and mount it on the manipulator of the stereotaxic frame.
14. Make 12° angle by rotating the manipulator (*see* Fig. 1).
15. Move the guide cannula to the hole. Insert it slowly into the brain by lowering the guide cannula for 1.2 mm.
16. Apply the dental cement to cover the base of the guide cannula and the bone screw enough to secure them. The surface of the skull should be dried by that time. If there is a part of skull exposed, apply some dental cement to cover it.
17. Wait until the dental cement is completely dried and firm. Remove the electrode holder from the stereotaxic adaptor. Remove the cap nut and pull out the stereotaxic adaptor from the guide cannula and replace it with the dummy probe and fasten the cap nut.
18. Release the mouse from the stereotaxic frame and house the mouse individually with food and water until the day of microdialysis.

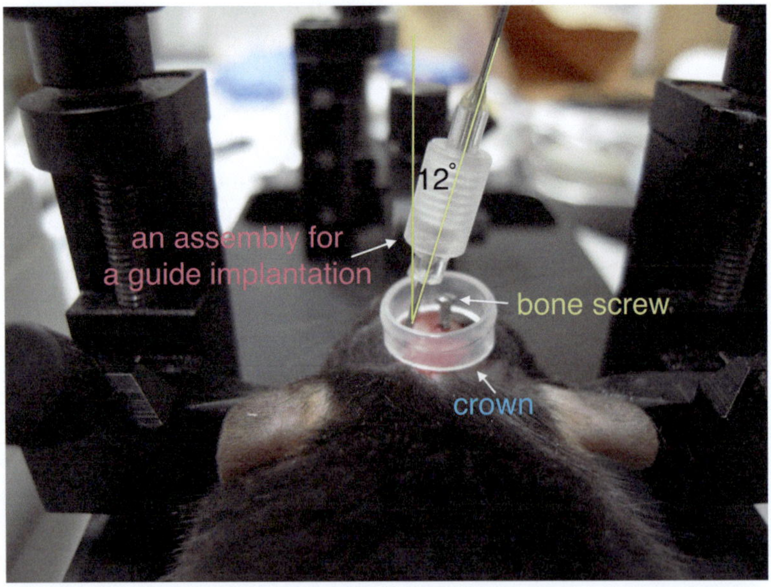

Fig. 1 Stereotaxic surgery for a guide cannula implantation. An assembly for the guide cannula implantation, which consists of a guide cannula, a cap nut, and a stereotaxic adaptor is inserted with 12°

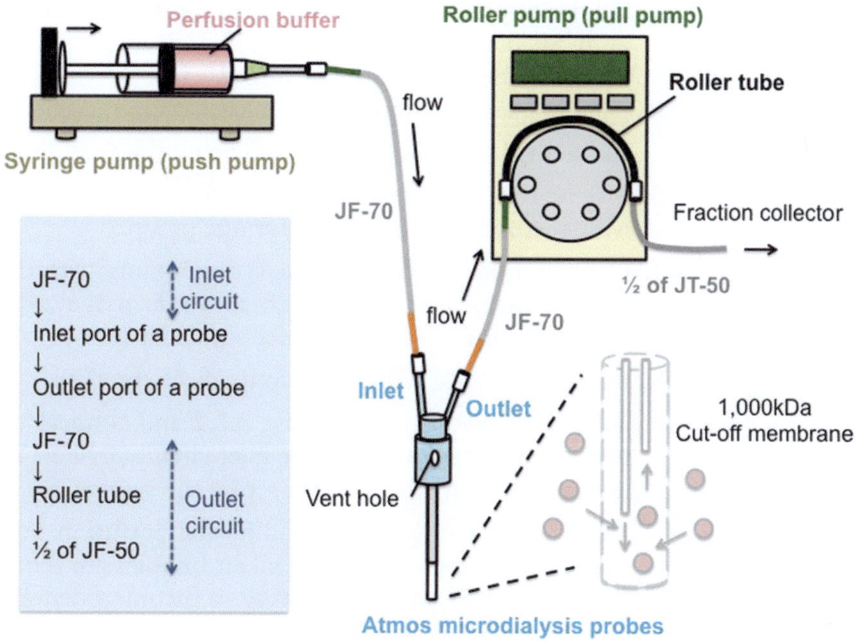

Fig. 2 The illustration of the microdialysis circuit construction. The order of tubing connections is shown in the *inset*

3.2 In Vivo Microdialysis in Freely Moving Mice

It has been recognized that the larger cutoff membranes become more susceptible to fluid loss through membranes due to accumulating internal pressure created by perfusion buffer entering the probes.

To avoid this issue, Atmos microdialysis probes with 1000 kDa cutoff membranes from Eicom microdialysis have a unique pressure cancelation system. There is a vent hole at the head of the probe, which circumvents back-pressure by equalizing the inner pressure with outside atmospheric pressure. Because the pressure will not be built up in these probes because of this structure, the system has to be operated by a push-pull mode. Although a syringe pump and a peristaltic pump are used as a push and a pull pump respectively in this chapter, a peristaltic pump can be used as both push and pull pumps if it has more than two channels.

3.2.1 Prepare the Perfusion Buffer

1. Dilute 30% BSA solution to 4% with 1×aCSF on the day of microdialysis.
2. Filter the solution through 0.1 μm syringe filter unit (*see* **Note 3**).

3.2.2 Construction of the Circuit for Microdialysis

1. Make connections of tubings in the order indicated in the inset of Fig. 2. At this point, use the connection needle instead of a probe for the connection between two JF-70 tubings (*see* **Note 4**).

3.2.3 Flushing the Circuit with Perfusion Buffer

1. Fill the disposable 3 ml syringe with perfusion buffer (4 % BSA in aCSF made in Subheading 3.2.1, **step 2**) and connect it with intramedic luer-stub adaptor.
2. Fill the perfusion buffer in the intramedic Luer-stub adaptor and get rid of air bubbles.
3. Mount this 3 ml syringe/intramedic Luer-stub adaptor filled with perfusion buffer on the syringe pump.
4. Contact a pusher block of the syringe pump with the plunger and manually press it enough to confirm that the solution comes out from intramedic luer-stub adaptor.
5. Fast-forward the syringe pump to check the flow.
6. Take the circuit in Subheading 3.2.2 and connect one end of FEP tubing (JF-10-70) of the inlet circuit (*see* Fig. 2) with the intramedic Luer-stub adaptor set on the syringe pump.
7. Starting the syringe pump and fill the perfusion buffer in the entire circuit. Make sure that all air bubbles are removed from the circuit before the connecting to the microdialysis probe.

3.2.4 Quality-Check of Probes: Prepare the Microdialysis Probe

1. Confirm that there is no leak of solution from microdialysis probe by performing quality-check procedure by following the manufacturer's instruction.
2. Fill a disposable 1 ml syringe with distilled water and connect it to the outlet of an Atmos probe using a byton tube.
3. Cover the vent holes with the fingers and gently infuse distilled water to the probe.
4. Confirm that water appears from the inlet and there is no leakage around microdialysis membrane.

3.2.5 Activation of Probes

1. Submerge the microdialysis membranes in 70 % ethanol for 1 s.
2. Infuse water to the probe with a syringe again.

3.2.6 Connecting a Probe to the Circuit

1. Once the circuit is filled with the perfusion buffer, stop the syringe pump.
2. Replace the connection needle with the activated microdialysis probe by the following procedure:
3. Put the cap nut on the probe so that both inlet/outlet ports stick out.
4. Remove the connection needle and connect the free end of JF-70 tubing in the outlet circuit to the outlet port of microdialysis probe.
5. Connect the free end of JF-70 tubing of inlet circuit to the inlet port of the microdialysis probe.

3.2.7 Start the Perfusion

1. Mount the roller tube in the outlet circuit by placing it on the tubing guide of the roller pump (*see* **Note 5**).
2. Pull the knob of the roller pump so that roller tube is in contact with the roller and secured.
3. Start the syringe pump at 10 μl/min and then start the roller pump at 9.5–9.8 μl/min (*see* **Note 6**). If air bubbles are generated, make sure to get rid of them by the time of the probe insertion.

3.2.8 Insertion of a Probe

1. Anesthetize the mouse implanted with the guide cannula and transfer to a microdialysis cage.
2. Apply the mouse collar around the neck.
3. Loosen the cap nut and remove the dummy probe.
4. Take off an Atmos probe from the protective sleeve and slowly insert it into the brain through the hole of the guide cannula.
5. Tighten the cap nut.
6. Tether the mouse with the collar to Raturn systems and turn the system on.
7. Keep running both the syringe pump and the roller pump for at least 1 h.
8. Make sure the tubing will not interfere with the sensor of Raturn systems.

3.2.9 Start Sampling ISF

1. Stop the roller pump and then the syringe pump.
2. Set the flow rate of syringe pump 20% faster than the roller pump. For example, if the desired flow rate is 1.0 μl/min, operate the syringe pump at 1.2 μl/min, the roller pump at 1.0 μl/min.
3. Start the syringe pump first and then the roller pump.
4. Put sample collection tubes in the fraction collector and set up the free end of ½ of JT-50 tubing in the outlet circuit on the refrigerated fraction collector and start the sample collection. The fraction collector should be refrigerated in advance. *See* Fig. 3 for an example of lab setup for microdialysis experiments.
5. House the mouse with water and food during the experiment.

3.2.10 Analyze ISF Samples to Determine Tau Concentration

1. Tau concentration in ISF can be measured by ELISA (*see* **Notes 7** and **8**).

3.3 Calculation of Actual Concentration of Tau

ISF collection by microdialysis needs a continuous flow of the perfusion buffer. Therefore there is not enough time for ISF tau to reach the perfect equilibrium with the perfusion buffer. As a consequence, tau concentration in the dialysate is only a fraction of tau

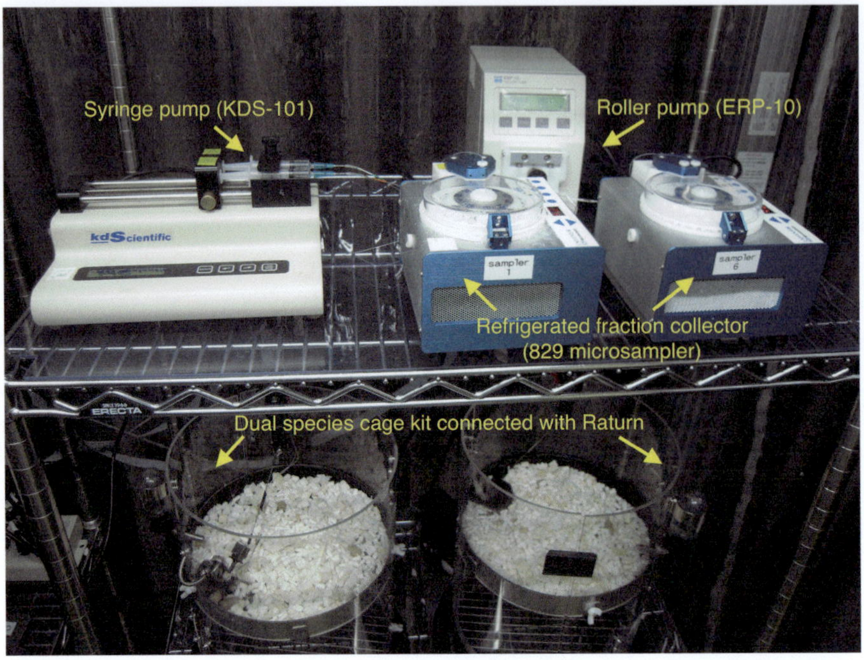

Fig. 3 An example of lab setup for microdialysis. The free-moving system (Raturn), a syringe pump, a peristaltic pump, and refrigerated fraction collectors are shown

that is actually present in ISF. To estimate the actual concentration of exchangeable tau in ISF, a zero flow method is often applied [5]. This method determines tau concentration by varying the flow rates. There is an inverse relationship between the flow rate and tau concentration. This correlation can be plotted in *X*–*Y* scatter plot with an exponential curve. The theoretical zero flow can be obtained by extrapolating the curve back to 0 flow rate, which represents the perfect recovery of tau protein. Since the microdialysis only allows the collection of extracellular tau that is small enough to pass through microdialysis membranes, note that the tau concentration at the theoretical zero flow will represent the concentration of "exchangeable tau" not "total extracellular tau".

4 Notes

1. Raturn systems equip the sensor that turns the cage in an opposite direction to the mouse movement to prevent microdialysis circuits from being twisted. Liquid Swivel with tether (Eicom) can be alternatively used for free-moving instead of Raturn systems.
2. Verification of correct probe placement will be encouraged by histology such as cresyl violet staining.

3. The addition of BSA at 4% significantly improves the recovery of tau protein compared to 0.15% BSA (the author's unpublished observation).
4. Microdialysis circuits are reusable when the maintenance is appropriately performed. Fill both lines with 1% liquid bleach and let it sit for 30 min. And flush both circuits extensively with distilled water.
5. To extend the life-time of a roller tube, remove it from the peristaltic pump so that it will not be compressed when the roller pump is not being used.
6. It is highly recommended to calibrate the peristaltic pump speed beforehand by measuring the volume of liquid pumped for certain length of time. Peristaltic pump speed is proportional to the flow rate.
7. Tau concentration right after microdialysis starts tends to be unstable probably due to acute local injury to the surrounding tissue caused by the probe insertion. To obtain stable baseline, it is recommended to exclude at least the first 9-h samples from analysis. In addition to such acute injury, a probe implantation could also have chronic effects on functional or metabolic status in the brain [13]. Thus the typical microdialysis experimental window for tau collection is set from 9 to 48 h from implantation.
8. There are cases where the recovery of tau declines for various reasons such as clogging of the membrane. In order to ensure the consistency of probe function, the simultaneous measurement of internal control substances such as glucose, lactate, or urea will be encouraged [6, 14].

Acknowledgments

The author thanks Dr. David M. Holtzman and Dr. John R. Cirrito for technical suggestions and helpful discussion on the initial tau microdialysis study. This work was supported in part by "Grant-in-Aid for Scientific Research on Innovative Areas (Brain Protein Aging and Dementia Control) (15H01552) from MEXT".

References

1. Karch CM, Jeng AT, Goate AM (2012) Extracellular Tau levels are influenced by variability in Tau that is associated with tauopathies. J Biol Chem 287:42751–42762
2. Pooler AM, Phillips EC, Lau DHW, Noble W, Hanger DP (2013) Physiological release of endogenous tau is stimulated by neuronal activity. EMBO Rep 14:389–394
3. Chai X, Dage JL, Citron M (2012) Constitutive secretion of tau protein by an unconventional mechanism. Neurobiol Dis 48:356–366
4. Bright J, Hussain S, Dang V, Wright S, Cooper B, Byun T, Ramos C, Singh A, Parry G, Stagliano N, Griswold-Prenner I (2015) Human secreted tau increases amyloid-beta production. Neurobiol Aging 36:693–709

5. Yamada K, Cirrito JR, Stewart FR, Jiang H, Finn MB, Holmes BB, Binder LI, Mandelkow E-M, Diamond MI, Lee VM-Y, Holtzman DM (2011) In vivo microdialysis reveals age-dependent decrease of brain interstitial fluid tau levels in P301S human tau transgenic mice. J Neurosci 31:13110–13117
6. Yamada K, Holth JK, Liao F, Stewart FR, Mahan TE, Jiang H, Cirrito JR, Patel TK, Hochgrafe K, Mandelkow E-M, Holtzman DM (2014) Neuronal activity regulates extracellular tau in vivo. J Exp Med 211: 387–393
7. Emmanouilidou E, Elenis D, Papasilekas T, Stranjalis G, Gerozissis K, Ioannou PC, Vekrellis K (2011) Assessment of α-synuclein secretion in mouse and human brain parenchyma. PLoS One 6, e22225
8. Ulrich JD, Burchett JM, Restivo JL, Schuler DR, Verghese PB, Mahan TE, Landreth GE, Castellano JM, Jiang H, Cirrito JR, Holtzman DM (2013) In vivo measurement of apolipoprotein E from the brain interstitial fluid using microdialysis. Mol Neurodegener 8:13
9. Takeda S, Hashimoto T, Roe AD, Hori Y, Spires-Jones TL, Hyman BT (2013) Brain interstitial oligomeric amyloid β increases with age and is resistant to clearance from brain in a mouse model of Alzheimer's disease. FASEB J 27:3239–3248
10. Magnoni S, Esparza TJ, Conte V, Carbonara M, Carrabba G, Holtzman DM, Zipfel GJ, Stocchetti N, Brody DL (2012) Tau elevations in the brain extracellular space correlate with reduced amyloid-β levels and predict adverse clinical outcomes after severe traumatic brain injury. Brain 135:1268–1280
11. DeVos SL, Goncharoff DK, Chen G, Kebodeaux CS, Yamada K, Stewart FR, Schuler DR, Maloney SE, Wozniak DF, Rigo F, Bennett CF, Cirrito JR, Holtzman DM, Miller TM (2013) Antisense reduction of tau in adult mice protects against seizures. J Neurosci 33:12887–12897
12. Cirrito JR, May PC, O'Dell MA, Taylor JW, Parsadanian M, Cramer JW, Audia JE, Nissen JS, Bales KR, Paul SM, DeMattos RB, Holtzman DM (2003) In vivo assessment of brain interstitial fluid with microdialysis reveals plaque-associated changes in amyloid-beta metabolism and half-life. J Neurosci 23:8844–8853
13. Benvenisten H, Hüttemeiern PC (1990) Microdialysis—theory and application. Prog Neurobiol 35:195–215
14. Schwetye KE, Cirrito JR, Esparza TJ, Mac Donald CL, Holtzman DM, Brody DL (2010) Traumatic brain injury reduces soluble extracellular amyloid-β in mice: a methodologically novel combined microdialysis-controlled cortical impact study. Neurobiol Dis 40:555–564

Chapter 18

Proximity Ligation Assay: A Tool to Study Endogenous Interactions Between Tau and Its Neuronal Partners

Alexis Bretteville, Florie Demiautte, and Julien Chapuis

Abstract

Tau is a microtubule associated protein (MAP) that is expressed in neurons of the central nervous system. Tau proteins are deregulated in a group of pathologies, including Alzheimer's disease, commonly called tauopathies. Therefore intensive research has been conducted to understand both the regulation of Tau and its involvement in neuronal cellular pathways. Since its originally described interactor tubulin, Tau has been described to interact with several other proteins, including tyrosine kinases (Src, Fyn, Lck) and Phospholipase C-γ. In this chapter, we describe the use of proximity ligation assay as a versatile method to study the endogenous interaction of Tau with these different neuronal partners and use the recently identified Tau interactor Bin1 as case study.

Key words Protein–protein interaction, Protein complexes, Alzheimer's disease, Neurodegeneration

1 Introduction

Tau is a microtubule associated protein (MAP) that is expressed as six isoforms in neurons of the central nervous system. Polymerization and stabilization of microtubules are its main described functions [1, 2], which are mediated through interaction of Tau with tubulin via the binding of three or four microtubule binding domains, depending on the Tau isoform considered (for review [3]). Recently though, new Tau functions have been suggested since Tau has been described to interact with other partners such as tyrosine kinases (Src, Fyn, Lck) [4], Phospholipase C-γ [5–7], actin and the bridging integrator 1 (Bin1) [8, 9]. While Tau interactions with other proteins have been identified, their cellular roles as well as their regulations are far less elucidated. However, Tau phosphorylation, which is the main Tau posttranslational modification, has been implicated in the regulation of Tau interaction with other partners and is highly deregulated during the course of neurodegeneration in AD [10].

Caroline Smet-Nocca (ed.), *Tau Protein: Methods and Protocols*, Methods in Molecular Biology, vol. 1523,
DOI 10.1007/978-1-4939-6598-4_18,

Protein–protein interaction studies are rather challenging in a cellular context and usually involve both molecular engineering to fuse fluorescent tags to the proteins of interest and/or particular imaging techniques, such as fluorescence resonance energy transfer (FRET). One bias of these methodologies is that it requires the administration of exogenous constructs usually expressed at high levels in the cell, making it impossible to study endogenous protein interactions at physiological expression levels. On the other hand, immunofluorescent staining combined with confocal imaging analyses are widely used to assess protein co-localizations as a reflection of potential interactions of proteins at endogenous levels. However, due to the high expression levels of Tau in neurons, concluding that Tau interacts with another protein based on co-localization with conventional confocal approaches is rather difficult. Therefore, further confirmations, obtained with immunogold electronic microscopy studies for example, are required. To override these limitations, proximity ligation assay (PLA) has recently been developed as a method that allows assessment of the presence of two proteins in close proximity with a resolution up to 28 nm, without requiring tenuous protein engineering or particular imaging techniques (*see* **Note 1**). Indeed, PLA is a methodology that uses a particular set of secondary antibodies coupled to oligonucleotide probes that, if close together, could be connected by the addition of small linear DNA strands. This process leads to the formation of a circular DNA which is then amplified by rolling cycle amplification (RCA) and, finally, stained with fluorescent complementary DNA probes (Fig. 1) [11]. We recently successfully used this methodology to study the interaction of Tau with Bin1 as well as its subsequent regulation by phosphorylation (*see* **Note 2**) [9]. Here, we review the steps used to positively identify Tau neuronal interactors using Bin1 as a study example.

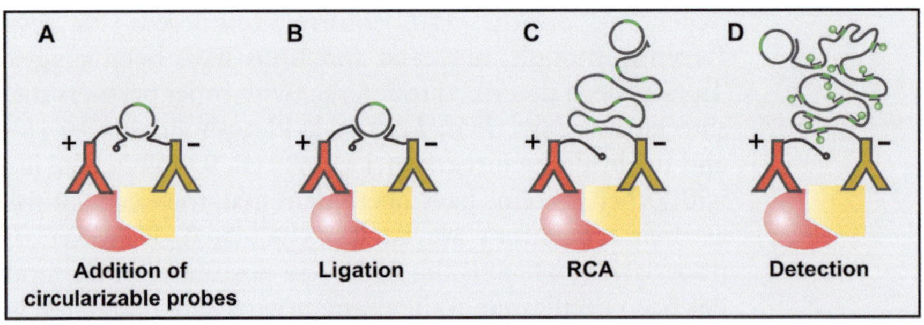

Fig. 1 Schematic representation of proximity ligation assay principle theory. (**a**) If two target proteins interact, the two specific proximity probes (+ and –), are close enough to form an open circular structure when linear connector oligonucleotides are added. (**b**) Those probes are then covalently joined by enzymatic DNA ligation to close the circular DNA structure. (**c**) The rolling cycle amplification is initiated using one of the proximity probes as a primer to amplify a tag sequence (*green*). (**d**) This tag sequence is secondly detected by hybridation of complementary fluorescence-labeled oligonucleotides

2 Materials

All solutions are prepared with ultrapure water (Purelab Ultra DV35, 18 MΩ). The reagent dilution buffer used is 1× phosphate buffer saline (PBS), pH 7.4 without calcium and magnesium: add 1 volume of a 10× PBS stock solution (Thermofisher, Life Technologies) to nine volumes of water.

2.1 Neuronal Coverslip Preparation

1. Rat primary neurons are obtained from P0 new born rats and cultured for at least 14 days in vitro (DIV) on borosilicate coverslips (Marienfeld, 18 mm N°1) coated with 0.01% (w/v) poly-D-lysine in a 12-well plate, as previously described [9].
2. Neuron fixation solution: add 0.25 volumes of 16% paraformaldehyde (EMS) to 0.1 volumes of 10× PBS and complete with 0.65 volumes of ultrapure water to prepare a fixating solution of 4% para-formaldehyde.
3. Neuron permeabilization solution: add 0.25 volumes of Triton X-100 solution (Sigma-Aldrich) to 100 volumes of 1× PBS.
4. Neuron immunofluorescence blocking solution: add 2.5 g of bovine serum albumin (BSA) (Interchim) to 100 ml of 1× PBS to prepare a 2.5% (w/v) BSA solution in PBS.
5. Neuron immunofluorescence blocking solution is used for primary antibody dilution.
6. 1× PBS solution is used as washing buffer.

2.2 Proximity Ligation Assay: Antibodies and Staining Solutions

1. PLA kit: the PLA kit is developed by O-link solution (Sigma-Aldrich) and consists of Duolink® In Situ PLA® Probe Anti-Mouse MINUS, Duolink® In Situ PLA® Probe Anti-Rabbit PLUS, and Duolink® In Situ Detection Reagents Green. Probes are stored at +4 °C and the detection kit is kept at −20 °C and protected from light.
2. Primary antibodies (*see* **Note 3**): Polyclonal rabbit anti-Tau antibody (Total Tau; purchased from Dako) is used at a final concentration of 0.0155 g/l. Monoclonal mouse anti-Bin1 99D antibody (Bin1; purchased from Millipore) is used at a final concentration of 0.02 g/l. Polyclonal rabbit anti-Bin1 SH3 (Bin1-SH3; purchased from Abcam) is used at a final concentration of 0.0025 g/l. Monoclonal anti-phospho-Tau pS396-404 (PHF1), pS202 (CP-13), pT231 (RZ-3) antibodies are a kind gift from Peter Davies and are used at a final 1/1000 dilution from stocks. Commercial antibodies are stored as recommended by the manufacturer and pTau antibodies are kept at +4 °C.
3. Secondary antibodies: Polyclonal goat anti-mouse and anti-rabbit IgGs (H + L) Alexa Fluor® 555 antibodies (Thermofisher, Life Technologies) are stored at +4 °C, as recommended by the manufacturer and are used at a 1/400 dilution from stock.

4. Nucleus staining solution: add one volume of Hoechst 33258 stock solution (Thermofisher, Life Technologies) to 10,000 volumes 1× PBS, to prepare a 1× Hoechst staining solution (1 μg/ml).

2.3 Picture Acquisition

1. Fluoromount aqueous mounting solution (Sigma-Aldrich) is used to mount coverslips on Menzel-Gläser superfrost plus microscope slides (Thermofisher).
2. Pictures are acquired with a confocal microscope (LSM 710, Zeiss) using an oil-immersed 40× objective and processed using ZEN 2012 software (Zeiss).

3 Methods

3.1 Day 1

1. Remove the culture medium by aspiration and wash the cells twice with 1 ml of 1× PBS pre-warmed at +37 °C.
2. Fix the cells on coverslips (diameter: 18 mm) by adding 500 μl/well of freshly prepared Neuron fixation solution and incubate for 20 min at room temperature (*see* **Note 4**).
3. Wash the cells 2 × 5 min with 1× PBS at room temperature.
4. Permeabilize the cells by adding 500 μl/well of neuron permeabilization solution and incubate for 10 min at room temperature.
5. Wash the cells 2 × 5 min with 1× PBS.
6. Block samples by adding 500 μl/well of a freshly prepared blocking solution and incubate coverslips for at least 30 min at room temperature.
7. For each coverslip, prepare 40 μl primary antibody solution at the appropriate dilution.
8. Prepare a humidity chamber using a box containing a paper towel saturated with water over-topped with a Parafilm layer (*see* **Note 5**).
9. Drop primary antibody solution onto the Parafilm layer.
10. Pick up each coverslip and remove the excess of blocking solution by tapping the coverslip edges on a filter paper.
11. Flip over the coverslip (cell side down) onto the appropriate primary antibody solution.
12. Place the humidity chamber for overnight incubation at +4 °C without agitation.

3.2 Day 2

1. Wash the cells 3 × 5 min with 1× PBS.
2. For each coverslip, prepare 40 μl PLA probes according to manufacturer's instructions.

3. Prepare a clean humidity chamber by changing the Parafilm layer.
4. Drop the previously prepared PLA probe solution onto the Parafilm layer.
5. Flip over the coverslip (cell side down) onto the PLA probe solution.
6. Place the humidity chamber at +37 °C for a 1 h incubation.
7. Wash the cells 3 × 5 min with 1× PBS.
8. For each coverslip, prepare 40 μl ligation solution according to manufacturer's instructions.
9. Prepare a clean humidity chamber by changing the Parafilm layer.
10. Drop 40 μl of previously prepared ligation solution onto the Parafilm layer.
11. Flip over the coverslip (cell side down) onto the ligation solution.
12. Place the humidity chamber at +37 °C for a 30 min incubation.
13. Wash the cells 3 × 5 min with 1× PBS.
14. For each coverslip, prepare 40 μl amplification solution according to manufacturer's instructions.
15. Prepare a clean humidity chamber by changing the Parafilm layer.
16. Drop 40 μl amplification solution onto the Parafilm layer. Protect from light.
17. Flip over the coverslip (cell side down) onto the amplification solution.
18. Place the humidity chamber at +37 °C for a 100 min incubation. Protect from light.
19. Wash the cells 3 × 5 min with 1× PBS. Protect from light.

For PLA co-staining with conventional immunofluorescence (Fig. 2d), perform **steps 20–25** otherwise go directly to **step 26** (*see* **Notes 6** and 7).

20. For each coverslip, prepare 40 μl of the appropriate secondary fluorescent antibody solution. Protect from light.
21. Prepare a clean humidity chamber by changing the Parafilm layer.
22. Drop the previously prepared secondary antibody solution onto the Parafilm layer.
23. Flip over the coverslip (cell side down) on the secondary antibody solution.
24. Place the humidity chamber at room temperature for a 45 min incubation. Protect from light.
25. Wash the cells 3 × 5 min with 1× PBS. Protect from light.

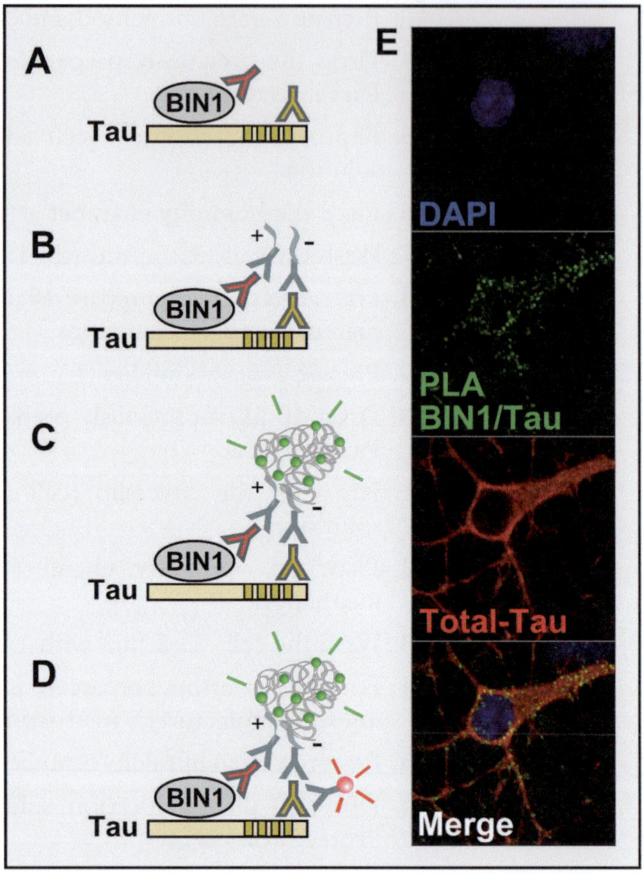

Fig. 2 Example of Proximity Ligation Assay experiment applied to detect interaction between Tau and Bin1. (**a**) Primary antibodies against Bin1 and Tau are used for staining. (**b**) Specific proximity (+ and –) probes are added to detect primary antibodies. (**c**) The Bin1-Tau complexes are tagged by fluorescence labeled oligonucleotides after RCA reaction. (**d**) A complementary Alexa Fluor-secondary antibody is finally added to visualize Total-Tau in neurons. (**e**) Representative PLA (*green*) and Total Tau (*red*) staining in primary neuron cultures. Note that PLA staining appears as puncta corresponding to the putative Bin1-Tau complexes

26. Counterstain coverslips in a 1× Hoechst solution and incubate at room temperature for 5 min. Protect from light.
27. Wash the cells 3 × 5 min with 1× PBS. Protect from light.
28. Mount coverslips with 20 μl Fluoromount: Drop 10 μl Fluoromount on both the coverslip (side with cells) and on the microscope slide. Flip the coverslip over onto the Fluoromount drop on the microscope slide.
29. Air dry 30 min at room temperature. Protect from light.
30. Store slides at +4 °C, in the dark, until image acquisition (*see* **Note 3**).

4 Notes

1. PLA staining using Duolink secondary probes, as presented in this protocol, has a resolution of 28 nm at best. This latter corresponds to the use of two primary antibodies and two secondary antibodies with an average length of 7 nm each. This resolution can be increased by using Duolink Kits which allow the fusion of oligo-nucleotide probes directly on the primary antibodies of interest, thus providing a 14 nm resolution. Nonetheless, even at the best resolution, PLA enables studying the presence of two proteins in the same complex, but assessing the existence of a direct or indirect interaction between these proteins will require additional in vitro methodologies such as glutathione S-transferase pull-down or nuclear magnetic resonance studies.
2. Even if the PLA staining can be used as a quantitative method, it is highly dependent on the nature of the interaction. Indeed, analyses of fine-tuned biologically labile interactions are rather difficult in fixed material and require live imaging. Therefore, only drastic interaction modulations, such as the effect of tau phosphorylation on BIN1–Tau interaction, can be robustly quantified (as presented in Fig. 3b).
3. Primary antibody choice is a fundamental aspect for PLA to assure biologically relevant results. Therefore, prior to PLA, choose your primary probes carefully and establish the appropriate dilutions by performing conventional immunofluorescences, to guaranty both sensitivity and specificity. Apply the optimized primary antibody conditions directly to PLA.
4. We have observed variations of PLA staining depending on the fixation solutions used. Therefore, precautions should always be taken, when modifying fixation protocols for any technical reason.
5. As a more technical note, due to the mandatory +37 °C incubation for PLA, the use of a well humidified humidity chamber during the whole experiment is a critical point to prevent coverslip dehydration. Also, make sure that the Parafilm layer is completely flat to obtain a homogeneous incubation.
6. The association of PLA with conventional staining on the same sample, as presented in Fig. 2b, allows simultaneous observation of Tau interacting with its partners and total-Tau. However, it has to be noted that the PLA proximity probes and Alexa Fluor® secondary antibodies can potentially compete with each other for the binding on the same primary antibody. Therefore, performing PLA with and without conventional immunofluorescence is needed and must be compared to avoid possible interferences.

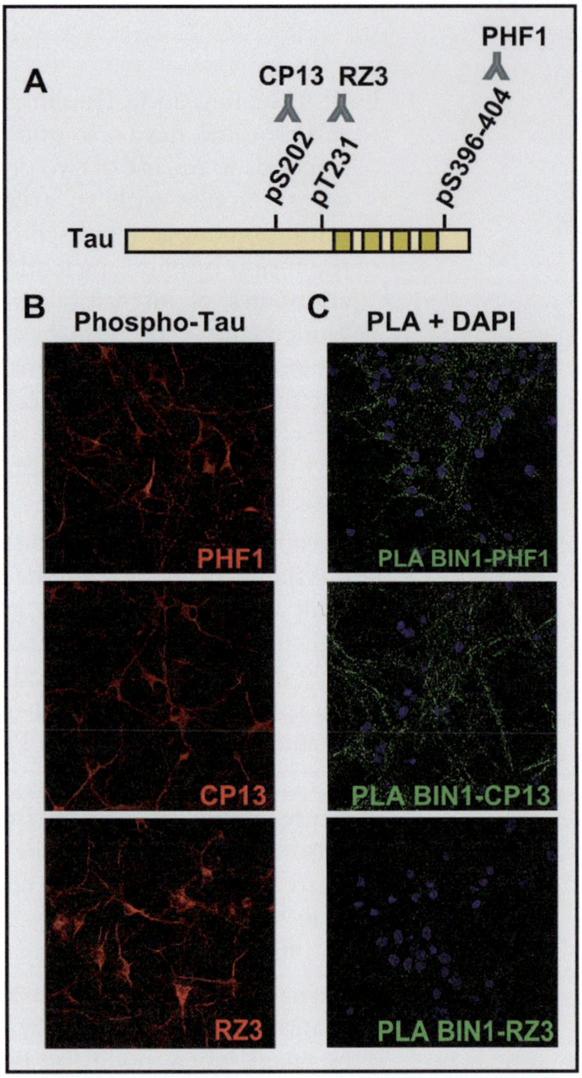

Fig. 3 Proximity Ligation Assay experiment to assess the impact of Tau phosphorylation on its interaction with Bin1. (**a**) Schematic representation of antibodies used against different Tau phosphorylated sites. (**b**) Conventional immunofluorescence staining for phospho-Tau (*red*) obtained with the same antibodies as used in PLA (**c**). (**c**) PLA combining an anti-Bin1 and three anti-phospho Tau antibodies. Note that the PLA signal is not observed with antibody RZ3 (targeting pThr231) suggesting that strong modulations of Tau–Bin1 interaction [9] could be quantified by PLA

7. As a general recommendation, and particularly in the establishment of the first PLA staining, one should care to perform appropriate negative controls by either removing one or the other secondary probe and by performing PLA staining without primary antibodies. Positive controls such as PLA staining between Tau and tubulin can be added as well [9].

Acknowledgments

We thank the BioImaging Center Lille-Nord de France (BICeL) facility (Lille, France). We also thank the Albert Einstein College of Medicine of Yeshiva University (New York, USA) for the kind gift of Tau-phosphorylation-dependent antibodies. This work is funded by the French government's LABEX DISTALZ (development of innovative strategies for a transdisciplinary approach to Alzheimer's disease) and the US Alzheimer's association (IIRG-06-25487).

References

1. Weingarten MD, Lockwood AH, Hwo SY et al (1975) A protein factor essential for microtubule assembly. Proc Natl Acad Sci U S A 72:1858–1862. doi:10.1073/pnas.72.5.1858
2. Cleveland DW, Hwo SY, Kirschner MW (1977) Purification of tau, a microtubule-associated protein that induces assembly of microtubules from purified tubulin. J Mol Biol 116:207–225
3. Sergeant N, Delacourte A, Buee L et al (2005) Tau protein as a differential biomarker of tauopathies. Biochim Biophys Acta 1739:179–197. doi:10.1016/j.bbadis.2004.06.020
4. Lee G, Newman ST, Gard DL et al (1998) Tau interacts with src-family non-receptor tyrosine kinases. J Cell Sci 111(Pt 21):3167–3177
5. Hwang SC, Jhon DY, Bae YS et al (1996) Activation of phospholipase C-gamma by the concerted action of tau proteins and arachidonic acid. J Biol Chem 271:18342–18349
6. Jenkins SM, Johnson GV (1998) Tau complexes with phospholipase C-gamma in situ. Neuroreport 9:67–71
7. Reynolds CH, Garwood CJ, Wray S et al (2008) Phosphorylation regulates tau interactions with Src homology 3 domains of phosphatidylinositol 3-kinase, phospholipase Cγ1, Grb2, and Src family kinases. J Biol Chem 283:18177–18186. doi:10.1074/jbc.M709715200
8. Chapuis J, Hansmannel F, Gistelinck M et al (2013) Increased expression of BIN1 mediates Alzheimer genetic risk by modulating tau pathology. Mol Psychiatry 18:1225–1234. doi:10.1038/mp.2013.1
9. Sottejeau Y, Bretteville A, Cantrelle F-X et al (2015) Tau phosphorylation regulates the interaction between BIN1's SH3 domain and Tau's proline-rich domain. Acta Neuropathol Commun 3:58. doi:10.1186/s40478-015-0237-8
10. Hanger DP, Anderton BH, Noble W (2009) Tau phosphorylation: the therapeutic challenge for neurodegenerative disease. Trends Mol Med 15:112–119. doi:10.1016/j.molmed.2009.01.003
11. Söderberg O, Gullberg M, Jarvius M et al (2006) Direct observation of individual endogenous protein complexes in situ by proximity ligation. Nat Methods 3:995–1000. doi:10.1038/nmeth947

Chapter 19

Finding *MAPT* Mutations in Frontotemporal Dementia and Other Tauopathies

Carol Dobson-Stone and John B.J. Kwok

Abstract

Sanger sequencing is a classic technique in molecular genetics to detect single nucleotide DNA variants in genomic DNA. Here we describe the detection of *MAPT* mutations by polymerase chain reaction amplification of patient genomic DNA followed by bidirectional Sanger sequencing. Exon trapping is a technique whereby genomic DNA covering the exon of interest and flanking intronic sequence is cloned into the intron of an expression vector and transfected into human cell lines. RNA is extracted and splicing products are examined by reverse-transcriptase PCR and agarose gel electrophoresis. We outline the application of this technique to assess the effect of novel DNA variants on the splicing efficiency of *MAPT* exon 10, a common mechanism of disease for pathogenic *MAPT* mutations.

Key words Polymerase chain reaction, Sanger sequencing, Heterozygote detection, Exon trapping, pSPL3

1 Introduction

Since *MAPT* was identified in 1998 as the cause of disease in several families with hereditary dementia and parkinsonism [1], over 50 different mutations have been identified [2]. *MAPT* encodes the microtubule-associated protein tau, a major component of intraneuronal inclusions in Alzheimer's disease and roughly half of frontotemporal lobar degeneration cases [3]. The neuropathology associated with *MAPT* mutations varies substantially: intraneuronal inclusions may resemble Pick bodies or the neurofibrillary tangles of Alzheimer's disease, and glial pathology reminiscent of progressive supranuclear palsy or corticobasal degeneration may also be present [2]. Clinically, patients with *MAPT* mutations most often show typical frontotemporal dementia with a mixture of behavioral and personality changes, cognitive impairment and parkinsonism; however, *MAPT* mutations have also been observed in patients with mild late-onset

Caroline Smet-Nocca (ed.), *Tau Protein: Methods and Protocols*, Methods in Molecular Biology, vol. 1523, DOI 10.1007/978-1-4939-6598-4_19, © Springer Science+Business Media New York 2017

parkinsonism only, dementia with epilepsy, or pure progressive supranuclear palsy or corticobasal syndrome phenotypes [4].

Six tau isoforms are expressed in adult human brain, the result of alternative splicing of *MAPT* exons 2, 3 and 10 (Fig. 1). Alternative splicing of exon 10 gives rise to isoforms containing either three microtubule-binding domains (3R tau) or four microtubule-binding domains (4R tau) [5]. 3R and 4R tau differ in their interaction strength with microtubules [6] and the concentrations of 3R and 4R tau are roughly equal in normal adult brain [7]. About half of all reported *MAPT* mutations, including two of the three most common mutations (N279K and IVS10+16C>T), increase the relative concentration of exon 10-containing transcripts, resulting in an overproduction of 4R tau relative to the 3R isoform [1, 8].

Whole-exome sequencing (WES) is rapidly becoming the gold standard for assessing the cause of disease in Mendelian disorders and identifying novel disease genes [9]. However, due to the relatively high cost of WES, frontotemporal dementia patients with a strong family history of disease are still routinely screened for mutations in known common disease genes, including *MAPT*, before inclusion in WES cohorts. In this chapter, we describe a method for mutation screening of point mutations in *MAPT* by polymerase chain reaction (PCR) amplification of *MAPT* exons and flanking intronic sequence (Subheading 3.1), followed by purification (Subheading 3.2), bidirectional direct Sanger sequencing (Subheading 3.3), and sequence analysis (Subheading 3.4) of amplicons.

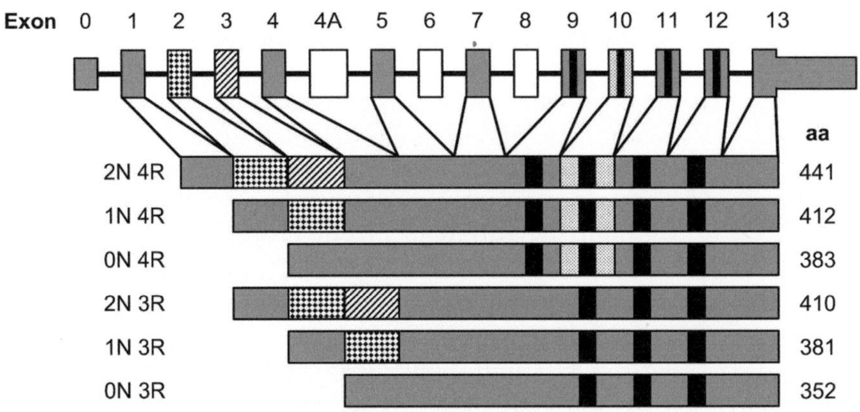

Fig. 1 *MAPT* splice variants in adult central nervous system (CNS). Schematic of *MAPT* exons with their traditional notation, denoting exons present in all isoforms (*grey*), exons that are not present in adult CNS splice isoforms (*white*), and alternatively spliced exons 2 (*checked*), 3 (*striped*), and 10 (*dotted*). Microtubule-binding domains are depicted as *vertical black bars*. The six isoforms present in human brain are depicted below. Alternative splicing of exons 2 and 3 results in proteins with 0, 1, or 2 N-terminal inserts (0N, 1N, 2N), while exon 10 alternative splicing generates proteins with either 3 or 4 microtubule-binding repeats (3R or 4R)

If a novel DNA variant is identified, functional analysis may be necessary to determine if it is likely to have a pathogenic effect on the *MAPT* gene. The particular assays used will of course depend on the nature of the variant identified, but one of the most common effects of *MAPT* mutations is differential splicing of exon 10. Exon trapping is a common technique used to assess gene splicing [10, 11]. In this technique, a genomic fragment is cloned into the intron of an expression vector, e.g., pSPL3. Exons encoded in the genomic fragment will be spliced into the transcript encoded on the expression vector, assuming that the splice signals in this region remain intact. Differential splicing of the transcript can be observed by reverse-transcriptase PCR of RNA extracted from cells transfected with wild-type and mutant pSPL3 constructs and comparing the relative yields of products containing or excluding the exon in question. We describe how to prepare DNA constructs and perform the exon-trapping assay in Subheadings 3.5 and 3.6, respectively.

2 Materials

Prepare all solutions using ultrapure water (prepared by purifying deionised water to attain a sensitivity of 18 MΩ cm at 25 °C) and molecular biology grade reagents. Prepare and store all reagents at room temperature, unless indicated otherwise. Follow local waste disposal regulations when disposing of waste materials.

2.1 PCR

1. PCR master mix (per reaction): 2.5 μL 10× PCR Buffer, 0.75 μL 50 mM $MgCl_2$ (both supplied with polymerase enzyme), 2 μL 2.5 mM dNTP mix (Promega Corporation), 0.5 μL 10 μM forward primer, 0.5 μL 10 μM reverse primer, 1.25 μL dimethyl sulfoxide (DMSO), 0.1 μL 5 U/μL Platinum Taq DNA polymerase (Invitrogen), nuclease-free water to 24 μL (*see* **Note 1**).
2. Agarose gel: 1–2.5% (w/v) agarose, 1× Tris-acetate EDTA (TAE) buffer (40 mM Tris pH 8.0, 20 mM acetic acid, 1 mM EDTA). Measure out TAE buffer into a conical flask. Add agarose, swirl to mix and microwave on full power in 30 s increments until agarose is completely dissolved. Swirl after each 30 s heating to avoid superboiling of gel solution. Cool agarose with stirring until flask is hand hot (~50 °C) and add 1× RedSafe Nucleic Acid Stain (iNtRON Biotechnology) for qualitative gels or 1× SYBR Safe DNA Gel Stain (Invitrogen) for preparative gels. Pour into gel tray and place comb in, moving bubbles away to the edge of the gel with a pipette tip. Once set, gel can be wrapped in cling film and stored in the dark at 4 °C.

3. 100 bp DNA Ladder (New England Biolabs): dilute to 100 ng/μL in 1× sample buffer (supplied with ladder), 1× Tris EDTA (TE) buffer (10 mM Tris pH 8.0, 1 mM EDTA). Store at 4 °C for up to 3 months, or long-term at −20 °C.
4. ExoSAP master mix (per reaction): 0.25 μL 10× PCR buffer, 1.75 μL nuclease-free water, 0.5 μL ExoSAP-IT (Affymetrix). Keep on ice.
5. PCR plate and adhesive PCR Plate Seal (ThermoFisher Scientific).

2.2 Sanger Sequencing

1. Sequencing master mix (per reaction, using BigDye Terminator v3.1 Cycle Sequencing Kit, Applied Biosystems): 4 μL 5× Sequencing Buffer, 1 μL BigDye Terminator v3.1 Ready Reaction mix, 0.32 μL 10 μM primer, nuclease-free water to 18 μL. Keep on ice and protect from light (*see* **Note 2**).
2. 125 mM ethylene diamine tetraacetic acid (EDTA), pH 8.0. Prepare from 0.5 M, pH 8.0 stock solution. For 0.5 M stock: add 146.1 g EDTA to 800 mL water. Mix and adjust pH to 8.0 with NaOH. Adjust volume with water to 1000 mL and autoclave.
3. 3 M sodium acetate. Add 40.8 g sodium acetate to 70 mL water. Mix and adjust pH to 5.2 with glacial acetic acid. Adjust volume with water to 100 mL and autoclave.
4. Microplate and TopSeal-A 96-well microplate seal.

2.3 Exon Trapping Construct

1. PCR master mix, agarose gel and DNA 100 bp Ladder as detailed in Subheading 2.1.
2. Primers MAPTex10trapF (5′-CGAGCAAGCAGGCGGG TCCAG-3′), MAPTex10trapR (5′-GTACGCACTCACA CCACTTCC-3′), SP6 (5′-ATTTAGGTGACACTATAG-3′), SD2 (5′-GTGAACTGCACTGTGACAAGCTGC-3′), SA4 (5′-CACCTGAGGAGTGAATTGGTCG-3′), *MAPT* exon 10 forward (Table 1).
3. QIAquick PCR Purification Kit (QIAGEN).
4. pGEM-T Easy Vector System (Promega).
5. Chemically competent *E. coli* One Shot TOP10 Competent Cells (Invitrogen).
6. Luria Broth (LB) medium. Dissolve 5 g tryptone, 2.5 g yeast extract, 5 g NaCl in 400 mL water. Adjust to 500 mL with water and autoclave. Add ampicillin to 100 μg/mL just before use. Once autoclaved, open only next to a flame for reasons of sterility.
7. LB—ampicillin agar plates. Prepare as for LB medium, with the addition of 7.5 g Bacto agar before autoclaving. Cool to ~50 °C before adding ampicillin to 100 μg/mL and pouring into ~20 100 mm petri dishes, filling each plate 1/2–2/3 full.

Table 1
Oligonucleotide primers used for amplification of *MAPT* exons

Exon (TN)	Exon (UN)	Expn	Forward primer	Reverse primer	Amplicon size (bp)
0	1	Adult (nc)	NA	NA	
1	2	Adult	GAGGCTCCTCTCTCTTCA	TTTAAAATTTCTTTGCAATTCTG	320
2	3	Adult	ACTAGCTGGTTAAGAGGCACT	GTGTGGGAGGACAATCGTAT	648
3	4	Adult	CTACTTCAGGGCTGCTTTCT	AGCTACTCACAACCCAAGATT	385
4	5	Adult	GAAGGTACAGAGAGCTTGGTTT	TAGAGCTCAGGTCCAAATGAT	242
4A	6	Fetal	NA	NA	
5	7	Adult	GACCTATCCCAGAGAAATCC	AATGAATCAGACATAAAGCACA	225
6	8	Fetal	NA	NA	
7	9	Adult	AGTCTTAGCCACGTTTTGAGT	CAGGGGTTATTCATTCTCAGT	338
8	10	Fetal	NA	NA	
9	11	Adult	GTCTCTGTCCCTAAGCAAAA	AAGAACTGAACTTCCTTGAAGA	677
10	12	Adult	GTCACTCATCGAAAGTGGAG	AAGTGTACGCACTCACACCA	255
11	13	Adult	CACCCTTAATAAAATAATCCCA	TTCTGAAGTCTGGAGCAGTT	266
12	14	Adult	GTAGACTGCAGACCTCATGTAA	AGCCTTTGTCCACTCAACTT	274
13	15	Adult	CGATAGAGCAAGACCCTGTC	GTGATTAACCAATTAACCGAAC	484

TN traditional notation, *UN* UCSC notation, *Expn* transcript expression, *nc* noncoding

Pour the plates next to a flame. Replace covers and allow plates to cool at room temperature overnight. Store plates at 4 °C in a polythene bag, upside down and protected from light.

8. Plasmid preparation kits: Wizard Plus SV Minipreps DNA Purification System (Promega) and Plasmid Midi Kit (QIAGEN).
9. Sequencing master mix as detailed in Subheading 2.2.
10. pSPL3 plasmid (Invitrogen—discontinued). This plasmid is available from the authors, or contact any group who have published using this construct.
11. *Eco*RI restriction enzyme, calf intestinal alkaline phosphatase and T4 DNA ligase (all from New England Biolabs).
12. 1 kb DNA Ladder (New England Biolabs). Dilute to 100 ng/μL in 1× sample buffer (supplied with ladder), 1× TE. Store at 4 °C for up to 3 months, or long-term at −20 °C.
13. QIAquick Gel Extraction Kit (QIAGEN).
14. Safe Imager™ 2.0 Blue-Light Transilluminator.

2.4 Exon Trapping Assay

1. Phosphate-buffered saline (PBS), e.g., HyClone Dulbecco's PBS without Calcium, Magnesium (GE Healthcare Life Sciences). PBS-EDTA: Add 5 mL 0.5 M EDTA to 500 mL PBS (i.e., final concentration 5 mM EDTA). Prepare in a tissue culture hood and store at 4 °C.
2. HEK medium: 1× minimal essential medium (MEM) with 1× Hanks' Salts, 10% (v/v) fetal bovine serum (FBS), 2 mM L-glutamine, 0.057% sodium bicarbonate. To 500 mL MEM Hanks' Balanced Salts (Gibco), add 50 mL FBS, 5 mL 200 mM L-glutamine (both from GE Healthcare Life Sciences), and 3.8 mL 7.5% (w/v) sodium bicarbonate. Prepare in a tissue culture hood and store protected from light at 4 °C.
3. SK-N-MC medium: 1× Dulbecco's modified Eagle medium (DMEM):Ham's F12 nutrient mixture (1:1), 10% (v/v) FBS, 2 mM L-glutamine, 1× nonessential amino acids (NEAA), 0.057% sodium bicarbonate. To 500 mL DMEM/F12 1:1 (GE Healthcare Life Sciences), add 50 mL FBS, 5 mL 200 mM L-glutamine, 5 mL NEAA (Gibco), and 3.8 mL sodium bicarbonate. Adjust pH to 7.4 with ~800 μL 1 M NaOH. Prepare in a tissue culture hood and store protected from light at 4 °C.
4. SK-N-MC and HEK 293 cell lines (both from American Type Culture Collection, VA, USA).
5. Lipofectamine® 2000 and Opti-MEM I Reduced Serum Medium (Gibco).
6. RNeasy Mini Kit (QIAGEN) and Superscript III First-Strand Synthesis System (Invitrogen).

7. Exon-trapping PCR master mix (per reaction): 2 μL 10× PCR Buffer, 0.6 μL 50 mM $MgCl_2$, 1.6 μL 2.5 mM dNTP mix, 2 μL 10 μM SD2 primer, 2 μL 10 μM SA4 primer, 0.2 μL 5 U/μL Platinum Taq DNA polymerase, nuclease-free water to 18 μL. Keep on ice.
8. Agarose gel and 100 bp DNA Ladder as detailed in Subheading 2.1.
9. ChemiDoc system and Quantity One Analysis Software.

3 Methods

Perform all procedures at room temperature unless otherwise stated.

3.1 Polymerase Chain Reaction (PCR) Amplification of MAPT Exons

1. A typical *MAPT* mutation screen will assess exons 1–4, 5, 7, 9–13 of *MAPT*, i.e., all exons expressed in the adult human central nervous system (Fig. 1). Table 1 lists the primers required to amplify these exons.
2. Quantitate genomic DNA (gDNA), e.g., by OD260 absorbance in a spectrophotometer, and make up 10 ng/μL working stocks of each sample to be amplified. Make up working stocks at least 2 h before use, ideally the day before (*see* **Note 3**).
3. Assemble all plasticware and pipette tips to be used for the PCR in a PCR hood and UV irradiate for 5 min. Perform **steps 4** and **5** in the PCR hood (*see* **Note 4**).
4. For each exon to be amplified, calculate the number of reactions required (no. of test gDNA samples + 1 gDNA sample from a healthy control individual + 1 negative (no-DNA) control + 10% for pipetting error). Prepare PCR master mix accordingly. Vortex, briefly spin down to collect, and store on ice until required.
5. Aliquot out 24 μL master mix into separate wells of a 96-well V-bottomed microplate. Transfer plate to a lab bench.
6. Add 1 μL (i.e., 10 ng) gDNA working stock from each sample to the relevant well(s). If multiple exons are to be amplified for a sample, use a new tip each time to avoid contamination of the gDNA stock. For the negative control, add 1 μL of sterile H_2O. Place a high-adhesion seal on microplate (*see* **Note 5**).
7. Briefly spin down microplate in a bench-top centrifuge to collect, and place in a PCR thermocycler. Run the following cycling conditions: 94 °C 2 min; a "touchdown" of eight cycles (94 °C for 30 s, 70 °C with −2 °C per cycle for 30 s, 72 °C for 30 s); 25 "standard" cycles (94 °C for 30 s, 56 °C for 30 s, 72 °C for 30 s); 72 °C for 10 min (*see* **Note 6**).

8. Place a 2% agarose TAE gel in an agarose gel electrophoresis tank containing sufficient 1× TAE buffer to cover the wells. Briefly spin down microplate to collect and remove seal from microplate. Mix 5 μL of each sample with 0.5 μL 10× agarose loading buffer and load on the gel alongside 3 μL 100 bp DNA Ladder (*see* **Notes** 7 and **8**).
9. Subject gel to electrophoresis: 20 min at 100 V in a ~15 cm electrophoresis tank should be long enough to resolve PCR products and give a rough estimate of size.
10. Visualize gel using a UV transilluminator and check that each reaction has produced an amplicon of the expected size and that there is no amplicon present in the negative control (*see* **Note 9**).

3.2 Purification of PCR Amplicons

1. On ice, prepare sufficient ExoSAP master mix for each PCR amplicon to be purified (+10% to allow for pipetting error) and aliquot out 2.5 μL per well into a clean 96-well V-bottomed microplate on ice (*see* **Note 10**).
2. Add 5 μL of PCR amplicon, mix by pipetting, apply high-adhesion seal and briefly spin down microplate to collect.
3. Incubate 37 °C for 30 min, inactivate enzymes at 80 °C for 15 min and cool on ice (*see* **Note 11**).

3.3 Preparation of MAPT Amplicons for Sanger Sequencing

1. For each amplicon, set up two sequencing reactions: one with the forward primer and one with the reverse primer. On ice, prepare sufficient sequencing master mix for each sample to be sequenced (+10% to allow for pipetting error) and aliquot 18 μL per well into a V-bottomed 96-well microplate on ice (*see* **Note 12**).
2. Add 2 μL of the relevant purified amplicon to the wells, seal with adhesive PCR seal, briefly spin down to collect.
3. Place microplate into a thermocycler and run sequencing program as follows: 96 °C 1 min; 30 cycles of 96 °C for 10 s, 50 °C for 5 s, 60 °C for 4 min. Briefly spin down microplate to collect (*see* **Note 13**).
4. Precipitate samples by addition of 2 μL 125 mM EDTA pH 8.0, 2 μL 3 M sodium acetate, 50 μL absolute ethanol and mix by pipetting. Leave plate at room temperature for 15 min (*see* **Note 14**).
5. Spin microplate at 3000 ×*g* at 4 °C for 30 min. As soon as centrifuge has stopped spinning, remove plate, gently remove adhesive seal, invert plate on a paper towel and spin at 185 ×*g* for 10–15 s (*see* **Note 15**).
6. Add 70 μL freshly made 70% ethanol to wash cell pellets, spin microplate at 1650 ×*g* at 4 °C for 15 min, immediately proceed to next step.

7. Gently remove adhesive seal, invert plate on a paper towel and spin at 185 × *g* for 1 min (*see* **Note 15**).
8. Leave microplate in a dark place for ~10 min to ensure wells are dry before resealing plate with a low-adhesion plate seal. Wrap plate in foil to protect from light and store at −20 °C until samples are ready to be sequenced.
9. Run samples on a capillary sequencer. We send our samples to a facility that runs the samples on an Applied Biosystems 3730 DNA Analyzer.

3.4 Strategy for Sequence Analysis of Electropherograms

1. Download the published reference sequence for the amplicon of interest from the UCSC Genome Browser (https://genome.ucsc.edu/) and use this to generate a reference sequence file (*see* **Note 16**).
2. For each exon to be analyzed, assemble all forward and reverse sequence files, together with the reference sequence file, within a sequencing program (*see* **Note 17**).
3. Check that sequence data adequately covers the exon for each individual plus at least 25 bp of flanking intronic sequence.
4. Examine all deviations from the reference sequence by visual examination of the electropherograms at the relevant positions. An example of a true heterozygous variant is shown in Fig. 2 (*see* **Note 18**).

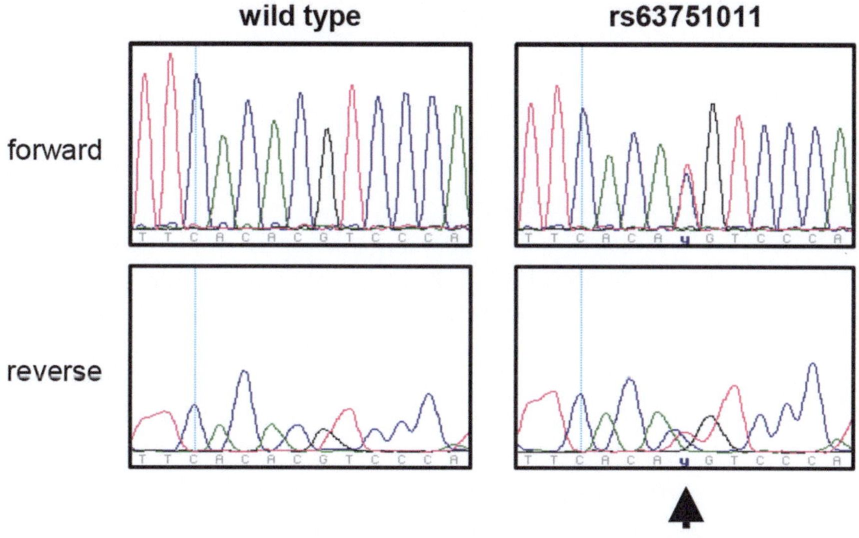

Fig. 2 *MAPT* mutation detection by Sanger sequencing. Sequence electropherograms are presented for forward and reverse sequencing reactions of exon 10 in a wild-type individual (*left*) and a heterozygote for rs63751011, the IVS10+16C>T mutation (*right*). Location of the mutation is *arrowed*. Note that the double peak is more easily identifiable in the forward sequence than the reverse sequence

5. Check all heterozygous positions against the single nucleotide polymorphism dbSNP track of the UCSC Genome Browser (http://genome.ucsc.edu/cgi-bin/hgGateway) and the AD & FTD Mutation Database (http://www.molgen.ua.ac.be/ADMutations/default.cfm?MT=1&ML=0&Page=ADMDB) as a first guide to whether the variant is a benign polymorphism or a pathogenic mutation (*see* **Note 19**).
6. For unknown or suspected pathogenic variants, PCR and sequencing reactions should be repeated on a separate genomic DNA dilution to confirm the presence of the variant in the proband, plus any available family members to determine whether the variant segregates with disease. For a novel suspected pathogenic variant, it is best practice to screen a large ethnically matched cohort ($n \geq 500$) of healthy elderly individuals to determine its frequency in the population (*see* **Note 20**).

3.5 Exon Trapping—Construction of pSPL3 Constructs

1. Perform PCR of *MAPT* exon 10 plus flanking intronic sequence (59 bp 5′, 70 bp 3′) from the gDNA sample harboring the suspected mutation as described in Subheading 3.1, using primers MAPTex10trapF & MAPTex10trapR.
2. Purify amplicons using, e.g., the QIAquick PCR Purification Kit according to manufacturer's instructions, eluting in 30 μL nuclease-free water.
3. Quantitate purified amplicons using a spectrophotometer, ligate ~10 ng amplicon into 50 ng pGEM-T Easy vector according to manufacturer's instructions and transform into chemically competent *E. coli* according to manufacturer's instructions, selecting transformants on LB-ampicillin agar plates.
4. Inoculate eight single colonies from the test transformation plate each into 4 mL LB-ampicillin medium and perform a small-scale plasmid preparation according to manufacturer's instructions.
5. Perform Sanger sequencing of each clone as detailed in Subheading 3.3, using oligonucleotide primer SP6, 2 μL BigDye and 200 ng plasmid per reaction. Determine the orientation of the insert relative to the SP6 and T7 ends of pGEM-T Easy and whether the insert contains wild-type or mutant sequence and confirm that no additional mutations have been introduced by PCR amplification (*see* **Note 21**).
6. Set up the following digests for one clone containing mutant insert and one containing wild-type insert. In separate reactions, digest 2 μg of each clone and 2 μg of pSPL3 plasmid, each with 40 U *Eco*RI and 1× NEBuffer *Eco*RI in a 50 μL total reaction. Incubate overnight at 37 °C.

7. Add 1 U calf intestinal alkaline phosphatase to the pSPL3 digest and incubate at 37 °C for 30 min. Inactivate all digests by incubation at 65 °C for 10 min (*see* **Note 22**).
8. Load samples on agarose 1× TAE gels containing 1× SYBR Safe DNA Gel Stain alongside appropriate size standard: a 2% agarose gel with 100 bp DNA Ladder for the pGEM-T-exon10 digests and a 1% agarose gel with 1 kb DNA Ladder for the pSPL3 digest. Subject to electrophoresis at 100 V until the digested bands are approximately halfway down the gel (*see* **Note 23**).
9. Place gel on a blue light transilluminator and check that the digested bands are of the correct size: ~240 bp for the *MAPT* exon 10 insert bands, ~6.0 kb for the pSPL3 vector band. Excise the *MAPT* insert and pSPL3 vector bands with a scalpel blade, using a new blade for each excision and placing the gel slice from each band into a separate 2 mL microcentrifuge tube.
10. Purify digested DNA from the agarose gel slices, e.g., using the QIAquick Gel Extraction Kit according to manufacturer's instructions, eluting in 30 μL nuclease-free water.
11. Quantitate DNA with a spectrophotometer and reduce volume by evaporation in a vacuum pump if necessary to increase DNA concentration (*see* **Note 24**).
12. Set up ligations between each digested insert and the digested pSPL3 vector. Combine the following in a total volume of 10 μL in a microcentrifuge tube on ice: 6 ng insert, 50 ng pSPL3, 1 μL 10× T4 DNA ligase buffer, 1 μL T4 DNA ligase. In addition, set up a vector-only ligation as a negative control (i.e., omit insert from the ligation). Incubate overnight at 4 °C (*see* **Note 25**).
13. Transform into competent *E. coli* according to manufacturer's instructions, selecting on LB-ampicillin agar plates at 37 °C overnight.
14. Draw and number 16 sections on the back of a pre-warmed LB-ampicillin agar plate. Inoculate eight pSPL3-exon10 mutant and eight pSPL3-exon10 wild-type colonies from the transformation plates, each into 4 mL LB-ampicillin medium. After each inoculation, use the inoculation tip to streak a separate section of the pre-warmed plate. Incubate the plate and cultures overnight at 37 °C and perform a small-scale plasmid preparation as in **step 4** (*see* **Note 26**).
15. Digest the plasmids to check for presence of insert. Combine 2 μL of each miniprep with 10 U (0.5 μL) *Eco*RI and 1× *Eco*RI buffer in a total volume of 30 μL and incubate at 37 °C for 2 h. Load 20 μL of each digest alongside 1 kb and 100 bp size

standards on a 1.5% agarose TAE gel and subject to electrophoresis. Clones containing insert will yield bands at ~6.0 kb and ~240 bp.

16. Perform Sanger sequencing of each clone as detailed in Subheading 3.3, using oligonucleotide primers SD2 and *MAPT* exon 10 forward, 2 μL BigDye and 200 ng plasmid per reaction. Confirm that the insert is in the correct orientation, contains wild-type or mutant sequence as expected and that ligation junctions are intact (*see* **Note 27**).
17. Use a streaked colony from one pSPL3-exon 10 mutant clone and one pSPL3-exon 10 wild-type clone for medium-scale plasmid preparation of pSPL3-exon 10 constructs with, e.g., the QIAGEN Plasmid Midi Kit according to manufacturer's instructions. Before processing the overnight culture, remove 810 μL for preparation of glycerol stocks (*see* **Note 28**).

3.6 Exon Trapping Assay

Steps 1–7 should be performed in a tissue culture hood using aseptic technique to avoid microbial contamination of mammalian cells. PBS, PBS-EDTA and culture media should be pre-warmed to 37 °C unless otherwise stated.

1. Grow SK-N-MC and HEK 293 cells to maximal confluency in 75 cm^2 flasks.
2. Prepare 100 ng/μL dilutions of each DNA construct in sterile TE buffer.
3. Aspirate media, wash cells with 5 mL PBS and lift cells off flask by incubation with 5 mL PBS-EDTA for 5 min at 37 °C. Transfer cells to 15 mL Falcon tubes, subject tubes to centrifugation at 200 × *g* for 5 min and resuspend the cell pellets in 5 mL media (*see* **Note 29**).
4. Count cells using a hemocytometer and prepare a sufficient quantity of cell suspension in media for each cell line (two wells per DNA construct used + 10% to allow for pipetting error). Allow ~2×10^5 SK-N-MC cells per well or ~1×10^5 HEK cells per well, in 1 mL total volume per well.
5. Aliquot out SK-N-MC and HEK cell suspensions to a 12-well plate (1 mL per well). Leave cells to recover at 37 °C for 16–24 h before transfection (*see* **Note 30**).
6. For each cell line set up duplicate transfections per DNA construct using Lipofectamine 2000, according to manufacturer's instructions. Incubate cells for 48 h at 37 °C (*see* **Note 31**).
7. Remove media from wells and lift cells off by incubation with 1 mL PBS-EDTA per well for 5 min at 37 °C. Transfer cells to a 1.5 mL microcentrifuge tube and spin at 200 × *g* for 5 min. Wash cell pellets with 500 μL ice-cold PBS.

8. Extract total RNA using, e.g., the RNeasy Mini Kit according to manufacturer's instructions, eluting in 100 μL nuclease-free water.
9. Perform cDNA synthesis from 3 μg each RNA sample using, e.g., the Superscript III First-Strand Synthesis System according to manufacturer's instructions, using oligo(dT) primer.
10. Perform PCR as detailed in Subheading 3.1, **steps 3–7** with the following modifications: prepare exon-trapping PCR master mix and aliquot out 18 μL per well + 2 μL each cDNA sample; cycling conditions as follows: 95 °C 2 min, then 38 cycles of 92 °C for 30 s, 58 °C for 30 s, 72 °C for 30 s.
11. Load samples alongside 100 bp DNA Ladder on 2.5 % agarose 1× TAE gel and subject to electrophoresis until bands are halfway down the gel.
12. Visualize and quantitate bands on a gel documentation system. Expected band sizes are 270 bp for the exon 10 inclusion product and 177 bp for the exon 10 exclusion product (Fig. 3). If a variant leads to a significant increase in the exon 10 inclusion product relative to wild-type, it is an indication that the variant could be pathogenic (*see* **Note 32**).

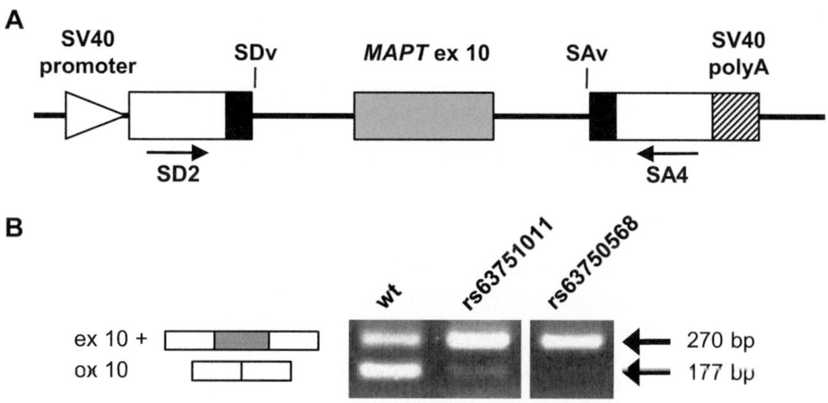

Fig. 3 Exon trapping experiment to determine effect of *MAPT* exon 10 splice region mutations. (**a**) Schematic of pSPL3-MAPTex10 construct. pSPL3 contains partial exons from the rabbit beta-globin gene (*white rectangles*) and splice donor (SDv) and acceptor (SAv) sequences from the HIV *tat* gene (*black*). A multiple cloning site in the *tat* intron allows the introduction of *MAPT* exon 10 and flanking intronic sequence. RT-PCR primers SD2 and SA4 are indicated by *arrows*. (**b**) Agarose gel of exon trapping RT-PCR products from wild-type (wt) and mutant pSPL3-MAPTex10 constructs. Bands corresponding to products containing (ex 10+) or omitting (ex 10–) *MAPT* exon 10 are *arrowed*. Mutations rs63751011 (IVS10+16C>T) & rs63750568 (S305S) affect splicing of exon 10, as evidenced by differential intensity of 10+ and 10– bands relative to wild type

13. If desired for confirmation of identity, or if additional bands are observed, individual bands can be excised, DNA purified and samples sequenced with SD2 and SA4 primers as detailed in Subheading 3.5, **steps 9** and **10** and Subheading 3.3.

4 Notes

1. Most of the premix can be prepared at room temperature. As with most enzymatic procedures, the enzyme is the most temperature-sensitive (and expensive) reagent and so should be added last, and the mix kept on ice once enzyme is added until ready for aliquoting. *MAPT* exon primer sequences and PCR amplicon sizes are listed in Table 1.
2. We find that a combination of 18 μL sequencing master mix and 2 μL PCR product is appropriate for most amplicons. The amount of PCR product can be adjusted up or down as desired for particularly low- or high-yield PCRs, adjusting the volume of master mix accordingly to keep a total volume of 20 μL.
3. Do not vortex gDNA to resuspend as it is likely to fragment; instead flick the tube to mix or pipette up and down gently using a pipette tip with the end cut off. Store gDNA samples at 4 °C until ready to be used. As long as gDNA is sufficiently pure (OD260/OD280 ratio ~1.85), it is stable in water for several weeks at 4 °C. Long-term storage of gDNA working stocks should be at −20 °C.
4. PCRs are highly susceptible to contamination with nucleic acids from gDNA stocks or products of previous PCRs. To minimize this contamination, use aerosol-resistant pipette tips and change gloves often. If access to a PCR hood is not available, use separate designated work areas and pipettes for pre- and post-amplification steps.
5. To avoid evaporation of well contents during thermocycling, use a high-adhesion PCR seal, and ensure that it makes contact with the top of all wells by, e.g., rubbing over each well with the backing paper from the seal. These plate seals are designed to be used in thermocyclers with a heated lid system that makes tight contact with the top of the plate: if this is not your setup, use foil or adhesive seals that are applied with a heat sealer. If PCRs cannot be run immediately, the plate containing the assembled reagents can be stored on ice, at 4 °C, or for a few days at −20 °C if need be.
6. If PCR will not be collected immediately from the thermocycler, add a 10 °C hold step to the program. Most PCRs can be held at this temperature overnight or over the weekend if necessary.

7. The glue on the adhesive PCR seals is relatively strong and the microplate is liable to tip up when removing the seal and spray the contents of the wells. The easiest way to avoid this is to place the plate into a microplate stand or empty pipette tip box, hold two sides of the plate with one hand and gently peel away the seal from one corner with the other hand.
8. The microplate with the remaining PCR product can be resealed and stored at 4 °C until the purification step or long term at −20 °C.
9. We find that the *MAPT* PCRs work reliably with the premix components and cycling conditions provided. The most common reason for absence of amplification in our experience (assuming that the premix was correctly assembled) is impure or degraded gDNA template. Adding extra gDNA to the reaction setup can mitigate this problem. If this does not work, it may be that a contaminant in the gDNA preparation is inhibiting the amplification, in which case diluting the gDNA working stock 1/2–1/10 before use may give success. Conversely, presence of an amplification product in the negative control is indicative of nucleic acid contamination and the amplicons from this PCR should be discarded. UV irradiate the plasticware and pipettes to be used and use new stocks of the PCR reagents. If in doubt as to the source of the contamination, replace everything except the polymerase enzyme—this can be replaced if the PCR is still contaminated on the second attempt.
10. Plate purification using, e.g., the MultiScreen $PCR_{\mu 96}$ Filter Plate (Merck Millipore Corporation, Germany), can be employed instead to remove unincorporated dNTPs and primers from the sequencing reaction. However, we find ExoSAP-IT treatment to lead to superior sequencing results for a similar cost.
11. Purified amplicons can be stored at 4 °C overnight or long term at −20 °C.
12. Bidirectional sequencing is recommended for mutation detection, as it is uncommon but not impossible for heterozygotes to be missed by the sequencing software in one direction, due to a low second peak. The fluorescent dye terminator mix is light sensitive so keep in the dark as much as possible.
13. If the sequencing reaction will not be collected immediately from the thermocycler, add a 10 °C hold step to the program. The reaction can be held at this temperature overnight if necessary. We also get good results for most amplicons using the following cycling conditions: 94 °C 2 min; 38 cycles of 94 °C for 30 s, 56 °C for 30 s, 72 °C for 1 min 15 s; 72 °C for 5 min.
14. One can make a premix of EDTA and sodium acetate if desired, but do not make more than a few minutes beforehand as the EDTA will precipitate. It is important to keep the incubation

step to 15 min. Incubation for much less than 15 min will result in loss of very short extension products. Longer times may increase the precipitation of unincorporated dye terminators.

15. If the samples are left too long after the centrifuge stops spinning in **steps 5** and **6**, the pellets will start to detach from the bottom of the plate and fall out when the plate is inverted. To avoid this, either hold the centrifuge at 3000 × *g* until you are ready to proceed to the next step, or spin the plate for an additional 5 min before removing the supernatant. The supernatant can lead to corrosion if spilled in the centrifuge. To avoid this, we invert the plate quickly over a sink and blot on a paper towel before placing in centrifuge buckets lined with two to three layers of paper towel cut to fit the buckets. The centrifuge bowl should also be wiped down with a damp paper towel after use.

16. Most sequencing programs include the facility to create a sequence file from text. We use the DNASTAR Lasergene SeqMan Pro package (DNASTAR Inc.), which includes the EditSeq program for this purpose.

17. We employ the default SeqMan Pro parameters, except that heterozygotes are called when the second peak is at least 30 % of the height of the main peak.

18. There are several online guides to distinguishing true heterozygous peaks from sequencing artifact, including this website: https://seqcore.brcf.med.umich.edu/sites/default/files/html/interpret.html. Most true heterozygotes will show a double peak in both the forward and reverse sequences, although the height of the second peak may vary (Fig. 2). If in doubt, repeat the PCR and sequencing reactions for the sample in question.

19. Note that several pathogenic mutations are listed in dbSNP so one must check the rs entry for each variant to determine its classification. Other useful databases of DNA variants are the 1000 Genomes Browser (http://www.ncbi.nlm.nih.gov/variation/tools/1000genomes/), the Exome Sequencing Project (http://evs.gs.washington.edu/EVS/) and the Exome Aggregation Consortium Browser (http://exac.broadinstitute.org/). Note though that if a variant is present at very low frequency in these databases this does not exclude its pathogenicity; conversely, absence of a variant in databases of normal variation does not prove its pathogenicity. Determining whether a novel variant is truly pathogenic will ultimately rest on its functional analysis and observation of segregation with disease in the patient's family, if possible.

20. We recommend screening at least 500 normal people (i.e., at least 1000 chromosomes) to exclude rare polymorphisms with a minor allele frequency of >0.001. IVS10+29G>A (rs63751443) was initially thought to be a pathogenic mutation [12] but was

subsequently found to be a rare polymorphism [13], with a minor allele frequency of ~0.006 in the 1000 Genomes GBR population.

21. Select mutant and wild-type clones in the same orientation so that the pGEM-T vector sequence that remains in the final constructs will be identical.
22. Dephosphorylation of the digested vector ends is necessary to avoid self ligation of the vector, which would otherwise lead to a high background of incorrect clones.
23. Use of a DNA stain that is visible in blue light avoids the use of UV light, which can nick DNA in preparative gels and is a carcinogen. If UV transillumination must be used when excising bands from gels, every caution must be taken to minimize exposure to DNA, skin and eyes, including switching the light box off between band excisions and wearing a UV face shield, long-sleeved lab coat and gloves.
24. Rather than quantitate DNA after sample evaporation, we calculate final DNA concentration by measurement of sample volume before and after evaporation.
25. Ligation reactions can be kept for several days at 4 °C before transformation, if necessary.
26. After incubation, store the plate with the 16 streaked colonies at 4 °C until needed for inoculation of starter cultures for medium-scale plasmid preparations (**step 17**). Wrap the edge of the plate with parafilm to stop the agar drying out. The plate can be stored in this way for up to a month.
27. Ligation using a single restriction site means that approximately half of the clones will contain the insert in the incorrect orientation. The complete sequence and a restriction map of pSPL3 can be downloaded from the Addgene Vector Database (https://www.addgene.org/vector-database/4202/).
28. Glycerol stocks should be prepared for long-term storage of transformed bacterial cultures in case more construct is required for subsequent experiments. To prepare, combine 810 μL bacterial culture with 190 μL sterile 80% glycerol in a sterile 2 mL cryotube, snap freeze on dry ice and store at −80 °C. To revive the bacteria, scrape a small amount (~10 μL) from the top of the frozen stock with an inoculation loop, streak on to a pre-warmed LB-ampicillin agar plate and incubate overnight at 37 °C. Use a single colony from this plate for plasmid preparation.
29. Give the flask a sharp tap against the side of a lab bench to help lift the cells off the flask.
30. Gently agitate the cell suspension before aliquoting and between aliquots to help keep cell numbers equal between

wells. The next day, visualize each well under a microscope to check that cells have adhered in sufficient numbers to achieve ~80 % confluency and that there is no microbial contamination before proceeding with transfection.

31. We transfect using 1 μg DNA construct and 2 μL Lipofectamine 2000 per well.

32. We routinely perform at least $n = 5$ experiments, which is sufficient to allow nonparametric statistical analysis of difference if necessary. Another consideration when concluding whether a variant is pathogenic is effect size. All verified *MAPT* exon 10 splicing mutations were found to lead to >95 % change in inclusion of exon 10 [12].

References

1. Hutton M, Lendon CL, Rizzu P et al (1998) Association of missense and 5′-splice-site mutations in tau with the inherited dementia FTDP-17. Nature 393:702–705
2. Ghetti B, Oblak AL, Boeve BF et al (2015) Invited review: frontotemporal dementia caused by microtubule-associated protein tau gene (MAPT) mutations: a chameleon for neuropathology and neuroimaging. Neuropathol Appl Neurobiol 41:24–46
3. Josephs KA, Hodges JR, Snowden JS et al (2011) Neuropathological background of phenotypical variability in frontotemporal dementia. Acta Neuropathol 122:137–153
4. Kertesz A (2003) Pick's complex and FTDP-17. Mov Disord 18:S57–S62
5. Goedert M, Spillantini MG, Jakes R et al (1989) Multiple isoforms of human microtubule-associated protein tau: sequences and localization in neurofibrillary tangles of Alzheimer's disease. Neuron 3:519–526
6. Goedert M, Jakes R (1990) Expression of separate isoforms of human tau protein: correlation with the tau pattern in brain and effects on tubulin polymerization. EMBO J 9:4225–4230
7. Hong M, Zhukareva V, Vogelsberg-Ragaglia V et al (1998) Mutation-specific functional impairments in distinct tau isoforms of hereditary FTDP-17. Science 282:1914–1917
8. D'Souza I, Poorkaj P, Hong M et al (1999) Missense and silent tau gene mutations cause frontotemporal dementia with parkinsonism-chromosome 17 type, by affecting multiple alternative RNA splicing regulatory elements. Proc Natl Acad Sci U S A 96:5598–5603
9. Rabbani B, Tekin M, Mahdieh N (2014) The promise of whole-exome sequencing in medical genetics. J Hum Genet 59:5–15
10. Apel TW, Scherer A, Adachi T et al (1995) The ribose 5-phosphate isomerase-encoding gene is located immediately downstream from that encoding murine immunoglobulin kappa. Gene 156:191–197
11. Buckler AJ, Chang DD, Graw SL et al (1991) Exon amplification: a strategy to isolate mammalian genes based on RNA splicing. Proc Natl Acad Sci U S A 88:4005–4009
12. Stanford PM, Shepherd CE, Halliday GM et al (2003) Mutations in the tau gene that cause an increase in three repeat tau and frontotemporal dementia. Brain 126:814–826
13. Pickering-Brown SM, Baker M, Gass J et al (2006) Mutations in progranulin explain atypical phenotypes with variants in MAPT. Brain 129:3124–3126

Part V

Cellular and In Vivo Models of Tau Physiopathology

Chapter 20

Tracking Tau in Neurons: How to Grow, Fix, and Stain Primary Neurons for the Investigation of Tau in All Developmental Stages

Hans Zempel and Eva-Maria Mandelkow

Abstract

Primary neurons have proved to be an invaluable tool for the investigation of Tau in the context of neuronal development and neurodegeneration. Culturing neurons usually is time consuming and requires multiple feeding steps and media exchanges, and either the use of proprietary media supplements or tedious preparation of complex media. Here we describe a relatively cheap and easy cell culture procedure based on a commercially available neuronal culture supplement (NS21) of known composition, as well as basic fixation techniques. Further, we demonstrate a staining technique that can be carried out in pre-coated hydrophobic multi-well plates, which minimizes antibody consumption and allows fast and convenient processing of samples for immunofluorescence microscopy of endogenous Tau in primary neurons. We also provide a protocol that allows cryopreservation of fixed cells for years without loss of Tau stainability.

Key words Primary neuronal cell culture, NS21, Immunostaining, Tau, Polarity

1 Introduction

The microtubule-associated protein Tau is sorted into the axon of mature neurons [1]. In the central nervous system (CNS), Tau also comes in four different isoforms in rodents and in six different isoforms in humans, the relative amounts of these isoforms are developmentally regulated [2, 3]. Further, Tau is phosphorylated at different sites, which also depends on the developmental stage, pathological processes and compartmentalization (for review *see* [4]). Thus it is of outmost importance to study Tau in a neuronal and cellular context. Primary rodent forebrain neurons have proved to be an important tool to study Tau during development and also for modeling pathological processes [5]. However, culturing primary neurons according to common protocols is time consuming and—especially when one aims to investigate mature, synaptically active,

Caroline Smet-Nocca (ed.), *Tau Protein: Methods and Protocols*, Methods in Molecular Biology, vol. 1523,
DOI 10.1007/978-1-4939-6598-4_20, © Springer Science+Business Media New York 2017

and well-polarized neurons—requires multiple feeding steps with expensive media supplements of proprietary formulations.

Here we describe a protocol developed for convenient long-term culturing of primary neurons that requires only two steps (plating, followed by only one feeding 3–4 days later) and is based on the neuronal supplement NS21 [6], which is commercially available for half the price or less than more commonly used supplements, and whose formulation is known in detail. The NS21-based culturing method with the modification as outlined below has proved to be cost-efficient, reproducible, and generates cultures that achieve sufficient maturity to investigate Tau distribution, isoform expression, and posttranslational modification in the context of neuronal development and degeneration.

Of note, this culturing method allows for low density cultures that are amenable to microscopy analysis also of endogenous Tau. Here we describe a standard fixation and staining procedure suitable for the investigation of Tau and phosphorylation of Tau in primary neurons. This procedure makes use of hydrophobic multi-well plates which reduces antibody consumption but still allows for fast and convenient processing (i.e., blocking, staining, washing) of samples for immunofluorescence microscopy.

2 Materials

2.1 Cell Culture Media

Prepare all solutions that come in contact with the cells or the multi-well plates in a sterile manner and in sterile conditions. For washing solution we use sterile PBS (Sigma) and for cell solutions sterile Ca^{2+}/Mg^{2+}-free HBSS (Sigma) and 1 × Trypsin (Sigma).

1. Coating solution: prepare a stock solution of 10 mg/ml poly-D-lysine in ddH_2O, make suitable aliquots (1–5 ml) and store at −20 °C. Directly before use, thaw and dilute to 50 μg/ml in sterile PBS. Do not store dilution.
2. Plating media: we use Neurobasal media (Nb; Gibco) as base, but other companies (MACS) worked fine too. It is possible to use DMEM, but with inferior results. Add 1 % heat-inactivated horse serum (HS), 1× penicillin–streptomycin mixture (PS; Sigma), 1× L-glutamine, 1 mM sodium pyruvate (Sigma), and 1× NS21 (PanBiotech). Store at 4° for up to 1 month.
3. Feeding media: Nb medium, add L-glutamine, PS and NS21, but no HS, no pyruvate. Store at 4° for up to 1 month. Immediately before use, add 0.5 μM cytosine β-D-arabinofuranoside (AraC; Sigma).
4. Multi-well plates and coverslips: we use preferentially 12-well cell culture plates (Corning) and 12 mm coverslips (VWR), but upscaling is easily possible. Down-scaling to 24-well plates is possible too, but cells tend to not grow longer than 6 weeks.

2.2 Fixation Solutions

1. Silanizing solution for preparing hydrophobic multi-well plates: 2% dimethyldichlorosilane in 1,1,1-trichlorethane (both Sigma). Store sealed at room temperature under a fume hood.
2. Fixation solutions: standard fixative is 3.7% of formaldehyde (FA) in PBS, i.e., 1:10 dilution in PBS of 37% formaldehyde (Sigma). Alternatively, cell shrinkage is usually slightly less if one uses a 1:5 dilution of 37% FA (i.e., a 7.4% solution) and mixes this solution with conditioned medium (i.e., growth medium conditioned by the culture to be fixed) one to one (just add on top of the culture, *see* Subheading 3). Always prepare fresh, do not store.
3. Storing solution for fixed cells: storing solution for fixed cells consists of a 1:1 mixture of PBS and Glycerol (Sigma), supplemented with 0.1% of formaldehyde.

2.3 Blocking and Staining Solutions

1. Blocking and permeabilizing solution (comprises nuclear stain): 5% bovine serum albumin, 0.5% Triton X-100, and if desired Hoechst (Sigma, 1:10,000 dilution of the 1 mg/ml stock) as nuclear counterstain.
2. Antibodies and f-actin counterstain: dilute antibodies directly in PBS. Add fluorescently labeled (e.g., with CF405) phalloidin (VWR) to the secondary antibody mixture for f-actin counterstaining, dilute 1:100. Use the following dilutions for primary antibodies: total Tau (K9JA, Dako, 1:500); pTau-epitope present in axon (PHF1, generously provided by P. Davies, AEC, NY, dilute 1:200); pTau-epitope present in dendrites (12E8, Elan, dilute 1:300). Secondary fluorescent antibodies from VWR (dilution of 1:200; CF488, CF555, CF647, donkey, low cross-reaction) work fine.
3. Washing solution: all washing steps are carried out with PBS if not stated otherwise.
4. Mounting medium: we use Polymount (Polysciences), but any mounting media will do.

3 Methods

3.1 Prepare Cell Culture Plates for Neurons

1. Put 1 coverslip in each well of a 12-well culture plate.
2. Add 1 ml of coating solution (50 μg/ml poly-D-lysine in PBS) into each well.
3. Incubate plate at 37 °C overnight, e.g., in a humidified cell culture incubator, plates must not dry.
4. Wash carefully three times with PBS (*see* **Note 1**). Plates can be stored after washing in the last washing solution for up to 3 weeks at 4 °C.

3.2 Preparation of a Single Cell Solution

1. Dissect rat embryos at day 18.5 or mice embryos at day 16.5 of gestation for preparation of forebrain neurons
2. Dissect out the complete brain under sterile conditions, carefully remove meninges, brain stem, olfactory bulb, thalamus, also separate cortex from hippocampus if desired (note that mixed cultures are feasible too). Keep the dissected tissue always in cold (4 °C or on ice) HBSS during the dissection.
3. Add 1× Trypsin (0.1 ml for each hippocampus, and 0.25 ml for each cortex half) and incubate for 8 min at 37 °C.
4. Inactivate trypsin with 10% horse serum (HS) in HBSS by at least doubling the volume (i.e., if you have 5 ml of trypsinized cells, add at least 5 ml of 10% HS in HBSS).
5. Remove trypsin–HS mixture and dilute cells with approximately 0.25 ml per hippocampus and 2 ml per cortex with HBSS for cell counting.

3.3 Plating of Neurons

1. When the single cell solution has been obtained, dilute it to 200,000/ml (cortical neurons) or 100,000/ml (hippocampal neurons) in pre-warmed (37 °C) plating medium (*see* **Note 2**).
2. Plate 1.5 ml of the cell suspension into each well of a poly-D-lysine coated and washed 12-well plate (*see* **Note 3**).
3. Transfer plate immediately into a normal humidified cell culture incubator, which is kept at 37 °C and 5% CO_2.

3.4 Feeding and Growing of Neurons

1. Grow rat neurons for 3 days and mouse neurons for 4 days after plating in a normal humidified cell culture incubator at 37 °C and 5% CO_2.
2. Right before addition of growth medium, add fresh AraC (*see* **Note 4**).
3. Double the volume of the cell culture medium by adding growth medium. Do not replace medium, only add (*see* also **Note 5**). For example, if you plated 300,000 cells in 1.5 ml of plating medium in a 12-well plate, add 1.5 ml growth medium to obtain a final volume of 3 ml. Cells can now be grown up to 2 months, depending on what kind of analysis of Tau is desired (*see* **Note 6**).

3.5 Silanizing Multi-well Plates for Convenient Immunostainings with Low Antibody Consumption

Conduct all steps under a fume hood.

1. Add 2–4 ml of silanizing solution to the first well, so that the bottom and the first few mm of the wall are wetted. Reaction is instantaneous so there is no need to incubate.
2. Suck out silanizing solution and transfer to the next well.
3. Repeat **steps 1–2** until the whole plate has been wetted with silanizing solution. Silanizing solution can be used several times until used up.
4. Dry plate overnight under a fume hood. Plastic will turn whitish-turbid.

3.6 Fixing and Staining Neurons

At the desired stage of Tau development, neurons can then be treated (*see* **Note 7**) and fixed for immunofluorescence analysis.

1. Remove cell culture medium completely and immediately add fixation solution (*see* **Note 8**).
2. Incubate for 1 h at room temperature or for 30 min at 37 °C.
3. Wash three times with PBS.
4. Add blocking/permeabilizing solution for 5 min at room temperature.
5. Wash twice with PBS.
6. Transfer coverslips to PBS-filled silanized multi-well plate, take care that cells are on the upper side of the coverslip (i.e., do not flip coverslips).
7. Remove PBS completely, one well at a time.
8. Add 45 μl of the primary antibody mixture on top of the coverslip, one well at a time. The silanization of the multi-well plate will prevent the antibody solution from flowing off the coverslip. Go back to **step 7** for the next well and repeat until all coverslips are processed.
9. Incubate for 16 h at 4 °C.
10. Wash three times with PBS.
11. Remove PBS completely, one well at a time.
12. Add 45 μl of the second antibody mixture on top of the coverslip, one well at a time. The silanization of the multi-well plate will prevent the antibody solution from flowing off the coverslip. Go back to **step 11** for the next well and repeat until all coverslips are processed.
13. Incubate for 1 h at 37 °C (*see* **Note 9**).
14. Wash three times with PBS.
15. Mount coverslips on microcopy glass slides upside down with a drop (5 μl) of mounting medium. Take care that coverslips are mounted as plane as possible.
16. Allow mounting medium to dry/polymerize overnight at 4 °C before imaging. Microscopy slides can be stored at 4 °C for at least 5 years with little decrease in fluorescence intensity.

3.7 Optional: Cryopreservation of Fixed Cells

After fixation, but before permeabilization, fixed cells can be stored for up to 2 years in storing solution at –20 °C. For this, wash fixed cells once with PBS, then add sufficient volume to completely immerse cells in storage solution (e.g., 0.5 ml into a 24-well plate, or 1 ml into a 12-well plate). Storage solution will not freeze at –20 °C, so when the time comes, one can simply take out the multi-well plate in which the coverslips are kept out of the freezer, wash them once with PBS, and go to **step 3.6.3**.

4 Notes

1. Coating of plates requires at least 3 h, but results are poorer (i.e., cells do not grow as well and not as long) if coating is done for less than 12 h. Washing needs to be done with great caution, as common preparations of poly-D-lysine have trace amounts of toxic substances that prevent growth of neurons.
2. Cell Culture medium must not be oxygenated. Usually, cell culture medium (also the recommended Nb medium) contains phenol red as a pH indicator, color must be bright red, but not purple or yellow. If color is other than bright red, pre-incubate medium in 5 % CO_2 until color turns bright red.
3. Cell density can be easily adjusted by changing the volume of the solution for plating, but this cell/volume ratio has proven to be robust. We use the same plating solution also for 24- and 6-well plates, but with different volumes (e.g., roughly half for 24-well plates, double for 6-well plates, but best conditions need to be determined for each well size and desired cell density). For variations see e.g. [7].
4. AraC is rather unstable, it will not resist several freeze–thaw cycles. Prepare a 500× stock and make small aliquots which then can be stored at −20 °C.
5. Primary neurons are very sensitive to medium replacement, especially at later stages of development. While it apparently is convenient to double the medium after a few days of culture, already after a week neurons do not support medium replacement. With this protocol neurons are usually fine for up to 2 months, but if medium replacement becomes necessary (e.g., for high-density long-term cultures) do not replace more than 20% of the medium. In this case omit the AraC of the growing medium.
6. Establishment of polarized distribution of Tau and differential Tau phosphorylation starts at around 3 div, and is usually completed within 12 div [8]. Mature Tau isoforms start to appear at 7–14 div. If one wants to study Tau in the context of adult-like neurotransmission and age-dependent neurodegeneration, cultures should be maintained 14–21 days or more.
7. Treatment of primary neurons with cell culture compatible (i.e., nontoxic) substances is not trivial, the following points must be taken into consideration: (1) Primary neurons, especially after weeks of culturing, do not support medium replacement. Even fresh and completely supplemented growth medium is insufficient to sustain growth. Instead, primary neurons need to be kept in "conditioned medium" (CM), which means that the growth medium has been conditioned by the neuron (or alternatively astrocytes) for several days. Thus, treatments of primary neurons should be conducted simply by adding substances to the CM and not by replacing the CM.

(2) To save valuable substances, it is possible to reduce the amount of CM, e.g., in a 12-well plate from 2 to 0.5 ml for a limited time (1–3 days is usually fine, but this depends on the age of the culture, the older the more difficult it gets). Replacing more than 10% of CM with PBS, or adding fresh growth medium or other non-neuronal cell culture medium will influence Tau distribution and phosphorylation (see for example ref. [9]). (3) Tau distribution and phosphorylation is also temperature sensitive [10], so good care has to be taken to treat samples fast and to maintain them at 37 °C.

8. The standard fixation solution is 3.7% FA in PBS, but slightly better results can be obtained by using double concentrated fixation solution (i.e., 7.4% FA in PBS) which is added on top of the cell culture volume in an amount to reach a final concentration of 3.7% FA. For example if cells are grown in 1 ml medium, add 1 ml of 7.4% FA. This not only gives slightly better fixation results, but also allows for faster processing and reduces cellular stress (as media is not removed from cells, cells are not exposed to oxygen, etc.). Alternatively, 3.7% FA can also be supplemented with 4% sucrose.
9. The incubation step with the secondary antibody mixture is quite robust and can be 1 h at 37 °C, 3 h at room temperature and overnight at 4 °C, but must be kept constant for reproducible results.

Acknowledgments

The authors would like to acknowledge Bianca Helling, who tested the NS21 formulation, Julia Luedtke for stimulating discussions and excellent technical expertise, and Eckhard Mandelkow for support. We are grateful for funding from DZNE and MPG.

References

1. Binder LI, Frankfurter A, Rebhun LI (1985) The distribution of tau in the mammalian central nervous system. J Cell Biol 101:1371–1378
2. Bullmann T, Holzer M, Mori H, Arendt T (2009) Pattern of tau isoforms expression during development in vivo. Int J Dev Neurosci 27:591–597. doi:10.1016/j.ijdevneu.2009.06.001
3. Goedert M, Spillantini MG, Potier MC, Ulrich J, Crowther RA (1989) Cloning and sequencing of the cDNA encoding an isoform of microtubule-associated protein tau containing four tandem repeats: differential expression of tau protein mRNAs in human brain. EMBO J 8:393–399
4. Zempel H, Mandelkow E (2014) Lost after translation: missorting of Tau protein and consequences for Alzheimer disease. Trends Neurosci 37:721–732. doi:10.1016/j.tins.2014.08.004
5. Zempel H, Luedtke J, Kumar Y, Biernat J, Dawson H, Mandelkow E, Mandelkow E-M (2013) Amyloid-β oligomers induce synaptic damage via Tau-dependent microtubule severing by TTLL6 and spastin. EMBO J 32:2920–2937. doi:10.1038/emboj.2013.207
6. Chen Y, Stevens B, Chang J, Milbrandt J, Barres BA, Hell JW (2008) NS21: re-defined and modified supplement B27 for neuronal cultures. J Neurosci Methods 171:239–247. doi:10.1016/j.jneumeth.2008.03.013

7. Fath T, Ke YD, Gunning P, Götz J, Ittner LM (2009) Primary support cultures of hippocampal and substantia nigra neurons. Nat Protoc 4:78–85. doi:10.1038/nprot.2008.199
8. Li X, Kumar Y, Zempel H, Mandelkow E-M, Biernat J, Mandelkow E (2011) Novel diffusion barrier for axonal retention of Tau in neurons and its failure in neurodegeneration. EMBO J 30:4825–4837. doi:10.1038/emboj.2011.376
9. Zempel H, Thies E, Mandelkow E, Mandelkow E-M (2010) Abeta oligomers cause localized Ca(2+) elevation, missorting of endogenous Tau into dendrites, Tau phosphorylation, and destruction of microtubules and spines. J Neurosci 30:11938–11950. doi:10.1523/JNEUROSCI.2357-10.2010
10. Bretteville A, Marcouiller F, Julien C, El Khoury NB, Petry FR, Poitras I, Mouginot D, Lévesque G, Hébert SS, Planel E (2012) Hypothermia-induced hyperphosphorylation: a new model to study tau kinase inhibitors. Sci Rep 2:480. doi:10.1038/srep00480

Chapter 21

Tracking Tau in Neurons: How to Transfect and Track Exogenous Tau into Primary Neurons

Hans Zempel, Julia Luedtke, and Eva-Maria Mandelkow

Abstract

Primary neurons have proved to be an essential tool for investigating neuronal polarity in general and polarized Tau distribution in particular. However, mature primary neurons are notoriously difficult to transfect with nonviral vectors and are very sensitive both to cytoskeletal manipulation and to imaging. Common nonviral transfections require the use of a monolayer of supportive glia or high density cultures, both of which complicate imaging. Here, we provide a simple nonviral transfection method enabling transfection of Tau to achieve expression levels comparable to endogenous Tau. This allows to investigate specific effects on, e.g., distribution and transport of Tau, without grossly affecting other cytoskeleton-based parameters such as microtubule density or microtubule-based transport.

Key words Tau, Transfection, Primary neuronal cell culture, Polarity, Neurodegeneration

1 Introduction

Primary neuronal cultures from the forebrain, especially from the developing cortex and hippocampus, are widely used for the investigation of neuronal cell maturation, including synaptogenesis and cell polarity. Using low density cultures, differential axodendritic and synaptic distribution of the protein in question can be investigated with high magnification power light and fluorescence microscopes. Unfortunately, primary neurons are difficult to grow to maturity and are notoriously difficult to transfect with increasing age using nonviral vectors. On the other hand, generating viral vectors is time consuming, and titer control is often unprecise, making it difficult to compare different constructs and to avoid overexpression.

Here, we describe a protocol based on liposomal transfection, which is compatible with a standardized, reproducible and economical culturing procedure for primary neurons (*see* Chapter 20). Avoiding overexpression is particularly important for the protein Tau, as even wildtype-Tau can over-stabilize and bundle microtubules, and impair microtubule-based transport, leading to

Caroline Smet-Nocca (ed.), *Tau Protein: Methods and Protocols*, Methods in Molecular Biology, vol. 1523,
DOI 10.1007/978-1-4939-6598-4_21, © Springer Science+Business Media New York 2017

loss of spines and other detrimental effects (see, e.g., [1, 2]). As Tau can interact with the microtubule as well as with the actin cytoskeleton, overexpression can induce a wide range of cytoskeletal disturbances, possibly masking more subtle signaling or pathology-related effects [2–6]. Only when lower expression levels are used, one can investigate effects caused by mutated or (pseudo)phosphorylated Tau [2, 7, 8]. Making use of this protocol, we achieve a transfection efficiency of 30%, but avoid strong overexpression. Importantly, this transfection method can be used with cultures of several weeks of age, even in cultures without support glia, and at suitable cell densities for microscopy-based single cell analysis. This protocol favors expression in neurons vs. glial cells and also allows handling in the same multi-well plate/dish in which the neurons are growing, without transferring coverslips.

2 Materials

2.1 Cell Culture

Primary neurons are grown as described elsewhere (*see* Chapter 20). The age of neurons is of little importance, but expression (not so much transfection) efficiency reduces with days in culture (*see* **Note 1**).

2.2 Plasmids

We use CMV-promotor-based plasmids. Commonly used plasmids for Tau are, e.g., pRK5-EGFP-Tau with and without the P301L mutation (both available from Addgene, *see* also [8]). As a transfection marker, we use tdTomato (available from Addgene, deposited by M. Davidson), which is very bright and at the expression levels of this protocol well tolerated by the cells. We recommend using very bright fluorescent proteins as transfection markers, because of the low expression efficiency of achieved by this protocol.

2.3 Transfection Reagents

All reagents should be handled under sterile conditions.

1. Lipofectamine2000 (L2K, Invitrogen).
2. Unsupplemented Neurobasal (Nb) medium (Gibco), preheated at 37 °C and preincubated with 5% CO_2 (*see* **Note 2**).
3. Sufficient amounts of conditioned medium (CM). CM refers to the growth medium in which the neurons are grown. CM should be from the same neuronal culture to be transfected.

3 Methods

Primary neurons should be kept in a humidified incubator at 37 °C with 5% CO_2. These cells do not tolerate medium exchange without supporting glia; thus all steps should be conducted with conditioned medium if not indicated otherwise. Volumes and amounts are given for 12-well plates, but the same volumes can be used for 24-well plates. Volumes and amounts should be doubled for 6-well plates.

3.1 Preparation of Primary Neurons for Transfection

1. Reduce the amount of conditioned medium per well to 0.5 ml before you start the preparation of the transfection mix (*see* Subheading 3.2). Do not discard the aspirated conditioned medium.
2. Place neurons back into the incubator for the time of the preparation of the transfection mix.
3. Keep conditioned medium at 37 °C under 5% CO_2 for later use, if left over it can be stored at –20 °C (*see* **Note 3**).

3.2 Preparation of Transfection Mix

1. DNA-mix: Mix 0.5 μg of each DNA (*see* **Note 4**) and 30 μl of fresh, unsupplemented Nb medium.
2. L2K-mix: Mix 1.5 μl of L2K with 30 μl fresh, unsupplemented Nb medium.
3. Incubate L2K-mix for 5 min.
4. Combine DNA-mix and L2K-mix, which is referred to as the transfection mix.
5. Incubate the transfection mix for 30 min at room temperature (*see* **Note 5**).

3.3 Transfection and Washing Procedure

1. Right before transfection, take cells with adjusted amount of conditioned medium (0.5 ml, *see* **step 1** of Subheading 3.1) out of the incubator.
2. Add the transfection mix dropwise on top of the conditioned medium of the well to be transfected, from about 1.5 to 2 cm distance.
3. After all wells to be transfected were treated, gently swirl the whole plate three to five times.
4. Place the cells immediately back into the incubator (*see* **Note 6**).
5. Incubate cells for 3 h with the transfection mix in the incubator, time can be adjusted to 1–6 h depending on the desired expression level (*see* **Note 7**).
6. Wash cells gently two times with pre-warmed (37 °C), 5% CO_2 pH-equilibrated and unsupplemented Nb medium (*see* **Note 2**).
7. Place pre-warmed (37 °C) conditioned medium, preferentially from the same culture (*see* Subheading 3.1), back onto the cells (*see* **Note 2**).
8. Place cells immediately back into the incubator.
9. Allow for at least 2 days for young (7 div) and for 4 days for old (18 div) cells for moderately expressing proteins, such as wild-type Tau protein (*see* **Note 8**).

3.4 Optional: Increasing Expression Efficiency

The following steps can be modified to increase the expression efficiency. Modifications will result in faster and higher protein expression (but *see* **Note 8** for adverse effects of high expression).

Alternative **step 4** of Subheading 3.2. DNA-mix: Mix 2 μg (instead of 0.5 μg) of each DNA (*see* **Note 4**) and 30 μl of fresh, unsupplemented Nb medium.

Alternative **step 1** of Subheading 3.1. Reduce the amount of conditioned medium per well to 0.4 ml (instead of 0.5 ml) before you start the preparation of the transfection mix (*see* Subheading 3.2). Do not discard the aspirated conditioned medium. Add 0.1 ml per well of fresh, unsupplemented Nb medium (*see* **Note 9**).

4 Notes

1. Expression efficiency of this protocol is good, but does not compare to other protocols (especially viral based). Thus, if one wants to test the transfection efficiency, one should investigate it with a bright fluorescent protein without or with little physiological or pathological activity in neurons. Expression of an untagged fluorescent protein such as tdTomato can be observed as soon as 1 day after transfection in young (up to 14 div) neurons. In older cells (14 div) expression is still good and relatively stable for 1 week, thus if one wants to investigate mature neurons of an age around 14–18 div, transfection at an age of 10–14 div is suitable. Transfection of cells 18 div and older is possible and feasible, but in this case cells should be allowed to express the protein for 5–6 days. Importantly, it has to be taken into account that independent of the age of the neurons, cells that highly express any transfected protein already after 1–2 days will die after 3–5 days simply due to overexpression stress. Thus, analyzing only brightly fluorescent cells 1–2 days after transfection will yield biased results, because at that time only soon-to-die cells will express the transfected proteins in high amounts.
2. All media that come in direct contact with the cells should always be maintained at 37 °C and should be equilibrated with 5% CO_2. Nb medium contains the pH-sensitive dye phenol red, which should be bright red. If the medium is purple, do not use it. Nb and CM will change their pH within minutes under normal atmospheric conditions (i.e., without 5% CO_2) and become purple. In that case, place them in an incubator diffused with 5% CO_2 and until medium is again bright red. Sufficient experience with these media should be acquired before using media without phenol red.
3. Conditioned medium can be stored at −20 °C for future use. This is particularly useful when this transfection protocol is

used for multi-well plates with small volumes, as then the aspirated amount is insufficient for further processing and growing of cells. Before use, heat conditioned medium to 37 °C and keep it under 5 % CO_2 until the medium color becomes bright red.

4. Several DNAs (we use up to 3 with very good results) can be mixed together at different ratios, depending on the desired expression efficacy. If co-transfection is desired, DNAs have to be mixed prior to mixing the DNA-mix with the L2K-mix. Expression efficacy can easily be adjusted by changing the amount of DNA. As fluorescent proteins not tagged to functional proteins have usually very high expression efficiencies, we often reduce the DNA amount to half the suggested amount (i.e., 0.25 μg for one-well of a 12-well plate). For Tau protein, 0.5 μg per well of a 12-well plate seems to be a good compromise between visibility with decent (eGFP, YFP, Citrine) fluorescent proteins and sufficiently low expression in order to not disturb physiological cellular functions.
5. Incubation time of the transfection mix can be varied from 20 to 45 min, but best results are obtained after 30 min.
6. The time of adding the transfection mix to the multi-well plate should be kept below 3 min, as the reduced amount of conditioned medium quickly alkalizes under common laboratory airflow clean benches. It is important to add the transfection mix in a dropwise manner, thus transfecting all wells of a 24-well multiwell plate manually is challenging but still possible.
7. Transfection time can be varied to adjust the expression level. 3 h-transfection time works for all ages of neurons and is particularly suited for expression of Tau protein to near physiological levels in already polarized and synaptically active primary neurons aged 10–14 div. In case of very young neurons (3–7 div), which are easily transfected and express high amounts of the transfected protein, we decrease transfection time to 1 h, in case of old neurons (>15 div), which are more difficult to transfect and express only low levels of the transfected protein, we increase the transfection time up to 6 h. We do not observe increased cell death up to 6 h of transfection time.
8. In case of wildtype Tau protein, depending on the type of analysis, expression should be at least 2–3 days in mature primary neurons, due to two reasons: (a) Tau protein propagates very slowly into the axon, in 10 div cortical neurons after 24 h of expression it has not even reached the growth cone (i.e., the end of the axon), which means that at least for some Tau proteins a certain amount of time (>1 day) is required to obtain homogenous or physiological-like distribution throughout the whole cell [9]. (b) It is certainly possible to attain sufficient expression even after 1 day of transfection, but cells expressing

such high amounts of protein will soon succumb due to unspecific high expression-related toxicity. In this case, well-designed controls are necessary if, e.g., one wants to investigate specific Tau protein-related toxicity.

9. The ratio of fresh Nb versus conditioned medium can be increased up to 1:1 for cells younger than 10 div, but will negatively affect cell viability, especially for older cells.

Acknowledgments

The authors thank Eckhard Mandelkow for support. We are grateful for funding from DZNE and MPG.

References

1. Thies E, Mandelkow E-M (2007) Missorting of tau in neurons causes degeneration of synapses that can be rescued by the kinase MARK2/Par-1. J Neurosci 27:2896–2907. doi:10.1523/JNEUROSCI.4674-06.2007
2. Zempel H, Luedtke J, Kumar Y, Biernat J, Dawson H, Mandelkow E, Mandelkow E-M (2013) Amyloid-β oligomers induce synaptic damage via Tau-dependent microtubule severing by TTLL6 and spastin. EMBO J 32:2920–2937. doi:10.1038/emboj.2013.207
3. Biernat J, Wu Y-Z, Timm T, Zheng-Fischhöfer Q, Mandelkow E, Meijer L, Mandelkow E-M (2002) Protein kinase MARK/PAR-1 is required for neurite outgrowth and establishment of neuronal polarity. Mol Biol Cell 13:4013–4028. doi:10.1091/mbc.02-03-0046
4. Sharma VM, Litersky JM, Bhaskar K, Lee G (2007) Tau impacts on growth-factor-stimulated actin remodeling. J Cell Sci 120:748–757. doi:10.1242/jcs.03378
5. Whiteman IT, Gervasio OL, Cullen KM, Guillemin GJ, Jeong EV, Witting PK, Antao ST, Minamide LS, Bamburg JR, Goldsbury C (2009) Activated actin-depolymerizing factor/cofilin sequesters phosphorylated microtubule-associated protein during the assembly of Alzheimer-like neuritic cytoskeletal striations. J Neurosci Off J Soc Neurosci 29:12994–13005. doi:10.1523/JNEUROSCI.3531-09.2009
6. Zempel H, Mandelkow E (2014) Lost after translation: missorting of Tau protein and consequences for Alzheimer disease. Trends Neurosci 37:721–732. doi:10.1016/j.tins.2014.08.004
7. Frandemiche ML, De Seranno S, Rush T, Borel E, Elie A, Arnal I, Lanté F, Buisson A (2014) Activity-dependent tau protein translocation to excitatory synapse is disrupted by exposure to amyloid-beta oligomers. J Neurosci 34:6084–6097. doi:10.1523/JNEUROSCI.4261-13.2014
8. Hoover BR, Reed MN, Su J, Penrod RD, Kotilinek LA, Grant MK, Pitstick R, Carlson GA, Lanier LM, Yuan L-L, Ashe KH, Liao D (2010) Tau mislocalization to dendritic spines mediates synaptic dysfunction independently of neurodegeneration. Neuron 68:1067–1081. doi:10.1016/j.neuron.2010.11.030
9. Li X, Kumar Y, Zempel H, Mandelkow E-M, Biernat J, Mandelkow E (2011) Novel diffusion barrier for axonal retention of Tau in neurons and its failure in neurodegeneration. EMBO J 30:4825–4837. doi:10.1038/emboj.2011.376

Chapter 22

Image-Based Analysis of Intracellular Tau Aggregation by Using Tau-BiFC Cell Model

Sungsu Lim, Dohee Kim, Dong Jin Kim, and Yun Kyung Kim

Abstract

Abnormal tau aggregation is a pathological hallmark of neurodegenerative disease classified as tauopathy. Preventing tau aggregation becomes an important therapeutic strategy to cure tau-mediated neurodegeneration. Here, we describe a method to investigate intracellular tau aggregation by using a recently developed tau aggregation cell-based model named tau-BiFC. High-throughput and high-contents screening method for quantifying intracellular tau aggregation would expedite the discovery of drugs that inhibit tau aggregation.

Key words Tau protein, Tau aggregation, Tau-BiFC cells, Tau phosphorylation, Forskolin, Cell image analysis, High content imaging system (Operetta™), Harmony® software

1 Introduction

Tau is a neuron-specific microtubule-associated protein (MAP) that binds microtubule and promotes microtubule assembly [1]. When pathologically hyperphosphorylated, tau loses its binding affinity to microtubules and becomes insoluble aggregates called neurofibrillary tangles (NFTs) [2, 3]. The accumulation of hyperphosphorylated tau and the aggregates in the brain is the pathological hallmark of multiple neurodegenerative diseases including Alzheimer's disease and frontotemporal dementia [4–6]. Once tau aggregates, microtubule is destabilized causing neuronal degeneration. Moreover, tau aggregates itself have gain of toxicity inducing neuronal cell death [2, 7]. Due to the pathological significance, preventing tau aggregation becomes an important therapeutic strategy to cure tau-mediated neurodegenerative disorders.

Recently, we have developed a cell-based sensor, named tau-BiFC (Bimolecular Fluorescence Complementation) as a tool for investigating intracellular tau aggregation [8–11]. To visualize tau aggregation in real time, nonfluorescent N- and C-terminal compartments of Venus protein are fused to full-length tau and

Caroline Smet-Nocca (ed.), *Tau Protein: Methods and Protocols*, Methods in Molecular Biology, vol. 1523,
DOI 10.1007/978-1-4939-6598-4_22, © Springer Science+Business Media New York 2017

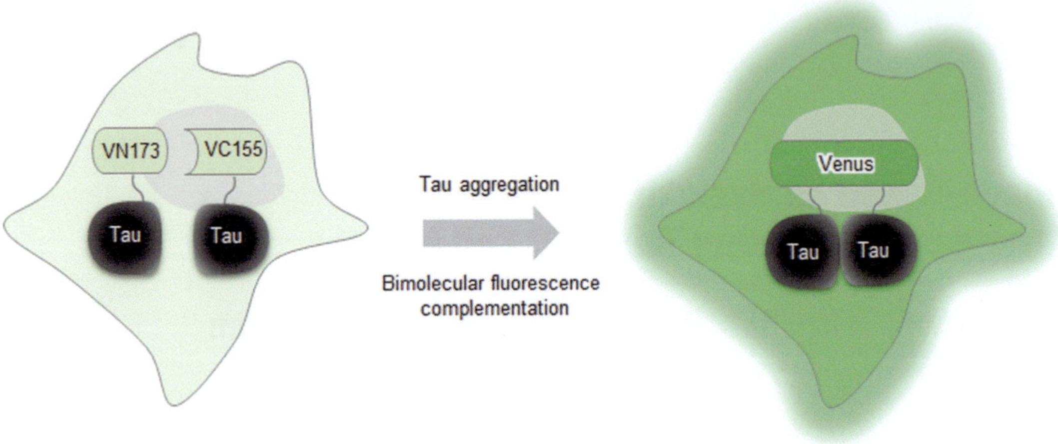

Fig. 1 Tau-BiFC cell model. Nonfluorescent N- and C-terminal compartments of Venus protein are fused to full-length tau. Venus fluorescence turns on only when tau assembles

co-expressed in a cell. Only when tau assembles, the Venus fluorescence turns on as an indication of tau assembly (Fig. 1). The tau-BiFC technique allows for monitoring and quantifying intracellular tau aggregation from the early stage of tau assembly. Here, we describe a method to investigate intracellular tau aggregation by using tau-BiFC cell model. For the investigation, here we use Forskolin, a well-known activator of tau phosphorylation, to tau-BiFC cells. Then, the intensity of tau-BiFC fluorescence is monitored and quantified automatically by using a high content imaging system (Operetta®).

2 Materials

1. Cell line: HEK293 Tau-BiFC cells.
2. Growth medium: Dulbecco's Modified Eagle's Medium (DMEM) that supplemented with 10% fetal bovine serum (FBS), 100 units/ml penicillin and 100 μg/ml streptomycin and 100 μg/ml Geneticin (G418 sulfate solution); store at 4 °C.
3. Balanced salt solution: Dulbecco's Phosphate Buffered Saline (DPBS) pH 7.3 ± 0.3; store at 4 °C.
4. Dissociation reagent: Trypsin-EDTA solution contains 0.05% Porcine Trypsin (1:250), 0.53 mM EDTA·4Na, and Phenol Red in HBSS without Ca^{2+} and Mg^{2+}; store at 4 °C.
5. Cell counting: 0.4% Trypan Blue solution and hemocytometer.
6. Plastic ware: Cell-culture dishes, 15 ml disposable conical centrifuge tubes, and 50 ml reservoir with sloping bottom.
7. 8-channel multichannel 30–300 μl pipettor.

8. 384-well μClear™ Polystyrene Microplates (Greiner).
9. Compound stock solution: 100 mM Forskolin [10] in dimethyl sulfoxide (DMSO); store at −20 °C.
10. High content imaging system: Operetta™ (PerkinElmer™).
11. High content imaging and analysis software: Harmony® 3.1 software (PerkinElmer™).

3 Methods

3.1 Cell Culture Preparation

For tau-BiFC assay, cell passages should not be higher than 40. Tau-BiFC response will be reduced over continuous passaging. Therefore, it is important to freeze a large number of aliquots of the cells at an early passage, and pull out a fresh vial of cells from the early passage stock before the current batch of cells hits their passage limit. In addition, for the optimum results, the cells should be passaged 2 days before the experiment to promote good proliferation and cell physiology.

All cell culture procedures are carried out in a laminar flow hood.

1. Remove and discard the culture medium from the culture dishes of tau-BiFC cells using pipette with a vacuum suction.
2. Wash cells with 5 ml of pre-warmed DPBS. Then, collect the DPBS and discard.
3. Add 1.5 ml of the pre-warmed Trypsin-EDTA solution to the side of the dishes. Then, incubate the culture dishes at room temperature for approximately 2–3 min.
4. When ≥90% of the cells have detached, add 3 ml of pre-warmed growth medium to the cell suspension. Transfer the suspended cells to a new 15 ml conical tube and centrifuge at 200 × *g* for 2 min. Discard the dissociation reagent solution.
5. Resuspend the cell pellet in 4 ml of pre-warmed growth medium. Make uniform cell suspension by pipetting. HEK293 cells tend to grow in clumps. One cells clump, it is difficult to analyze the tau-BiFC fluorescence properties per a cell (*see* **Note 1**).
6. Count the number of viable cells/ml and calculate the total cell number using hemocytometer with Trypan Blue [12]. Dilute cells in 20 ml of pre-warmed growth medium to prepare (7×10^4 cells/ml) cell suspension (*see* **Note 2**).
7. Dispense 50 μl of the cell suspension to each well of a 384-well plate using multichannel pipette (*see* **Note 3**).
8. Incubate the 384-well plate in 37 °C incubator overnight. For phase contrast and stability of cells, 24 h after plating cells on the 384-well plate are suitable for compound treatment.

3.2 Compound Treatment

For small-molecule screening with tau-BiFC cells, it is advisable to keep final DMSO concentration as low as possible, preferably below 1 % v/v. It is also useful to keep DMSO concentrations constant in serial dilutions of stock compounds in pre-warmed growth medium, to eliminate DMSO toxicity effects.

All procedures of compound treatment are carried out in a laminar flow hood.

1. Warm a growth medium in a 37 °C water bath. Dilute a stock compound to working concentration (30 μM Forskolin) in pre-warmed growth medium. If the compound is not premixed with growth medium, the compound sinks quickly to the bottom of a well.
2. Add 10 μl of the mixture (compound and growth medium) to a 384-well plate (*see* **Note 4**). After the compound treatment, the total volume of medium is 60 μl per a single well of the 384-well plate.
3. Incubate the cell plate in 37 °C incubator. Forskolin-induced tau aggregation appears usually 12 h after the treatment and shows the maximum effect at 48 h.

3.3 Cell Imaging

At the various time points, the entire 384-well plate of tau-BiFC cells is automatically imaged by using Operetta® High Contents Image System (PerkinElmer™). However, to get optimal results, parameter values should be adjusted depending on tau-BiFC cell condition.

1. Set up experimental parameters of tau-BiFC assay using Harmony® 3.1 software (PerkinElmer™). For tau-BiFC cell imaging, we select a 10 x objective and tau-BiFC channel (Excitation: 460–490 nm; Emission: 500–550 nm) (*see* **Note 5**).
2. On the *Run experiment* screen, insert a plate name and start. During the measurement, we have to check the images of tau-BiFC cells (Fig. 2).

3.4 Cell Image Analysis and Evaluation

Images of tau-BiFC cells are analyzed using Harmony® 3.1 software.

1. After the measurement, select the images to be evaluated. Since a separation of a single cell depends on BiFC fluorescence during the image analysis, the detected cells are tau-BiFC positive cells (Fig. 3).
2. Select three *Lists of output* with parameters (*see* **Note 6**): (A) number of tau-BiFC positive cells, (B) average intensity, and (C) cell area with mean values. Then, set up *Formula output* (A*B*C) to calculate total tau-BiFC response per a well. Formula output indicates integrated intensity of BiFC on tau-BiFC cells.
3. On the *Evaluation* screen, select wells and fields to be evaluated and start. After the evaluation, make a graph of total tau-BiFC response (Fig. 4).

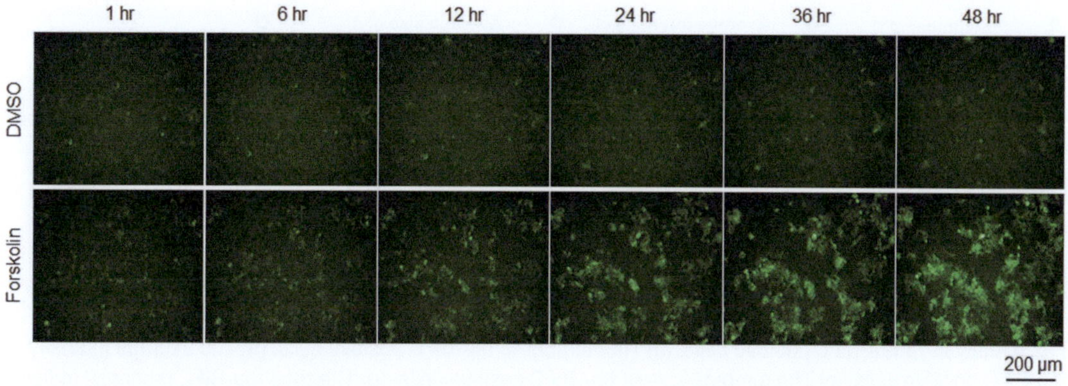

Fig. 2 Representative images of tau-BiFC cells upon the treatment of DMSO or Forskolin. Upon the treatment of Forskolin, the intensity of tau-BiFC fluorescence increased time-dependency. Scale bar, 200 μm

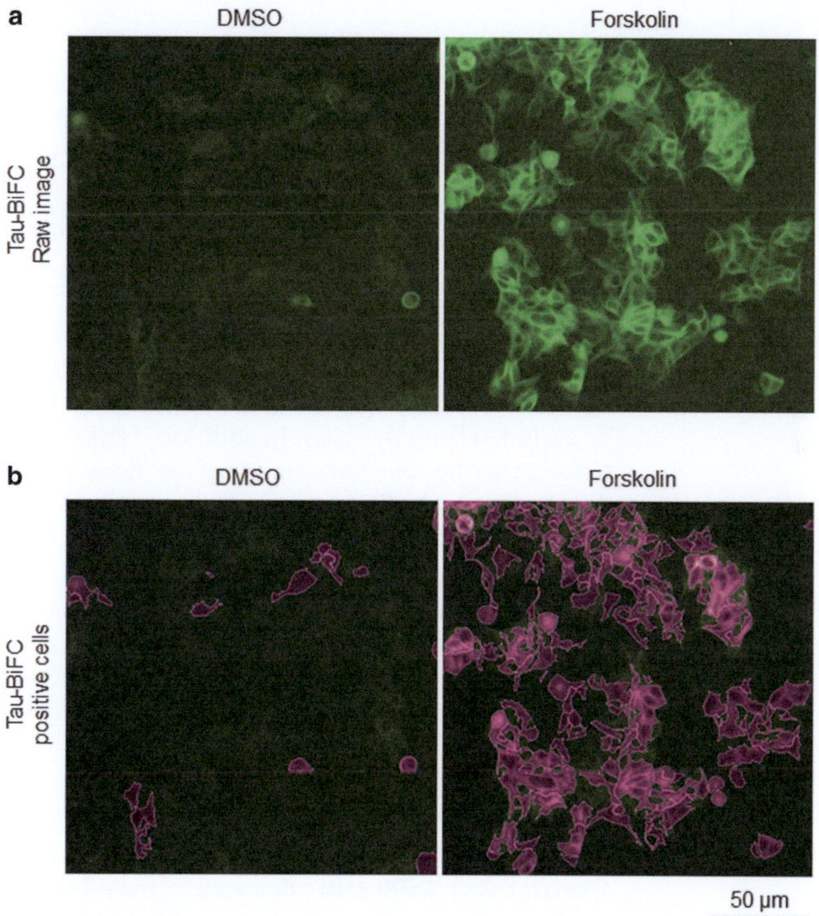

Fig. 3 Selection of tau-BiFC positive cells. (**a**) The images of tau-BiFC cells after 36 h of the treatment of compounds. (**b**) Tau-BiFC positive cells were selected automatically by using a Harmony® software for further analysis. Scale bar, 50 μm

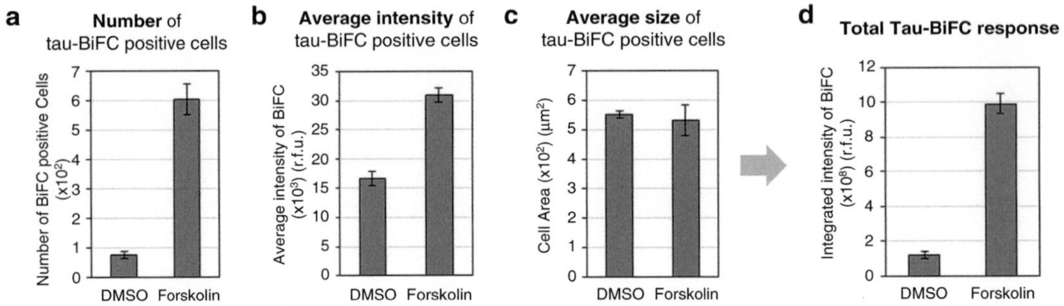

Fig. 4 Analysis of tau-BiFC positive cells. (**a**) The number of tau-BiFC positive cells. (**b**) The average intensity of tau-BiFC positive cells. (**c**) The average size of tau-BiFC positive cells. (**d**) The total tau-BiFC response indicating tau aggregation

4 Notes

1. Before plating cells on a 384-well plate, pay extra attention to make sure cells to be separated as a single unit. To avoid cell clumping, it is recommended to use 37 °C pre-warmed medium and pipetting the cells gently before plating.
2. Seeding 3500 cells per a well of a 384-well plate is the most optimal condition for tau-BiFC screening. We use a 384-well plate μClear™.
3. While dispensing cells into the 384-well plate, be careful not to create air bubbles. Once air bubbles form on the bottom of a well, it is difficult to get rid of the bubbles, and the well should be excluded from the assay. In addition, HEK293 cells are very sensitive to bubble bursting. For the homogeneous cellular distribution in the 384-well plate, do not tilt or shake the plate while or after dispensing cells.
4. When we do screening using 384-well plates, we avoid using the first two and the last two columns of a 384-well plate. Due to uneven temperature and humidity within a plate, the 384-well plate assay often shows edge-well effects. To evaluate edge effects, assign positive and negative controls of the assay located on both edges in a palindromic conformation.
5. For tau-BiFC cell imaging, we usually select a 10 x objective to observe the largest number of cells at once. The optimal cell number for a calculation of BiFC fluorescence intensity is approximately 1500 cells per a well of the 384-well plate.
6. For the identification of total tau-BiFC positive cells with BiFC fluorescence, we usually select the parameters as hereafter. A: number of tau-BiFC positive cells (*Channel*: tau-BiFC). B: Average intensity with mean value (*Channel*: tau-BiFC). C: Cell with area values (Area > 25 μm²).

Acknowledgment

This work was supported by the National Research Council of Science & Technology (NST) grant by the Korea government (MSIP) (No. CRC-15-04-KIST) and the Brain Research Program through the National Research Foundation of Korea (NRF) funded by the Ministry of Science, ICT & Future Planning (NRF-2016M3C7A1913845).

References

1. Reddy PH (2011) Abnormal tau, mitochondrial dysfunction, impaired axonal transport of mitochondria, and synaptic deprivation in Alzheimer's disease. Brain Res 1415:136–148. doi:10.1016/j.brainres.2011.07.052
2. Gendron TF, Petrucelli L (2009) The role of tau in neurodegeneration. Mol Neurodegener 4:13. doi:10.1186/1750-1326-4-13
3. Mukrasch MD, Biernat J, von Bergen M, Griesinger C, Mandelkow E, Zweckstetter M (2005) Sites of tau important for aggregation populate {beta}-structure and bind to microtubules and polyanions. J Biol Chem 280:24978–24986. doi:10.1074/jbc.M501565200
4. Johnson GV, Hartigan JA (1999) Tau protein in normal and Alzheimer's disease brain: an update. J Alzheimers Dis 1:329–351
5. Buée-Scherrer V, Buée L, Leveugle B, Perl DP, Vermersch P, Hof PR, Delacourte A (1997) Pathological tau proteins in postencephalitic parkinsonism: comparison with Alzheimer's disease and other neurodegenerative disorders. Ann Neurol 42:356–359. doi:10.1002/ana.410420312
6. Buée Scherrer V, Hof PR, Buée L, Leveugle B, Vermersch P, Perl DP, Olanow CW, Delacourte A (1996) Hyperphosphorylated tau proteins differentiate corticobasal degeneration and Pick's disease. Acta Neuropathol 91:351–359
7. Brandt R, Hundelt M, Shahani N (2005) Tau alteration and neuronal degeneration in tauopathies: mechanisms and models. Biochim Biophys Acta 1739:331–354. doi:10.1016/j.bbadis.2004.06.018
8. Han DH, Na H-K, Choi WH, Lee JH, Kim YK, Won C, Lee S-H, Kim KP, Kuret J, Min D-H, Lee MJ (2014) Direct cellular delivery of human proteasomes to delay tau aggregation. Nat Commun 5:5633. doi:10.1038/ncomms6633
9. Lim S, Haque MM, Nam G, Ryoo N, Rhim H, Kim YK (2015) Monitoring of Intracellular Tau Aggregation Regulated by OGA/OGT Inhibitors. Int J Mol Sci 16:20212–20224. doi:10.3390/ijms160920212
10. Tak H, Haque MM, Kim MJ, Lee JH, Baik J-H, Kim Y, Kim DJ, Grailhe R, Kim YK (2013) Bimolecular fluorescence complementation; lighting-up tau-tau interaction in living cells. PLoS One 8:e81682. doi:10.1371/journal.pone.0081682
11. Kim D, Lim S, Haque MM, Ryoo N, Hong HS, Rhim H, Lee D-E, Chang Y-T, Lee J-S, Cheong E, Kim DJ, Kim YK (2015) Identification of disulfide cross-linked tau dimer responsible for tau propagation. Sci Rep 5:15231. doi:10.1038/srep15231
12. Strober, W. (2001) Trypan blue exclusion test of cell viability. Curr Protoc Immunol Ed John E Coligan Al Appendix 3, Appendix 3B. doi:10.1002/0471142735.ima03bs21

Chapter 23

FRET and Flow Cytometry Assays to Measure Proteopathic Seeding Activity in Biological Samples

Jennifer L. Furman and Marc I. Diamond

Abstract

Transcellular propagation of protein aggregates—or seeds—is increasingly implicated as a mechanism for disease progression in many neurodegenerative disorders, including Alzheimer's disease and the related tauopathies. While neuropathology generally originates in one discrete brain region, pathology progresses as disease severity advances, often along discrete neural networks. The stereotypical spread of tau pathology suggests that cell-to-cell transfer of toxic protein aggregates could underlie disease progression, and recent studies implicate seeding as a proximal marker of disease, as compared to standard histological and biochemical analyses. Commonly used metrics for protein aggregation detection, however, lack sensitivity, are not quantitative, and/or undergo subjective classification. Here, we describe a FRET and flow cytometry cell-based assay that allows for rapid and quantitative detection of protein aggregates from human and rodent biological specimens.

Key words FRET, Flow cytometry, Protein aggregation, Spread, Tau, Synuclein, Huntingtin, Prion, Neurodegeneration, Mouse model

1 Introduction

The accumulation and progression of toxic protein aggregates is a prominent feature of many neurodegenerative disorders, including a panel of tauopathies, synucleinopathies, prionopathies, and polyglutamine diseases such as Huntington's. While each disorder has distinct clinical and pathological characteristics, all feature the deposition of insoluble protein, especially as amyloid, or fibrillar assemblies. This accumulation of aggregates is associated with the demise of otherwise healthy brain networks (reviewed in [1–5]).

The microtubule-associated protein tau (MAPT) is not aggregated in healthy brain tissue. It stabilizes microtubules in vitro, while its function within the cell is not as well understood [6]. In disease, however, tau self-aggregates into fibrillar assemblies and instead adopts a toxic role. In contrast to accumulation of amyloid beta, insoluble tau correlates very well with cognitive decline in

Caroline Smet-Nocca (ed.), *Tau Protein: Methods and Protocols*, Methods in Molecular Biology, vol. 1523, DOI 10.1007/978-1-4939-6598-4_23, © Springer Science+Business Media New York 2017

Alzheimer's disease (AD) [7–9]. A recent study demonstrated that tau seeding activity is the earliest detectable marker of tauopathy in a mouse model [10]. This implicates tau seeding as a causal event in disease progression and highlights the importance of establishing an assay that sensitively and quantitatively detects toxic aggregates. Numerous methods for the detection of aberrant protein aggregation and/or seeding activity are currently available, but each has limitations. For example, thioflavin assays generally require high-purity starting material, making them incompatible with brain-derived specimens that contain large amounts of contaminating protein [11]. Other methods, such as protein misfolding cyclic amplification (PMCA)—while reported as compatible with human AD brain homogenates [12]—show relatively large variability between experiments, which ultimately limits their utility. Further, brain levels of soluble vs. insoluble tau may vary considerably, and are difficult to quantify. Immunohistochemistry, the most widely used assay for the detection of aberrant pathology, is only qualitative and analyses may be discrepant from pathologist to pathologist. Here, we describe a novel cell assay based on the use of flow cytometry to measure Forster resonance energy transfer (FRET). This method enables sensitive and specific quantification of tau aggregates in biological samples.

We engineered monoclonal HEK 293T cell lines that constitutively express Tau repeat domain (RD) containing the disease-associated P301S mutation tagged to either CFP or YFP. In the absence of tau aggregates, endogenous Tau RD-CFP/YFP proteins do not aggregate. Upon the introduction of tau seeds to the cells, however, Tau RD-CFP/YFP forms intracellular amyloids that are observable by microscopy (Fig. 1a vs. 1b), and flow cytometry sensitively quantifies the resultant FRET signal (Fig. 1c). By combining the stability of a monoclonal cell line, the specificity of FRET, and the sensitivity of flow cytometry, this technique is ideal for evaluating the presence of tau aggregates in biological samples or for quantifying the therapeutic efficacy of compounds that may decrease tau aggregation.

2 Materials

1. Cell Culture Medium: Dulbecco's Modified Eagle's Medium, 10% Fetal Bovine Serum; 1% Penicillin/Streptomycin, 1% Glutamax.
2. Homogenization buffer: 1× Tris-buffered saline (20 mM Tris, 0.9% NaCl, pH 7.4), Complete Protease Inhibitors (EDTA-free; Roche).
3. Flow cytometry buffer: 1× Hank's Balanced Salt Solution, 1% Fetal Bovine Serum, 1 mM EDTA.
4. Dulbecco's Phosphate-buffered saline (DPBS).

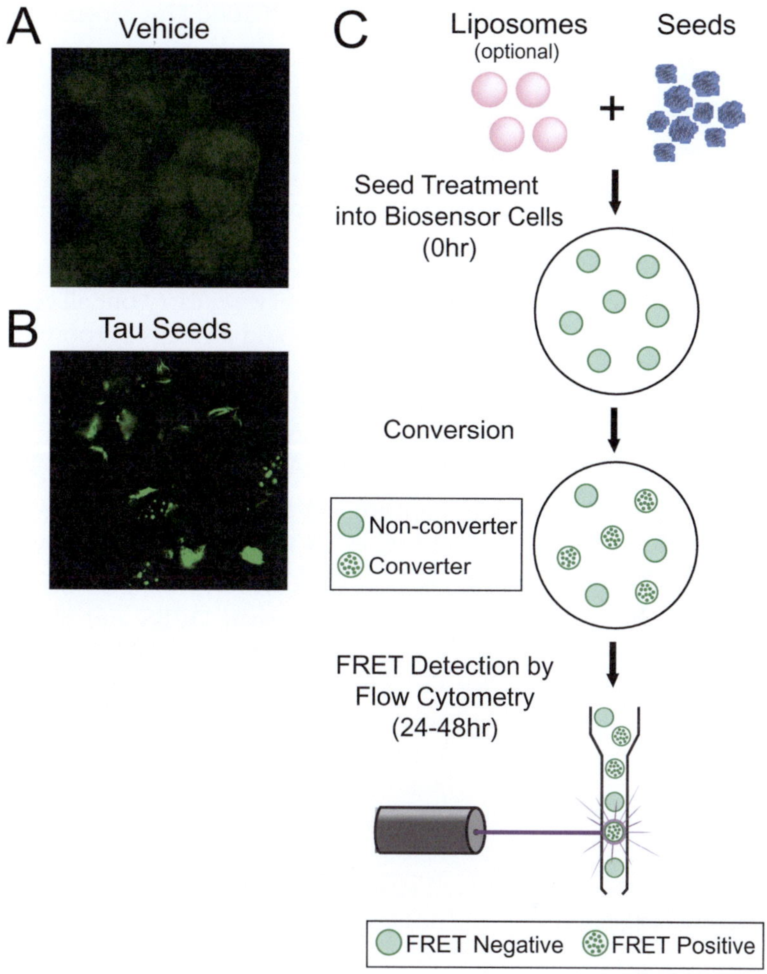

Fig. 1 Detection of Tau Seeding Activity with FRET Flow Cytometry. Monoclonal HEK 293T cells expressing Tau-RD-CFP/YFP generally maintain Tau-RD in an unaggregated state (**a**), but when cells are treated with a seed-containing biological sample, they display prominent inclusions (**b**). Cells are harvested 24–48 h after treatment and analyzed for tau seeding on a single cell basis using FRET flow cytometry (**c**)

5. Trypsin–EDTA 0.05 %.
6. Tissue culture-treated flat bottom 96-well plates.
7. Round bottom 96-well plates.
8. Bench Centrifuge.
9. Microcentrifuge.
10. Multichannel Pipet.
11. Analytical Balance.
12. Ultrasonic Homogenizer.
13. Opti-MEM Reduced Serum Medium (Gibco).

14. Lipofectamine® 2000 (Invitrogen).
15. Flow Cytometer with FRET compatible lasers and filters (*see* **Note 1**).
16. FlowJo Data Analysis Software.

3 Methods

Unless stated otherwise, carry out all cell culture procedures under sterile conditions and all steps at room temperature.

3.1 Culturing Biosensor Cells

1. Tau RD-CFP/YFP cells—hereafter referred to as "biosensor cells"—are available for purchase at American Type Culture Collection (ATCC; ATCC® CRL-3275™).
2. Culture cell lines in Cell Culture Medium. Incubate cells at 37 °C, 5 % CO_2, and ≥80 % relative humidity.
3. Passage cells when they are ~80–90 % confluent by quickly rinsing with warm DPBS and dissociating cells from culture vessel using Trypsin–EDTA 0.05 % for 3–5 min at 37 ° C. Quench the dissociation reaction with warm culture medium, transfer suspension to a conical tube, and centrifuge cells at 1000 × *g* for 5 min. Aspirate the supernatant, and resuspend the pellet in warm culture medium. Subculture at the desired confluency (*see* **Note 2**) until further use.

3.2 Preparing Biological Material

1. Prepare homogenization buffer by dissolving one tablet of protease inhibitors (*see* **Note 3**) into 10 mL of 1× Tris-buffered saline. Once dissolved, store on ice.
2. Weigh frozen brain tissue (*see* **Note 4**), transfer to an appropriately sized conical tube, and add a volume of homogenization buffer such that the total concentration is 10 % weight/volume.
3. In a cold room (4 ° C) (*see* **Note 5**), use a probe sonicator to mechanically dissociate tissue (*see* **Notes 6** and **7**). Sonicator settings may vary depending on the specific instrument used. However, as a starting point, deliver 75 W, 500 ms pulses. Continue pulses until the tissue is completely dissolved and has formed a homogenous lysate (*see* **Notes 8** and **9**).
4. Spin down lysates at 21,000 × *g* for 15 min at 4 ° C. Transfer supernatant to a clean tube, taking care not to disturb the pellet. Aliquot supernatant and store at −80 ° C until further use (*see* **Note 10**).

3.3 Replating Biosensor Cells

1. Harvest cells as described in Subheading 3.1. Following suspension of the pellet, count the cells to determine cell density.
2. Make a master mix of each cell line, such that one well of a 96-well plate contains 35,000 cells in 130 μL culture medium

(*see* **Note 11**). Include at least one well each of: HEK 293 T cells, Tau RD-CFP cells, and Tau RD-YFP cells (all of which will be used for flow cytometer and analysis setup). Use the remainder of the plate for biosensor cells, which are treated with biological samples.

3. Using a multichannel pipet, dispense 130 μL of cell suspension into each well of a flat bottom, tissue culture-treated 96-well plate. To ensure even spread of cells, slowly dispense the volume, and place the pipet tip in the center, but not pressed against the bottom of the well.
4. Let cells settle by leaving the plate undisturbed at room temperature for 10 min. Transfer plate to an incubator overnight.

3.4 Treatment of Biosensor Cells (See Note 12)

3.4.1 Liposome-Mediated Treatment of Biosensor Cells (See Note 13)

1. In one tube ("A"), mix liposome reagent with warm reduced serum medium, such that the total volume is 10 μL, and flick gently to mix. Incubate at room temperature for at least 5 min.
2. In another tube ("B"), mix biological sample with warm reduced-serum medium, such that the total volume is 10 μL (*see* **Notes 14** and **15**).
3. Combine contents of Tube "A" and Tube "B," flick gently to mix, and quickly centrifuge to collect the volume at the bottom of the tube. Incubate transduction complex at room temperature for at least 20 min and up to 2 h.
4. To each well containing biosensor cells, gently pipet 20 μL transduction complex to the side of the well, making sure to avoid disturbing the cell layer.
5. Return plate to incubator for 24–48 h.

3.4.2 Physiological Treatment of Biosensor Cells (See Notes 16 and 17)

1. In one tube, combine biological sample with warm culture medium such that the total volume is 20 μL. Add entire volume to the side of the well, making sure to avoid disturbing the cell layer.
2. Return plate to incubator for 24–48 h.

3.5 Harvesting Cells for FRET Flow Cytometry (See Note 18)

1. 24–48 h after treatment, aspirate culture medium with a multichannel pipet. Dissociate cells (*see* **Note 19**) from plate by incubating in 50 μL Trypsin-EDTA 0.05% for 5 min at 37 °C. Quench the reaction with 150 μL cold culture medium.
2. Using a multichannel pipet, transfer entire volume (200 μL) from each well to a round bottom 96-well plate. Pellet cells by centrifuging at 1000 × *g* for 5 min.
3. Aspirate supernatant, and resuspend cells in 50 μL 2% paraformaldehyde. Incubate plate at room temperature for 10 min, then pellet cells by centrifuging at 1000 × *g* for 5 min (*see* **Note 20**).
4. Aspirate supernatant, and resuspend cells in 200 μL flow cytometry buffer (*see* **Note 21**).

3.6 FRET Flow Cytometry

1. Choose a flow cytometer with laser lines and filter sets that are compatible with CFP/YFP FRET detection (*see* **Note 1**), such as the Miltenyi VYB or BD Biosciences LSR Fortessa.
2. Use HEK 293 T cells to define the cell population (SSC-A vs FCS-A bivariate plot) and then to exclude doublet events (FSC-H vs. FSC-A bivariate plot). Set voltages such that no signal is detectable on the CFP, YFP, or FRET filters.
3. Compensate (*see* **Note 22**) CFP bleed through into the FRET and YFP filters. This can be performed manually or in an automated fashion, depending on the flow cytometer used. For compensation, three cell lines are employed: HEK 293 T (negative), Tau RD-CFP (single color control), and Tau RD-YFP (single color control).
4. Following setup, highlight the wells of interest, and flow the remainder of the cells.

3.7 Data Analysis (Fig. 2)

1. Using flow cytometry software, define the cell population (Fig. 2a) and exclude doublet events (Fig. 2b), as performed during setup.
2. Using YFP single-positive cells, correct for YFP spillover into the FRET channel by constructing a "False FRET" gate (FRET-A vs. YFP-A bivariate plot), similar to that described previously [13, 14]. Draw the gate such that FRET-positive events are excluded (Fig. 2c).
3. Using empty liposome-treated cells or media-only cells (*see* **Note 23**), define a FRET gate (FRET-A vs. CFP-A bivariate plot). Draw the gate along the slope of the cell population, such that ~99 % of events fall outside the gate (*see* **Note 24**) (Fig. 2d).
4. For each well, record the percent FRET-positive events and the median fluorescence intensity (MFI) of FRET-positive events (Fig. 2d–f).
5. Calculate the product of the two parameters described in **step 4** of Subheading 3.7 to obtain the Integrated FRET Density, our recommended output measure (*see* **Note 25**).

4 Notes

1. Laser excitation and filter detection wavelengths will vary across flow cytometers, and thus the compatibility of a specific instrument with FRET flow cytometry will need to be empirically tested. The flow cytometer used during development of the assay has the following laser and filter settings: CFP is measured

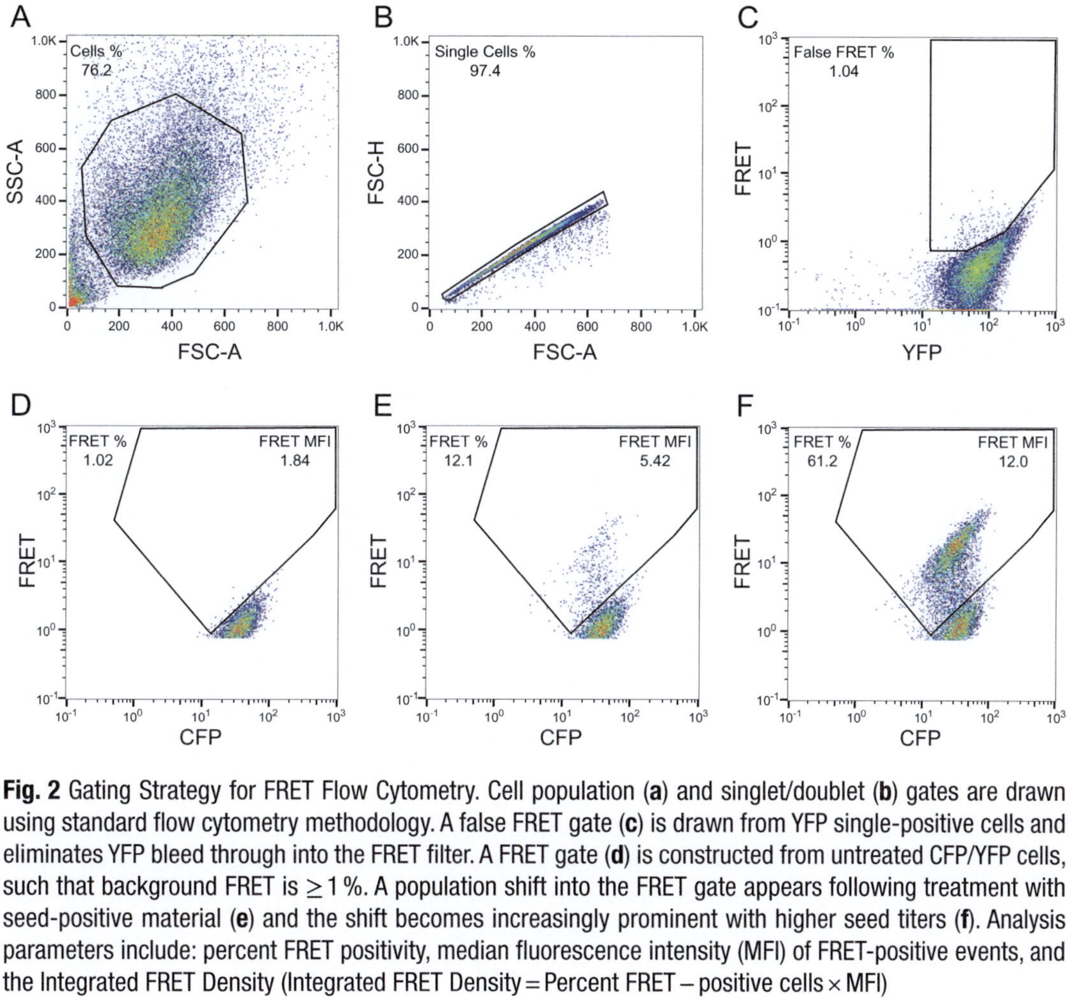

Fig. 2 Gating Strategy for FRET Flow Cytometry. Cell population (**a**) and singlet/doublet (**b**) gates are drawn using standard flow cytometry methodology. A false FRET gate (**c**) is drawn from YFP single-positive cells and eliminates YFP bleed through into the FRET filter. A FRET gate (**d**) is constructed from untreated CFP/YFP cells, such that background FRET is $\geq 1\,\%$. A population shift into the FRET gate appears following treatment with seed-positive material (**e**) and the shift becomes increasingly prominent with higher seed titers (**f**). Analysis parameters include: percent FRET positivity, median fluorescence intensity (MFI) of FRET-positive events, and the Integrated FRET Density (Integrated FRET Density = Percent FRET − positive cells × MFI)

by exciting with a 405 nm laser and detecting with a 450/50 nm filter. YFP is measured by exciting with a 488 nm laser and detecting with a 525/50 nm filter. FRET is measured by exciting with a 405 nm laser and detecting with a 525/50 nm filter.

2. Suggested passaging is between 1:3 and 1:10. Generally, a split of 1:3 will have cells in a nearly confluent state ~24 h later. A split of 1:10 will have cells in a nearly confluent state ~48 h later.
3. Ensure that protease inhibitors are EDTA-free. If not, cell viability will be compromised following treatment of biosensor cells.
4. Detection of tau aggregates with FRET flow cytometry is compatible with human and mouse tissue [10]. While additional

biological samples have not currently been tested, it is expected that other tissues would be compatible with the assay.

5. If it is not possible to store the probe sonicator in the cold room, samples may be homogenized at room temperature if tubes are kept on ice during the procedure.
6. Empirical testing shows that probe sonication is greater than 10 times more sensitive in isolating seeding activity from biological specimens as compared to other commonly used homogenization techniques, such as handheld homogenization. Thus, it is strongly recommended that probe sonication be employed, especially if seeding activity is not abundant or if the difference in seeding activity across samples is a parameter of interest.
7. Although currently unconfirmed, there is a possibility that during probe sonication, biological materials may be aerosolized. Thus, use proper personal protective equipment (including goggles, mask, gown, and gloves) to ensure exposure to tau seeds is prevented.
8. The number of pulses to deliver will vary depending on the starting size of the biological source (i.e., a larger chunk of tissue will take longer to dissociate). If desired, to decrease homogenization time, a sample may be pulverized or chopped into smaller pieces prior to mixing with homogenization buffer.
9. While homogenizing, take care to avoid foaming of the sample. This can commonly occur if a) the overall volume is too low or b) the sample becomes overheated. To avoid these problems, do not use a volume <250 μL and store the tube on ice while homogenizing.
10. Freeze/thaw cycles of biological samples reduce seeding efficacy. Make small aliquots of lysates following homogenization to avoid the need for freeze/thawing.
11. Cell number per well can be adjusted to fit alternative treatment timelines. 35,000 cells per well is recommended for treatment ~18 hours after replating. Increase or decrease this number for earlier or later treatment times, respectively.
12. The ideal cell confluency at the time of treatment is 60–65 %. If cells are more confluent, they are less likely to internalize tau aggregates, which will decrease sensitivity. If cells are less confluent, application of biological sample may be toxic.
13. Liposome-mediated transduction of tau aggregates is the most sensitive treatment method. This is recommended when evaluating samples that may contain low amounts of seed material, when determining maximum seeding potential, or when comparing tissues from animals that have been treated with compounds. Run each sample in triplicate, if possible.

14. The optimal volume of biological sample will vary in an experiment-dependent manner. Variables to take into consideration include: abundance of tau seeds in the sample and potential toxicity to cells. Previously tested amounts of protein range from 1 to 30 μg.
15. The appropriate negative control for this treatment method is using an "empty liposome" condition, where Tube "A" and Tube "B" are mixed, but Tube "B" contains only reduced serum medium. A slight shift in fluorescence properties is observed in cells exposed to reduced serum medium and liposomes, as compared to cells only cultured in their normal medium.
16. Application of tau seeds in the absence of liposomes is recommended when evaluating samples for their ability to undergo physiological uptake into cells [14–16], or when pretreating tau seeds with compounds to evaluate their potential for inhibiting uptake and/or seeding [14].
17. If evaluating the therapeutic efficacy of a compound to block uptake and/or seeding, it is recommended that the biological sample be preincubated with compound prior to treating cells. Appropriate preincubation times will vary by compound, although previously tested time frames range from 3 to 24 h.
18. The remaining steps do not have to be conducted under sterile conditions.
19. A DPBS rinse is not performed at this step, as it may result in cell lifting and subsequent cell loss. Dead cells or debris can be removed with proper gating strategies while performing flow cytometry.
20. Cell fixation prior to flow cytometry is not required, but is highly recommended because it provides a cleaner and more consistent signal.
21. If cells are fixed, they do not have to be run immediately. However, it is recommended that they be run as soon as possible to prevent cell clumping, which will affect setup and analysis parameters.
22. Compensation is not required for obtaining preliminary or approximate data points. However, it is highly recommended, as CFP bleed through into the FRET channel may lead to exaggeration of FRET positivity and/or high background. For the most sensitive and quantitative results, apply compensation.
23. Empty liposome-treated cells are used if biological samples are transduced into cells. Media-only cells are used if tau seeds were applied in the absence of liposomes.
24. Establish gating such that the "background" level of FRET is ≥ 1 %, as this parameter will ultimately be used as a multiplying

factor in analysis. A value less than 1 % will lead to underestimation of FRET signal.

25. Individually, both percent FRET-positive events or MFI of FRET-positive events can provide important information about seeding activity in a sample. We recommend reporting the product of these values—Integrated FRET Density—as it provides a measure that more wholly represents the abundance of aggregation in a population of cells as well as the degree of aggregation within an individual cell.

Acknowledgments

We thank Brandon B. Holmes for his experimental insight and contributions to the assay development. This work was supported by the Tau Consortium (M.I.D.); National Institutes of Health 1R01NS071835 (M.I.D.) and 1F32NS087805 (J.L.F.).

References

1. Aguzzi A, Lakkaraju AK (2016) Cell biology of prions and prionoids: a status report. Trends Cell Biol 26(1):40–51. doi:10.1016/j.tcb.2015.08.007
2. Goedert M, Falcon B, Clavaguera F, Tolnay M (2014) Prion-like mechanisms in the pathogenesis of tauopathies and synucleinopathies. Curr Neurol Neurosci Rep 14(11):495. doi:10.1007/s11910-014-0495-z
3. Goedert M, Clavaguera F, Tolnay M (2010) The propagation of prion-like protein inclusions in neurodegenerative diseases. Trends Neurosci 33(7):317–325. doi:10.1016/j.tins.2010.04.003
4. Holmes BB, Diamond MI (2012) Cellular mechanisms of protein aggregate propagation. Curr Opin Neurol 25(6):721–726. doi:10.1097/WCO.0b013e32835a3ee0
5. Ross CA, Poirier MA (2004) Protein aggregation and neurodegenerative disease. Nat Med 10(Suppl):S10–S17. doi:10.1038/nm1066
6. Weingarten MD, Lockwood AH, Hwo SY, Kirschner MW (1975) A protein factor essential for microtubule assembly. Proc Natl Acad Sci U S A 72(5):1858–1862
7. Lee VM, Goedert M, Trojanowski JQ (2001) Neurodegenerative tauopathies. Annu Rev Neurosci 24:1121–1159. doi:10.1146/annurev.neuro.24.1.112124/1/1121
8. Arriagada PV, Growdon JH, Hedley-Whyte ET, Hyman BT (1992) Neurofibrillary tangles but not senile plaques parallel duration and severity of Alzheimer's disease. Neurology 42(3 Pt 1):631–639
9. Giannakopoulos P, Herrmann FR, Bussiere T, Bouras C, Kovari E, Perl DP, Morrison JH, Gold G, Hof PR (2003) Tangle and neuron numbers, but not amyloid load, predict cognitive status in Alzheimer's disease. Neurology 60(9):1495–1500
10. Holmes BB, Furman JL, Mahan TE, Yamasaki TR, Mirbaha H, Eades WC, Belaygorod L, Cairns NJ, Holtzman DM, Diamond MI (2014) Proteopathic tau seeding predicts tauopathy in vivo. Proc Natl Acad Sci U S A 111:E4376–E4385. doi:10.1073/pnas.1411649111
11. Santa-Maria I, Perez M, Hernandez F, Avila J, Moreno FJ (2006) Characteristics of the binding of thioflavin S to tau paired helical filaments. J Alzheimers Dis 9(3):279–285
12. Meyer V, Dinkel PD, Rickman HE, Margittai M (2014) Amplification of tau fibrils from minute quantities of seeds. Biochemistry 53(36):5804–5809. doi:10.1021/bi501050g
13. Banning C, Votteler J, Hoffmann D, Koppensteiner H, Warmer M, Reimer R, Kirchhoff F, Schubert U, Hauber J, Schindler M (2010) A flow cytometry-based FRET assay

to identify and analyse protein-protein interactions in living cells. PLoS One 5(2), e9344. doi:10.1371/journal.pone.0009344

14. Furman JL, Holmes BB, Diamond MI (2015) Sensitive detection of proteopathic seeding activity with FRET flow cytometry. J Vis Exp 106:e53205. doi:10.3791/53205

15. Holmes BB, DeVos SL, Kfoury N, Li M, Jacks R, Yanamandra K, Ouidja MO, Brodsky FM, Marasa J, Bagchi DP, Kotzbauer PT, Miller TM, Papy-Garcia D, Diamond MI (2013) Heparan sulfate proteoglycans mediate internalization and propagation of specific proteopathic seeds. Proc Natl Acad Sci U S A 110(33):E3138–E3147. doi:10.1073/pnas.1301440110130144011O

16. Mirbaha H, Holmes BB, Sanders DW, Bieschke J, Diamond MI (2015) Tau trimers are the minimal propagation unit spontaneously internalized to seed intracellular aggregation. J Biol Chem 290(24):14893–14903. doi:10.1074/jbc.M115.652693

Chapter 24

In Vivo Imaging of Tau Aggregates in the Mouse Retina

Christian Schön and Jochen Herms

Abstract

Aggregates of hyperphosphorylated tau can be observed in the human brain affected by various neurodegenerative disorders. The development of a noninvasive technique for the visualization of these protein accumulations is a promising task. In the following protocol, we describe a method to image fibrillar tau in the retina of a transgenic mouse model of frontotemporal dementia with parkinsonism linked to chromosome 17 (FTDP-17). This technique can be highly valuable for the preclinical in vivo testing of approaches that target tau aggregation.

Key words Tauopathies, Tau protein, Confocal scanning laser ophthalmoscopy, P301S mutant mice, FSB

1 Introduction

A common feature of neurodegenerative disorders is the pathological accumulation of protein aggregates within the central nervous system [1]. One of these proteins is the microtubule associated protein tau (MAPT). During disease progression, the hyperphosphorylation of tau in neurons leads to so-called fibrillar tangles. Human diseases that show this neuropathological hallmark are therefore called tauopathies. It remains a hot topic in current research to which extent the accumulation of fibrillar tau protein causes or influences the progression of tauopathies. On the other hand, it is widely accepted that the accumulation of hyperphosphorylated tau protein correlates with the severity of neurodegenerative disorders like Alzheimer's disease [2]. Therefore, it would be of great value to visualize the progression of tau accumulation in the living being. First in vivo studies to visualize tau aggregates and the consequences of pathological tau accumulations to neuronal physiology were already successfully conducted in mouse models of tauopathies [3, 4]. However, the visualization of changes on cellular level in the central nervous system of mammals is still challenging and was so far only achieved by invasive means. In this protocol, we want to present the details of a method that allows for the noninvasive imaging of tau aggregates in P301S transgenic

Caroline Smet-Nocca (ed.), *Tau Protein: Methods and Protocols*, Methods in Molecular Biology, vol. 1523,
DOI 10.1007/978-1-4939-6598-4_24,

mice, a model of tauopathies [5]. These mice accumulate hyperphosphorylated tau in neurons of the central nervous system including the retina. During this process, single cells in the retinal ganglion cell layer (GCL) develop fibrillar aggregates of tau protein that can be stained with the fluorophore (E, E)-1-fluoro-2,5-*bis*(3-hydroxycarbonyl-4-hydroxy)styrylbenzene (FSB) [6, 7]. Our approach focuses on the tangle formation in the retina—a part of the central nervous system that is easily accessible for noninvasive ophthalmologic examinations. Using confocal scanning laser ophthalmoscopy (cSLO) in combination with the fluorophore FSB binding to fibrillar tau, we were able to monitor the aggregation of fibrillar tau in cells of the GCL. It is important to mention that we found no clear evidence for fibrillar tau aggregates in the retina of patients with Alzheimer's disease or FTDP-17. Therefore, this technique seems not to be directly translatable for diagnostic approaches in humans. However, we see a big potential of this method for preclinical tests of therapeutic interventions targeting tau aggregation. The biggest advantage of our approach lies in the option to follow up individual animals over several months. Important aspects and details of the experimental procedure are described in the following chapter.

2 Materials

2.1 Setup and Materials to Adapt the cSLO for Mice Experiments

1. The scanning of the mouse retina was performed with a modified Spectralis HRA + OCT system (Heidelberg Engineering) equipped with three different excitation lasers (450, 488, and 820 nm) and an integrated set of emission filters (LP 458 nm, BP 550/49 nm, and BP 617/73 nm) (*see* **Note 1**).
2. The Spectralis HRA + OCT can be easily adapted for the use on mice (Fig. 1) by attaching a 78 diopter slit lamp lens (Volk) in front of the camera and 100 diopter contact lenses (Roland Consult) to the mouse eyes [8]. Alternatively, one can consider a combination of the contact lenses and a 55° noncontact lens (Heidelberg Engineering).
3. For fixation of the mouse, we designed a custom-made mouse holder that allowed the tilting of the mouse head. This mouse holder was placed on a micromanipulator for shifting the mouse body in horizontal and vertical directions.
4. We used homozygous mice expressing human mutant P301S tau [9].

2.2 Reagents and Pharmaceuticals

1. Anaesthesia cocktail: 4 ml Ketamine (10%; WDT), 1 ml Xylazine (2% Rompun; Bayer Healthcare) and 24 ml NaCl (0.9%) or a multiple of that mixture.
2. 1% FSB (Merck) in DMSO (*see* **Note 2**).

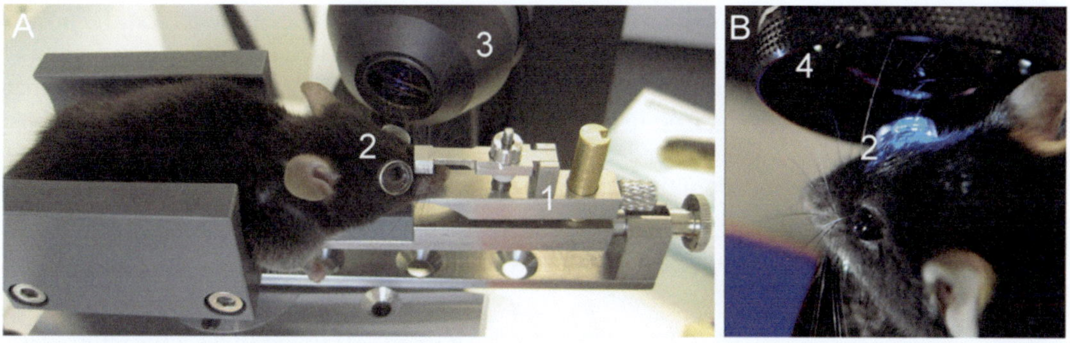

Fig. 1 Adapting the Spectralis HRA + OCT for examinations of the mouse eye. A custom-made mouse holder (*1*) was designed to align the animal in front of the camera lens. Different combinations of optical lenses are suitable to adapt the system (**a**, **b**). We combined 100 diopter mouse contact lenses (*2*) with a 55° noncontact lens (*3*) or a 78 diopter slit lamp lens (*4*), respectively

3. 2% Mouse Albumin (Merck) in Phosphate buffered-saline (PBS).
4. Tropicamid eye drops (Mydriadicum Stulln; Pharma Stulln GmbH).
5. 2% Methocel (OmniVision).
6. Eye ointment (e.g., Bepanthen).
7. Spectralis HRA + OCT system (Heidelberg Engineering, Germany) for examination of the mouse eye.
8. Heidelberg Eye Explorer software (HEYEX).

3 Method

In the following protocol, we present the experiment conducted in a well-established timeline. Of course, this timeline can be adapted depending on the demands of the experimentalist.

3.1 Pre-Examination (Day 1)

We recommend a pre-examination of the mouse retina to record spots of increased autofluorescence (*see* **Note 3**) and to evaluate the general morphology of the eye.

3.1.1 Mouse Preparation

1. Breed mice homozygous for P301S tau. Adult mice from 2 months of age develop fibrillar tau aggregates in the GCL that can be stained with FSB (*see* **Note 4**).
2. Apply 10 μl of the anaesthesia cocktail per gram of body weight by an intraperitoneal (i.p.) injection to adult mice (>15 g). This corresponds to a final concentration of 0.14 mg/g Ketamine and 0.01 mg/g Xylazine. Under normal conditions, mice will receive an adequate level of anaesthesia after 5–10 min and will wake up after approximately 1 h (*see* **Note 5**).

3. While waiting for the animal to fall asleep, switch on the power supply and scanning box and start the Heidelberg Eye Explorer software. Create a definite "patient" file for the animal that can be used for subsequent experiments. Additionally, mouse contact lenses can be cleaned. A simple cleaning equipment contains standard dish detergent and lint free tissues.
4. As soon as mice are unconscious, add one drop of Tropicamid to the eyes and wait for 2 min until pupils are completely dilated.
5. Add one drop of 2 % Methocel to the contact lenses and place them on the mouse eye. Center the lenses to the pupils if they slide down. The Methocel will keep the lenses at the right position after a few seconds.
6. Fix the animals head to the custom-made mouse holder.

3.1.2 Scanning Procedure

1. Start the scan in high resolution mode with a 30° or 55° field of view preference (*see* **Note 6**).
2. Align the animal by tilting the mouse holder and moving the micromanipulator to a position where you can clearly see the retina with the optic nerve head best possible centered to the displayed picture (*see* **Note 7**).
3. Adjust the brightness with the supplied touch panel.
4. For imaging of the mouse fundus, we used three different combinations of excitation lasers and emission filters:
   - Setting 1: Excitation laser: 820 nm //Emission filter: LP 458 nm.
   - Setting 2: Excitation laser: 450 nm //Emission filter: LP 458 nm.
   - Setting 3: Excitation laser: 488 nm //Emission filter: BP 617/73 nm.
5. Focus the confocal system to the GCL. The retinal nerve fiber layer is best displayed in setting 1. Spots of increased autofluorescence will appear when using settings 2 and 3.
6. Obtain images of the mouse fundus using the setting 1–3. Readjust the focus when changing between the three settings. Take advantage of the average recording mode of the system. An average of 30 frames will significantly improve the signal-to-noise ratio of your images.
7. Also consider checking the general retinal morphology by optical coherence tomography. This allows for the exclusion of retinas with morphologic abnormalities.
8. Save images and exit the scanning program.

3.1.3 Procedure After Scanning

1. After scanning, contact lenses were removed by adding a drop of water.
2. Care has to be taken to cover the eyes with eye ointment.
3. Place the mice on a heating plate (37 °C) and wait until they wake up (*see* **Note 8**).

3.2 FSB Application (Day 7)

After one week, mice will have recovered from the first anesthesia on the first day of the experiment. Fibrillar tau aggregates can now be stained with one single i.p. injection of FSB.

1. Freshly mix the cocktail containing the fluorophore FSB. For this purpose, prepare solutions of 2 % mouse albumin in PBS and 1 % FSB in DMSO (*see* **Note 9**). Vortex until the solutions appear clear. Carefully dissolve the 1 % FSB solution 1:10 in the 2 % mouse albumin solution and vortex right before the injection.
2. Apply 10 μl per gram of body weight of the FSB cocktail by an i.p. injection to achieve a final concentration of 10 mg/kg FSB.

3.3 Baseline Examination (Day 8)

24 h after systemic injection of the FSB solution, fibrillar tau aggregates will be stained with the fluorophore and can be examined by cSLO (*see* **Note 10**)

1. Repeat the steps mouse preparation and scanning procedure like described above (*see* Subheading 3.1).
2. In order to compare successive examinations to the pre-examination, it is important to scan at every time point the same area of the retina. This can be achieved by aligning the field of view to the retinal blood vessel system best visible in setting 1.
3. Again, spots of increased autofluorescence will appear at the scanner settings 2 and 3. However, additional distinct fluorescent spots will appear after FSB injection. These spots are only visible at 450 nm excitation and LP458 emission (setting 2) but not in other settings. These additional fluorescent dots are cells of the GCL layer stained by the fluorophore FSB (Fig. 2).
4. Count FSB-positive cells within the selected field of view to determine a baseline for subsequent examinations.

3.4 Post-Examinations (Day 30)

1. After the baseline examination, mice can be treated with various compounds directed against the hyperphosphorylation or aggregation of fibrillar tau aggregates.
2. We recommend a subsequent post-examination 1 month after the baseline examination (*see* **Note 11**). Therefore, execute the experiment like described before (*see* Subheading 3.1).
3. This procedure can be repeated various times.

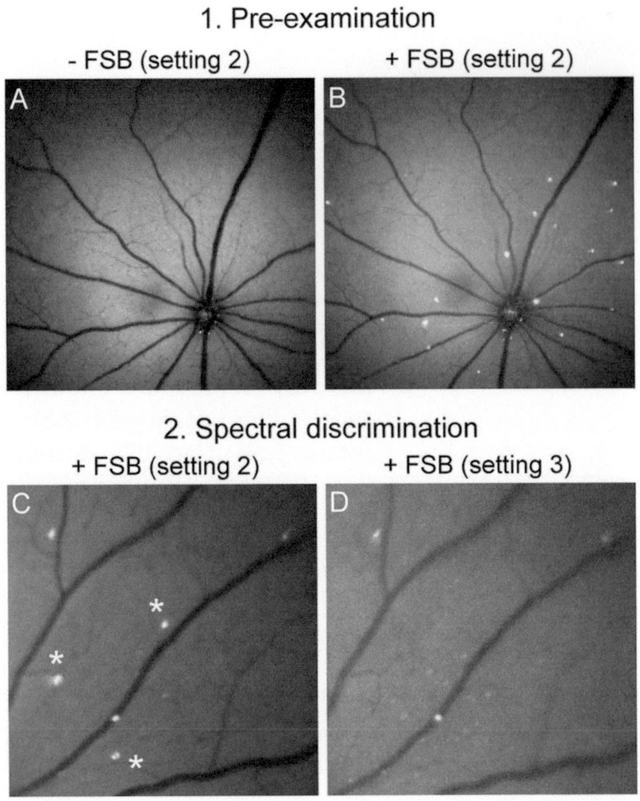

Fig. 2 Identification of FSB-positive cells in the retina of P301S tau mutant mice. Firstly, compare pictures before and after the application of FSB (**a**, **b**). Secondly, obtain images using the settings 2 and 3 of the cSLO for spectral discrimination (**c**, **d**). FSB-positive cells (*asterisks*) will appear only after the application of FSB (**b**, **c** and **d**) and are only visible at 450 nm excitation and the emission filter set to LP458 (setting 2)

4. Quantify the observed FSB-positive cells in treated and untreated groups and check for statistical significant differences in the formation kinetics of fibrillar tau aggregates.

4 Notes

1. Due to the highly specific exchange of lasers and filter sets, the described cSLO is hardly comparable to the commercially available system. However, different devices for FSB excitation can be considered. When choosing a scanning system, it is important to keep in mind that FSB may only be excitable with short wavelength light sources [10]. For example, we were only able to excite FSB with a 450 nm laser, but not at 488 nm excitation.

2. In our hands, FSB is so far the only successfully tested dye for the in vivo staining of fibrillar tau aggregates. Furthermore, it is important to mention that we tested FSB from two different companies (Merck Millipore and Dojindo molecular technologies). Both dyes could be applied to stain tau aggregates ex vivo in retinal whole mounts of P301S transgenic mice. However, only the fluorophore from Merck was detectable by cSLO in in vivo examinations at an excitation of 450 nm.
3. With proceeding age, mice accumulate spots of increased autofluorescence in the retina due to the accumulation of lipofuscin [11]. Due to their size and the spectral properties of lipofuscin [12], these spots can be easily misinterpretation as fibrillar tau aggregates stained by FSB. Therefore, it is critical to carefully discriminate these spots of increased autofluorescence from the FSB positive cells by a pre-examination and by their spectral properties.
4. Mice homozygous for P301S mutant tau will develop a neurological phenotype dominated by a severe paraparesis at the age of 5–7 months.
5. Consider a lower dosage of anaesthesia for mice lighter than 15 g. Mice heavier than 30 g can be difficult to anaesthetize and it might be necessary to apply a higher dosage. If mice won't reach a deep level of anaesthesia at all, keep in mind that the animals do not have to be deeply anaesthetized for the scanning procedure.
6. In general, we observed no influence of the room light turned on or off on the obtained picture quality. However, we recommend turning of the light. Firstly, the experimentalist will be able to see more subtle changes in the imaged fluorescence signal. Secondly, the mice pupils are dilated and therefore the eyes are more vulnerable to bright light exposure.
7. Finding the right position of the mouse eye in front of the camera can take some time and is a matter of practice. The infrared laser offers the highest signal-to-noise ratio of the Spectralis HRA + OCT system and is therefore the best setting for finding the mouse retina in the beginning of the scanning procedure.
8. Littermates that undergo the experimental procedure can be kept together in one cage. Male and female mice tolerate the stress due to the anaesthesia quite well and we did not observe an enhanced level of aggression. However, keep in mind that mice waking up will stumble around and will remove the ointment from the eyes of their littermates. If this happens the eyes dry out and cataracts can develop.
9. FSB dissolved in DMSO and albumin dissolved in PBS can be stored at −20 °C for at least 1 month without loss of quality.

Getting FSB in solution is the most critical step of the protocol. Carefully check that the fluorophore is dissolved before the injection. Without the mouse albumin, FSB will precipitate in the solution.

10. Adhere to the timing of 24 h between FSB injection and imaging. This is the first reliable time point that allows for the discrimination of background fluorescence and FSB-positive cells. The cells stained with the fluorophore can be observed up to 1 month after a single i.p. injection of FSB.

11. The formation of fibrillar tau aggregates in the retina of P301S mice progresses steadily. However, the appearance of novel FSB-positive cells is a quite rare event (for details please compare quantitative date in [5]). Therefore, shorter observation periods than 1 month are not reasonable.

References

1. Goedert M (2015) Neurodegeneration. Alzheimer's and Parkinson' diseases: the prion concept in relation to assembled Abeta, tau, and alpha-synuclein. Science 349 (6248): 1255555
2. Braak H, Alafuzoff I, Arzberger T, Kretzschmar H, Del Tredici K (2006) Staging of Alzheimer disease-associated neurofibrillary pathology using paraffin sections and immunocytochemistry. Acta Neuropathol 112(4):389–404
3. de Calignon A, Fox LM, Pitstick R, Carlson GA, Bacskai BJ, Spires-Jones TL, Hyman BT (2010) Caspase activation precedes and leads to tangles. Nature 464(7292):1201–1204
4. Hoffmann NA, Dorostkar MM, Blumenstock S, Goedert M, Herms J (2013) Impaired plasticity of cortical dendritic spines in P301S tau transgenic mice. Acta Neuropathol Commun 1:82
5. Schön C, Hoffmann NA, Ochs SM, Burgold S, Filser S, Steinbach S, Seeliger MW, Arzberger T, Goedert M, Kretzschmar HA, Schmidt B, Herms J (2012) Long-term in vivo imaging of fibrillar tau in the retina of P301S transgenic mice. PLoS One 7(12):e53547
6. Gasparini L, Crowther RA, Martin KR, Berg N, Coleman M, Goedert M, Spillantini MG (2011) Tau inclusions in retinal ganglion cells of human P301S tau transgenic mice: effects on axonal viability. Neurobiol Aging 32(3): 419–433
7. Velasco A, Fraser G, Delobel P, Ghetti B, Lavenir I, Goedert M (2008) Detection of filamentous tau inclusions by the fluorescent Congo red derivative FSB [(trans, trans)-1-fluoro-2,5-bis(3-hydroxycarbonyl-4-hydroxy) styrylbenzene]. FEBS Lett 582(6):901–906
8. Seeliger MW, Beck SC, Pereyra-Munoz N, Dangel S, Tsai JY, Luhmann UF, van de Pavert SA, Wijnholds J, Samardzija M, Wenzel A, Zrenner E, Narfstrom K, Fahl E, Tanimoto N, Acar N, Tonagel F (2005) In vivo confocal imaging of the retina in animal models using scanning laser ophthalmoscopy. Vision Res 45(28):3512–3519
9. Allen B, Ingram E, Takao M, Smith MJ, Jakes R, Virdee K, Yoshida H, Holzer M, Craxton M, Emson PC, Atzori C, Migheli A, Crowther RA, Ghetti B, Spillantini MG, Goedert M (2002) Abundant tau filaments and nonapoptotic neurodegeneration in transgenic mice expressing human P301S tau protein. J Neurosci 22(21): 9340–9351
10. Sato K, Higuchi M, Iwata N, Saido TC, Sasamoto K (2004) Fluoro-substituted and 13C-labeled styrylbenzene derivatives for detecting brain amyloid plaques. Eur J Med Chem 39(7):573–578
11. Xu H, Chen M, Manivannan A, Lois N, Forrester JV (2008) Age-dependent accumulation of lipofuscin in perivascular and subretinal microglia in experimental mice. Aging Cell 7(1):58–68
12. Delori FC, Dorey CK, Staurenghi G, Arend O, Goger DG, Weiter JJ (1995) In vivo fluorescence of the ocular fundus exhibits retinal pigment epithelium lipofuscin characteristics. Invest Ophthalmol Vis Sci 36(3):718–729

Chapter 25

In Vivo Hyperthermic Stress Model: An Easy Tool to Study the Effects of Oxidative Stress on Neuronal Tau Functionality in Mouse Brain

Alban Chauderlier, Lucie Delattre, Luc Buée, and Marie-Christine Galas

Abstract

Oxidative damage is an early event in neurodegenerative disorders such as Alzheimer disease. To increase oxidative stress in AD-related mouse models is essential to study early mechanisms involved in the physiopathology of these diseases. In this chapter, we describe an experimental mouse model of transient and acute hyperthermic stress to induce in vivo an increase of oxidative stress in the brain of any kind of wild-type or transgenic mouse.

Key words Hyperthermic stress, Heat stress, Oxidative stress, Brain, Neuron, Tau

1 Introduction

Many neurodegenerative disorders, such as Alzheimer disease (AD), are associated with a disturbance of the cellular redox balance. From the early stages of AD, elevated oxidative damage to proteins, lipids, and nucleic acids occurs in vulnerable neurons [1–4]. To study the effects of increased oxidative stress in the brain is essential to decipher early mechanisms involved in the physiopathology of these diseases.

In this context, a new in vivo model was necessary to test the impact of an increase of oxidative stress on Tau biology in mouse brain. Hyperthermic stress (HS) is known to be a strong inducer of oxidative stress [5–7]. Here, we describe an vivo mouse model of transient and acute hyperthermic stress based on the rat model previously described by Papasozomenos et al. [8]. This method offers the advantage of being noncytotoxic [7, 9] and easily useable with any kind of mouse, wild-type or transgenic. This protocol has been successfully used to induce in the brain the dephosphorylation and nuclear translocation of Tau, to discover the key role of Tau in DNA and RNA protection in vivo [7] and its deregulation in early stages

Caroline Smet-Nocca (ed.), *Tau Protein: Methods and Protocols*, Methods in Molecular Biology, vol. 1523,
DOI 10.1007/978-1-4939-6598-4_25, © Springer Science+Business Media New York 2017

of Tau pathology [10]. Strikingly, it is an easy mouse model to induce the formation of toxic Tau oligomers in vivo [10].

The mice are anesthetized and maintained at 37 °C to avoid anesthesia-related hypothermia and subsequent Tau hyperphosphorylation as previously described [11]. Then, the mice are maintained at 37 °C (control condition) or heat stressed by being placed at 44 °C for 20 min. Mice are either sacrificed right after the heating step or after an allowed recovery time at room temperature.

2 Materials

2.1 Equipment

1. Weighing scale.
2. Heating blanket.
3. Heating lamp.
4. Surgical supplies and dissecting tools.
5. Incubator at 44 °C.
6. Rectal Thermometer.
7. Graduated syringes.

2.2 Animals and Reagents

1. Mouse (*see* **Note 1**).
2. Ketamine (100 mg/kg) and Xylazine (10 mg/kg) diluted in 1× Phosphate Buffer Saline.
3. Vaseline lubricant.
4. Physiological serum.

3 Method

1. Weigh the mice to adapt the volume of anesthetics to inject.
2. Anesthetize the mice by intraperitoneal injection of Ketamine (100 mg/kg) and Xylazine (10 mg/kg) diluted in 1X Phosphate Buffer Saline.
3. Put back the mice in the cage, while waiting for the first effects of anesthesia to reduce the animals' stress (*see* **Note 2**).
4. Remember to record the number associated with each mice to distinguish them.
5. Place the mice on the heating blanket, under the heating lamp to maintain the mice in a 37 °C environment for 30 min to avoid anesthesia-related hypothermia and related Tau hyperphosphorylation [11].

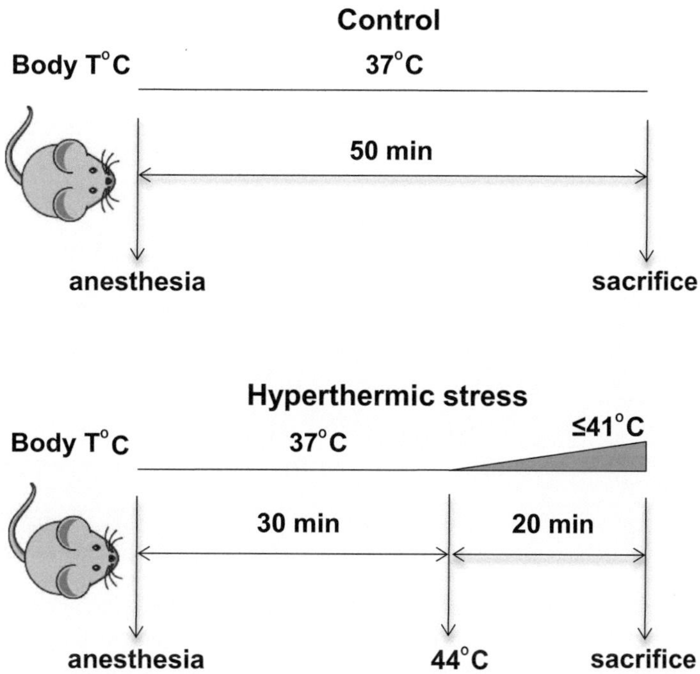

Fig. 1 Schematic representation of the hyperthermic stress (HS) experiment. After anesthesia, mice are maintained in a 37 °C environment for 30 min. Then, mice are placed in 37 or 44 °C environment, respectively, for control and HS group (*see* **Note 3**)

6. Maintain the mice at 37 °C (control group) or put the mice in an incubator heated at 44 °C (heat-stressed group) for 20 min (Fig. 1).
7. To directly study the effects of HS, sacrifice the mice by cervical dislocation right after temperature exposure. In this case, maintain the mice at 37 °C until the end of the anesthesia (*see* **Note 4**). After awakening, the mouse is kept at room temperature and sacrificed when needed.
8. Dissect the mice to extract the brain.
   - To perform RNA extraction, tissues are snap-frozen in liquid nitrogen and then stocked at −80 °C.
   - To prepare lysate or fractionation, tissues are placed on ice [7].
   - To carry out immunohistochemistry, tissues are fixed in paraformaldehyde 4% (*see* **Note 5**).

Under those conditions, the HS induces noncytotoxic oxidative damage (Fig. 2) [7].

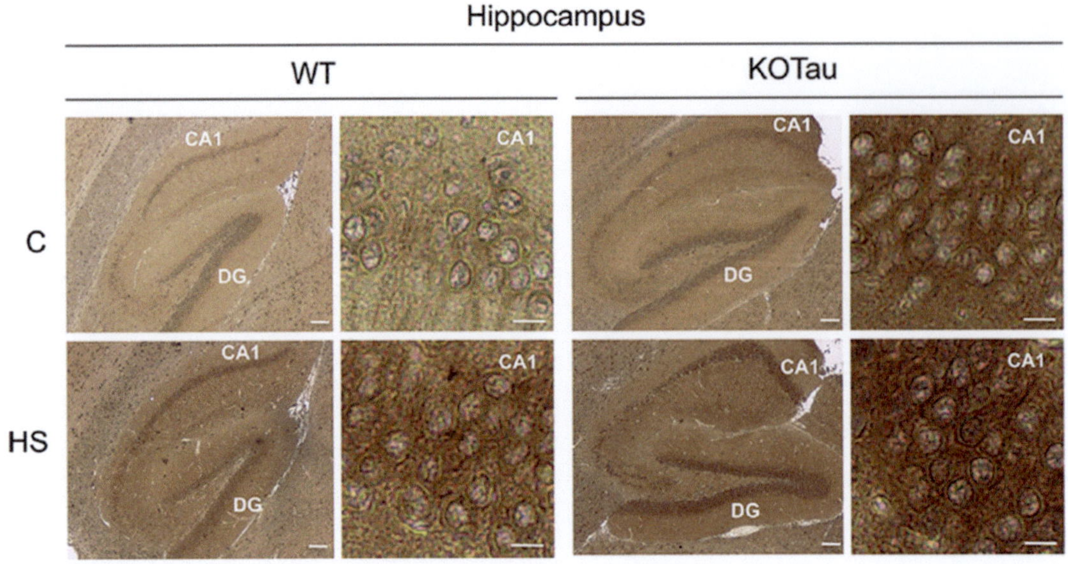

Fig. 2 Hyperthermia generates oxidative stress in WT and KO-Tau hippocampi. Protein carbonyl groups are biomarkers of oxidative stress. Carbonyl-modified protein generated by oxidative stress was visualized using immunolabeling after reaction with 2,4-dinitrophenylhydrazine (DNPH) in sagittal sections of the hippocampus from WT and KO-Tau mice under control (C) or heat stress (HS) conditions. HS increased DNPH staining in WT and KO-Tau sections. *Scale bars* indicate 200 or 10 μm (zoom)

4 Notes

1. Any kind of wild-type and transgenic mouse can be used. We used KO-Tau and littermate WT mice to demonstrate the nucleic acid protective function of Tau in the cortex and the hippocampus [7]. Concerning the effect of Tau pathology on the nucleic acid protective function, we used Thy-Tau22 and littermate WT mice [10].
2. The anesthetic depth is monitored by observing the loss of palpebral reflex and the toe-pinch withdrawal reflex.
3. Monitor the rectal temperature of the mice every 10 min using a rectal thermometer and a lubricant (Fig. 1). The body temperature must not exceed 41 °C to prevent animal death. Record the evolution of rectal temperature. Physiological serum is used to hydrate animals' eyes.
4. It is important to maintain the mice at 37 °C during the awakening hours to avoid hypothermia.
5. Protein oxidation was analyzed using an OxyIHC oxidative stress-detection kit (Millipore) (Fig. 2). According to the manufacturer's direction, fixation was performed in Methacarn. Nucleic acid integrity was analyzed using Comet assay (DNA) and TUNEL (DNA and RNA).

Acknowledgments

This work was supported by French government funding from the Agence Nationale de la Recherche MALZ EPITAUDNA grant and by LabEx (Excellence Laboratory), DISTALZ (Development of Innovative Strategies for a Transdisciplinary, Approach, to, Alzheimer's, Disease),(INSERM((Institut National de la santé et de la recherche médicale), LMCU (Lille Métropole Communauté urbaine), Région Nord/Pas de Calais, and FEDER (Fonds Européen de Développement Economique et Régional).

References

1. Nunomura A, Perry G, Aliev G et al (2001) Oxidative damage is the earliest event in Alzheimer disease. J Neuropathol Exp Neurol 60:759–767
2. Nunomura A, Tamaoki T, Motohashi N et al (2012) The earliest stage of cognitive impairment in transition from normal aging to Alzheimer disease is marked by prominent RNA oxidation in vulnerable neurons. J Neuropathol Exp Neurol 71:233–241
3. Nunomura A, Perry G, Aliev G et al (2008) Nucleic acid oxidation in Alzheimer disease. Free Radic Biol Med 44:1493–1505
4. Bradley-Whitman MA, Lovell MA (2015) Biomarkers of lipid peroxidation in Alzheimer disease (AD): an update. Arch Toxicol 89: 1035–10447
5. Flanagan SW, Moseley PL, Buettner GR et al (1998) Increased flux of free radicals in cells subjected to hyperthermia: detection by electron paramagnetic resonance spin trapping. FEBS Lett 431:285–286
6. Bruskov VI, Masalimov ZK, Chernikov AV (2002) Heat-induced generation of reactive oxygen species in water. Dokl Biochem Biophys 384:181–184
7. Violet M, Delattre L, Tardivel M et al (2014) A major role for Tau in neuronal DNA and RNA protection in vivo under physiological and hyperthermic conditions. Front Cell Neurosci 18(8):84
8. Papasozomenos SC (1996) Heat shock induces rapid dephosphorylation of tau in both female and male rats followed by hyperphosphorylation only in female rats: implications for Alzheimer's disease. J Neurochem 66:1140–1149
9. Sultan A, Nesslany F, Violet M et al (2011) Nuclear tau, a key player in neuronal DNA protection. J Biol Chem 286:4566–4575
10. Violet M, Chauderlier A, Delattre L et al. (2015) Prefibrillar Tau oligomers alter the nucleic acid protective function of Tau in hippocampal neurons *in vivo*. (under revision)
11. Planel E, Richter KE, Nolan CE et al (2007) Anesthesia leads to tau hyperphosphorylation through inhibition of phosphatase activity by hypothermia. J Neurosci 27:3090–3097

Chapter 26

Identification of Tau Toxicity Modifiers in the *Drosophila* Eye

Pierre Dourlen

Abstract

Drosophila is a powerful model to study human diseases thanks to its genetic tools and ease of screening. Human genes can be expressed in targeted organs and their toxicity assessed on easily scorable external phenotypes that can be used as readout to perform genetic screen of toxicity modifiers. In this chapter, I describe how to express human Tau protein in the *Drosophila* eye, assess protein expression by western blot, assess Tau toxicity by quantifying the size of the Tau-induced rough eye, and perform a genetic screen of modifiers of Tau toxicity in the *Drosophila* eye.

Key words Neurodegeneration, Drosophila, Tau, Neuron, Eye, Alzheimer, Modifier screen

1 Introduction

Drosophila melanogaster has been a model organism for more than a century [1]. First used to dissect the principles of inheritance, researchers gained from all the genetic tools they had developed initially to study many other biological processes. Although different at the level of the body plan, *Drosophila* has many similarities with mammals and human at the cellular and genetic levels. Most of the cell signaling pathways are conserved in *Drosophila*. 75 % of human disease-associated genes have a functional *Drosophila* homolog. Therefore, *Drosophila* mutants of these homologs can model many human diseases, like neurodegenerative diseases [2]. Another strategy consists of expressing human transgene in *Drosophila*. For neurodegenerative diseases, a classical approach is to express the proteins that aggregate and constitute the histopathological hallmark of the disease. Thus, several *Drosophila* models of Alzheimer's disease express Tau, a protein that aggregates into neurofibrillary tangles in the brain of patients [3].

In *Drosophila*, the main method to express a transgene is the UAS/Gal4 system [4] (Fig. 1a). It is a binary system in which the transgene is inserted downstream of a yeast sequence called Upstream

Caroline Smet-Nocca (ed.), *Tau Protein: Methods and Protocols*, Methods in Molecular Biology, vol. 1523, DOI 10.1007/978-1-4939-6598-4_26, © Springer Science+Business Media New York 2017

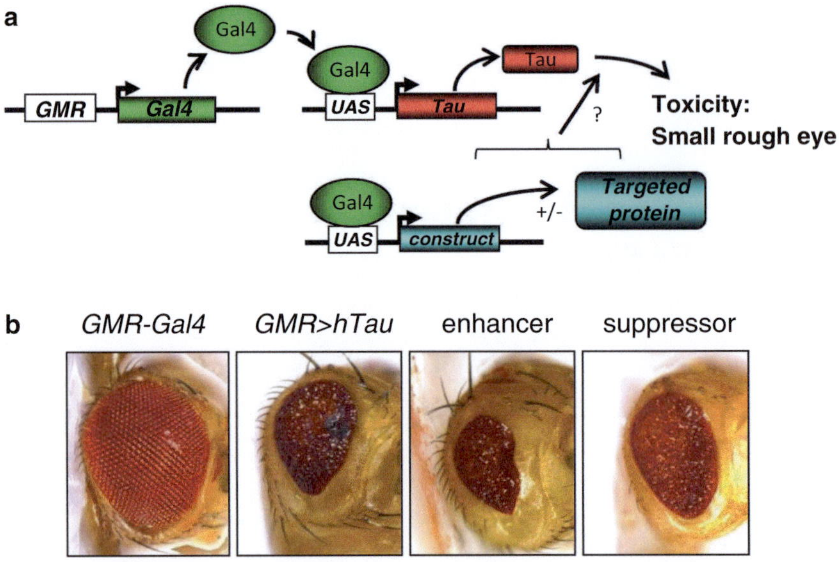

Fig. 1 Principle of a modifier screen. (**a**) Scheme of the UAS/Gal4 system enabling Tau expression and the co-expression of a screened construct. The goal of the screen is to determine if the construct modifies Tau-associated toxicity. (**b**) Images of fly eyes illustrating a control eye, the Tau-induced small rough eye, and enhancing versus suppressing effects on Tau toxicity

Activating Sequence (UAS). Transcription of the transgene is activated when the yeast Gal4 transcription factor binds to the UAS sequence. Gal4 is produced by a second construct called the driver under the control of a time- and tissue- specific promoter. Large collections of drivers and UAS-transgene have been generated and when crossed together, allow the expression of many transgenes nearly anywhere and at any time of the fly life cycle.

Several UAS-Tau constructs have been generated to study Tau isoforms (constructs expressing 0N3R, 0N4R, 2N4R isoforms), Tau phosphorylation (phosphodeficient and phosphomimetic Tau construct in which phosphorylation site is mutated into alanine and glutamate respectively), Tau mutation (Fronto-Temporal Dementia-related TauR406W, TauV337M and TauP301L mutated constructs), and Tau degradation (protease-resistant and truncated constructs) [3]. Using specific drivers, Tau neurotoxicity and dysfunction have been studied thanks to several readouts, the small rough eye phenotype (GMR driver), the loss of notal bristle (Eq driver), reduced lifespan and lethality (Elav and Appl drivers), axonal transport defects (D42 and OK6 drivers), and the loss of olfactory learning and memory (C492 and C772 drivers) [3]. The choice of a readout results from a trade-off between the ease and speed of scoring a phenotype and its biological or pathogenic relevance. The eye has the advantage of being easily accessible, loss of the eye is not lethal and the eye is composed of around 800 repetitive units that are scorable. Therefore, the eye external morphology has been successfully used in screens for modifiers of Tau pathology and other neurodegenerative diseases [5–9]. The concept of such screens is to co-express Tau in the eye with loss-of-function or gain-of-function constructs targeting a set of genes and to determine if the co-expression enhances or suppresses Tau toxicity in the eye (Fig. 1b). If the construct modifies Tau toxicity, the targeted gene likely interacts with Tau.

In this chapter, I describe how to express human Tau2N4R in the *Drosophila* eye with the UAS/Gal4 system. I present how to assess Tau expression in the fly eye by western blot. I detail how to take images of a fly eye and quantify its size. Finally, I explain how to perform a modifier screen and present the results of DIAP1 and p35, two published suppressors of Tau toxicity [10].

2 Materials

1. Drosophila stocks can be ordered from stock centers (http://flybase.org/wiki/FlyBase:Stock_Collections) or by sending requests to colleagues after stocks have been published in scientific journals. The UAS-mCD8GFP, Cs, UAS-DIAP1, UAS-p35, and GMR-Gal4 stocks come from the Bloomington Drosophila Stock Center. The UAS-Tau2N4R and UAS-Tau2N4R/Cyo;GMR-Gal4 stocks come from two publications [11, 12].
2. Fly stocks are reared in small vial filled to a fifth with fly food. Flies lay eggs on the food. Larvae eat and live in the food and then metamorphose into pupae outside of the food on the side of the vial. Adult flies hatch from pupae. Several fly food recipes have been developed (http://flystocks.bio.indiana.edu/Fly_Work/media-recipes/media-recipes.htm). Fly food, vials, and plugs are commercially available (https://geneseesci.com/).
3. Common equipment of *Drosophila* lab such as stereomicroscopes, CO_2 anesthesia apparatus, and incubators.
4. 1.5 mL microcentrifuge tubes.
5. Forceps Dumont #5.
6. LDS buffer: Dilute 25 μL of NuPAGE® LDS Sample Buffer (4×) (ThermoFisher scientific) and 10 μL of NuPAGE® Sample Reducing Agent (10×) (ThermoFisher scientific) in 65 μL deionized water. Vortex and aliquot 30 μL in 1.5 mL microcentrifuge tubes for each condition.
7. A heating block.
8. Upper and lower chamber running buffer: Add 50 mL of NuPAGE® MOPS SDS Running Buffer (20×) (ThermoFisher

scientific) to 950 mL of deionized water to prepare the lower chamber running buffer. Add 500 μL of NuPAGE® Antioxidant (ThermoFisher scientific) to 200 mL 1× running buffer to prepare the upper chamber running buffer.

9. XCell SureLock® Mini-Cell gel running tank (ThermoFisher scientific).
10. NuPAGE® Novex® 4–12% Bis-Tris Protein Gels, 1.0 mm, 10 wells (ThermoFisher scientific).
11. A power supply.
12. SeeBlue® Plus2 Pre-stained Protein Standard (ThermoFisher scientific).
13. A Trans-Blot® Turbo™ Transfer System (Biorad).
14. A Trans-Blot® Turbo™ Mini Nitrocellulose Transfer Pack (Biorad).
15. Ponceau red staining solution: in a bottle add in the following order, 50 mL deionized water, 3 mL glacial acetic acid, 0.2 g of ponceau red, qsp 100 mL with deionized water. Store at room temperature.
16. A shaker.
17. TNT solution (10×): Dissolve 36.32 g Tris base and 163.6 g NaCl in 1.5 L deionized water. Add 10 mL Tween 20. Adjust to pH 8.0 with HCl and add deionized water to qsp 2 L. Store at 4 °C. Dilute ten times with deionized water to get TNT solution (1×).
18. SuperBlock™ T20 (TBS) Blocking Buffer (ThermoFisher Scientific).
19. Primary antibodies: anti human Tau rabbit polyclonal antibody (Dako), anti-actin rabbit polyclonal antibody (Sigma-Aldrich).
20. Secondary Goat Anti-Rabbit IgG (H+L)-HorseRadish Peroxydase Conjugate (Biorad).
21. Chemiluminescent detection reagent: mix 5 mL of detection reagent 1 and 5 mL of detection reagent 2 of the ECL Western Blotting Detection Reagent (GE healthcare Life Sciences).
22. Plastic bag.
23. A blot imager, like the ChemiDoc™ XRS+ System with Image Lab™ Software (Biorad).
24. The open source Fiji software (http://fiji.sc/Fiji).
25. A quarter round 15×15 mm.
26. A macroscope, like Leica Z16 Apo Macroscope.

3 Methods

3.1 UAS-Gal4 Mediated Expression of HumanTau in the Drosophila Eye

1. Select the human Tau construct that has to be expressed in the Drosophila eye [3]. In this chapter, I use the UAS-Tau2N4R construct expressing the 2N4R isoform of human Tau. Select also a Gal4 driver and a control line. In this chapter, I use the GMR-Gal4 driver that is specifically expressed in the eye from development until adulthood and the CantonS (Cs) line.
2. Collect virgin female from the GMR-Gal4 line to cross them with male of the UAS-Tau2N4R and Cs lines (*see* **Notes 1** and **2**).
3. Cross 6–8 virgin GMR-Gal4 females with the same number of UAS-Tau2N4R males and of Cs males in two new vials (*see* **Notes 3** and **4**).
4. Incubate the vial for 10 days at 25 °C (*see* **Note 5**). Remove the parents after 7 days to avoid a mix between the parents and the progeny.
5. 10 days later, the progeny starts hatching. Anesthetize the progeny of each vial with CO_2 and observe the eye.

3.2 Assessment of Human Tau Expression in the Drosophila Eye by Western Blot

1. Protein extracts are prepared from the *Drosophila* heads of the progeny.
2. Proteins are separated by electrophoresis using the Novex® NuPAGE® SDS-PAGE Gel System and transferred onto nitrocellulose membrane using the Trans-Blot® Turbo™ Transfer System. Finally, they are detected and quantified with antibodies and chemiluminescence.
3. Anesthetize the progeny of one cross. With two forceps, dissect ten fly heads by holding the thorax with one forceps and cutting the connection between the head and the thorax with the other. Put the heads in a 1.5 mL microcentrifuge tube filled with 30 μL of LDS buffer (*see* **Note 6**).
4. Crush the head with a pestle for 30 s to 1 min. Centrifuge at 8,000 × *g* for 10 min at 4 °C. Retrieve the supernatant in a new vial (*see* **Note 7**).
5. Repeat the same steps with the progeny of the other cross.
6. Heat the samples at 85 °C for 10 min.
7. Remove the comb and the white tape near the bottom of the precast gel cassettes. Place the gel in the XCell SureLock® Mini-Cell gel running tank. Use the plastic buffer dam in place of the second gel cassette to form the upper buffer chamber. Lock into place with the gel tension wedge. Fill the upper chamber with the upper chamber running buffer. Wash the wells by aspirating and expelling the buffer in each well. Check

the tightness of seal of the upper chamber and fill the lower chamber with the lower chamber running buffer.

8. Load 7 μL of sample per well and the molecular weight standard (*see* **Note 8**).
9. Place the XCell SureLock® Mini-Cell lid on the buffer core. Connect the electrode cords to the power supply and run for around 50 min at 200 V constant until the dye front gets out of the gel.
10. After completion of the electrophoresis, turn off the power supply, disconnect the electrodes, and remove the gel cassette from the running tank.
11. The cassette is constituted of two plates, one shorter at the level of the wells, and another one with a slot where is located the foot of the gel. Place the cassette with the shorter plate facing up. Insert the gel knife in between the two plates at the three bonded sides and gently pry open the cassette. The gel should stick to the slotted plate. Remove the wells at the top and return the gel stuck on the slotted plate back on the shorter plate. Unstick the gel from the slotted plate by pushing the gel foot out of the slot with the gel knife.
12. Open one cassette of the Trans-Blot® Turbo™ Transfer System and a Trans-Blot® Turbo™ Mini Nitrocellulose Transfer Pack. Place the buffered-saturated nitrocellulose membrane and the bottom (anode) ion reservoir stack in the cassette. If necessary, expel any trapped air bubbles with a blot roller. Hold the gel by the foot and place it on the nitrocellulose membrane (*see* **Note 9**). Expel any trapped air bubbles with a lot of care to avoid damaging the gel with the blot roller. Place the top (cathode) ion reservoir stack on the gel. Use the blot roller to remove any trapped air bubbles in the assembled blotting sandwich. Close the cassette with its lid.
13. Insert the cassette in one of the two bays of the Trans-Blot® Turbo™ instrument. Select the mixed molecular weight mini program that transfers proteins at 1.3 A, 25 V for 7 min.
14. Once the transfer is finished, remove the cassette, open it, and disassemble the blotting sandwich. Retrieve the nitrocellulose membrane and place it in a tray.
15. Cover the membrane with ponceau red staining solution (around 10 mL depending on the size of the tray). Incubate for 1 min at room temperature under mild stirring. Retrieve the ponceau red staining solution (*see* **Note 10**) and wash the membrane twice in deionized water for 1 min at room temperature under mild stirring. Remove the last wash and take a picture of the ponceau red-stained membrane (Fig. 2a). Ponceau red stains all the proteins and allows assessing the quality of the electrophoresis and transfer, for example, at the

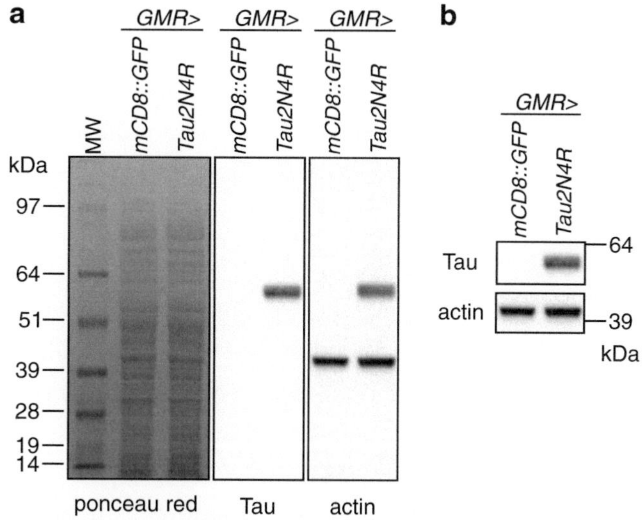

Fig. 2 Western blot analysis of human Tau expression in the *Drosophila* eye. Protein extracts from adult head of flies expressing the mCD8::GFP control construct and human Tau2N4R were analyzed by western blot. (**a**) Ponceau red staining of the blot (*left panel*, *MW* molecular weight standard), immunodetection of human Tau (*middle panel*) and actin (*right panel*), used as a loading control. (**b**) Final figure

level of the amount of loaded proteins, the sharpness of the separated bands, and the presence of air bubble during the transfer (*see* **Note 11**).

16. Wash the membrane 3 × 5 min in TNT solution to remove the ponceau red and block the membrane for 30 min in 5 % skimmed milk TNT solution at room temperature under mild stirring.
17. Wash the membrane 3 × 5 min in TNT solution.
18. Incubate the membrane overnight at 4 °C in Superblock T20 buffer, in which the primary Tau antibody is diluted at 1:10,000 (*see* **Note 12**).
19. Wash the membrane 3 × 10 min in TNT solution.
20. Incubate the membrane for 2 h at room temperature in 5 % skimmed milk TNT solution, in which the secondary anti-rabbit HRP-conjugate antibody is diluted at 1:5000.
21. Wash the membrane 3 × 10 min in TNT solution.
22. Incubate the membrane in the chemiluminescent detection reagent for 1 min.
23. Place the membrane in a plastic bag and then in the blot imager. Take images every 5 s for 5 min. Stop when one pixel of the image gets saturated. Keep the last image before the first image with saturated pixels. Take an image under white light to see the border of the membrane and the molecular weight standard. Merge the two images to localize the molecular

weight standard on the chemiluminescent image (Fig. 2a, middle panel).

24. Wash the membrane 3 × 5 min in TNT solution.
25. Incubate the membrane overnight at 4 °C in Superblock T20 buffer, in which the primary anti-actin antibody is diluted at 1:2000.
26. Wash the membrane 3 × 10 min in TNT solution.
27. Incubate the membrane for 2 h at room temperature in 5% skimmed milk TNT solution, in which the secondary anti-rabbit HRP-conjugate antibody is diluted at 1:5000.
28. Wash the membrane 3 × 10 min in TNT solution.
29. Incubate the membrane in the chemiluminescent detection reagent for 1 min.
30. Place the membrane in a plastic bag and then in the blot imager. Take images every 10 s for 5 min. Stop when one pixel of the image gets saturated. Keep the last image before the first image with saturated pixels. Take an image under white light to see the border of the membrane and the molecular weight standard. Merge the two images to localize the molecular weight standard on the chemiluminescent image (Fig. 2a, right panel) (*see* **Notes 13** and **14**). Using Fiji, crop the images around the band of interest (crop command in the Image menu after having selected the region with the rectangle selection tool) and make the final figure (Fig. 2b) (*see* **Note 15**). While the protein loading control is equal between the two conditions, as shown by the actin staining, the GMR>Tau2N4R condition exhibits a strong overexpression of Tau. Similarly with phosphorylation specific Tau antibodies, Tau phosphorylation can also be studied in *Drosophila*.

3.3 Assessment of Human Tau Toxicity by Eye Size Measurement

1. Anesthetize the progeny of one cross. With two forceps, dissect ten fly heads by holding the thorax with one forceps and cutting the connection between the head and the thorax with the other (*see* **Note 16**).
2. Align the head on a quarter round with double-sided adhesive tape (Fig. 3a). Carefully orient the head such that the right eye points to the top.
3. Place the quarter round under the objective of the macroscope. Define a z stack of ten images. Set the first plane and the last plane (Fig. 3a). Take the pictures. In each of the ten pictures (Fig. 3b), only a part of the eye and head is focused. The software retrieves the focused part of the ten images to reconstitute a fully focused final image (Fig. 3b, lower panel).
4. Shift the quarter round laterally to take pictures of the ten heads.

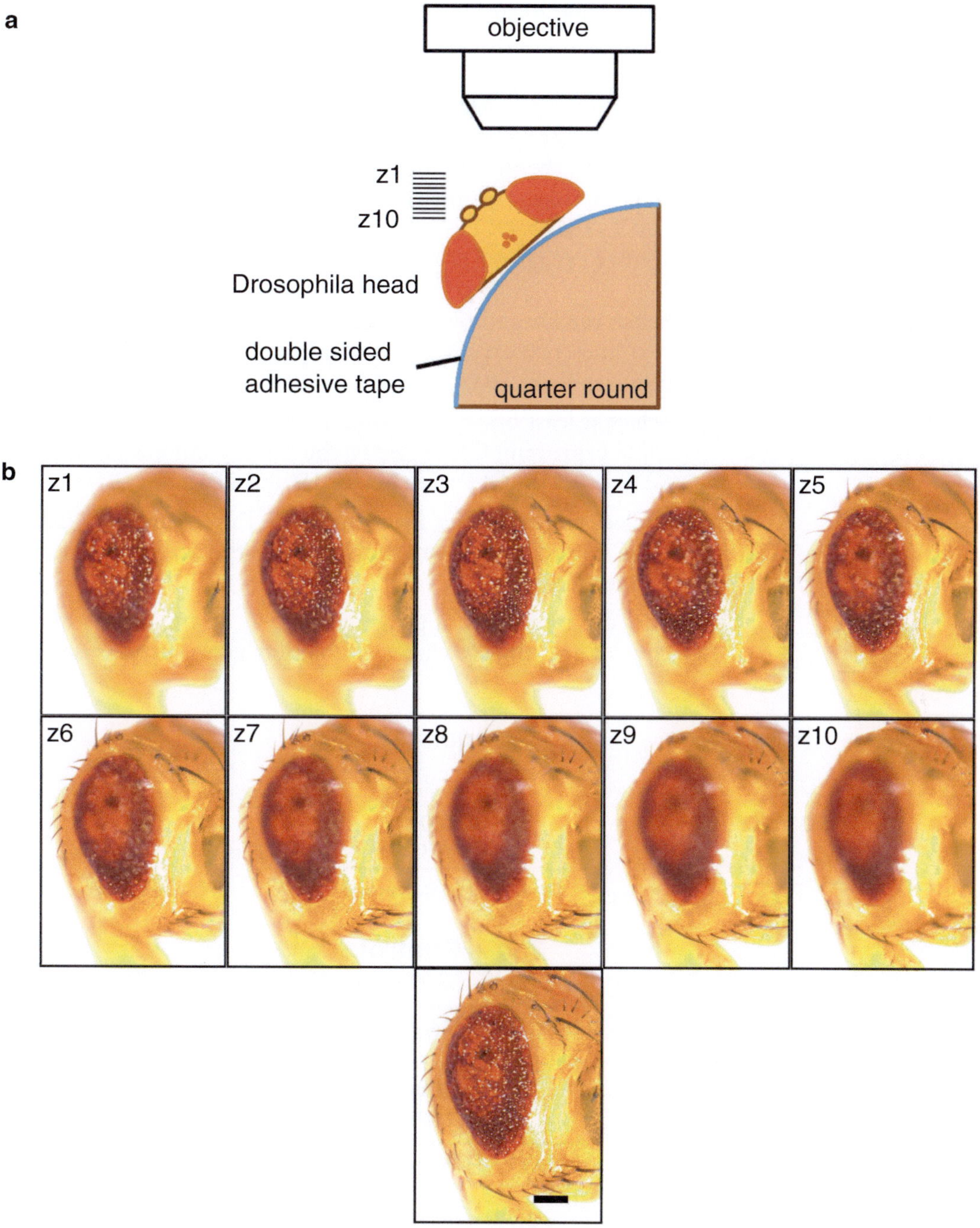

Fig. 3 Image acquisition of a *Drosophila* eye. (**a**) Experimental assembly. The *Drosophila* head is stuck on a quarter round with double-sided adhesive tape and oriented to have one eye pointing toward the objective of the macroscope. (**b**) Stack of ten focal plan images covering the thickness of the eye from which the fully focused final image (*lower panel*) is reconstituted (scale bar = 0.1 mm)

5. Open each image in Fiji. In the Image>Transform submenu, rotate and flip the image as required. With the polygon selection tool, make a selection around the eye. In the Analyse menu, click the Measure command. A new window with the measurement opens. Measure all the images and then copy and paste the measurements in a spreadsheet to make a graph and perform statistical analysis (*see* **Note 17**). The GMR>Tau2N4R eye is rough and around half the size of the control eye (Fig. 5).

3.4 Genetic Screening of Tau Toxicity Modifier

1. Select the lines to be screened. In this chapter, I have selected two lines overexpressing two caspase inhibitors, DIAP1 and p35 (*see* **Note 18**).
2. Select at least one control line. In this chapter, I have selected a line expressing GFP fused to the transmembrane segment of CD8 (mCD8::GFP) (*see* **Notes 19** and **20**).
3. Collect virgin females from the UAS-Tau2N4R/Cyo; GMR-Gal4 line to cross them with male of the screened lines (*see* Subheading 3.1).
4. Sort the four types of progeny (Fig. 4). In the cross with the control line, the Cyo flies (with curled wings) express only the

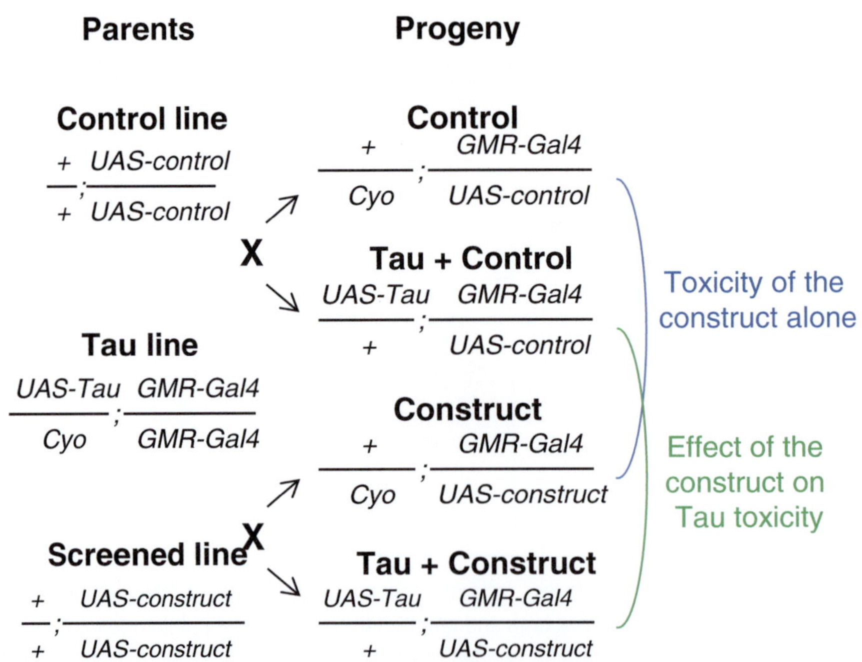

Fig. 4 Crossing scheme of the modifier screen. The Tau-expressing line is crossed with a control line and a screened line. Their progeny expresses either the control construct alone, the screened construct alone, Tau and the control construct, or Tau and the screened construct. Comparison of the progeny indicates the effect of the screened construct alone on the eye and its effect on Tau toxicity

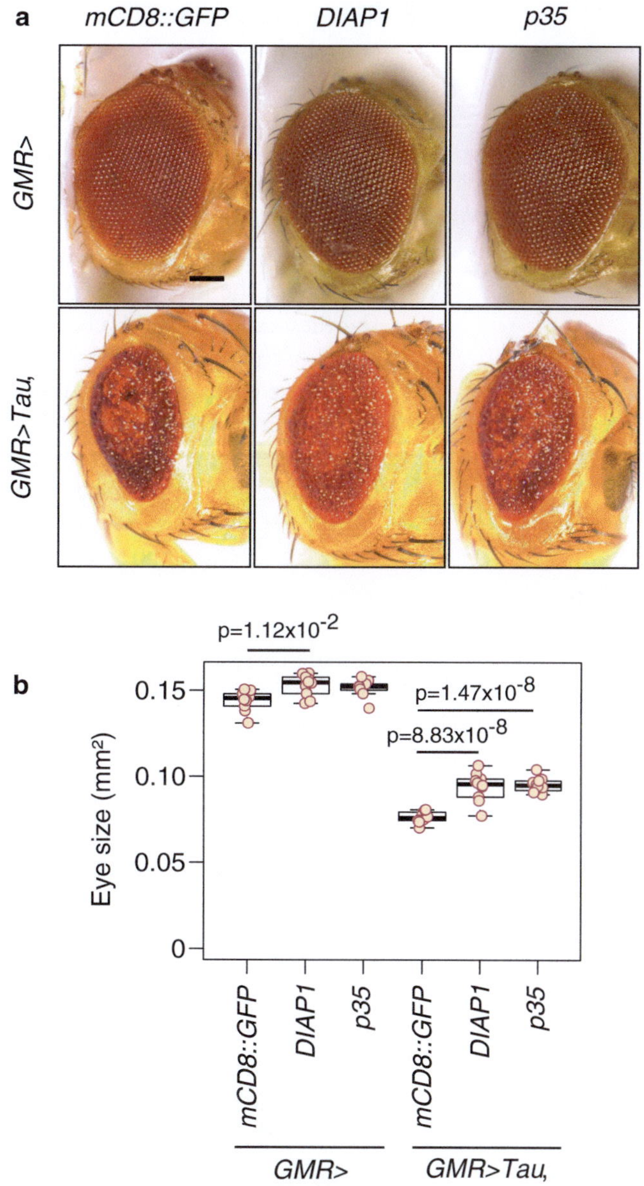

Fig. 5 Screening of two constructs targeting DIAP1 and p35 on Tau toxicity in the *Drosophila* eye. Two constructs expressing DIAP1 and p35 were crossed with a GMR>Tau line and a GMR-Gal4 line. The mCD8::GFP construct is used as a control. (**a**) Representative images of the eye of the progeny (scale bar = 0.1 mm). (**b**) Quantification of the eye size ($n = 10$, ANOVA, Tukey post-hoc)

control construct whereas the non-Cyo flies (straight wings) express both the control construct and Tau. In the cross with the screened line, the Cyo flies (with curled wings) express only the screened construct whereas the non-Cyo flies (straight wings) express both the screened construct and Tau.

5. Take pictures of ten flies of the four types of progeny and measure their eye size (Fig. 5a, b).
6. Compare the size of the flies expressing Tau and the screened construct with the flies expressing Tau alone (Fig. 4). If the size is bigger, the construct suppresses Tau toxicity. To the opposite, a smaller size indicates an enhancement of Tau toxicity. To know if the effect is just additive or if there is an interaction between Tau and the target of the screened construct, compare the size of the flies expressing the screened construct alone with the flies expressing the control only. If the effect is similar, the effect of the screened construct on Tau toxicity is likely additive, otherwise there is an interaction (*see* **Notes 21**). In Fig. 5, co-expression of DIAP1 and p35 with Tau suppresses Tau toxicity. The expression of DIAP1 alone also results in slightly bigger eye but with a smaller fold change and significance.

4 Notes

1. Female can store sperm. They have to be selected before they are fertilized by sibling males in their vial of origin. Otherwise, the progeny will be a mix from males with different genotypes.
2. After hatching, young females and males do not mate until around 8 h at 25 °C and 18 h at 18 °C. To collect virgin females, the fly vial is emptied in the morning at 9 am for example and carefully checked to avoid that any fly remains in the vial. Then the vial is incubated at 25 °C for several hours. Before 5 pm, females that have hatched are still virgin and have to be collected. The vial is again emptied at 5 pm and incubated at 18 °C during the night to collect virgin females next morning before 11 a.m. Young flies can also be recognized a few hours after hatching based on their depigmented soft cuticle and the presence of a black spot visible on the ventral side of the abdomen, called meconium.
3. If the males are fit, the number of males can be reduced to a third of the total number of flies.
4. Females from the UAS-Tau2N4R and Cs lines can also be crossed to GMR-Gal4 males. In that case, one has to collect virgin females from two stocks instead of one. In addition, using males or females may not be symmetric when the construct is inserted on the sexual chromosomes, the first pair of chromosomes in *Drosophila*.
5. The duration of the life cycle of *Drosophila* depends on temperature. From an egg to an adult, it lasts 10 days at 25 °C and longer or shorter at lower or higher temperature. To avoid hatching of the progeny during the weekend, it is convenient

to perform crosses on a Friday at 25 °C. Hatching of the progeny will start 10 days later on a Monday.

6. The protocol does not include any quantification of the protein extract. If the same number of heads, of the same sex (female tends to be bigger than males) are put in the same volume of buffer, the range of protein concentration should be the same in the different conditions. If the protein concentration happens to be too different on the following blot, a protein extraction in RIPA buffer, followed by a protein assay and a mixing with LDS buffer, may be preferred.
7. Samples can be stored at −80 °C at this step to continue the protocol later.
8. Avoid a symmetric loading, such as the samples in the middle with the standard on each side, to be able to place the right side of the gel on the membrane when transferring the protein on the nitrocellulose membrane.
9. The gel is very fragile and should be handled with care. When placing the gel, if the loading of the samples was asymmetric, check the orientation of the gel to have the right order of the sample on the nitrocellulose membrane and not the opposite.
10. The ponceau red can be reused as many times as it still stains the proteins.
11. While proteins are stained, one can cut the excess of membrane to be able to use a smaller tray and smaller amount of reagents such as antibodies in the following steps. The band of the molecular weight can also be highlighted with a ball-point pen.
12. After the incubation, the antibody solution can be stored at 4 °C to be reused several times. Azide can be added to the solution for a better conservation.
13. The band of Tau is still detected on the image (Fig. 2a, right panel). To remove previous immunodetection, the membrane can be subjected to a stripping procedure.
14. The membrane can be dried in between two sheets of absorbent paper and stored in a plastic bag for putative future reuse.
15. The bands can be quantified with the commands of the submenu Analyse>Gels. Image processing can also be performed by the software of the blot imager.
16. Take only males or females because females are bigger than males. Mixing males and females increases variability between flies of the same condition.
17. If the macroscope does not include the scale bar in the output image, it is possible to define the scale by taking a picture of an object with a known size. To this aim, a cell counting chamber, like a Malassez cell counting chamber or a Thoma cell counting chamber, can be useful.

18. Several collections of fly lines with constructs covering most of the genome are available (http://flybase.org/wiki/FlyBase:Stock_Collections). For loss-of-function constructs, RNAi libraries have been developed in Vienna (VDRC, http://stockcenter.vdrc.at/control/main), Kyoto (NIG, http://www.shigen.nig.ac.jp/fly/nigfly/) and Harvard (TRiP, http://www.flyrnai.org/TRiP-HOME.html). They differ mainly in the type of constructs (long double-stranded hairpin, short hairpin microRNA) and the insertion (random or targeted). The sequence specificity and expression is guaranteed but not their efficiency. Some mutant collections, especially generated by random insertion of P-element, like the BDGP and Bruinfly collections (http://www.fruitfly.org/p_disrupt/index.html, http://www.bruinfly.ucla.edu/) are also available. Random insertion of P-element has also been used to generate gain-of-function collections like the EP and EY collections (http://flystocks.bio.indiana.edu/Browse/in/misexp_top.htm, http://flypush.imgen.bcm.tmc.edu/pscreen/). These P-elements carry UAS sequences that potentially allow expression of nearby endogenous genes. Of note, these UAS insertions can also behave as loss-of-function constructs by blocking transcription of the gene if they are inserted in the opposite orientation. A collection of gain-of-function UAS-cDNA constructs has also started to be generated (http://flyorf.ch/). To find *Drosophila* orthologs of a list of human genes, one can use the DIOPT tool available online (http://www.flyrnai.org/cgi-bin/DRSC_orthologs.pl).
19. Because it is used as the reference, the selection of the control is important. The genetic background can affect the phenotype. Therefore, it is advised to retrieve control from the same collection. For example, the TRiP center has generated a construct expressing a hairpin against mCherry (http://www.flyrnai.org/TRiP-CTR.html). If such constructs are not available, constructs targeting genes that are not expressed in the eye, like the ovary specific *hunchback* gene, can be chosen as controls.
20. For large-scale screens (from hundreds to thousands of screened lines), up to 50 lines can be reasonably analyzed per week for two persons. The controls have to be rescreened each week.
21. If the expression of the screened construct alone affects the size of the eye, it may be due to an interaction with the Cyo balancer chromosome which carries many mutations. To rule it out, cross again the line with a GMR-Gal4 driver alone and check the effect on the eye size.

References

1. Bellen HJ, Tong C, Tsuda H (2010) 100 years of Drosophila research and its impact on vertebrate neuroscience: a history lesson for the future. Nat Rev Neurosci 11:514–522
2. Lessing D, Bonini NM (2009) Maintaining the brain: insight into human neurodegeneration from Drosophila melanogaster mutants. Nat Rev Genet 10:359–370
3. Gistelinck M, Lambert J-C, Callaerts P et al (2012) Drosophila models of tauopathies: what have we learned? Int J Alzheimers Dis 2012:970–980
4. Brand AH, Perrimon N (1993) Targeted gene expression as a means of altering cell fates and generating dominant phenotypes. Development 118:401–415
5. Shulman JM, Imboywa S, Giagtzoglou N et al (2014) Functional screening in Drosophila identifies Alzheimer's disease susceptibility genes and implicates Tau-mediated mechanisms. Hum Mol Genet 23:870–877
6. Shulman JM, Feany MB (2003) Genetic modifiers of tauopathy in Drosophila. Genetics 165:1233–1242
7. Shulman JM, Chipendo P, Chibnik LB et al (2011) Functional screening of Alzheimer pathology genome-wide association signals in Drosophila. Am J Hum Genet 88:232–238
8. Blard O, Feuillette S, Bou J et al (2007) Cytoskeleton proteins are modulators of mutant tau-induced neurodegeneration in Drosophila. Hum Mol Genet 16:555–566
9. Ambegaokar SS, Jackson GR (2011) Functional genomic screen and network analysis reveal novel modifiers of tauopathy dissociated from tau phosphorylation. Hum Mol Genet 20: 4947–4977
10. Jackson GR, Wiedau-Pazos M, Sang T-K et al (2002) Human wild-type tau interacts with wingless pathway components and produces neurofibrillary pathology in Drosophila. Neuron 34:509–519
11. Chatterjee S, Sang T-K, Lawless GM et al (2009) Dissociation of tau toxicity and phosphorylation: role of GSK-3beta, MARK and Cdk5 in a Drosophila model. Hum Mol Genet 18:164–177
12. Chapuis J, Hansmannel F, Gistelinck M et al (2013) Increased expression of BIN1 mediates Alzheimer genetic risk by modulating tau pathology. Mol Psychiatry 18:1225–1234

Chapter 27

Regulation of Neurotrophic Factors During Pathogenic Tau-Aggregation: A Detailed Protocol for Double-Labeling mRNA by In Situ Hybridization and Protein Epitopes by Immunohistochemistry

Katharina Schindowski-Zimmermann

Abstract

Alzheimer's disease (AD), most tauopathies, and other neurodegenerative diseases are highly associated with impaired neurotrophin regulation and imbalanced neutrophin distribution. Tau phosphorylation occurs at different sites of the tau protein and some phospho-epitopes are associated with normal ageing (like tau phosphorylated at Ser202/Thr205 detected by the antibody clone AT8) and others are highly associated with AD (abnormally phosphorylated tau at Thr212/Ser214 detected by the antibody clone AT100). Neurotrophins are crucial for the survival and maintenance of distinct neuronal population; therefore, their supply is essential for a healthy brain. Though their importance is well known, their analysis in tissue is not trivial and needs careful consideration. Here, a detailed a protocol is presented, how to combine in situ hybridization (ISH) with immunohistochemistry (IHC) to analyze neurotrophins during tau neuropathology and the results were confirmed by immunological methods. In addition, the preparation of the riboprobes is presented step-by-step. Since there are growing evidences for the relevance of neurotrophic factor distribution in the pathogenesis of AD, this technique is one useful tool to investigate the underlying mechanisms and therapeutic intervention.

Key words Double staining, In situ hybridization, Immunohistochemistry, Immunofluorescence, Riboprobe synthesis, Neurotrophins, Tauopathy, Dementia, Tau hyperphosphorylation

1 Introduction

Alzheimer's disease (AD) is the most common form of dementia in the elderly and is characterized by global cognitive decline including a progressive loss of memory, orientation, and reasoning. The microtubule-associated tau protein is hyperphosphorylated in AD and forms intracellular neurofibrillary tangles, one of the neuropathological hallmarks of AD. Even before their occurrence hyperphosphorylation of tau causes a destabilization of microtubules and thereby an impairment of axonal transport [1, 2]. In AD, NFT formation and neurodegeneration start in the basal forebrain.

Caroline Smet-Nocca (ed.), *Tau Protein: Methods and Protocols*, Methods in Molecular Biology, vol. 1523, DOI 10.1007/978-1-4939-6598-4_27, © Springer Science+Business Media New York 2017

Neurons in general, but in particular basal forebrain cholinergic neurons are dependent on trophic supply with neurotrophins. Nerve growth factor (NGF) and brain-derived-neurotrophic-factor (BDNF) are the major neurotrophic factors of the neurotrophin (NT) family. NGF displays a restricted target population and promotes survival and functioning to basal forebrain neurons. Being synthesized in the cortex and hippocampus, it is then retrogradely transported along the axons to cell bodies [3, 4]. BDNF promotes neuronal survival, differentiation, and synaptic plasticity in cortical, hippocampal, and basal forebrain neurons [5]. It is highly expressed and is distributed widely throughout the CNS, specifically in the hippocampal formation, cerebral cortex, and amygdala. After protein synthesis from its mRNA, BDNF is antero- and retrogradely transported to ipsi- and contralateral subfields of the hippocampus, the medial septum, the horizontal and vertical limbs of the diagonal band of Broca, several cortical areas, and the striatum [6]. Several studies suggest a loss axonal transport causing a loss of neurotrophic factor supply as the root for the degeneration of the basal forebrain cholinergic neurons in AD [7]. In AD, an imbalance of BDNF and NGF was observed with decreased levels of BDNF [8–10] and high levels of NGF [10, 11] in hippocampus and cortex. In addition, the NGF concentration in the basal forebrain is substantially reduced [12]. The reduced regulation of BDNF plays a key role in the AD pathogenesis. From experimental data it appears that many symptoms and hallmarks of AD like impaired memory, deficits in synaptic transmission and neurodegeneration are related to the loss of BDNF [13, 14]. Very interestingly, such imbalance was also observed in non-AD tauopathies like Pick's disease [15]. One rational of the present study was to find out, if such imbalance can be observed in a tau-transgenic mouse model [16].

To uncover if such imbalance is related to axonal transport and/or to a specific pathology a neuroanatomical determination of neurotrophins in distinct brain regions is needed as a whole brain homogenates will not give any details about distribution or transport. The dissection of the brain into distinct regions without contamination of different neuronal populations may cause relevant hurdles as the brain's size does definitively matter. As this is still manageable for human brains it is rather though for rodent brains (though not impossible [17]). Moreover, neurotrophins are expressed and produced in rather small quantities, very often below the detection limit of a classical protein analysis with Western Blot or ELISA. As neurotrophic proteins are stored in vesicles, transported along axons and secreted in the target area, their detection is more complex than most other proteins. Even dissected brain areas only show a part of the reality since very often neuronal projections are cut in different pieces (as it is between hippocampus and cortex and in particular for the projections to the different parts of the basal forebrain). Quantitative RT-PCR may overcome the limited detection of protein analytics, but still needs homogenates, which does not provide any information about the neuroanatomical distribution or neuronal populations. Though laser capture microdissection and single-cell RT-PCR is a sophisticated method, only a few laboratories are able and equipped to perform this method [18]. Therefore, classical histology is a powerful tool for the analysis of regional and neuroanatomical differences as brain regions and projections are easy to identify for a trained scientist. But as neurotrophins are expressed in low abundances, also here their histological detection with antibodies may be difficult or needs a powerful laser-scanning microscope. In situ hybridization (ISH) is a practical and very cost-effective method that can be implemented in more or less every laboratory. It is commonly used to analyze neurotrophins in tissue sections [19]. As with ISH the mRNA is detected and hence stains the cell bodies, no analysis of protein transport processes is possible. Nevertheless, for neurotrophins it is considered a standard method as it gives detailed answers about the expressed levels, the neuroanatomical site of expression, and the neuronal population. In addition ISH can be combined with IHC and/or counterstains. The protocol presented here is based on the method of Uwe Ernsberger [20].

BDNF promotes survival and function of hippocampal and cortical neurons, it is essential for synaptic transmission, learning, and memory. The THY-Tau22 mouse model shows neurodegeneration and cell loss in CA1 and impaired spatial learning and memory [16]. Moreover, the tau-transgenic mice display hyperphosphorylation of tau on several AD-relevant tau epitopes like AT100 and AT8. A hallmark of neurodegenerative disease is an imbalance of neurotrophic factors that appears to be the consequence of impaired axonal transport. Therefore, the regulation of neurotrophin mRNA and protein was analyzed in hippocampus and cortex of THY-Tau22 mice during tau pathology using a double staining method that combines ISH and IHC. The aim of the present study was to investigate if this can be linked to altered BDNF levels as reported for AD.

Using this protocol, it could be shown that tau-trangenic mice displayed an AD-like dysregulation of BDNF- and Tropomyosin receptor kinase B (TrkB)-mRNA, while neurotrophin-3 (NT-3) and TrkC were not altered. BDNF mRNA and protein levels are reduced by 6 months in tau transgenic animals. Interestingly, the loss of BDNF correlates negatively with tau abnormally phosphorylated (AT100) in the affected neurons. Histological analysis of BDNF and NGF protein levels revealed an imbalance in the tau-transgenic mouse model like it is reported from AD post mortem brain. BDNF protein levels were clearly reduced while some hippocampal and cortical neurons accumulated NGF-immunoreactivity (ir). Interestingly, most of these neurons co-localize with AT100-ir, but not AT8 indicating a crucial role of abnormal phosphorylated tau in this process. Immunoblotting confirmed significantly increased pro-NGF during increasing tau pathology. However, in contrast to data from AD, an increased number of DG neurons expressing NGF mRNA was observed that appear not to be directly associated with NGF protein accumulation.

1.1 Expression Pattern of BDNF and NT-3 and Their Receptors TrkB and TrkC

ISH using antisense riboprobes of BDNF, NT-3, TrkB, and TrkC were performed on coronal brain sections of 15-month-old THY-Tau22 and their littermate controls. ISH with sense controls showed very weak background without any specific staining. In Fig. 1a representative photos of the hippocampal-specific staining with the riboprobes BDNF, TrkB, NT-3, and TrkC are shown. BDNF expression can be found in all areas of the hippocampus: CA1, CA2, CA3, and DG. Also cells of the hilus and the subiculum were stained but have not been quantified. In THY-Tau22 mice the intensity of BDNF expression was largely diminished and since

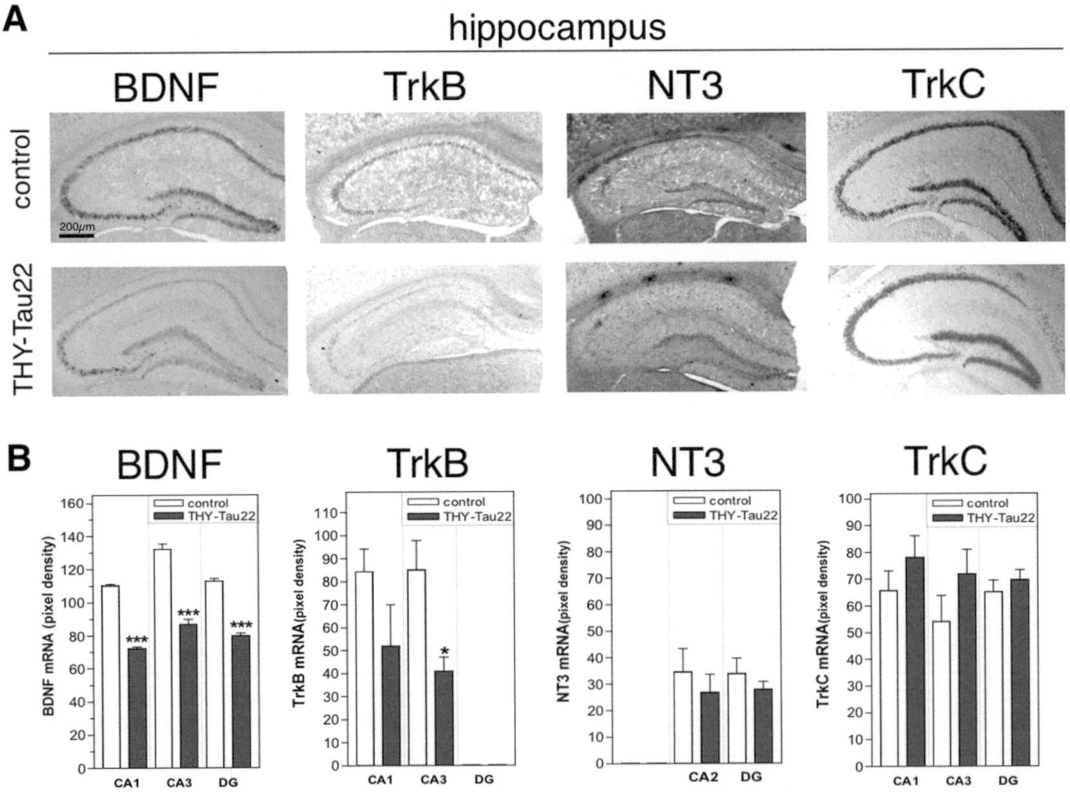

Fig. 1 Expression of BDNF, TrkB, NT-3, and TrkC. (**a**) Representative sections of the hippocampal area from wild-type control animals (*upper panel*) and tau-transgenic mice (*lower panel*; all 12–15 months old) are shown. Sections were processed for ISH with antisense probes for BDNF, TrkB, NT-3, and TrkC. (**b**) The pixel density of the stained areas was quantified with ImageJ. BDNF and TrkC were decreased in tau-transgenic mice compared to their age-matched littermates, while NT-3 and TrkC were not altered (*n*= 5 animals per group)

those animals suffer from neurodegeneration [16] also the number of BDNF mRNA-positive cells was reduced. When the pixel density of the hippocampal areas CA1, CA3, and DG was quantified a highly significant reduction by >30% was observed (Fig. 1b).

TrkB is a BDNF receptor and hence responsible for mediating BDNF activity. The TrkB antisense probe used here covers nearly whole coding sequence of the full-length TrkB. Even in wild-type controls TrkB mRNA staining was rather low (Fig. 1a) and decreased up to 50% in tau-transgenic mice (Fig. 1b). When the present expression levels are compared to BDNF and TrkB expression in the human brain during aging, a similar expression is observed [21]. In the present study, BDNF and TrkB ISH have not been performed in parallel with an internal standard. Therefore, it is not possible to compare the quantitative data of BDNF and TrkB pixel density. Nevertheless, a comparable tendency is observed, which BDNF shows a stronger expression in hippocampus than TrkB as it is observed in the human brain. Very interestingly, rather no TrkB expression was observed in the DG of wild-type controls and transgenes (Fig. 1a, b), though other studies show TrkB mRNA [22] and protein [23] in comparable amounts like in CA1. Only few studies show a decreased TrkB mRNA density in DG [24].

ISH with the NT3 antisense riboprobe showed expression in DG and CA2 and reproduced a pattern described previously [25]. No difference between THY-Tau22 and controls was detected.

The expression of the NT-3 receptor TrkC was rather strong in hippocampus in all hippocampal areas. As for NT-3, no difference between tau-transgenic mice and wild-types was evident. Though some sections of THY-Tau22 mice showed a tendency for higher TrkC levels, this did not reach statistical significance.

1.2 Cortical and Hippocampal Loss of BDNF in Tau Transgenic Mice

In AD post mortem tissue, decreased levels of BDNF in hippocampus and cortex are well documented. As BDNF is highly implicated in learning and memory, the observed decrease of BDNF mRNA was analyzed in more detail. In normal control animals, BDNF was widely expressed in hippocampal (Fig. 2a) and cortical (Fig. 2b) neurons. Some neurons display even high levels of BDNF mRNA in non-transgenic animals mainly in the CA3 and DG. In THY-Tau22, most of hippocampal areas expressed less BDNF (Fig. 2a). In the area that strongly displays cell loss (CA1), the reduction of BDNF was already present at 6 months, hence long before the onset of neurodegeneration of pyramidal neurons (data not shown). However, even in aged animals with severe cell loss in CA1 seen as reduced thickness of the pyramidal cell layer, the loss of BDNF was present in the remaining pyramidal neurons by reduced staining intensities (Fig. 1b). This can be interpreted that THY-Tau22 do not or are not able to compensate their BDNF loss by upregulating BDNF mRNA in the remaining cells.

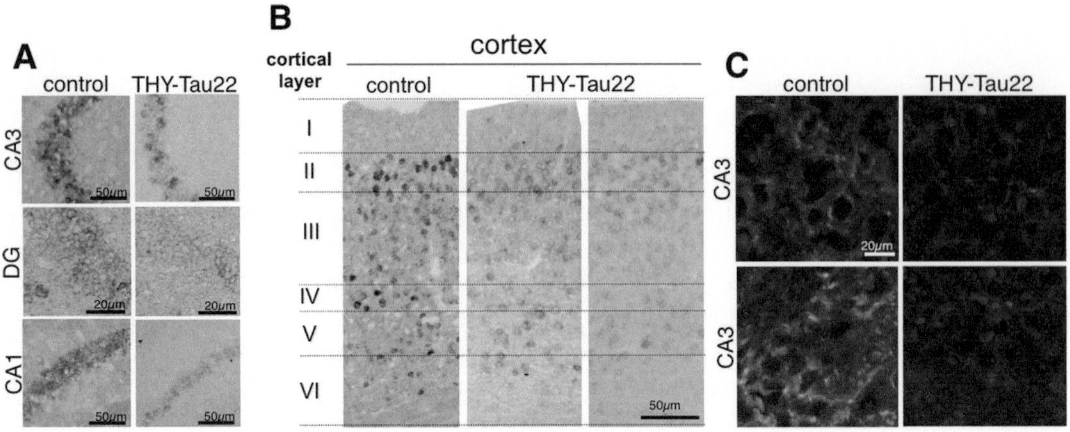

Fig. 2 Decreased levels of BDNF mRNA and BDNF protein in THY-Tau22. (**a**) Higher magnification of the hippocampal areas CA3, DG, and CA1. 15-month-old THY-Tau22 showed decreased staining intensity of BDNF-mRNA compared to age-matched littermate controls. In CA1 a neuronal loss is also evident. (**b**) BDNF-mRNA was reduced in 15-month-old THY-Tau22 in layer II–VI in the primary cortex. Layer I does hardly show any BDNF expression even in control animals. (**c**) In concordance to the decrease in BDNF mRNA, BDNF-ir (*green color*) was also reduced in CA3 of 15-month-old animals compared with littermate age-matched controls. DAPI (*blue color*) was used as a nuclear counterstain

The loss of BDNF is not limited to the hippocampus. The cortex of THY-Tau22 mice shows also a clearly decreased signal after ISH with BDNF antisense probes (Fig. 2b). The distribution of wild-type's BDNF expression in the different cortical layers is comparable to the distribution in the human cortex [21].

Immunofluorescence (IF) confirmed the BDNF ISH data and revealed that BDNF protein levels were also decreased in CA3, CA1 (Fig. 2c), and DG (data not shown).

To elucidate the relevance of phospho-tau in BDNF regulation, sections were double stained for BDNF mRNA (using ISH) and phospho-tau detected with AT100 ir. AT100 detects phosphorylation of tau in the proline-rich region at Thr212/Ser214 and is highly implicated in the pathogenesis of AD. The loss of BDNF in THY-Tau22 correlates to tau pathology in CA1 (Fig. 3a) and CA3 [26]: the more neurons displayed AT100-ir the less BDNF mRNA was expressed. In the cerebral cortex the interaction between AT100-ir and BDNF mRNA was similar. At a higher magnification a strong negative correlation between BDNF mRNA and phospho-tau can be observed in the cortical cell bodies: The more BDNF mRNA is detectable, the less AT100-ir is observed and vice-a-versa. The phospho-epitope AT270 (tau; pThr181; Fig. 3e) showed a comparable relation like AT100, while this was less obvious for AT8 (tau; Ser202/Thr205; Fig. 3c) and AT180 (tau; pThr231; Fig. 3d). The THY-Tau22 mouse model shows an AD-like progression of tau pathology, starting early in subiculum, entorhinal cortex, basal forebrain, CA1 and

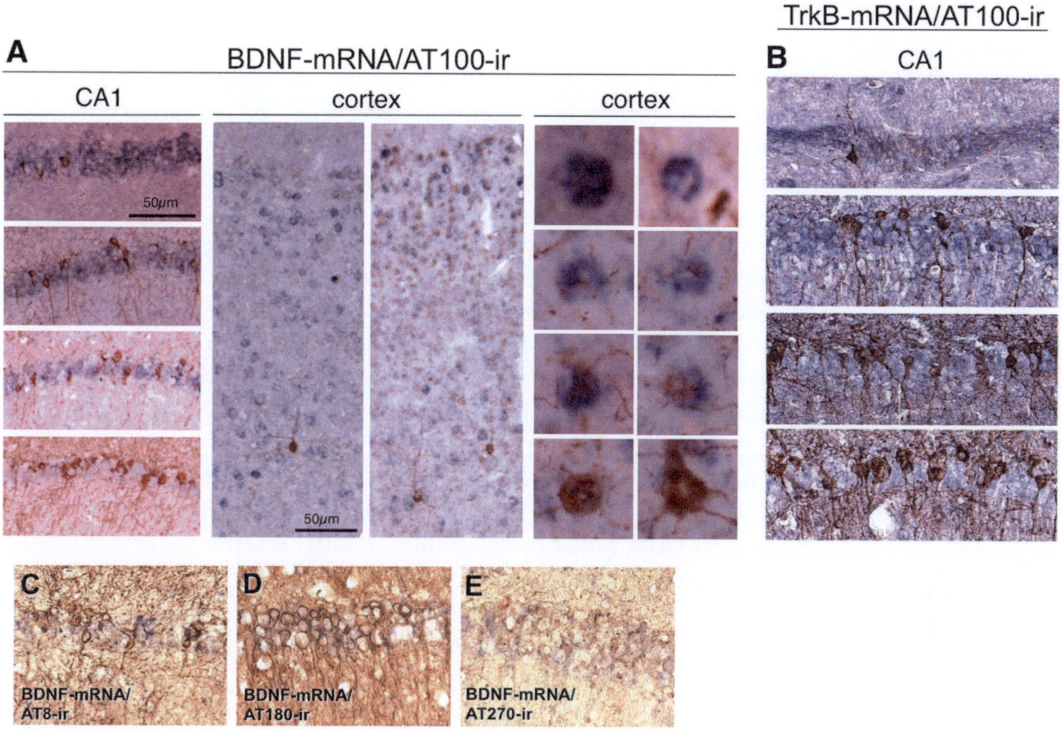

Fig. 3 Decreased BDNF and TrkB mRNA in THY-Tau22 is associated with phospho-tau specific epitopes AT100-ir. (**a**) Double labeling of BDNF-mRNA (*blue color*) and AT100-ir (*brown color*) showed a decrease of BDNF expression in areas with high AT100-ir in CA1 and cortex. In cortical neurons a very clear negative correlation could be observed when using a higher magnification: the stronger the AT100-ir was the lower was the amount of BDNF-mRNA and vice-a-versa. Animals used were between 3 and 15 months old. (**b**) Double labeling of TrkB-mRNA (*blue color*) and AT100-ir (*brown color*) showed a similar pattern as for BDNF. However, the correlation here was not as strong as for cortical BDNF loss. Animals used were between 3 and 15 months old. (**c–e**) Double labeling of BDNF-mRNA (*blue color*) and AT8-ir (**c**), AT180-ir (**d**), and AT270-ir (**e**; *all in brown color*) in 14-month-old THY-Tau22. A similar correlation as for AT100 appears to be evident in AT270 (**e**), but less for AT8 (**c**) and AT180 (**d**)

the perforant path and spreading to CA3, amygdala, basal forebrain, other cortices, thalamus and DG with aging (unpublished data). Most hippocampal fiber tracts display hyperphosphorylated tau in axons rather early by 3 months. This may be a consequence of BDNF mediating tau dephosphorylation [27]. However, little is understood yet about the role of tau on BDNF regulation. No differences in hippocampal and striatal BDNF levels were detected in APP23 mice but an increase in cortex [28]. In addition, this model displayed an increase of BDNF in plaque-associated glial cells [29].

As BDNF regulates the expression of its receptor TrkB, a loss of TrkB may be seen whenever BDNF is reduced. Coherent to the double stain with BDNF, also TrkB showed a negative correlation with AT100-ir (Fig. 3b).

1.3 Increase in the Number of NGF-Positive Neurons in Hippocampus and Cortex of THY-Tau22 Mice

Pyramidal and dentate granule neurons and some interneurons express NGF regulated by neuronal activity. Among glial cells, NGF is produced by astrocytes and microglia. Cerebral insults markedly upregulate neuronal and glial NGF expression. The increase of cortical and hippocampal NGF levels is well described in AD. However, whether it is related to tau pathology is still not clear. Therefore, the expression of NGF mRNA was examined in the tau transgenic mouse model. In non-transgenic controls, distinct cells displaying NGF mRNA were detected in the polymorph and granular DG, in the hilus, and in the CA1 and CA3 regions at all ages investigated (Fig. 4a). In THY-Tau22 an elevated number of cells in the granular cell layer of the DG and hilus expressing NGF were detected by 3 months up 17 months (Fig. **4**a). The majority of these cells displayed clearly the morphology of granule cells. Other hippocampal areas and the cerebral cortex displayed no changes in NGF mRNA expression at young ages up to 15 months (data not shown), similar to what is reported in AD brains. A quantification of the number of NGF mRNA positive DG neurons per mm^2 revealed an increase of 41 % in 15-month-old THY-Tau22 compared to age-matched littermate controls (Fig. **4**b). While NGF expression in these tau-transgenic mice is coherent to studies in human AD brains for most neuroanatomical areas, it is not reported that NGF mRNA is altered in DG in AD. Though there was a low deviation and high reproducibility among the analyzed brains, the total amount of NGF mRNA-positive cells was rather low with regard to the total number of hippocampal neurons. As previously shown, a cell cycle reentry can be observed in THY-Tau22 mainly in the DG [30]. It cannot be excluded that this is linked to the enhanced NGF mRNA observed here. Nevertheless, this model displays tau pathology and no evident signs of amyloid deposition. AD is an interplay of both pathologies, and amyloid pathology appears also to play a role in NGF regulation [31].

To explore the relevance of tau in this scenario, sections were double stained for NGF mRNA and AT100. In several NGF expressing DG neurons a co-localization (shown in Fig. **4**c as arrows) with the phospho-epitope was found, but only in few NGF expressing neurons in CA3, CA1 (Fig. **4**c) or cortex (data not shown). Hence, no clear correlation between NGF mRNA and neurofibrillary pathology could be determined. It should be noted that high magnifications are needed to adequately determine co-localization of ISH and IHC since both methods produce a precipitated dye that may cover each other.

IF was used to compare with the NGF mRNA data. While in non-transgenic controls NGF-ir was usually located with weak granular fluorescence in compartments of the soma (Fig. **5**a, e), in THY-Tau22 NGF-ir accumulated to high fluorescence and filled

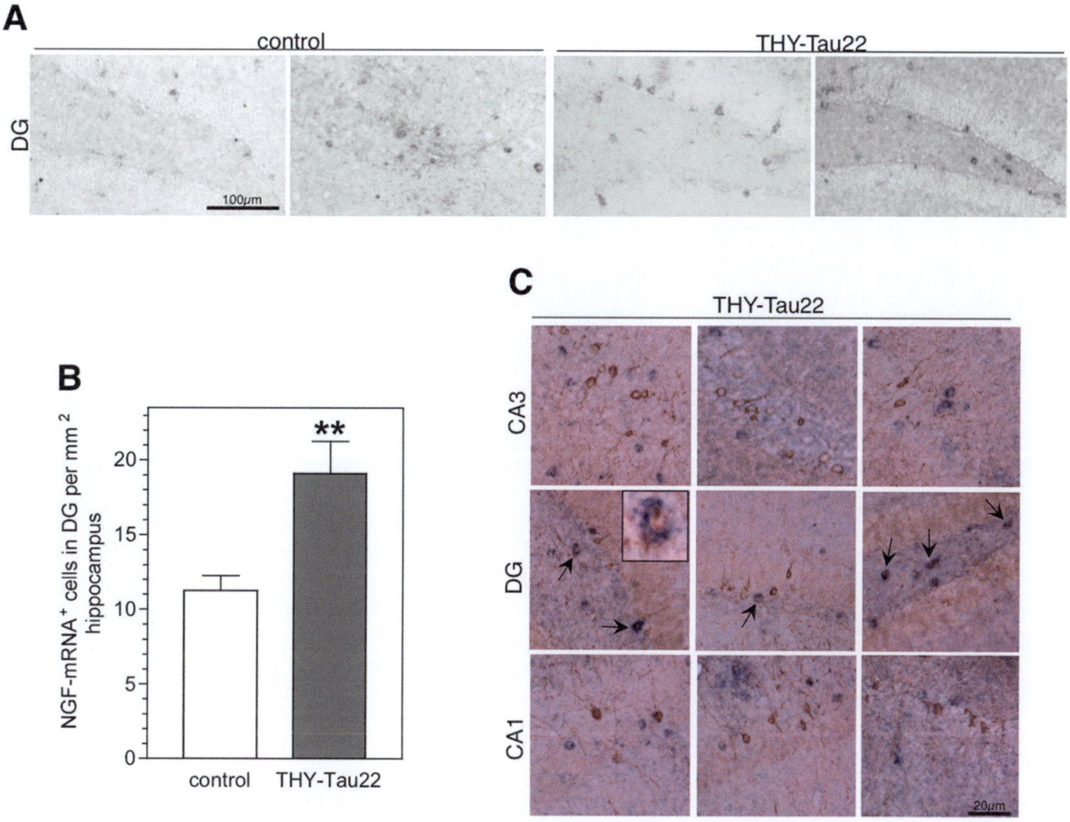

Fig. 4 Increase of NGF-mRNA in DG and co-localization with AT100-ir in THY-Tau22. (**a**) ISH detected an increased number of NGF expressing cells in the granular cell layer and the hilus of the DG in aged THY-Tau22 (here at 15 months). (**b**) Semiquantification of the density of NGF expressing cells in DG showed a significant increase by over 40 % in THY-Tau22 (**$p < 0.01$). (**c**) Double labeling with AT100-ir (*brown color*) and NGF-mRNA (*blue color*) showed co-localization in some cells of DG (high magnification in *inset*) in THY-Tau22 mice at 15 months (see *arrows*). In CA3 and CA1, such co-localization was a very rare event. In addition, the number of NGF expressing neurons was not altered in CA3 and CA1 between controls and transgenic mice

the whole cell body in cells in CA3, DG (Fig. 5a), entorhinal cortex (Fig. 5b), and CA1 (Fig. 5e). In addition, the tau-transgenic mouse model displayed weak NGF-ir in glial cells in the hippocampal hilus (Fig. 5c), which was not observed in controls (data not shown). As reported previously, adult THY-Tau22 mice display a strong gliosis in the hilus region [16]. Interestingly, astroglial NGF production was shown to induce tau hyperphosphorylation [32] and NGF also activates the tau kinase GSK3beta [33]—a vicious circle in AD pathogenesis. It should be noted that glial cells were identified by morphology and not by staining of glial-specific markers.

Immunoblotting of NGF confirmed the histological data (Fig. 5d). Pro-NGF (32 kDa) was the predominant form in mouse brain homogenates, while the mature NGF protein (14 kDa) was too close to the detection limit to give acceptable results (data not

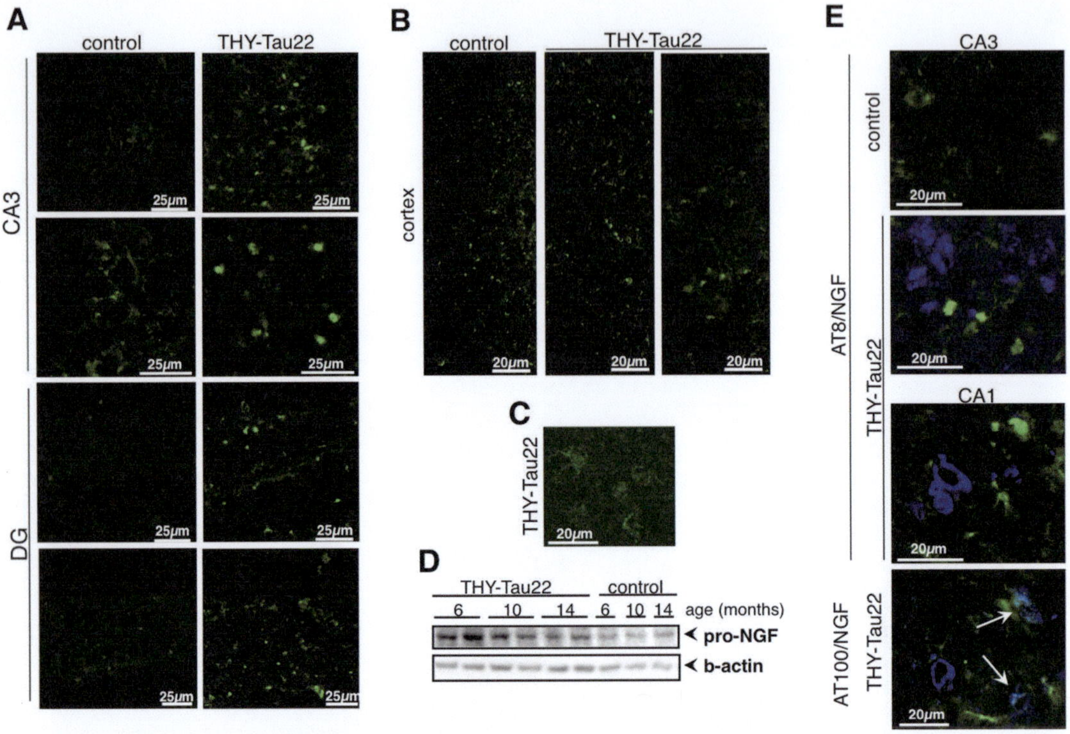

Fig. 5 NGF protein accumulation in CA3, DG and cerebral cortex in THY-Tau22 mice. (**a**) NGF-ir was detected in hippocampal CA3 and DG in controls and tau-transgenic animals at 12–15 months. However, in THY-Tau22 some hippocampal neurons in CA3 and DG showed strong accumulation of NGF-ir in the cell bodies. (**b**) Accumulation of NGF-ir in cells of the cerebral (entorhinal) cortex of 15-month-old THY-Tau22 mice. (**c**) NGF-ir glial cells were detected in the hilus in tau-transgenic animals at 15 months. However, NGF-ir was less intensive than in neurons. (**d**) Immunoblot analysis revealed an increase of pro-NGF levels in THY-Tau22. (E) NGF-ir (in *green*) was found in the vicinity of AT8-positive cells (in *blue*) in the CA3 region and the CA1 region at 15 months in tau-transgenic mice. However, no co-localization of NGF-ir accumulation with AT8-ir was detected in any section. By contrast, AT100-ir (in *blue*) was detected in CA1 in aged THY-Tau22 mice that was frequently co-localized with NGF-ir accumulation (see *arrows*)

shown). The levels of pro-NGF were increased in THY-Tau22 compared to controls, similar to AD [15, 34, 35]. The antibody used here was the same used for IF. Therefore, it cannot be excluded that the NGF-ir observed in hippocampal and cortical neurons of THY-Tau22 mice is caused predominantly by pro-NGF-ir.

Finally, to investigate the impact of tau phosphorylation on NGF accumulation, sections were double stained for phopho-tau (in blue) and NGF (in green; Fig. 5e). Interestingly, the site of tau hyperphosphorylation appears to influence the observed NGF accumulation in CA3 and CA1: no co-localization of AT8 and NGF was observed, while frequently neurons accumulating NGF co-localize with AT100. However, highly NGF-ir cells were detected in the proximity to AT8-ir. Interestingly, any co-localization with phospho-tau in the DG, in which the number of neurons showing hyperphosphorylated tau in their perykarya was rather low, was not observed. AT100 detects abnormal phorphorylated tau considered typical for AD. As demonstrated previously, AT100-ir increased in THY-Tau22 mice in an age-dependent manner and was linked to the occurrence of aggregated tau and filaments [16]. No AT100 is detectable in control animals while AT8 was observed in controls in aging. Aggregated tau may interfere with transport processes and as a possible consequence of impaired retrograde transport, NGF is gathered in the target neuron.

However, tau is only one prominent piece in the mosaic of AD. Also other AD models have so far implicated a relevance of NGF in AD. A transgenic mouse (AD11), which produces antibodies against NGF and thereby causing a lack of free and bioavailable NGF, acquires an age-dependent pathology reminiscent of human AD, including amyloid plaques, hyperphosphorylated tau in cortical and hippocampal neurons, extensive cortical cell loss, behavioral deficits, and impairment in short- and long-term potentiation [36–38]. In a mouse model of trisomy 16 (Ts65Dn), failed retrograde transport of NGF was observed [39–41]. Interestingly, these mice show increased levels of Abeta but no hyperphosphorylated tau, thus, there are data showing reduced levels of NGF in the basal forebrain and—in aged animals—also in the hippocampus [42].

The role of APP and Abeta on NGF is still not clearly elucidated. NGF is reduced in hippocampus and striatum but not in cortex of APP23 [28]. Excitingly, cognitive performance correlated inversely with hippocampal NGF levels in trained APP23 animals. Thus, impaired NGF transport causing cholinergic degeneration was linked to increasing APP levels in the Ts65Dn trisomic model [43] without deficits in general transport.

The importance of neurotrophins in neurodegenerative dementia is now widely accepted. An NGF gene trial in AD patients [44] showed by postmortem neurohistology that neurons of the degenerating AD brain were able to respond to NGF with axonal sprouting, cell hypertrophy, and activation of functional markers. Sprouting persists for 10 years after gene transfer [45, 46]. No adverse events were reported and, unfortunately, no effects on cognition were discussed.

Though, it is well established that the transport of neurotrophins is impaired in most neurodegenerative diseases [2] there is still a lot of work to do until this can translate to therapeutic strategy to prevent or treat AD. Co-localization methods as presented here for ISH and ICH are a powerful tool to uncover molecular mechanisms.

2 Materials

2.1 Transgenic Mice

The THY-Tau22 mouse line was developed by Sanofi/France by Dr. Nathalie Touchet and Dr. Günter Tremp in collaboration with Dr. Luc Buée (Inserm, UMRS1172). A cDNA coding for the 412 amino acid 4-repeat isoform of human tau mutated at G272V and P301S was inserted into a Thy1.2-expression vector. Adult to aged mice (25–40 mg) of both sex were used throughout this study. This construct was microinjected into a C57Bl6/CBA background. All experiments on animals were in accordance with standards for the care and use of laboratory animals and with French and European Community rules.

2.2 Solutions and Reagents for Generation of cDNA Templates, PCR, Ligation, and In Vitro Transcription Reactions

1. Trizol (LifeTechnologies/ThermoFisherScientific).
2. Nuclease-free, sterile water.
3. 20–40 U/μl RNase-inhibitor (RNasin®/Promega).
4. 5× M-MLV buffer (Promega).
5. 100 pmol/μl dT_{15} primer.
6. 10 mM dNTP (New England Biolabs).
7. 200 U/μl M-MLV reverse transcriptase (Promega).
8. 10× PCR buffer : 10× *Taq* Polymerase buffer.
9. 25 mM $MgCl_2$.
10. 100 pmol/μl of each forward and reverse primers (Table 1).
11. 5 U/μl *Taq* polymerase.
12. 1% agarose gel: dissolve 1 g of agarose in 100 ml TAE (Tris-Acetate-EDTA) buffer and heat until agarose is dissolved. Add dye for nucleic acids like ethidium bromide or Roti®-GelStain (Roth).
13. QIAquick DNA Gel Extraction kit (Qiagen).
14. pGEM-T vector (Promega).
15. 2× ligation buffer (Promega).
16. 1–3 U/μl T4 DNA ligase (Promega).
17. KRX competent cells for blue/white screening (Promega).
18. LB-ampicillinplatessupplementedwithX-gal(5-bromo-4-chloro-3-indolyl-β-D-galactopyranoside).
19. Miniprep and Midiprep DNA extraction kits.
20. QIAquick PCR purification kit (Qiagen).
21. Digoxigenin (DIG) RNA Labeling Kit (Roche).
22. 10 μg/μl t-RNA (from baker's yeast; Roche).
23. Mini Quick Spin RNA column (Roche).

Table 1
Primer combinations used to generate cDNA templates for riboprobes

Template	Forward primer	Reverse primer	Tm [°C]	length [bp]
NGF	GTGCCTCAAGCCAGTGAAAT	TCTATCCGGATGAACCTCCA	60.2	870
BDNF	GCGCCCATGAAAGAAGTAAA	CTTATGAATCGCCAGCCAAT	60.0	647
NT-3	GATCCAGGCGGATATCTTGA	TTTCTCGACAAGGCACACAC	59.8	648
TrkB	CGGCACATAAATTTCACACG	CTTGGCCATCAGGGTGTAGT	55.3	741
TrkC	AAGTAACCGGCTCACCACAC	GTGGGCTTTTTGAAGAGCAG	57.3	702

2.3 Solutions and Buffer for ISH

1. Hybridization buffer: 50% formamide (v/v); 10% dextrane sulfate (w/v); 1 mg/ml yeast RNA; 1× Denhardts; 5 mM EDTA; 12.6 mM Tris, pH 7.5; 185 mM NaCl; 5 mM Na_2HPO_4; 0.5 mM NaH_2PO_4.
2. SSC/formamide: 1× SSC (Saline-Sodium Citrate buffer); 50% formamide; 0.1% Tween 20.
3. MABT: 100 mM maleic acid; 150 mM NaCl; 0.1% Tween 20; adjust to pH 7.5.
4. MABT/ChS: 20% heat-inactivated chicken serum in 1× MABT.
5. AP (alkaline phosphatase) buffer: 100 mM Tris, pH 9.5; 50 mM $MgCl_2$; 100 mM NaCl; 0.24 mg/ml levamisole (Abcam); 0.1% Tween 20.
6. Nitro blue tetrazolium/(5-Brom-4-chlor-1H-indol-3-yl) dihydrogenphosphate (NBT/BCIP) solution (Roche).
7. anti-DIG alkaline phosphatase conjugate (Roche) diluted 1:1700 in MABT/ChS.
8. Kaiser's Glycerol gelatine (Merck).

2.4 Solutions, Buffer, and Materials for Tissue Preparation, IHC, Immunofluorescence (IF), and Western-Blot (WB)

1. Rompun™ and Ketanest™ anesthetics.
2. 0.9% NaCl.
3. 4% paraformaldehyde (PFA).
4. 30% sucrose.
5. Cryostat.
6. Superfrost® Plus slides (Menzel-Glaser).
7. 1× PBS: 137 mM NaCl; 2.7 mM KCl; 10 mM Na_2HPO_4; 2 mM KH_2PO_4; adjust to pH 7.4.
8. PBS-T: 1× PBS supplemented with 0.05% Triton X-100.
9. Tau phospho-specific antibodies; *see* Table 2 for dilutions in PBS-T.

Table 2
Tau phospho-specific antibodies used for the double stainings

Antibody	Specificity	Species	Dilution	Source
AT8	Tau; $pSer^{202}/pThr^{205}$	mouse	1:10,000	Innogenetics
AT100	Tau; $pThr^{212}/pSer^{214}$	mouse	1:2000	Innogenetics
AT180	Tau; $pThr^{231}$	mouse	1:500	Innogenetics
AT270	Tau; $pThr^{181}$	mouse	1:2000	Innogenetics

10. Biotinylated anti-mouse antibody (diluted 1:200 in PBS; Vector Labs).
11. VECTASTAIN ABC kit (Vector Labs).
12. 3,3′-diaminobenzidine (DAB; Sigma-Aldrich).
13. Anti-NGF or anti-BDNF (Santa Cruz Biotechnology) diluted at 1:100 in PBS-T for IF.
14. streptavidin-labeled fluorescein diluted at 1:100 in PBS (Vector Laboratories).
15. AMCA (aminomethylcoumarin acetate)-labeled anti-rabbit antibodies diluted at 1:200 in PBS (Vector Laboratories).
16. VectaShield (Vector Laboratories).
17. 4′,6-Diamidin-2-phenylindol (DAPI; Vector Laboratories).
18. Microscope equipped with a software-controlled digital camera.
19. Image J program (public domain, NIH).
20. SDS-PAGE gels (Invitrogen).
21. Nitrocellulose membranes (Invitrogen).
22. ECL chemoluminescence kit (GE Healthcare).
23. Anti-NGF antibody diluted at 1:1000, β-actin antibody (Santa Cruz Biotechnology) at 1:10,000 in PBS-T for WB.
24. GraphPad/Prism Software (GraphPad Software, Inc).

3 Methods

3.1 Tissue Preparation for Histology

1. Anesthetize mice deeply with Rompun™ and Ketanest™ and transcardially perfuse sequentially with 20 ml of 0.9% NaCl and 20 ml of 4% paraformaldehyde without any pressure (*see* **Note 1**).
2. Remove brains carefully from the skulls and transfer in a small vessel.

3. Postfix brains in 4% PFA overnight at 4 °C, and then place in 30% sucrose at 4 °C for cryoprotection until they sunk to the bottom, prior to be frozen on dry ice.
4. Cut serial coronal cryosections (20 μm) on a cryostat and mount on Superfrost slides. Slides are kept at −20 °C until use (*see* **Note 2**).

3.2 Generation of cDNA Templates for Riboprobes

The plasmids had been generated in the lab of Prof. Dr. Klaus Unsicker, Neuroanatomy, University of Heidelberg/Germany and have been kindly donated for this study.

1. Weigh a sterile empty microcentrifuge tube and note well its weight.
2. Sacrifice a mouse and remove the brain carefully from the skull.
3. Transfer the brain to the microcentrifuge tube, weigh again, and calculate the net weight of the brain.
4. Homogenize the brain in an appropriate amount of Trizol (1 ml per 50–100 mg tissue) and extract RNA according to manufacturer's instructions.
5. Prepare the following reaction to denature the secondary structure of the mRNAs:

1 μg	total RNA from mouse brain
0.1 μl	RNase-inhibitor RNasin®
H_2O added up to 10 μl	

6. Heat for 15 min at 65 °C, then place on ice.
7. To generate template cDNA from RNA add the following pre-mixed cocktail for first strand synthesis:

4 μl	5× M-MLV buffer
1 μl	dT_{15} primer
2 μl	dNTP
2 μl	M-MLV reverse transcriptase
1 μl	H_2O
10 μl	Denatured RNA from above

8. Incubate the reaction for 1 h at 37 °C, then heat for 10 min at 65 °C, and store at −20 °C until use.

9. Amplify templates from cDNA with the primer combinations described in Table 1 and perform per probe the following PCR reaction:

0.5 μl	cDNA
2.5 μl	10× PCR buffer
1.5 μl	$MgCl_2$
1.0 μl	dNTPs
0.5 μl	Forward primer (*see* Table 1 and **Note 3**)
0.5 μl	Reverse primer (*see* Table 1 and **Note 3**)
18.4 μl	H_2O
0.1 μl	*Taq* polymerase

Cycler program:

(a) 60 s at 94 °C.

(b) 30 s at 94 °C.

(c) 30 s at Tm (*see* Table 1).

(d) 60 s at 72 °C.

Go to step (b), repeat steps (b) to (d) 35 times.

(e) 600 s at 72 °C.

10. Analyze the reactions on a 1% agarose gel with an appropriate dye for nucleic acids (like ethidium bromide or Roti®-GelStain). Excise the appropriate bands from the agarose gel under UV radiation with a clean and sterile scalpel and transfer to a nuclease-free microcentrifuge tube.
11. Extract the DNA from the gel with a commercial kit (QIAquick Gel Extraction kit) according to the manufacturer's instructions and recover in a small amount (maximal 20 μl) of nuclease-free sterile H_2O.
12. Clone the PCR products in a pGEM-T vector with 3′ T-overhangs for insertion of PCR amplificates with 3′ A-overhangs generated by the *Taq* polymerase as follow:

5 μl	2× ligation buffer
1 μl	pGEM-T vector
30 ng	PCR product (extracted from agarose gel)
1 μl	T4 DNA ligase
H_2O added up to 10 μl	

13. Incubate 1 h at room temperature.

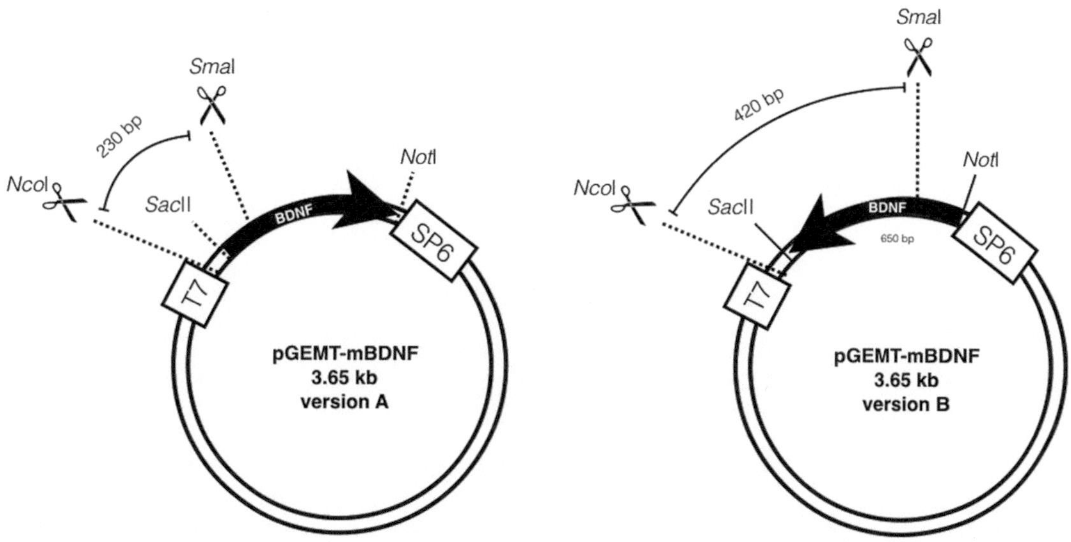

Fig. 6 Orientation of inserted PCR product in the vector pGEM-mBDNF. Version A: Sense of the murine BDNF (mBDNF) probe has inserted under T7 promoter. Restriction analysis with *Nco*I and *Sma*I results in a 230 bp and a 3420 bp fragment. Hence, an antisense probe is synthesized with the aid of the SP6 promoter and SP6 RNA polymerase subsequent to linearizing the vector with *Sac*II. Version B: Sense of the murine BDNF (mBDNF) probe has inserted under SP6 promoter. Restriction analysis with *Nco*I and *Sma*I results in a 420 bp and a 3230 bp fragment. Hence, an antisense probe is synthesized with the aid of the T7 promoter and T7 RNA polymerase subsequent to linearizing the vector with *Not*I

14. Transform competent cells according to standard procedures with ligation mix (blue-white screen is recommended) and plate on LB-ampicillin plates supplemented with X-gal.
15. Incubate the plates overnight at 37 °C.
16. Pick some white colonies for a mini prep according to standard procedures.
17. Digest the extracted plasmid DNA from the colonies with appropriate restriction enzymes and check the orientation of cloned PCR product to use the appropriate promoter (SP6 or T7) for sense and antisense probes (*see* **Note 4**, Fig. **6**, and Table 3).

18 Generate the final plasmids harboring the correctly inserted templates in larger amounts by a midi prep according to standard procedures.

3.3 Synthesis of DIG-Labeled Riboprobes for In Situ Hybridization

1. Linearize the plasmids appropriately for antisense probe and sense control, i.e., use a restriction enzyme that cuts behind the inserted PCR product. Perform digestion at least 1 h at 37 °C using standard procedures.
2. Purify the digest mix with the QIAquick PCR purification kit and elute with 30 μl nuclease-free H_2O.

3. Perform in vitro transcription with SP6 and T7 RNA polymerases to generate the appropriate antisense and sense riboprobes (*see* Table 3 and Fig. **6**) using the DIG RNA Labeling Kit according to the following scheme:

1 μg	Linearized purified DNA
2 μl	10× transcription buffer
2 μl	10× dNTPs
0.5 μl	RNase inhibitor
1 μl	RNA polymerase (either SP6 or T7)

4. Incubate for 2 h at 37 °C.
5. After the incubation, add 1 μl t-RNA and 2 μl DNase per reaction and additionally incubate for 15 min at 37 °C.

Table 3
Restriction analysis of cloned cDNA templates to decide if the inserted PCR product is orientated as version A or version B (*see* Fig. 6)

		Version A		Version B	
Insert	**Cut in insert**	**Digest with**	**Results in**	**Digest with**	**Results in**
NGF	*Nco*I (1×)	*Nco*I	3400 + 500 bp	*Nco*I	3500 + 370 bp
		Use for probe synthesis: *Sac*II/SP6 for antisense *Sac*I/T7 for sense		Use for probe synthesis: *Sac*I/T7 for antisense *Sac*II/SP6 for sense	
BDNF	*Sma*I (1×)	*Nco*I + *Sma*I	3420 + 230 bp	*Nco*I + *Sma*I	3230 + 420 bp
		Use for probe synthesis: *Sac*II/SP6 for antisense *Not*I/T7 for sense		Use for probe synthesis: *Not*I/T7 for antisense *Sac*I/SP6 for sense	
NT-3	*Sma*I (1×)	*Nco*I + *Sma*I	3420 + 230 bp	*Nco*I + *Sma*I	3230 + 420 bp
		Use for probe synthesis: *Nco*I/SP6 for antisense *Sac*I/T7 for sense		Use for probe synthesis: *Sac*I/T7 for antisense *Nco*I/SP6 for sense	
TrkB	*Not*I (1×)	*Not*I + *Xho*I	3700 + 70 bp	*Not*I + *Xho*I	3090 + 680 bp
		Use for probe synthesis: *Sac*II/SP6 for antisense *Not*I/T7 for sense		Use for probe synthesis: *Not*I/T7 for antisense *Sac*II/SP6 for sense	
TrkC	*Pst*I (2×)	*Pst*I	3100 + 340 + 270 bp	*Pst*I	3300 + 340 + 90 bp
		Use for probe synthesis: *Nco*I/SP6 for antisense *Not*I/T7 for sense		Use for probe synthesis: *Not*I/T7 for antisense *Nco*I/SP6 for sense	

The orientation is important to use the correct restriction enzyme to linearize the vector and the correct RNA polymerase promoter for probe synthesis

6. Purify the mixture using a mini Quick Spin RNA column according to the manufacturer's instructions and store at −80 °C until use.

3.4 ISH

Nonradioactive ISH on cryosections was modified from the protocol in [20].

1. Bring slides with cryosections and riboprobes to room temperature and transfer to a humid chamber that was bedded with some filter paper soaked with SSC/formamide.
2. Dilute the DIG-labeled riboprobes in hybridization buffer, heat to 70 °C for 10 min, and pipet on the slides (*see* **Note 5**).
3. During overnight incubation at 65 °C under high stringency conditions, cover the slides with glass cover slips to avoid evaporation.
4. Wash the slides the following day 2 × 30 min at 65 °C in a jar filled with SSC/formamide and 2 × 30 min at room temperature in MABT.
5. Block sections for 1 h with MABT/ChS before incubating in anti-DIG alkaline phosphatase conjugate diluted in MABT/ChS at room temperature overnight.
6. On the following day, wash slides 3 × 20 min in MABT and 2 × 10 min in AP buffer.
7. Start the color reaction by adding a NBT/BCIP solution according to the manufacturer's instructions. Incubate for 24 h at room temperature in a light-protected humid chamber. As the expression levels of neurotrophins (BDNF, NGF, and NT3) are rather low the color reaction is prolonged to 48 h (*see* **Note 6**).
8. Rinse all sections thoroughly with PBS to stop the dye development and either stain for phospho-tau (*see* Subheading 3.5) or mount on Kaiser's Glycerol gelatin (*see* **Note 7**).

3.5 Double-Labeling Protocol for ISH and IHC

1. For IHC with phospho-specific antibodies (*see* Table 2), use the sections that had been processed for ISH prior to Kaiser's glycerol embedding.
2. Add 100–200 μl of antibodies diluted in PBS-T to the slides (*see* **Note 8**).
3. Cover the slides with a plastic coverslip and incubate overnight in a humid chamber at 4 °C or room temperature.
4. Wash slides 4 × 5 min in PBS and incubate with a biotinylated anti-mouse antibody for 1–2 h at room temperature.
5. During the incubation, prepare the ABC solution of the VECTASTAIN ABC kit by mixing 20 μl of bottle A, 20 μl of bottle B and 1 ml PBS and incubate at least 30 min.

6. Wash the slides again 4 × 5 min in PBS and incubate with the ABC solution for 30–60 min.
7. After another intensive wash, incubate the slides with DAB according to manufacturer's instructions until the color developed (approximately 5–7 min).
8. Rinse the slides very quickly with a large amount of H_2O to stop the color reaction.
9. As NBT/BCIP is not resistant against organic solvents, omit the dehydration step, and mount the slides in Kaiser's Glycerol gelatine.

3.6 IF

1. Dilute anti-NGF or anti-BDNF, and Tau phospho-specific antibodies in PBS-T, as indicated in Subheading 2.4.
2. Process slides like for IHC with regard to the primary and secondary antibody.
3. Detect secondary biotinylated anti-mouse antibodies used for Tau phospho epitopes with streptavidin-labeled fluorescein.
4. After an extensive wash in PBS, incubate with AMCA-labeled anti-rabbit antibodies for the detection of the neurotrophins.
5. Mount the slides in VectaShield containing DAPI for single-staining or without DAPI for double staining and store at 4 °C in the dark.

3.7 Semiquantitative Estimation of Cell Density and Pixel Density

Images were made using a software-controlled digital camera attached to a microscope and analyzed offline.

1. For analysis, take images using a 20× objective and count stained cells in a counting frame.
2. Measure the pixel density of stained cells on digitalized images using the Image J program and correct by background staining substraction (*see* **Note 9**).

3.8 Western Blot Analysis

1. Resolve whole brains homogenates on SDS-PAGE.
2. Blot onto Nitrocellulose membranes.
3. Incubate with appropriate antibodies.
4. Develop using the ECL chemoluminescence kit.

3.9 Data Analysis

Data were analyzed by Student's *t*-tests with Graphpad/Prism Software.

4 Notes

1. Perfuse the mice without any pressure to avoid damage to the tissue. For best results use a peristaltic pump or gravity pres-

sure to drive a prewash of physiological saline followed by the fixative. The complete removal of red blood cells is very important as they have a rather high autofluorescence and interfere with peroxidase used in the ABC kit.

2. Best ISH results are achieved with cryosections. However, paraffin sections may be used as well if appropriate RNase inhibition is assured during the whole procedure including embedding, sectioning, and storage. In addition, sections must not be too thin as this reduces the amount of mRNA per section if only parts of the nucleus are present.

3. The primers for the generation of riboprobes were selected by using bioinformatical tools like Primer3 and the software VectorNTI. All primers were designed to amplify cDNA from *Mus musculus*; however, some of them may be also used to synthesize probes for *Homo sapiens.*

4. The appropriate restriction enzymes to check the orientation of the cloned PCR product in this study are given in Table 3. After the successful cloning of the PCR product into the pGEMT vector, the orientation of the inserted sequence must be analyzed. Figure **6** shows the two possibilities of insertion (version A and version B) with BDNF as an example. Use the restriction enzymes given in Table 3 and analyze the digest fragment in an agarose gel. The fragment pattern obtained will show whether version A or B of the plasmid is present. According to the orientation of the insert, the correct restriction enzyme to linearize the vector prior to riboprobe synthesis is indicated in Table 3. The opened vector is than subjected to probe synthesis by using the appropriate RNA polymerase as given in Table 3.

5. When pipetting the probe on the sections avoid the formation of air bubbles.

6. To check the color intensity of the NBT/BCIP precipitate, the sections can be briefly inspected with a light microscope during the color reaction. Incubation can be prolonged or fresh NBT/BCIP may be added if the color is not sufficiently developed after 24 or 48 h, respectively. As long exposure to NBT/BCIP will lead to high background staining compare sections with antisense riboprobes to sense probes as negative control.

7. To monitor the quality of the ISH procedure, add a positive control with a riboprobe that was verified to give a strong staining signal. In the present study, an antisense probe of murine choline acetyltransferase (ChAT) was used [20]. A negative control with a sense probe should be used to evaluate background staining. Positive neurons should present a clear nuclear staining. A weak dotted cytosolic may be visible as

well. If no staining is observed for an antisense riboprobe although the positive shows a clear and specific staining pattern, repeat the preparation of the riboprobes. If needed, check the quality of the RNA, cDNA, and riboprobes used with agorose gel electrophoresis or PCR reactions with specific primers. If unspecific or unexpected staining patterns occur perform a sequence analysis of the cloned insert and verify it with the aid of the Basic Local Alignment Search Tool (BLAST; http://www.blast.ncbi.nlm.nih.gov). If appropriate, unexpected staining can be compared to published results (e.g., http://www.mouse.brain-map.org).

8. Best results are achieved when antibodies are used that give a cytosolic staining pattern. Antibodies with nuclear staining pattern should be avoided as they interfere with nuclear staining of the ISH.
9. Image J software can be downloaded at: http://rsb.info.nih.gov/ij/download.html.

Acknowledgments

This study was performed at Sanofi-Aventis, Vitry-sur-Seine/France, supported from a Marie Curie Fellowship from the European Community in collaboration with Dr. Thomas Rooney, Sanofi Sarl, Vitry-sur-Seine/France and Dr. Luc Buée, Inserm, Lille/France. A special thanks goes to Uwe Ernsberger for transfer of the ISH technique and to Klaus Unsicker for transfer of skills on neuroanatomy and on the importance of neuroanatomy for neuroscience.

References

1. Krüger L, Mandelkow EM (2015) Tau neurotoxicity and rescue in animal models of human tauopathies. Curr Opin Neurobiol 36:52–58
2. Schindowski K, Belarbi K, Buee L (2008) Neurotrophic factors in Alzheimer's disease: role of axonal transport. Genes Brain Behav 7(S1):43–56
3. Sofroniew MV, Howe CL, Mobley WC (2001) Nerve growth factor signaling, neuroprotection, and neural repair. Annu Rev Neurosci 24:1217–1281
4. Chowdary PD, Che DL, Cui B (2012) Neurotrophin signaling via long-distance axonal transport. Annu Rev Phys Chem 63:571–594
5. Siegel GJ, Chauhan NB (2000) Neurotrophic factors in Alzheimer's and Parkinson's disease brain. Brain Res Brain Res Rev 33:199–227
6. Murer MG, Boissiere F, Yan Q, Hunot S, Villares J, Faucheux B, Agid Y, Hirsch E, Raisman-Vozari R (1999) An immunohistochemical study of the distribution of brain-derived neurotrophic factor in the adult human brain, with particular reference to Alzheimer's disease. Neuroscience 88:1015–1032
7. Schliebs R, Arendt T (2011) The cholinergic system in aging and neuronal degeneration. Behav Brain Res 221:555–563
8. Michalski B, Fahnestock M (2003) Pro-brain-derived neurotrophic factor is decreased in parietal cortex in Alzheimer's disease. Brain Res Mol Brain Res 111:148–154
9. Garzon D, Yu G, Fahnestock M (2002) A new brain-derived neurotrophic factor transcript and decrease in brain-derived neurotrophic factor transcripts 1, 2 and 3 in Alzheimer's disease parietal cortex. J Neurochem 82:1058–1064

10. Hock C, Heese K, Hulette C, Rosenberg C, Otten U (2000) Region-specific neurotrophin imbalances in Alzheimer disease: decreased levels of brain-derived neurotrophic factor and increased levels of nerve growth factor in hippocampus and cortical areas. Arch Neurol 57:846–851
11. Goedert M, Fine A, Hunt SP, Ullrich A (1986) Nerve growth factor mRNA in peripheral and central rat tissues and in the human central nervous system: lesion effects in the rat brain and levels in Alzheimer's disease. Brain Res 387:85–92
12. Scott SA, Mufson EJ, Weingartner JA, Skau KA, Crutcher KA (1995) Nerve growth factor in Alzheimer's disease: increased levels throughout the brain coupled with declines in nucleus basalis. J Neurosci 15:6213–6221
13. Ferrer I, Marín C, Rey MJ, Ribalta T, Goutan E, Blanco R, Tolosa E, Martí E (1998) BDNF and full-length and truncated TrkB expression in Alzheimer disease. Implications in therapeutic strategies. J Neuropathol Exp Neurol 58:729–739
14. Patterson SL, Abel T, Deuel TA, Martin KC, Rose JC, Kandel ER (1996) Recombinant BDNF rescues deficits in basal synaptic transmission and hippocampal LTP in BDNF knockout mice. Neuron 16:1137–1145
15. Belrose JC, Masoudi R, Michalski B, Fahnestock M (2014) Increased pro–nerve growth factor and decreased brain-derived neurotrophic factor in non–Alzheimer's disease tauopathies. Neurobiol Aging 35:926–933
16. Schindowski K, Bretteville A, Leroy K, Begard S, Brion JP, Hamdane M, Buee L (2006) Alzheimer's disease-like tau neuropathology leads to memory deficits and loss of functional synapses in a novel mutated tau transgenic mouse without any motor deficits. Am J Pathol 169:599–616
17. Hideo H, Keiko T, Nobuyuki Y, Tsuyoshi M (2009) Dissection of hippocampal dentate gyrus from adult mouse. J Vis Exp 33:pii:1543
18. Keays KM, Owens GP, Ritchie AM, Gilden DH, Burgoon MP (2005) Laser capture microdissection and single-cell RT-PCR without RNA purification. J Immunol Methods 302:90–98
19. Wetmore C, Ernfors P, Persson H, Olson L (1990) Localization of brain-derived neurotrophic factor mRNA to neurons in the brain by in situ hybridization. Exp Neurol 109:141–152
20. Ernsberger U, Patzke H, Rohrer H (1997) The developmental expression of choline acetyltransferase (ChAT) and the neuropeptide VIP in chick sympathetic neurons: evidence for different regulatory events in cholinergic differentiation. Mech Dev 68:115–126
21. Webster MJ, Herman MM, Kleinman JE, Weickert CS (2006) BDNF and trkB mRNA expression in the hippocampus and temporal cortex during the human lifespan. Gene Expr Patterns 6:941–951
22. Nibuya M, Morinobu S, Duman RS (1995) Regulation of BDNF and trkB mRNA in rat brain by chronic electroconvulsive seizure and antidepressant drug treatments. J Neurosci 15:7539–7547
23. Micheli MR, Bova R, Laurenzi MA, Bazzucchi M, Zucconi GG (2006) Modulation of BDNF and TrkB expression in rat hippocampus in response to acute neurotoxicity by diethyldithiocarbamate. Neurosci Lett 410:66–70
24. Li Y, Luikart BW, Birnbaum S, Chen J, Kwon CH, Kernie SG, Bassel-Duby R, Parada LF (2008) TrkB regulates hippocampal neurogenesis and governs sensitivity to antidepressive treatment. Neuron 59:399–412
25. Smith MA, Makino S, Kvetnansky R, Post RM (1995) Stress and glucocorticoids affect the expression of brain-derived neurotrophic factor and neurotrophin-3 mRNAs in the hippocampus. J Neurosci 15:1768–1777
26. Schindowski K (2011) Decreased axonal transport during Alzheimer's-like spatial-temporal tau pathology leads to imbalance of NGF, cholinergic dysfunction, and decreased regulation of BDNF and TrkB. Alzheimers Dement 7:S560–S561
27. Elliott E, Atlas R, Lange A, Ginzburg I (2005) Brain-derived neurotrophic factor induces a rapid dephosphorylation of tau protein through a PI-3 Kinase signalling mechanism. Eur J Neurosci 22:1081–1089
28. Hellweg R, Lohmann P, Huber R, Kuhl A, Riepe MW (2006) Spatial navigation in complex and radial mazes in APP23 animals and neurotrophin signaling as a biological marker of early impairment. Learn Mem 13:63–71
29. Burbach GJ, Hellweg R, Haas CA, Del Turco D, Deicke U, Abramowski D, Jucker M, Staufenbiel M, Deller T (2004) Induction of brain-derived neurotrophic factor in plaque-associated glial cells of aged APP23 transgenic mice. J Neurosci 24:2421–2430
30. Schindowski K, Belarbi K, Bretteville A, Ando K, Buée L (2008) Neurogenesis and cell cycle-reactivated neuronal death during pathogenic tau aggregation. Genes Brain Behav 7(S1):92–100
31. Bruno MA, Leon WC, Fragoso G, Mushynski WE, Almazan G, Cuello AC (2009) Amyloid

β-induced nerve growth factor dysmetabolism in Alzheimer disease. J Neuropathol Exp Neurol 68:857–869
32. Saez ET, Pehar M, Vargas MR, Barbeito L, Maccioni RB (2006) Production of nerve growth factor by beta-amyloid-stimulated astrocytes induces p75(NTR)-dependent tau hyperphosphorylation in cultured hippocampal neurons. J Neurosci Res 84:1098–1106
33. Tatebayashi Y, Haque N, Tung YC, Iqbal K, Grundke-Iqbal I (2004) Role of tau phosphorylation by glycogen synthase kinase-3beta in the regulation of organelle transport. J Cell Sci 117:1653–1663
34. Peng S, Wuu J, Mufson EJ, Fahnestock M (2004) Increased proNGF levels in subjects with mild cognitive impairment and mild Alzheimer disease. J Neuropathol Exp Neurol 63:641–649
35. Fahnestock M, Michalski B, Xu B, Coughlin MD (2001) The precursor pro-nerve growth factor is the predominant form of nerve growth factor in brain and is increased in Alzheimer's disease. Mol Cell Neurosci 18:210–220
36. Capsoni S, Giannotta S, Cattaneo A (2002) Beta-amyloid plaques in a model for sporadic Alzheimer's disease based on transgenic anti-nerve growth factor antibodies. Mol Cell Neurosci 21:15–28
37. Houeland G, Romani A, Marchetti C, Amato G, Capsoni S, Cattaneo A, Marie H (2010) Transgenic mice with chronic NGF deprivation and Alzheimer's disease-like pathology display hippocampal region-specific impairments in short-and long-term plasticities. J Neurosci 30:13089–13094
38. Capsoni S, Brandi R, Arisi I, D'Onofrio M, Cattaneo A (2011) A dual mechanism linking NGF/proNGF imbalance and early inflammation to Alzheimer's disease neurodegeneration in the AD11 anti-NGF mouse model. CNS Neurol Disord Drug Targets 10:635–647
39. Cooper JD, Salehi A, Delcroix JD, Howe CL, Belichenko PV, Chua-Couzens J, Kilbridge JF, Carlson EJ, Epstein CJ, Mobley WC (2001) Failed retrograde transport of NGF in a mouse model of Down's syndrome: reversal of cholinergic neurodegenerative phenotypes following NGF infusion. Proc Natl Acad Sci U S A 98:10439–10444
40. Seo H, Isacson O (2005) Abnormal APP, cholinergic and cognitive function in Ts65Dn Down's model mice. Exp Neurol 193:469–480
41. Hunter CL, Isacson O, Nelson M, Bimonte-Nelson H, Seo H, Lin L, Ford K, Kindy MS, Granholm AC (2003) Regional alterations in amyloid precursor protein and nerve growth factor across age in a mouse model of Down's syndrome. Neurosci Res 45:437–445
42. Hunter CL, Bimonte-Nelson HA, Nelson M, Eckman CB, Granholm AC (2004) Behavioral and neurobiological markers of Alzheimer's disease in Ts65Dn mice: effects of estrogen. Neurobiol Aging 25:873–884
43. Salehi A, Delcroix JD, Belichenko PV, Zhan K, Wu C, Valletta JS, Takimoto-Kimura R, Kleschevnikov AM, Sambamurti K, Chung PP, Xia W, Villar A, Campbell WA, Kulnane LS, Nixon RA, Lamb BT, Epstein CJ, Stokin GB, Goldstein LS, Mobley WC (2006) Increased APP expression in a mouse model of Down's syndrome disrupts NGF transport and causes cholinergic neuron degeneration. Neuron 51:29–42
44. Tuszynski MH, Thal L, Pay M, Salmon DP, U HS, Bakay R, Patel P, Blesch A, Vahlsing HL, Ho G, Tong G, Potkin SG, Fallon J, Hansen L, Mufson EJ, Kordower JH, Gall C, Conner J (2005) A phase 1 clinical trial of nerve growth factor gene therapy for Alzheimer disease. Nat Med 11:551–555
45. Tuszynski MH, Yang JH, Barba D, U HS, Bakay RA, Pay MM, Masliah E, Conner JM, Kobalka P, Roy S, Nagahara AH (2015) Nerve growth factor gene therapy: activation of neuronal responses in Alzheimer disease. JAMA Neurol 72:1139–1147
46. Malkki H (2015) NGF gene therapy activates neurons in the AD patient brain. Nat Rev Neurol 11:548

Chapter 28

Pin1 Knockout Mice: A Model for the Study of Tau Pathology in Alzheimer's Disease

Asami Kondo, Onder Albayram, Xiao Zhen Zhou, and Kun Ping Lu

Abstract

Pin1 knockout in mice causes age-dependent neuropathy characterized by motor and behavioral deficits, tau hyper phosphorylation, tau filament formation, and neuronal degradation. Here, we describe the methods with mouse behavior test, immunostaining, and immunoblotting to detect many aspects of neurodegeneration in Pin1 knockout mice.

Key words Pin1, *Cis* P-tau, Alzheimer's disease, Behavior test, Immunohistochemistry, Immunoblotting

1 Introduction

Alzheimer's disease (AD), the most common form of dementia in older individuals, currently affects more than 5.4 million Americans and 30 million people worldwide [1]. As the baby-boom generation ages and life expectancy continues to grow, these numbers are expected to increase dramatically with some estimates projecting that by 2050, AD will affect ~130 million people worldwide at a cost of more than $1 trillion in the United States alone. However, there is no effective treatment yet [1]. Since AD may take more than a decade to develop, it is critical to understand the early disease mechanisms so that therapies can be developed to stop these early pathogenic events [2].

Tau hyper phosphorylation is an early event that precedes tangle formation in AD, but it remains unclear how tau phosphorylation becomes pathogenic and leads to dementia [3]. We have identified a unique proline isomerase, Pin1 that inhibits tauopathy in AD by converting the phosphorylated Thr231-Pro motif in Tau (P-tau) from *cis* to *trans* in AD cell and mouse models [4–13]. In human AD, Pin1 is inhibited by multiple mechanisms [5, 7, 14–16], whereas

---

Asami Kondo and Onder Albayram have contributed equally to this work.
Xiao Zhen Zhou and Kun Ping Lu have contributed equally to this work.

Caroline Smet-Nocca (ed.), *Tau Protein: Methods and Protocols*, Methods in Molecular Biology, vol. 1523,
DOI 10.1007/978-1-4939-6598-4_28, © Springer Science+Business Media New York 2017

the Pin1 genetic polymorphism that prevents its down regulation is associated with delaying AD age of onset [17]. In addition, Pin1 is located at a locus of associated with late-onset AD [18], P-tau appears early in pre-tangle AD neurons [19], and its cerebrospinal fluid level correlates with memory loss in mild cognition impairment and AD [20].

We focus on the methods we used to detect tauopathy in Pin1 knockout mouse model not only by mouse behavior test but immunostaining and immunoblotting.

2 Materials

2.1 Animals

Institutional Animal Care and Use Committees at Beth Israel Deaconess Medical Center have approved animal care and use for the experiments.

1. Young (2-month-old) and old (12-month-old) male $Pin1^{-/-}$ and WT mice on C57BL/6 background.
2. Animals are housed five per cage under a 12:12-h light/dark cycle, and maintained in standard conditions with food and water ad libitum.

2.2 Behavioral Studies

1. Morris water maze: a circular pool made of white PVC (83 cm diameter, 60 cm height), transparent Plexiglas platform (10 cm in diameter), water, heating lamp.
2. An automatic tracking system (EthoVision XT; Noldus).
3. Accelerating Rota-Rod: accelerating Rota-rod device (47650 Rota-Rod NG, Ugo-basile).
4. Balance beam: plastic beam (2 cm diameter, 80 cm length, 35 cm height), video recording device and tripod, timer.
5. Hind-limb clasping: video recording device and tripod, timer.

2.3 Perfusion of the Mouse

1. Forceps.
2. Micro-scissors.
3. Perfusion pump (e.g., Fisher Scientific, Variable-flow peristaltic Pumps).
4. 21 G butterfly needle.
5. PBS.
6. 4% paraformaldehyde (PFA) dissolved in PBS.
7. Mouse brain matrix.

2.4 Paraffin-Embedded Sections

1. Embedding cassettes.
2. 4% paraformaldehyde in PBS.
3. 70, 80, 95, 100% ethanol.

4. Xylene.
5. Paraffin wax (56–58 °C).
6. Microtome.
7. Poly-L lysine-coated glass slides.

2.5 Immunostaining

1. Histoclear.
2. 100, 95, 70 % ethanol.
3. 0.5 M sodium citrate, pH 6.0 pH should be adjusted using 1 M citric acid.
4. Autoclave instrument.
5. Fetal bovine serum.
6. 10 % Triton X-100: Triton X-100 solution should be prepared every experiment.
7. Cleaning wipes.
8. Hydrophobic marker.
9. Tris buffered saline (TBS).
10. Primary antibody: anti-*cis* pT231 tau monoclonal antibody, diluted 1:100 in TBS, 1 % fetal bovine serum, 0.1 % Triton X-100.
11. Secondary antibody: goat anti-mouse IgG antibody, dilute 1:100 in TBS, 0.01 % Triton X-100.
12. Glass cover slips (glass thickness should be matched with reflective index of mounting medium).
13. Mounting medium for fluorescein (e.g., Fluoromount-G, Southern biotech).
14. Humidified chamber.
15. Cold room (4 °C) or fridge.

2.6 Brain Lysate Preparation

1. RIPA buffer including protease inhibitor and phosphatase inhibitors: 50 mM Tris–HCl, pH 7.4, 150 mM NaCl, 2 mM EDTA, 1 % NP 40, 0.1 % SDS, 0.5 % Na-deoxycholate.
2. Pellet pestles Cordless motor and pestle (Sigma).
3. Microtube centrifuge.
4. 2× Sample loading buffer: 0.1 M Tris–HCl pH 6.8, 4 % beta-Mercapto-ethanol, 4 % SDS, 0.2 % Bromophenol blue, 20 % Glycerol.

2.7 SDS Polyacrylamide Gel

1. 40 % acrylamide solution.
2. 2 % Bisacrylamide solution.
3. 10 % Ammonium persulfate: solution in water (*see* **Note 1**).
4. 3 M Tris–HCl (pH 8.8).

5. 1 M Tris–HCl (pH 6.8).
6. 10% SDS.
7. *N,N,N,N′*-tetramethyl-ethylenediamine (TEMED).
8. One sets of gel casting glass and spacer.
9. Comb.
10. Electrophoresis unit.
11. Running Buffer: 250 mM Tris, 2 M Glycine, 1% SDS.
12. Protein marker for SDS-PAGE.

2.8 Immunoblotting

1. Rocking shaker.
2. PVDF transfer membrane.
3. Chromatography paper.
4. Semi-dry transfer unit.
5. 10× Transfer Buffer: 3% Tris, 1.9 M Glycine.
6. Methanol.
7. CBB staining solution: 40% Methanol, 10% Acetic acid, 0.25% (w/v) Coomassie brilliant blue.
8. Destaining solution: 50% Methanol, 10% Acetic acid.
9. Blocking reagent: 1% gelatin in TBST.
10. TBST: 0.06% Tween® 20 diluted in 1× TBS.
11. Bovin serum albumin (BSA).
12. 10% sodium azide.
13. Primary antibody: anti-*cis* pT231 tau monoclonal antibody, dilute 1:1000 in TBST, 3% BSA, 0.05% azide.
14. Secondary antibody: anti-mouse IgG antibody conjugated with HRP, dilute 1:5000 in TBST.
15. Enhanced chemiluminescence reagent: prepare the working solution by mixing equal volumes of Enhance reagent and Peroxide solution.
16. Membrane scanner with CDC camera or X-ray film and Film developer.

3 Methods

3.1 Morris Water Maze

The MWM is used for assessing hippocampal-dependent spatial and reference memory for rodents [21, 22].

1. Place mice in the tank facing the wall and given 90 s to find the platform, mount the platform, and remain on it for 5 s (*see* **Note 2**).

2. Guide mice that fail to mount the platform within the allotted time (90 s) to the platform and allow 10 s to become acquaint with its location (*see* **Note 3**).
3. In the first phase of the test, subject each mouse four consecutive sessions (the mouse is left for 1 min in a dark dry container between the sessions) daily over 6 days (*see* **Note 4**).
4. For probe trials, place mice in the tank with the platform removed and give 90 s to explore the tank.
5. For visible platform trials, use a red reflector to mark the top of the target platform.
6. Analyze the swim path with an automatic tracking system and calculate escape latencies, total path length, and swim speed.
7. Analyze data using two-way ANOVA (Injury and trials) followed by Bonferroni's correction.

3.2 Accelerating Rota-Rod Test

The accelerating rota-rod test is used for assessing sensorimotor coordination in 2- and 12-month-old Pin1$^{-/-}$ mice and age-matched wild-type controls [23].

1. Place mice in the rotating cylinder four sessions per day for 2 consecutive days.
2. Each session lasts a maximum of 10 min, during which the rotating rod accelerates from 4 to 40 rpm in the first 5 min of the trial and then remains at maximum speed for the remaining 5 min.
3. After each session, allow the mouse to rest for 5 min before the next session.
4. Clean the cylinder between subjects with 70% ethanol solution and dry before each session.
5. Record latency to fall and average over four sessions per day.

3.3 Balance Beam

The balance beam is used for assessing motor balance in 2- and 12-month-old Pin1$^{-/-}$ mice and age-matched wild-type controls [7, 8].

1. Divide the beam into 8 segments of 10 cm each with a marker.
2. Place mice in the center of the first segment and allow them to walk for 60 s for four consecutive sessions.
3. After crossing the beam, allow the mouse to rest for 1 min before the next session.
4. Record latency to cross the beam and average over four sessions.

3.4 Hin-limb Clasping

We control hind limb-clasping reflex in 2- and 12-month-old Pin1$^{-/-}$ mice and age-matched wild-type control (Fig. 1). Abnormal hind-clasping reflex is a marker of neurodegenerative disease progression in mouse [7, 8].

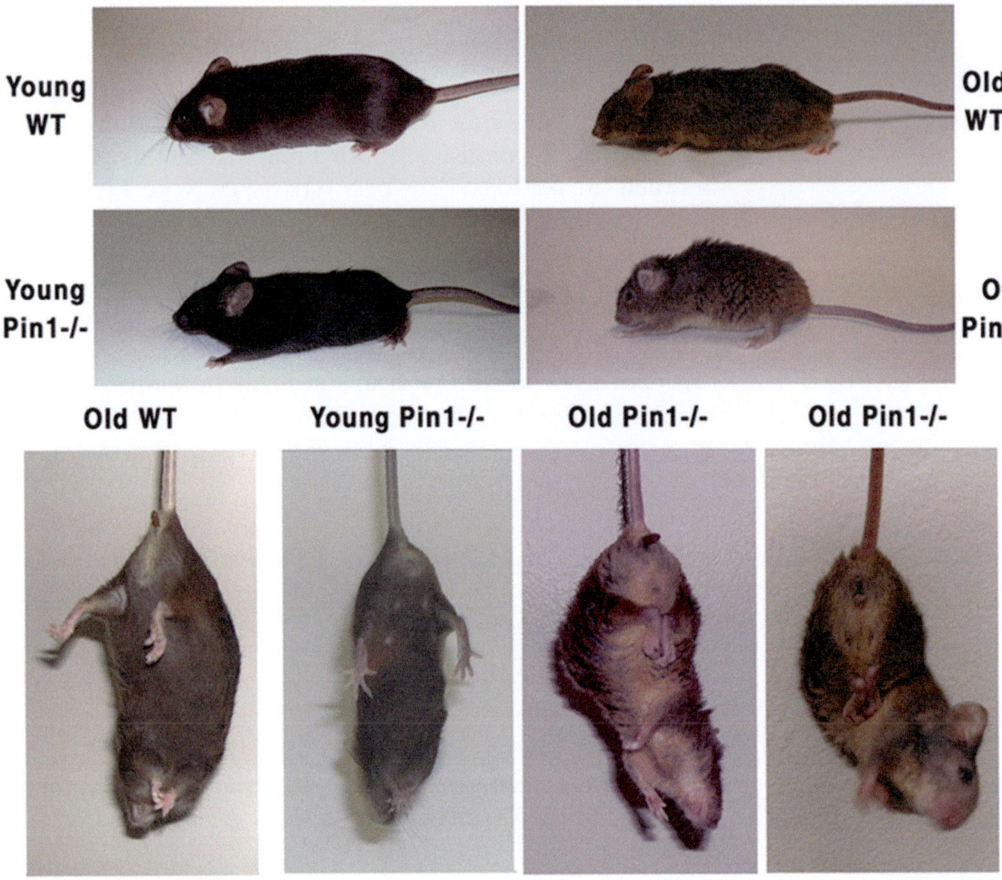

Fig. 1 Age-dependent abnormal hunched posture and hind-limb clasping reflexes in Pin1$^{-/-}$ mice (reproduced with the permission from Liou, Y. C. et al., 2003, Nature 424, 556–561)

1. Suspend the mouse by tail for 10 s for three sessions.
2. After each session, allow the mouse to rest for 1 min before the next one.
3. Monitor the hind-limb position and score from 0 to 3 depending on the severity of deficits. If the hind limbs are continually expanded outward, score is 0; if the hind limbs are flexed to the trunk and tightened to the bodies for 50% of the time suspended, score is 3.
4. Record performance and average over three sessions.

3.5 Harvesting Mouse Brain

1. Euthanize the mouse according to your institutional animal care and use the committee approved protocol.
2. Once the animal sedated, perform a midline laparotomy.
3. Make an incision at the diaphragm and side of ribs and hold the ribs by the hemostatic forceps to open the thoracic cavity and thus expose the heart.

4. Perfuse (for western blot) or perfuse and fix (for paraffin sections) at a physiological pressure using a perfusion pump.
5. Insert 21-G needle attached to the pump into the left ventricle of the mouse to start perfusion.
6. Make an incision at the right atrium.
7. Perfuse till the liver turns pale in color, which indicates circulating blood has been removed systemically.
8. Remove the brain and put it into the brain matrix.
9. Isolate the brain region on the brain matrix using a razor blade.
10. Dissect the brain and collect cortex and hippocampus for western blot.
11. Proceed to embedding in Subheading 3.6 or preparation of lysate in Subheading 3.8.

3.6 Preparation of Paraffin-Embedded Sections

1. Fix the harvested brain overnight at 4 °C using 4% PFA prepared in PBS. The fixative volume should be around 5 ml.
2. Wash the brain with PBS.
3. Store tissues in 70% ethanol till they are ready to be embedded in paraffin blocks.
4. Transfer tissues to embedding cassettes and dehydrate as follows:
   (a) 70% Ethanol, two changes, 1 h each.
   (b) 80% Ethanol, two changes, 1 h each.
   (c) 95% Ethanol, two changes, 1 h each.
   (d) 100% Ethanol, two changes, 1 h each.
   (e) Xylene, three changes, 1 h each.
   (f) Paraffin wax (56–58 °C), two changes, 1.5 h each.
5. Embed the tissues into paraffin blocks.
6. To section paraffin blocks, trim and mount the blocks onto the microtome.
7. Take care to trim the paraffin block that there is optional cutting surface area that includes the sample with small paraffin frame.
8. Cut the tissue in 3- to 10-μm thick slices using the microtome. Use a brush to gently hold the sections. If the microtome is adjusted properly, the paraffin sections should come out as a ribbon.
9. Place paraffin ribbons or a slice in 40–45 °C water bath using a second wash brush. As the sections expand on the water bath, the winkles on the sections vanish.

10. Gently scoop out the floating paraffin sections using microscope glass slides and position the section with the help of the brush.
11. Dry the sections overnight at 37 °C (lower temperature are better for subsequent antibody detection).
12. Sections are ready for immunostaining (*see* Subheading 3.7) or can be stored at room temperature for an extended period of time.

3.7 Processing of Paraffin-Embedded Brain Sections for Immunostaining

1. Deparaffinize the section as follows:
   (a) Histoclear for 5 min.
   (b) 100% ethanol for 2.5 min.
   (c) 95% ethanol for 2.5 min.
   (d) 70% ethanol for 2.5 min.
   (e) ddH_2O: two washes, 3 min each.
2. Retrieve masked antigen with autoclave instrument.
   (a) Autoclave the tissue in 20 mM sodium citrate (pH 6.0) for 20 min.
   (b) Cool down to room temperature in ddH_2O.
3. Wash sections with ddH2O for twice, 5 min each.
4. Outline the sections with a hydrophobic marker.
5. Block the nonspecific binding by incubating 30 min at room temperature with blocking solution (3% fetal bovine serum).
6. Gently remove excess liquid off the slide and wipe around the section.
7. Incubate the section for overnight at 4 °C with specific primary antibody diluted 1:100 in TBS, 1% fetal bovine serum, and 0.1% Triton X.
8. Rinse sections in TBS five times, 15 min each.
9. Incubate the sections for 90 min with secondary antibody diluted 1:300 in TBS and 0.1% Triton X.
10. Rinse sections in TBS three times, 20 min each.
11. Mount section with mounting media and cover grass.
12. Seal around the edges of the cover glass with nail polish and place slides horizontal and dry. Sections are ready to observe under the microscopy or can be stored in −20 °C freezer for several months.

3.8 Preparation of Brain Lysate

1. Weigh the tube including mouse brain sample and add RIPA buffer three times volume of brain weight.
2. Homogenize the brain with pellet pestle on ice.
3. Rotate for 15 min in cold room.
4. Centrifuge at 15,700 × *g* for 10 min in cold room.

5. Correct supernatant to a chilled tube.
6. Measure the protein concentration of the brain lysate with Coomassie protein assay.
7. Add same volume of 2× SB to the lysate and boil for 5 min.
8. The lysate is ready to proceed to run an acrylamide gel (*see* Subheading 3.9) or stored in −80 °C freezer.

3.9 Running a Gel

1. Set 10% acrylamide gel to the tank and add running buffer.
2. Load 10 μg of brain lysate to the well.
3. Run the gel at 120 V constant for 100 min.
4. The gel is ready to proceed to transferring step (*see* Subheading 3.10).

3.10 Transferring the Gel to the PVDF Membrane

1. Cut the PVDF membrane and filter paper.
2. Incubate the membrane in methanol for 30 s and put it in a transfer buffer and shake on the rocking shaker until use it.
3. Prepare transfer by soaking.
4. Take the gel out from the cassette and transfer to PVDF membrane with semi-dry transfer system at 250 mA for 1 h.
5. The membrane is ready to proceed to western blot (*see* Subheading 3.11).

3.11 Processing the PVDF Membrane for Western Blot

1. Stain the membrane with CBB staining buffer for 1–2 min on the shaker (*see* **Note 5**).
2. Wash the membrane with destaining buffer twice for 5 min each (*see* **Note 6**).
3. Wash the membrane with TBST buffer twice for 5 min each.
4. Block the membrane with blocking reagent for 1 h.
5. Wash the membrane with TBST buffer three times for 5 min each.
6. Prepare primary antibody solution in 3% BSA, 0.05% azide and incubate the membrane with antibody for 16 h in cold room.
7. Wash the membrane with TBST buffer for 1 h and change the buffer every 5–10 min.
8. Incubate the membrane with diluted secondary antibody conjugated to HRP for 50 min.
9. Wash the membrane with TBST buffer for 2 h changing the buffer every 10 min.
10. Put the membrane on the plastic bag and incubate with the enhanced chemiluminescence reagent for 30 s.
11. Detect the signal with the scanner with CDC camera or X-ray film system.

4 Notes

1. Prepare 10% APS in water and make small aliquots, for example 1 ml/tube, and store in −20 °C freezer.
2. The hidden platform remained at a fixed spatial location for the entire acquisition period.
3. Cognitive impairments in the MWM are independent of locomotor activity, as locomotor deficits do not affect swim speed [24].
4. Four points equally distributes along the wall served as starting points (designated as North, South, East, and West) and divides the area into four quadrants.
5. We do CBB staining because it is better to see the transfer step is going well before process further experiment.
6. The whole membrane turns to be blue after CBB staining. It is necessary to wash membrane with destaining buffer to see the protein band clearly.

Acknowledgments

This work is supported by NIH grants R01CA167677, R03DA031663, R01HL111430, R01AG029385, and R01AG046319, and Alzheimer's Association grant DVT-14-322623 to K.P.L., and BIDMC pilot grant and gift donations from the Owens Family Foundation to X.Z.Z. and K.P.L.

References

1. Alzheimer's Association (2016) Alzheimer's disease facts and figures. Alzheimer Assoc 12(4):459–509
2. Lu KP, Kondo A, Albayram O, Herbert MK, Liu H, Zhou XZ (2016) Potential of the antibody against cis–phosphorylated tau in the early diagnosis, treatment, and prevention of alzheimer disease and brain injury. JAMA Neurol. 73(11):1356–1362. doi:10.1001/jamaneurol.2016.2027
3. Ballatore C, Lee VM, Trojanowski JQ (2007) Tau-mediated neurodegeneration in Alzheimer's disease and related disorders. Nat Rev Neurosci 8(9):663–672
4. Lu KP, Hanes SD, Hunter T (1996) A human peptidyl-prolyl isomerase essential for regulation of mitosis. Nature 380:544–547
5. Lu PJ, Wulf G, Zhou XZ, Davies P, Lu KP (1999) The prolyl isomerase Pin1 restores the function of Alzheimer-associated phosphorylated tau protein. Nature 399:784–788
6. Zhou XZ et al (2000) Pin1-dependent prolyl isomerization regulates dephosphorylation of Cdc25C and tau proteins. Mol Cell 6:873–883
7. Liou Y-C et al (2003) Role of the prolyl isomerase Pin1 in protecting against agedependent neurodegeneration. Nature 424:556–561
8. Pastorino L et al (2006) The prolyl isomerase Pin1 regulates amyloid precursor protein processing and amyloid-beta production. Nature 440:528–534
9. Lu KP, Zhou XZ (2007) The prolyl isomerase PIN1: a pivotal new twist in phosphorylation signalling and human disease. Nat Rev Mol Cell Biol 8:904–916
10. Lim J et al (2008) Pin1 has opposite effects on wild-type and P301L tau stability and tauopathy. J Clin Invest 118:1877–1889

11. Lee TH, Pastorino L, Lu KP (2011) Peptidyl-prolyl cis-trans isomerase Pin1 in aging, cancer and Alzheimer's disease. Expert Rev Mol Med 13, e21
12. Driver JA, Zhou XZ, Lu KP (2015) Pin1 dysregulation helps to explain the inverse association between cancer and Alzheimer's disease. Biochim Biophys Acta 1850(10):2069–2076. doi:10.1016/j.bbagen.2014.12.025
13. Nakamura K et al (2012) Proline isomer-specific antibodies reveal the early pathogenic tau conformation in Alzheimer's disease. Cell 149:232–244
14. Chen CH et al (2015) Pin1 cysteine-113 oxidation inhibits its catalytic activity and cellular function in Alzheimer's disease. Neurobiol Dis 76:13–23
15. Lee TH et al (2011) Death associated protein kinase 1 phosphorylates Pin1 and inhibits its prolyl isomerase activity and cellular function. Mol Cell 42:147–159
16. Kim BM et al (2014) Death-associated protein kinase 1 has a critical role in aberrant tau protein regulation and function. Cell Death Dis 5, e1237
17. Ma SL et al (2012) A PIN1 polymorphism that prevents its suppression by AP4 associates with delayed onset of Alzheimer's disease. Neurobiol Aging 33:804–813
18. Wijsman EM et al (2004) Evidence for a novel late-onset Alzheimer disease locus on chromosome 19p13.2. Am J Hum Genet 75:398–409
19. Luna-Mun˜oz J, Chavez-Macias L, Garcia-Sierra F, Mena R (2007) Earliest stages of tau conformational changes are related to the appearance of a sequence of specific phospho-dependent tau epitopes in Alzheimer's disease. J Alzheimers Dis 12:365–375
20. Hampel H et al (2010) Total and phosphorylated tau protein as biological markers of Alzheimer's disease. Exp Gerontol 45:30–40
21. Albayram O et al (2011) Role of CB1 receptors on GABAergic neurons in brain aging. Proc Natl Acad Sci U S A 108:11256–11261
22. Kondo A et al (2015) Antibody against early driver of neurodegeneration cis P-tau blocks brain injury and tauopathy. Nature 523(7561): 431–436
23. Cao VY, Ye Y, Mastwal S, Ren M, Coon M, Liu Q, Costa RM, Wang KH (2015) Motor Learning Consolidates Arc-Expressing Neuronal Ensembles in Secondary Motor Cortex. Neuron 86(6):1385–1392
24. Vorhees CV, Williams MT (2006) Morris water maze: procedures for assessing spatial and related forms of learning and memory. Nat Protoc 1(2):848–858

Index

A

Caroline Smet-Nocca (ed.), *Tau Protein: Methods and Protocols*, Methods in Molecular Biology, vol. 1523,
DOI 10.1007/978-1-4939-6598-4, © Springer Science+Business Media New York 2017

U

V

W

MIX
Papier aus verantwortungsvollen Quellen
Paper from responsible sources
FSC® C105338

If you have any concerns about our products,
you can contact us on
ProductSafety@springernature.com

In case Publisher is established outside the EU,
the EU authorized representative is:
Springer Nature Customer Service Center GmbH
Europaplatz 3, 69115 Heidelberg, Germany

Printed by Libri Plureos GmbH
in Hamburg, Germany